Your Patient's ANATOMY

A CLINICAL VIEW OF HUMAN MORPHOLOGY

Stan R. Blecher, M.D., F.C.C.M.G.
Professor and Director
School of Human Biology
College of Biological Science
University of Guelph
Guelph, Ontario

APPLETON & LANGE
Norwalk, Connecticut/San Mateo, California

0-8385-9950-8

Notice: The author and publisher of this volume have taken care that
the information and recommendations contained herein are accurate and
compatible with the standards generally accepted at the time of publication.

Copyright © 1990 by Appleton & Lange
A Publishing Division of Prentice Hall

All rights reserved. This book, or any parts thereof, may not be used or
reproduced in any manner without written permission. For information,
address Appleton & Lange, 25 Van Zant Street, East Norwalk, Connecticut 06855.

90 91 92 93 94/10 9 8 7 6 5 4 3 2 1

Prentice-Hall International (UK) Limited, *London*
Prentice-Hall of Australia Pty. Limited, *Sydney*
Prentice-Hall Canada, Inc., *Toronto*
Prentice-Hall Hispanoamericana, S.A., *Mexico*
Prentice-Hall of India Private Limited, *New Delhi*
Prentice-Hall of Japan, Inc., *Tokyo*
Simon & Schuster Asia Pte. Ltd., *Singapore*
Editora Prentice-Hall do Brasil Ltda., *Rio de Janeiro*
Prentice-Hall, *Englewood Cliffs, New Jersey*

Library of Congress Cataloging-in-Publication Data

Blecher, Stan R.
  Your patient's anatomy.

  1. Anatomy, Human. I. Title. [DNLM: 1. Anatomy.
QS 4 B646y]
QM23.2.B54 1990   611   88-8051
ISBN 0–8385–9950–8

Acquisitions Editor: Peter M. Klamkin
Production Editor: Amanda D. Egan
Designer: Steven M. Byrum
Cover Designer: Michael J. Kelly

PRINTED IN THE UNITED STATES OF AMERICA

"All worthwhile works of men are done for the love of a beautiful woman." (Anonymous)

This book is for Hanne.

Contents

Preface vii

Introduction and User's Guide ix

Section 1: Visceral Structures of the Head and Neck 1

Section 2: Thorax and Abdomen 79

Section 3: Cranial Structures 181

Section 4: Somatic Structures of the Neck; Back and Upper Limb 223

Section 5: Lower Limb 279

Index 335

Preface

Medical students and students of the paramedical professions are normally required to learn facts concerning the basic sciences out of context of the clinical situation, very often by rote rather than by practical contact, and, for these reasons, largely without meaning. They are then unleashed in patient contact in the vain expectation, as Mager has stated in a different context,[1] that one can have learned to run forwards by being taught to run backwards. To the eternally repeated consternation of clinical teachers and students alike, students realize then that they have to unlearn the theoretically learned facts and relearn the basic sciences in correct context.

The idea presented here, of placing the living body at the center of anatomical basic science learning and of integrating dissectional and theoretical learning to that orientation, was conceived and germinated when I was a student and, subsequently, a junior faculty member in the Medical School of the University of the Witwatersrand, Johannesburg, South Africa. At the time it was considered by most of the medical teaching profession to be very radical. It was first tested in practice while I was working in Copenhagen, Denmark, where it also initially met with much scepticism from virtually all except the students. The pedagogical basis for this approach, and some of the early results, were described in a series of articles published mainly in Danish, but culminating with one in English.[2] Since those early days some changes have occurred on the international medical education scene. It appears that this idea's time may now have arrived.

I owe acknowledgments to several individuals and institutions. In particular my former mentor and colleague, Professor Phillip V. Tobias, Head of the Department of Anatomy in the University of the Witwatersrand, was a source of inspiration and strongly influenced my development.

Early drafts of parts of this book were prepared and published in Danish, for use by dental students at the Royal Dental College in Copenhagen. I gratefully acknowledge the expert advice I received from several members of that college.

The present form of the book was largely developed while I was a Professor of Anatomy in the Faculty of Medicine at Dalhousie University, Halifax, Nova Scotia. I thank Dr D. Graham Gwyn, Head of the Department of Anatomy, for facilitating my access to materials and services. I am also indebted to Mr Don Ferris for his excellent help in preparing prosections for photography, and to other anonymous prosectors. The majority of the illustrations originate from the Audio-Visual Division of the Faculty of Medicine at Dalhousie University. I thank the Director, Mr Anthony Gibson, and his staff, and in particular Mr Alan Floyd who produced most of the book's very beautiful photographs.

Some additional photographs were taken by Mr Frank Sasinek at Dalhousie University, and others by Mr Herb Rauscher and Ms Susan Hersey of Photographic Services, University of Guelph, Ontario, Canada. The lucid and artistic line drawings are by Ms Linda Wilson-Pauwels, of the Department of Art as Applied to Medicine, Faculty of Medicine, University of Toronto. These drawings are mainly based on sketches I have developed in my teaching, and are limited in number to only depict a few central concepts in the morphological approach to human structure.

For typing and word processing, electronic transfer of material, proofreading, and assistance with artwork I thank members of the office staffs of the Anatomy Department of Dalhousie University and the Word Processing Centre of that University, as well as Mrs Judy Cockburn, my present secretary, Mrs JoAnne Waechter, and various members of my laboratory staff. In particular my staunch and long-

---

[1] Mager RF: *Preparing instructional objectives.* Belmont, CA: Fearon, 1962.
[2] Blecher SR: Anatomy—The patient's or the book's? *Higher Education*, 1978, 7, 71–82.

suffering secretary of several years in the School of Human Biology, Mrs Marion Stillman, was of great assistance, and the school's expert computing technician, Mr Jim Hoare, was especially helpful in the last stages of preparation.

I also acknowledge the very helpful suggestions of three reviewers whose comments were passed on to me by the publisher.

Finally, I thank the Senior Medical Editor of Appleton & Lange, Mr Peter Klamkin, for facilitating production of this book with skill, insight, and a sense of humor.

Stan R. Blecher
Medical Doctor
Fellow of the Canadian College
   of Medical Geneticists
Guelph, 1989

Introduction and User's Guide

Traditionally, in the teaching of anatomy and other basic medical sciences, heavy emphasis is placed on theoretical knowledge and systematic classification of factual material. However, the anatomy which doctors, dentists, and other practitioners are required to know is that of their patients. An alternative to the traditional approach is to make the living body the center of focus for the study of the basic medical sciences.

By studying anatomical structures **first** on the living body and **subsequently** by dissection and at theoretical levels, this orientation can be achieved. The first contact with information is then in its correct context, and since the first contact tends to be didactically the most important, incorrect contexts and the inevitable misunderstandings that must accompany these are avoided. The written word, no matter how well formulated, can never describe an anatomical structure so accurately that the mental image created is entirely correct. Thus students are traditionally asked to learn anatomy first by theoretical description; then see the structures in dissection (and at this time correct the errors of the initial theoretical image); and then, some years later, discern the quite different nature of the structures as they really are in the living patient. In the opposite sequence, both dissection and theoretical description become methods of deepening understanding of morphological concepts already grasped in the context of the living body and the need to correct distractingly erroneous image is eliminated.

CLINICAL STUDY OF THE LIVING BODY

In the approach presented here, the student is urged to follow the instructions relating to study of the **living body** as completely and as conscientiously as possible. The instructions are formulated so that no prior knowledge of anatomy is required in order to execute them. Therefore, **no prior text study** is required and it is in fact strongly discouraged. The student will find that if approached in the sequence suggested, theoretical material will become more readily understandable and more easily retained than by any other method of learning. This approach will also best ensure success for the student in examinations in anatomy, a not entirely trivial consideration for some students.

The instructions themselves do not require prior preparation either, in the sense of studying them before performing the physical examinations described; however, they do require intense concentration while performing the actual exercise: their contents should be learned on the spot. This implies that the terms defined and introduced should be learned as they are encountered, just as one would normally try to learn terms by reading a text. Thus the actual session at which one performs the exercise of examining the structures according to the guidelines provided should be regarded as the primary learning session. Subsequent rereading of the guide is review rather than learning. To merely read through the instructions without performing them or to perform them frivolously with the intention of subsequently studying the instruction to learn its contents, would defeat the intention and purpose of this approach.

LABORATORY (PROSECTIONAL AND DISSECTIONAL) STUDY OF THE CADAVER

Study of a previously dissected specimen (a prosection) and actual, active dissection of specimens by the student in the anatomy laboratory are essential components of the follow-up of clinical study. The laboratory exercises described in this book have been specially devised in such a way that each clinical session of study can be followed and reinforced by laboratory study of the same topics as have already been examined in the living body. To achieve this, several new approaches have been employed.

One example concerns the anatomy of the oral cavity which, for reasons explained elsewhere, is the first organ studied on the living body. In dissectional anatomy this region is traditionally approached by dissection from the skin surface inwards, a method which would have made impossible the immediate laboratory follow up of the initial clinical learning experience. To make the oral cavity easily accessible at the first laboratory session of study, a new approach, from the median plane has been employed.

As explained in detail elsewhere[1], a consequence of placing extra emphasis on a conceptual, including clinical, approach, is that in this system anatomical details that are deemed to be less relevant are given lower priority in teaching and examining. The time made available for the teaching of anatomy in most present day medical schools continues to diminish, but often detailed dissection, and much theoretical knowledge, continues to be required of students. The dissection requirements of this course are much reduced and reflect the opinion that dissection can serve the purpose of increasing understanding of the living body and the purpose of training clinically relevant skills, but is not an end in itself.

NOTE ON THE FORMAT AND USE OF THIS BOOK AND OTHER STUDY AIDS

This book is intended as a **learning guide** for the study of human anatomy in the clinical context of the living body. It takes as its starting point the precept that, ultimately, all learning depends on direct contact with the reality of the object being studied. In anatomy, this reality is primarily the living body, and secondarily, for the purposes of study of aspects of anatomy that cannot be studied on living persons, the use of a cadaver. The instructions in this book have been formulated in such a way that, by following them conscientiously, the student can study the living body and the cadaver without any previous knowledge of the subject. This does not imply that this book embraces all that needs to be known. On the contrary, its major role is to present an **approach** to how anatomy should be learned. Some factual information has to be obtained elsewhere, by use of reference texts and atlases, in conjunction with practical study of the subject.

The book employs the principle of **programmed learning,** defined here to mean progress of the student in small, intelligible, and manageable steps, from the known to the unknown, with understanding of principles as the ultimate goal. The "known" starting point is the lay person's knowledge of the human body; in the process of moving toward the unknown, major conceptual principles are established and factual information placed in the context of these. One mechanism employed to ensure this linkage of facts to principles is to repeatedly refer the student back to earlier sections of the book, in order to place a new fact into a previously established context. Conscientious pursuit of the book's instructions implies that these cross-references should be consulted **every** time, **even** when it may appear to be unnecessary.

While the approach to study of the human body employed in this method is that of the clinical context, the approach to organization of knowledge into "conceptual principles" is that of morphology. The morphological approach to biological structure is concerned with deriving and understanding general rules governing form, including its development. This approach mediates the learning of both regional anatomy, essential in the surgical approach to disease, as well as the anatomy of the systems, which forms the basis of functional (nonsurgical or "internal") medicine.

The morphological approach encourages understanding, as opposed to rote learning. It is of value because facts learned on a background of understanding are better retained than rote-learned facts that have not been understood. More importantly, this approach provides training in problem solving. The ability to use basic science knowledge to solve problems in the clinical situation is among the most valuable skills physicians and other practitioners can acquire. The emphasis that the morphological approach places on developmental anatomy is not a mere academic exercise but rather a reflection of the clinician's real world in dealing with individuals who grow older, from conception and onward through adulthood.

The combination of these approaches to the study of human anatomy in one book and one course requires careful selection of priorities. For example, certain morphological considerations, as well as some practical ones, argue in favor of commencing the study of human anatomy with the thoracic region. This region demonstrates, the principle of segmentation of the body most clearly; and dissection of the skin is more easily learned here than, for example, in the head. On the other hand, deference to the overriding importance of the clinical approach, as well as other morphological factors, make it reasonable to start with the oral cavity. The oral cavity offers easy, immediate access to clinical examination of a region and of internal organs. The morphological principle of the devel-

---

[1] Blecher SR: Anatomy—The patient's or the book's. *Higher Education*, 1978; *7*, 71–82.

opment of the organism around the digestive tube paves the way for understanding of such concepts as the difference between visceral and somatic connections of the nervous system. The difficult skin dissection of the face is deferred, as explained previously, by approaching the oral cavity from a sagittal section.

The sessions describing study of the living body have been formulated for use either in a simulated clinical situation, in a formal class and with the guidance of an instructor, or for use by the student in his or her own time. In the latter situation, the exercises are ideally performed when a partner is available. For this purpose a fellow student is the most appropriate so that, working in pairs, each student can examine the other, thus reinforcing the learning process by repetition as in the class situation. If a fellow student is not available, a friend or spouse may serve the purpose. However, if no partner is available, excellent results can be obtained by self-examination using a mirror. If this procedure is employed, the student should bear in mind that a mirror image is being observed, and that left and right sides are thus interchanged.

Major **anatomical terms** and **concepts** are emphasized in the text when introduced for the first time. These should, as mentioned, be learned when encountered. Important terms and concepts are also emphasized elsewhere, to indicate their importance or for systematization . One reason for the emphasis is that they can readily be identified by scanning the text; after completing the session, they should be reidentified and their meaning checked. These highlighted terms thus serve as key words around which the student's learning is based. A measure of successful completion of the overall objectives given at the start of each session is to ensure, at the completion of the session, that all the structures have been identified, their names can be recalled, and conceptual understanding of the material is present. The emphasized key words should therefore be reviewed, and the student should test his or her knowledge of their meaning and, in the case of anatomical structures, their anatomical description in so far as it has been dealt with in the session. A good way of testing this and one's understanding of concepts is to describe and explain the material to a partner or fellow student.

Following completion of the clinical and dissectional sessions, the student should perform further study of anatomical specimens, review atlas illustrations, and pursue independent reading around topics introduced in the practical courses. The student should acquire expertise in the art of scanning a book, or a section of a book, chapter, or article, to extract the major conceptual content of it.

Students are encouraged to collaborate in their use of textbooks. Alternative texts, reference books, and atlases are recommended such that some students may buy one alternative and others another and fruitfully cooperate by mutual lending of different books. No one book can excel in all aspects, and the broadest education is obtained by the student who consults many sources of information.

Review of studied material is most usefully performed by the "challenge" method. This entails the active process of the student challenging him- or herself to explain and apply information, as opposed to the passive process of rereading or reexamining material. It is easy to get the impression, on rereading a passage of text or on looking at a structure, that one "knows" it since the mind does not, on rereading, distinguish sharply between information that is fully understood and information that is only vaguely familiar. Attempting to demonstrate understanding **before** reviewing the material clearly establishes which segment of it has been understood and which requires further study, and the mental process of the challenge plays an important role in subsequent retention of the material reviewed. In this process, collaboration with fellow students is an important skill to acquire.

The prime objective of the student is, however, to learn to organize and to take responsibility for his or her own personal learning processes. The major goal of any university education is to guide students to an understanding of, and a career-long commitment to, the ongoing nature of the learning process.

PRACTICAL INSTRUCTIONS FOR CLINICAL STUDY OF MORPHOLOGY

For the clinical study sessions, students should work in pairs, one student being the patient substitute (model or subject) and the other being the examiner. Although physical examination may take place in the standing, sitting, or lying position, all descriptions of anatomical structures assume that the person being examined (patient or model) is in the **anatomical position.** In this position the subject stands upright. A detailed description of the anatomical position will be given in Clinical Session 1.

In the clinical situation, the human body is mainly studied by the four techniques of **inspection** (visual examination), **palpation** (examination by feeling with the fingers), **percussion** (examination by tapping or beating on the surface of the structure to elicit audible and palpable vibrations, which can give information concerning internal structure of the object being percussed), and **auscultation** (listening, either by placing the ear directly on the object being examined or by listening with the aid of a special instrument such as a stethoscope for naturally occurring sounds emitted by an organ or structure).

By inspection, the general shape and contour, size, color, surface projections, and numerous other features can be observed. By palpation the consistency of observable structures can be determined, mobility can be ascertained, tenderness can be elicited, and deep structures not visible to inspection can be felt through surface coverings. Percussion is usually performed by placing a finger or fingers of one hand flat on the surface of the body and then tapping these fingers with the index finger of the other hand. This procedure creates a vibration that passes through the tissues being examined and allows detection by touch and by sound, of invisible, deep-lying structures. Air-containing or hollow organs produce a resonant note; an example of this is the lungs. Solid structures, such as the liver, produce a dull note. By auscultation, such normal sounds as the closure of the valves of the heart can be examined—disease in these structures can lead to abnormality of the sounds. As another example, diagnosis of dysfunction of the joint between the upper and lower jaws may be aided by stethoscopic examination of abnormal sounds produced by movement of this joint.

Other instruments may often be used in physical examination. For example, in the first section of your studies, concerned with the head and neck, the instruments will include probes, forceps, dental mirrors, spatulas (tongue depressors), and other instruments with which you will examine the interior of the **oral cavity** (mouth cavity) and other internal structures. In addition, you will examine external surfaces, such as the skin, and palpate structures through the skin, or mark them on the surface of the skin with skin markers.

It is essential when examining a patient, or a model, that high standards of hygiene are maintained. For the examination of **external structures,** such as the skin, the hands should always be carefully washed. When **internal structures,** such as the oral cavity, are examined, the hands should be scrubbed, using the full standard scrubbing technique as used by physicians and other clinical practitioners. This full scrubbing technique entails removal of all rings, wrist watches, and other items from fingers and wrists. Sleeves should be rolled up, and, using a sterilized nailbrush, the fingers, fingernails, hands, and wrists scrubbed thoroughly with soap and water under a running stream. No unwashed surface is touched once the hands have been scrubbed. Thus the taps at the handwash basin at which one scrubs one's hands should be turned off by using the elbows or, where available, a foot control. Ultimately the examiner puts on a pair of sterile gloves. In the instructions contained in this book, the student is frequently asked to examine a structure in a living subject (model), then observe an anatomical specimen, such as a skull or plastic model, and then return to examination of the living subject. In such cases, it is essential that the examiner rescrubs his or her hands prior to reexamination of the subject. Instruments for use in internal examination should be sterile at the start of the exercise. Instruments that are no longer sterile should be carefully kept apart from sterile instruments, in returning trays for recycling instruments.

The normal procedure, in the exercises described in this text, is, that students work in pairs, one student acting as the model or subject, and the other as the examiner. The examiner ensures that the patient is comfortably seated or reclining in an examination chair or couch, making any adjustments that may be necessary for adequate lighting. The examiner should then assemble the instruments to be used; if sterile instruments are required for the session to be studied, hands should be washed thoroughly and a sterile tray of instruments should be present. Thereafter, the student acting as model will normally read the instructions aloud from the instruction book, and the examiner will carry these instructions out by studying the appropriate structures. The student being examined can follow the activities of the examiner with the help of a hand mirror, which should be part of each student's equipment at all teaching sessions. When the examiner has completed the exercise, the examiner and model exchange positions and the exercise is repeated, with the second student now acting as examiner, collecting a new tray of sterile instruments and seeing all structures on the new model.

Aside from various practical advantages of performing clinical exercises in pairs, discussed here and elsewhere in this book, a major educational role which this system plays is for the student to experience the sensation of being a patient, and so gain insight into the need for practitioners to acquire the skills of care and consideration.

PRACTICAL INSTRUCTIONS FOR DISSECTIONAL STUDY OF MORPHOLOGY

The actual performance of careful, conscientious dissection is of crucial importance for students of the medical and paramedical professions. There are two main reasons for this. First, dissection is invaluable and irreplaceable as a learning aid for gaining insight into those aspects of anatomy which cannot be studied on the living individual. No amount of reading or study of line drawings can provide the same three-dimensional concept of form which actual examination of the living body and actual dissection of structures can give. Although study of prosections is an excellent and necessary supplement to dissectional study, the student who does not actually perform dissection will be deprived of an essential element in the learning process. Second, the manual skills which physicians and surgeons, dentists, physiotherapists, dental hygienists, and other health professionals require are largely based on the techniques of dissection. A student who fails to take advantage of the opportunity of mastering these skills under guidance and at a measured pace, which no future learning situation may offer, would be missing an important opportunity to develop a major skill of his or her profession.

Gloves should be worn when dissecting to protect the dissector from potential exposure to infectious material. The major objective of dissection, and the principle guideline is that organs of the body should be separated from each other, where possible in natural planes of separation between them. With respect to the internal organs of the body, this is most often best achieved by separating organs without the use of instruments (i.e., by **blunt dissection,** which means using only one's fingers). In the case of smaller structures and organs, finger dissection may not be possible. The use of instruments then becomes necessary. However, whether dissection is being done by the fingers or with instruments, the basic principle to follow is to seek the natural planes of cleavage that most often correspond to sites where connective tissue coverings separate organs from each other. Separation of structures in connective tissue planes is easier, and less damaging, than at other sites.

The instruments generally used for this purpose are a sharp scalpel and a pair of forceps. However, in many situations a pair of sharp pointed scissors may be better. Structures can be dissected out more completely, and with less danger of damage, if the scissors are used as a separating instrument rather than a cutting instrument. The scissors are held in the closed position (points of scissors approximated to each other) and inserted into the connective tissue plane being dissected. The scissors are then opened, thus separating components of the connective tissue from each other. This process, when repeated a few times, reveals and exposes the enclosed structures. This method is the ideal one for dissecting into a region in which the course of structures, such as arteries and nerves, is not visible.

Less frequently than **separating** structures, it may be required that one actually **dissect into** an organ, in order to study its internal structure. In this case, the cutting edge of the scissors or the cutting edge of a scalpel may be used. Often when cutting with a scalpel, it is useful to insert a pair of forceps deep to the anticipated scalpel incision. The incision is then made between the limbs of the forceps, thus protecting deeper tissues from damage.

Dissection of skin from deep-lying connective tissues requires a scraping motion, with the cutting edge of the scalpel facing the deep surface of the skin, and the blunt edge of the scalpel facing toward the tissues which the dissector wishes to preserve, such as superficial blood vessels or nerves. Note that a structure situated nearer to the body surface than a second structure is said to be **superficial** to that second structure. The second structure situated further from the surface is said to be **deep** to the first. The precise technique for superficial skinning is illustrated in Figure 5, page 16. Observe that the skin surface is held away from the deep-lying tissues at an obtuse angle, into which the point of the scalpel is inserted. The surface of the skin which has been dissected away from the deep-lying connective tissues should have a white and pitted appearance, as illustrated. Using this technique, the **subcutaneous tissue** (tissue lying deep to the skin) and its contained **superficial vessels** (arteries and veins) and **nerves** remain intact for subsequent study. Dissection of skin is usually more difficult than dissection of internal organs, but the method of handling the scalpel, as illustrated in Figure 5, applies to dissection of deeper organs as well as skin.

On occasion it may be appropriate to dissect the skin and subcutaneous tissue away in one layer, exposing the deep fascia and muscles immediately. The advantage of this technique is that it is faster than that described above. Its disadvantage is that superficial nerves and vessels cannot be seen and studied. A reasonable compromise is to dissect one side of the body by the one method and one by the other.

Great care should be exercised in the use of scalpels. Injuries caused by carelessness can be extremely serious. Elementary precautions should be observed, such as never to cut toward one's own or others' fingers, and never to have a scalpel in hand except when it is in use. Other instruments such as bone cutters, chisels, and so forth are more rarely needed, and special instructions on their use will be given in the laboratory when necessary.

It is an important precept of dissection that as much as possible of the structure of the specimen or cadaver should be conserved. The purpose is, of course, to allow the student to reconstruct the specimen, by replacing dissected structures into their natural position, for the purposes of review. The student should bear in mind that the more careful and conservative the dissection, the easier it will be to review the material for subsequent study.

Students should arrive in the anatomy dissection laboratory prepared for study of a region by having examined the appropriate region on the living body. Most of the terminology and a three-dimensional concept of where the structures are situated in the living body will have been acquired by the student in this way. This background will be reinforced and expanded, and the task of dissecting the relevant structures in the cadaver further elucidated and assisted by examination prior to actual dissection of prosected specimens in the laboratory. This may be facilitated by specimens being demonstrated by the course instructor, making use of television monitors, and by making available anatomical specimens for individual study by students prior to, and parallel with, their own dissection.

For reasons of conservation of your dissection for future study as well as for reasons of general hygiene, cleanliness, and respect for the human material you are dissecting, you are requested to treat the anatomical material with care. It should be handled gently and wrapped in a responsible manner after each and every use of the material according to prescribed guidelines which will be indicated to you in the laboratory.

Contents of Section 1

Visceral Structures of the Head and Neck

1. Morphology; Regional Anatomy; The Five Major Anatomical Regions; Systematic Anatomy; The Organ Systems of the Body; Visceral (Internal) Components of the Organism; Somatic (Outer, Body Wall) Components of the Organism
   Clinical Session 1 3
   Laboratory Session 1 8

2. Oral Cavity; Teeth; Gums; Vestibule of Oral Cavity; Oral Cavity Proper; Lips; Tongue; Facial Muscles
   Clinical Session 2 9
   Laboratory Session 2 21

3. Vestibule; Buccinator; Bony Structures of Vestibule; Bony Structures of Oral Cavity Proper; Gingiva; Floor of Mouth; Roof of Mouth
   Clinical Session 3 26
   Laboratory Session 3 32

4. The Masticatory (Chewing) Apparatus; Maxilla; Mandible; Masseter Muscle; Temporal Muscle; Medial Pterygoid Muscle; Lateral Pterygoid Muscle; Temporomandibular Articulation (Joint)
   Clinical Session 4 37
   Laboratory Session 4 42

5. Anterior Triangles of the Neck; Digastric Triangle; Submental Triangle; Carotid Triangle; Muscular Triangle; Submandibular Gland
   Clinical Session 5 49
   Laboratory Session 5 55

6. Nose; Nasal Cavity; Paranasal Sinuses; Nasal Portion of the Pharynx (Nasopharynx); Muscles of the Soft Palate
   Clinical Session 6 59
   Laboratory Session 6 63

7. Oropharynx; Palatine Tonsil; Waldeyer's Lymphatic Ring; Laryngopharynx; Larynx; Muscular Triangle; Strap Muscles; Thyroid Gland
   Clinical Session 7 68
   Laboratory Session 7 72

Clinical Session 1

■ **TOPICS**

Morphology
Regional Anatomy
The Five Major Anatomical Regions
Systematic Anatomy
The Organ Systems of the Body
Visceral (Internal) Components of the Organism
Somatic (Outer, Body Wall) Components of the Organism

Morphology is the study of form; it comprises the rules and generalizations governing the way living organisms are put together. This includes the way organisms **develop** and the way they **function,** both of which are related to an organism's structure as seen at a given moment of examination. The study of a particular organism's structure at a given moment of examination is called **anatomy.** A practitioner of medicine, dentistry, or any of the related health professions is concerned with patients at all stages and ages of development, and in normal as well as abnormal states of function. An understanding of how these and other factors determine anatomical **structure,** and are determined by it, is fundamental to rational clinical practice.

From this it will be clear that the study of morphology may be approached from various standpoints, including those of **developmental anatomy** and **functional anatomy.** In the following text, we will examine two further methods of viewing the structure of the human body for the purposes of study. These are **regional anatomy,** in which the body is regarded as being divided into various **regions;** and **systematic anatomy,** in which the body is considered as being comprised of various **organ systems.**

REGIONAL ANATOMY

In regional anatomy, the body is regarded as consisting of five major **regions,** each of which is again subdivided into numerous subregions. For the purpose of standardizing anatomical descriptions, the human body is always assumed to be in the **normal anatomical position.** Study this position in your model or subject. In the anatomical position, the subject stands upright, with the feet together. The arms are at the side of the body with the palms facing forward (i.e., with the thumbs facing out to the sides). The subject looks straight ahead, with the head in the so-called **Frankfurt plane.** In this plane, the lower border of the **orbit** (socket or hole in which the eyeball is situated) is on the same horizontal plane as the upper border of the **external auditory meatus** (the opening into the ear; with external indicating outer, auditory indicating concerned with sound, and meatus meaning canal).

In this position, the following directional parameters are employed (fig. 37, page 88). Structures which are situated higher or above (i.e., nearer to the top of the head) are called **superior.** Structures which are lower or below (i.e., nearer to the sole of the foot) are called **inferior.** Thus the head is superior to the chest, and the knees are inferior to the waist. The front surface of the body is said to be **anterior;** the back surface of the body is said to be **posterior.** Thus the navel is on the anterior surface of the body, and the **scapula** (shoulder blade) can be felt on the posterior surface of the body. Similarly the palms of the hands face anteriorly in the anatomical position, and the elbows face posteriorly in this position. Bear in mind (page 7) that irrespective of whether you are examining your subject in the sitting, supine, or any other position, the descriptive terms always refer to the body as if it were in the anatomical position.

Observe on a skeleton, or on a disarticulated (disconnected from the rest of the body) skull, two lines on the superior (top) surface of the skull. One of these lines runs in an **anteroposterior** direction (i.e., from anterior to posterior). This line represents the joint between a right-sided and a left-sided bone, and is called a **suture** (seam). The name of this particular suture is the **sagittal suture** (sagitta indicating arrow—the suture's name is a relic of the saga of William Tell).

A vertical plane which passes through the sagittal suture, and divides the human body into two equal halves, is known as the **sagittal plane.** This plane is also known as the **median plane.** Place your model in the anatomical position again, in order to study the meanings of the following directional terms. **Medial** indicates toward the median plane. The opposite of medial is **lateral.** Thus, in the anatomical position, the thumb points in a lateral direction. The little finger is the most medial of the fingers of each hand. A plane which is parallel to the sagittal, but somewhat lateral to it, is called **parasagittal** (para indicating beside).

Referring back to the superior surface of the skull, note the other suture that is found at right angles to the sagittal suture. This second suture is

3

known as the **coronal suture** (corona meaning crown or coronet). A vertical plane through the coronal suture, lying parallel to the front or anterior surface of the body, is known as the **coronal plane.** Such a plane is also called a **frontal plane.**

Finally, a plane in the remaining dimension, at right angles to both the frontal and median planes, is called the **transverse plane** of the body. This plane, also known as the **horizontal plane,** divides the body into superior and inferior parts.

The five major regions of the human body are defined in the following paragraphs.

Head and Neck
Palpate (meaning feel) on your model and yourself, the **clavicles** (collar bones). These arbitrarily delineate the head and neck region from the **thorax** on the anterior surface of the body. On the posterior surface, the border is defined as lying between the last of the seven **cervical vertebrae** (cervix meaning neck; vertebra meaning the segment or portion [separate bone] of the "spine" or "backbone"). The seventh cervical vertebra **(vertebra prominens)** can often be easily palpated through clothing, at the back of the neck. Confirm your observations of both the clavicles and the seventh cervical vertebra on a skeleton.

Palpate on your model and yourself the important landmark that demarcates the site where the head and neck overlap into one region, i.e. at the floor of the mouth. On the anterior surface, starting at the anterior, inferior tip of the chin, run your finger downward in the median plane, until you encounter a bony structure, the **hyoid bone** (tongue bone), just above the **thyroid cartilage** ("Adam's apple"). The Adam's apple is more prominent in males than in females. In both sexes, however, it is easy to palpate the superior edge of the thyroid cartilage, and, just above it, in the median plane, the hyoid bone. The portion of the hyoid bone palpable in the median plane is the anterior portion, or **body,** of the bone. See the hyoid bone on a skeleton. Now palpate the posterior portions of this horseshoe-shaped structure on your model using your index finger on the left side and your thumb on the right side of the model's neck. Palpate gently, as the posterior extremities of the horseshoe (known as the **greater horns)** are delicate. With your thumb and index finger on the hyoid bone, ask the model to stick out his or her tongue. What do you notice, and what do you conclude from this?

Palpate also the area of the chin anterior to (in front of) the body of the hyoid bone, as your model swallows. In so doing you are able to palpate the contraction of the **mylohyoid,** a flat **muscle** in the floor of the mouth. Identify this muscle in an anatomical specimen, noting its attachments to the hyoid bone and to the lower jaw or **mandible.**

Thorax
The thorax extends from the boundary with the neck superiorly, to the inferior end of the **sternum** (breast bone). This pointed inferior end (see skeleton) should be palpated on your model. Posteriorly, the border between the thorax and the abdomen is designated as being the twelfth **thoracic vertebrae.** Identify this vertebra on a skeleton, bearing in mind that the first seven vertebrae are **cervical,** and the following twelve are **thoracic.**

Abdomen
The abdomen is the portion of the body which extends inferiorly from the thorax. Posteriorly, the five **lumbar vertebrae** follow the thoracic. The abdomen encloses a space in which the **bowel** and other important organs are housed, and at its inferior extremity it includes a basin-shaped structure called the **pelvis** (basin). Observe, on the skeleton, the basin-shaped **bony pelvis** formed by right and left **hip bones** and, posteriorly, the five **sacral vertebrae** (fused to form the **sacrum)** and three to four **coccygeal vertebrae** (fused into the **coccyx).** The inferior portion of the pelvis, situated between the superior ends of the thighs, is known as the **perineum.** In the anterior half of the perineum the external **genital organs** are situated (external meaning outer; genital meaning to do with generation or reproduction). The posterior half of the perineum contains the external opening of the digestive tract, or the **anus** (meaning ring). By convention, the abdominal region includes the pelvis and perineum.

Upper Extremity
The upper extremity or upper limb is the portion of the body known, in lay terms, as the "arm." However, the term **arm** is used by anatomists to indicate the portion of the upper limb extending from the shoulder joint to the elbow joint; the portion from the elbow joint to the wrist is known as the **forearm.** Distal to (i.e., beyond) the forearm are the wrist, hand, and digits (fingers). The term **distal** refers to a point away from the center. The opposite of this is **proximal.** In the limbs it is assumed that the terms refer to the **trunk** (the thorax and abdomen) as the center of the body.

Lower Extremity
The lower extremity corresponds to the lay term "leg." The term **leg** is used by anatomists to indicate the part of the lower limb extending from the knee to the ankle. The portion of the lower limb extending

from the hip joint to the knee joint is known as the **thigh.** Distal to the leg are the ankle, foot, and digits (toes).

SYSTEMATIC ANATOMY

In systematic anatomy, the body is considered as comprising organ systems (i.e., groups of organs, each built of cells, which are integrated structurally and functionally, to operate in subserving a common purpose in the body's economy). Any subdivision of the body into organ systems is, of course, as arbitrary as is subdivision of the body into regions. It was seen previously that the head and neck merge into the thorax; and indeed the head and the neck overlap with each other at the hyoid bone. Similarly, the extremities are in continuity with, and merge into, the **trunk** (thorax and abdomen). In some contexts it may be convenient to consider the systems as having a certain existence of their own. Thus one may study the digestive system, which extends, as a continuous tube, from the mouth to the anus. However, no system has any true existence independent of other systems. For example, the **digestive system** receives a supply of nerves from the **nervous system,** and it receives its blood supply through the **cardiovascular system.**

Not only do systems integrate with each other, but they also, of course, extend through several regions. Thus, in the example given, the digestive system begins in the head and neck region, then passes through the thorax and the abdomen, to end at the inferior limit of the abdomen, i.e., the perineum. It is on the basis of an understanding of the above maxims concerning the integration of organ systems, and the arbitrary nature of subdivision of the body into organ systems or into regions, that any list of organ systems, like any list of regions, should be studied. Furthermore, depending on the interest and emphasis which authors wish to place on aspects of human structure and function, numerous different ways of subdividing the anatomy into organ systems have been described. The following list is therefore, like other lists, an arbitrary breakdown of the systematic anatomy and is not absolute.

Digestive System

In some ways the digestive system is of primary importance. In the evolutionary sense, the digestive tract is the first system that develops as such. Primitive organisms may, indeed, be considered to consist primarily of a **digestive tube,** (with an entrance for food intake at the front end and an exit for removal of wastes at the back end), surrounded by an outer wall or covering. Morphologically, it is helpful to think of the organism as consisting of such a tube, lined **internally** by a soft, moist membrane (continuous throughout the digestive canal with that lining the mouth); surrounded **externally** by certain outer coverings, the outermost of which is skin; and with a **middle** layer of structures in between. Embryologically, the inner lining develops from a layer called **endoderm** (endo indicating inner; derm indicating layer); the outer lining of skin develops from a layer called **ectoderm** (ecto meaning outer); and a middle layer called **mesoderm** (meso meaning middle) develops between these, and contributes developmentally to both internal and external structures.

Considered in this way, the organism can be viewed as comprising **internal organs** (related spatially and developmentally to the internal **digestive tube**) and **outer structures** (related to the outer **body wall**). Such inner structures are known as **visceral** organs, or **viscera,** and they differ in numerous important ways from the outer, or **somatic** structures derived from the body walls.

Identify, in the anatomical specimens provided, the digestive tube, and trace its course through the regions of the body.

Nervous System

The cells of the nervous system (nerve cells or **neurons**) are specialized for the function of conduction of an electrical impulse. Identify the following in the appropriate specimens, and visualize their situations and functions in yourself and your model.

The nervous system comprises a central portion, known as the **central nervous system (CNS),** and an outer portion, the **peripheral nervous system (PNS).** The CNS is a tubular structure that has an expanded front or top portion, the **brain,** situated in the skull (in close relation to the intake end of the digestive tube), and a thin, tapering portion, the **spinal cord,** that extends down within the **vertebral column** (bony "spine"). The PNS comprises **peripheral nerves.** Some of these connect to the brain, and are known as **cranial nerves** (cranium meaning the skull), while other peripheral nerves connect with the **spinal cord,** and are known as **spinal nerves** (figs. 49 and 50, page 104).

Like other cells, neurons contain nuclei. The portion of the nerve cell cytoplasm within which the nucleus is situated is called **the nerve cell body.** To carry out the above mentioned specialized function of conduction of an impulse, part of the cytoplasm of neurons is drawn out or extended from the nerve cell body as **nerve cell processes** or **fibers.** Bundles of nerve cell processes constitute the peripheral **nerves** of the PNS. The nerve cell processes of the CNS are also organized into bundles called **white matter.** Similarly the nerve cell bodies of both CNS and PNS are organized into clusters, called **gray matter** in the CNS and **ganglia** in the PNS.

The processes of some nerve cells carry **efferent (motor)** impulses centrifugally (out from the central nervous system to peripheral structures), while other nerve cells carry **afferent (sensory)** impulses along their processes (fibers), centripetally (in toward the CNS).

As stated previously in the discussion of the digestive system, the body may be morphologically regarded as consisting of inner or **visceral,** and outer or **somatic** components. The nervous system correspondingly supplies these two components of the organism in slightly differing ways. The outer body wall comprising, as it does, the organism's interface with the surrounding environment, conveys sensory impressions from the outside world into the CNS in such a way that they are perceived **consciously;** and impulses which are carried from the CNS to these outer structures are under **voluntary** control. This functional component of the nervous system, in both PNS and CNS, is known as the **somatic nervous system** (or **voluntary nervous system).** Examples of its functions are the sense of touch, and the voluntary ability to move one's limbs. Nerve supply of the viscera comprises the **visceral nervous system,** and since this component is neither **voluntary** nor **conscious,** but functions automatically, it is also known as the **autonomic** (automatic) **nervous system.** This, too, is represented in the CNS as well as in the PNS. Examples are the subconscious beating of the heart and the automatic regulation of its rate.

Respiratory System

Respiration (breathing; use of oxygen) is performed by the **lungs** and its associated organs. These organs develop directly as outgrowths of the digestive system, and are therefore also **viscera.** Identify the organs of this system in appropriate specimens and trace the passage of air from the exterior to the lungs.

Skeletal System

The skeleton comprises the so-called hard tissues of the body, which act, amongst other things, as a framework which gives support to other structures in the body. This framework is mainly situated in the outer wall of the body, and is thus somatic in origin. Study a skeleton and identify the components of each major region. What additional functions does the skeleton subserve?

Muscular System

Those muscles which contribute to determining the external shape of the human form comprise a portion of the somatic component of the body. Thus the muscles of the **thoracic** and **abdominal** body walls are clearly components of this outer wall structure. Less obvious is the fact that the extremities, both upper and lower, are, developmentally, outgrowths from the body wall. Thus muscles and bones of the extremities are also somatic structures, and, like the body wall muscles, limb muscles are supplied by **somatic** components of the **nervous system.** The somatic component of the muscular system, together with the skeletal system, are often regarded as comprising one system, the **musculoskeletal system.**

The visceral structures of the body also contain muscles of a different type. These are known as visceral muscles, and receive nerve supply from the **visceral** portion of the **nervous system.** Thus the muscular system may be thought of as having both somatic and visceral components.

Hematic System

The hematic system (hem meaning blood) comprises both the blood itself, as well as the blood forming organs, which include the **spleen, bone marrow,** and several other organs.

Immunolymphatic System

Immune reactions are body responses to foreign matter, such as infecting bacteria. The organism protects itself against such invasions in two main ways: "killer" cells attack the invading particles **(cellular immunity);** and other cells manufacture chemical substances known as **antibodies,** which circulate in the blood and neutralize the effects of the invading particles **(humoral immunity:** humor indicates fluid, in this case referring to the bloodstream). Both the killer cells themselves, as well as the cells that manufacture the antibodies, are known as **lymphocytes,** which accounts for the name of this organ system. Lymphocytes are produced in various organs. These include the small **lymph glands,** which are linked by **lymph vessels,** in which the fluid **lymph** flows, carrying lymphocytes in suspension. Other important organs of this system include the spleen, thymus, and bone marrow.

Lymphocytes are also blood cells and this underscores an overlap between this system and the hematic system. The lymphocytes reach the bloodstream through communications between the lymph vessels and blood vessels; thus there is also overlap between this system and the next.

Cardiovascular System

The cardiovascular system comprises the **heart** and the **blood vessels.** The function of this system is to circulate blood through the body, transporting, among other things, oxygen to the various tissues. Identify, in anatomical specimens, the heart and the

arteries (that carry blood away from the heart), **veins** (that carry blood to the heart), and depicted in appropriate specimens **capillaries** (that complete the circuit by carrying blood from arteries to veins).

Genital System
The genital system comprises the **internal** and **external sex organs**. Identify these in specimens of both sexes. In the male, identify the **penis** and **scrotum** (containing the **testes**) externally, and the **duct system** which passes internally from the testis on each side, and ultimately carries **sperm cells** to the penis. In the female identify the **vagina** which opens externally, the **clitoris** situated anterior to the vagina, the **uterus** which projects into the vagina, the **uterine tubes** on each side of the uterus, and the **ovaries** at the lateral end of each uterine tube, where the egg cell is released.

Renal System
This comprises the **kidneys** and associated ducts and organs, which should be observed in the specimens provided. This system is responsible for filtering the blood of impurities, and excreting them (**excretion** meaning removal of wastes to the exterior). Because of developmental and anatomical association between this system and the genital system, the two are often dealt with together in anatomy texts as the **urogenital system.** In the male, the urine is ultimately passed through the penis; in the female the urinary opening is just anterior to the vaginal opening.

Integumentary System
This system comprises the **integument** (skin) and its derivatives (organs derived from it). These include the **breast, hair,** and **fingernails.** The integument is clearly a somatic structure, and is, by virtue of this, an integral component of that portion of the nervous system that carries impulses into the organism from the outside environment. It is also that part of the body that is, in general, most easily accessible to inspection and other examination by the clinical practitioner, and is therefore of importance in diagnosis.

Endocrine–Metabolic System
Endocrine glands secrete their products into the bloodstream (endo indicating within; crine indicating to secrete or produce. Note the difference between **secretion,** defined here, and **excretion** as defined in the section on the Renal System). The endocrine glands produce secretions which influence the way cells function. **Metabolism** means **change,** and refers to the organism's uptake of nutrients, and change or **turnover** of these, to produce energy for bodily functions. The secretions of the endocrine glands, called hormones, influence these metabolic changes. The major endocrine glands include the pituitary, thyroid, parathyroid, and adrenal glands. Some organs have an endocrine portion as well as other, nonendocrine functions. These include the pancreas and the gonads (testis and ovary). Identify as many of these structures as possible on the specimens available.

Review the emphasized key words of the session you have just studied. You should emerge from this session with a clear understanding of the human body as comprised of integrated organ systems. You should also understand that the body and its organ systems are organized in such a way that some components are mainly situated internally, and comprise the **viscera,** while other components are mainly situated externally, and comprise a **somatic body wall** and its derivatives. Finally, you should understand how the body is arbitrarily divided into regions for the purpose of anatomical study.

Laboratory Session 1

In this, the first laboratory session in human anatomy, most students will be seeing and working with a **cadaver** (a dead human body) for the first time. Depending on the individual background and personality make-up of each student, this event will be experienced in different ways, and will have different significance for each individual. Notwithstanding that it may be a more intense experience for some than for others, every health professional student will and should give some thought to the question of the practitioner's relationship to life and death, and the use of cadavers in medical research and education. This confrontation will be discussed in the first laboratory session, and students are encouraged to bring any questions or concerns that they may have, to the attention of their teachers, whose own resolution of such confrontations may be of assistance.

In this laboratory session, the major objective is to review, in the context of cadaver (dissectional) anatomy, the concepts dealt with on the living body in your initial clinical session.

REGIONAL ANATOMY

Reidentify the five major regions of the body, on the cadaver. Within the head and neck region, palpate the hyoid bone, as you previously have done on the living body. Palpate also the clavicles on the cadaver, and visualize, with the aid of a skeleton, the border between the cervical and thoracic vertebrae (i.e., the boundary between the head and neck region and the thorax).

Reidentify by palpation on the cadaver the sternum and thoracic vertebrae, as well as the ribs, all of which constitute the thoracic region.

In the abdomen, palpate the lumbar, sacral, and coccygeal vertebrae and the bony pelvis.

Briefly observe on the skeleton, and palpate on the cadaver, the bony components of the upper and lower extremities.

SYSTEMATIC ANATOMY

Review, with the aid of the anatomical specimens provided, the organ systems described in pages 5 to 7, in the first clinical session. Ensure that you have understood the meanings of all the emphasized key words in Clinical Session 1. Identify on specimens, for example, the two main parts of the CNS, the brain and spinal cord; and the two main components of the PNS, cranial nerves and peripheral nerves. What do you understand by the terms visceral and somatic?

Clinical Session 2

■ **TOPICS**

Oral Cavity
Teeth
Gums
Vestibule of Oral Cavity
Oral Cavity Proper
Lips
Tongue
Facial Muscles

With the model comfortably seated or reclining in a suitable chair (e.g., a dentist's chair), and a light adjusted to focus into the model's mouth, commence your examination of the **oral cavity** by observing the teeth in the upper and lower jaws. Note initially that there are four different kinds of teeth (incisors, canines, premolars, and molars), differing in their shape and, correspondingly, in their function. What function could be ascribed to the different tooth types? Note which teeth have cusps and observe how many cusps upper and lower teeth have. Compare your findings on your model with the teeth of a disarticulated (separated from the skeleton) skull as you proceed with your observations.

Detailed study of the form of the individual teeth will not constitute part of this course in human anatomy. Rather, your immediate objective at this time is to observe the general arrangement of the teeth in a **dental arcade** (arch) in both upper and lower jaws with each arcade comprising two halves or **quadrants**. Study the way in which the dental arcade contributes to delineation of outer and inner portions of the oral cavity from each other. Note at this time that teeth, as seen on a disarticulated skull, can be seen to be lodged in sockets in the bones of the upper and lower jaws. In the model, this arrangement is obscured by surrounding soft tissue, comprising the **gingiva** (gum). Gently palpate through the gingiva, to feel the underlying bony structure of the jaw that you have identified on the skull. Each socket in which a tooth is lodged is known as an **alveolus** (depression or pocket), and the bony ridge in which the alveolae are situated is called the **alveolar process** (of the upper jaw, or **maxilla**) or the **alveolar part** (of the lower jaw, or **mandible**).

The oral cavity may, for the purposes of simplicity, be regarded as if it were a regular cube in shape. As such, in keeping with frequent usage in anatomical terminology, it may be described as if it were a room, having a roof, a floor, and walls. On this basis, identify the following boundaries of the oral cavity. The lateral wall on either side is formed by the inner surface of the **cheek**. More specifically, it is formed by the moistened, innermost layer of the cheek (i.e., its **mucous membrane** lining). Hollow organs of the body, such as the digestive tube, including the mouth, are generally lined on their inner surfaces by a closely packed layer of cells which face into the **lumen** (hollow space) of the tube. Such closely packed cells on a surface are known as **epithelial cells**. On their deep aspect (i.e., on the aspect which faces away from the lumen), these epithelial cells are usually surrounded by a layer of **connective tissue** in which some small **glands** are situated. A gland is an organ which manufactures a substance, to be put out (secreted) and used elsewhere. These glands empty their secretions (through small ducts which pass between the epithelial cells) onto the free surface lining the lumen. Such secretion is called **exocrine** (compare with **endocrine,** page 7). These secretions maintain the lining surface in a moist state, and the complex of a **moistened epithelial surface,** the thin **layer of connective tissue** deep to it, and the **glands contained therein,** is known as a **mucous membrane**. Note that the epithelial cells lining the mucous membrane of the entire digestive tract are derived from endoderm (page 5).

The anterior wall or boundary of the oral cavity is formed by the upper and lower **lips**. These will be studied in more detail in the following paragraphs. At this time, confirm that when the mouth is closed, the upper and lower lips together constitute an anteriorly situated wall in the frontal plane.

As indicated previously, the oral cavity is divided into an outer and inner portion by the **dental arch,** the **alveolar process of the maxilla (or alveolar part of the mandible),** and the associated **gingivae**. The outer portion of the oral cavity, thus demarcated, is horseshoe-shaped, with the two lateral portions, one on each side, connected by a transverse portion anteriorly. This horseshoe-shaped, outer portion of the oral cavity is called the **vestibule**. The lateral portions are bounded laterally by the mucous membrane of the cheek as their lateral boundary, and medially by the lateral surface of the gingivae and the lateral teeth (molars and premolars). The anterior portion, which extends across the midline, is bounded anteriorly by the posterior mucous membrane-lined surface of the lips. The posterior boundary of this portion of the vestibule is formed by the anterior surfaces of the front teeth (incisors) and the anterior surfaces of their associated gingivae. Note that within each quadrant (page 9) of the dental arch, direction is indicated by the terms **mesial** (meaning toward the medial incisor) and **distal** (meaning toward the third molar).

Study the superior and inferior boundaries of the lateral portions of the vestibule. Both roof and floor consist of **reflections** of mucous membrane which connect the mucous membrane of the cheek with the superior and the inferior gingivae respectively. These reflections are known as the superior and inferior **buccoalveolar sulci** (i.e., from bucca to gingival covering of alveolar bone: bucca indicates cheek; sulcus indicates groove). Anteriorly, in the transverse portion of the vestibule, the roof and floor are formed by similar mucous membrane-lined grooves connecting the inner surface of the upper and lower lips respectively to the gingivae of the upper and lower jaws. These two grooves are correspondingly known as the superior and inferior **labioalveolar sulci** (i.e., from labium [lip] to gingivae of the alveolar bone). Examine the extent of the vestibule and its boundaries, observing the features mentioned here.

The inner, and larger, portion of the oral cavity, demarcated from the vestibule by the dental arches, the alveolar bone, and the gingivae, is known as the **oral cavity proper**. With the mouth closed, the anterior boundary of the oral cavity proper is formed by the posterior surface of the anterior teeth, and the associated gingiva. The lateral boundary of this part of the oral cavity is similarly formed by the medial surfaces of the lateral teeth and the associated gingiva. These boundaries should be observed with the mouth open.

The roof of the oral cavity proper is formed by the **palate**. Observe the palate on the isolated skull, formed by a thin layer of bone. Most of this horizontal plate of bone is part of the **upper jaw** bone or **maxilla**. However, a small, posterior portion of the horizontal, bony palate is contributed by a separate bone, the **palatine bone**. Observe the junction of the **palatine process of the maxilla** and the **horizontal plate of the palatine bone** on the skull. Now palpate the palate on your living model, observing that the bony structures (the **hard palate**) can be palpated, through a mucous membrane, in the anterior two thirds of the palate. The posterior portion of the palate in the living model, however, contains no bony skeleton, and constitutes the **soft palate**. Palpate and inspect the junction of the hard and soft palates in your model.

The floor of the oral cavity proper consists of the **tongue** and the region visible around and below the tongue when the tongue is lifted. The tongue develops embryologically as an upgrowth from a more or less flat floor, which is why the entire surface of the tongue may be regarded as constituting the floor of the adult mouth. The tongue will be studied in more detail later in this section.

Finally, the remaining boundary of the oral cavity proper, the posterior, should now be studied. A posterior wall is mainly lacking, leaving free communication between the oral cavity and the next region of the digestive canal, the **pharynx**. The site of the communication is, however, indicated by a fold of mucous membrane that extends from the soft palate to the lateral edge of the tongue on either side. Find this fold, called the **palatoglossal fold**, or **palatoglossal arch** (fig. 1, page 12). The palatoglossal folds contain small muscles called the **palatoglossus muscles**. Gently palpate the arches, as your model moves his or her tongue, and appreciate the slight degree of tension within the fold which is created by the deep-lying muscles while the tongue moves.

Turn now to a more detailed study of the anterior boundary of the oral cavity, the lips. The upper and lower lips may each be divided into three regions, according to the superficial covering tissues. These regions are:

1. A portion covered by true skin.
2. A portion covered by modified skin (the red, or **prolabial** portion.) (Labium means lip.)
3. A portion covered by mucous membrane.

Study these three portions of the lip as follows:

Skin Covered Portion of the Lip

On your model define the borders of the **oral region**. The superior limit of this region is formed by the inferior border of the **base of the nose**. (The nose is pictured as being pyramidal in shape, with the apex, corresponding to the tip of the nose, pointing anteriorly, and the **base** facing posteriorly, flush with the face.) The oral region is limited superolaterally by the **nasolabial sulcus**, the skin furrow that runs inferolaterally from the nose to the "angles" (lateral ends) of the mouth opening (labium indicates lip; sulcus indicates groove or furrow). Laterally, the region may be bounded by a **buccolabial sulcus,** also known as a dimple and produced in some by smiling (bucca indicates cheek). Inferiorly, the oral region is limited by an imaginary horizontal line passing through the **mentolabial sulcus** (mentum indicates chin).

In the skin-covered portion of the upper lip, study a median, vertical groove, called the **philtrum,** which runs from the nose toward the opening of the mouth. It is bounded on either side by an elevated ridge, and ends inferiorly in a swelling, known as the **tubercle of the upper lip** (fig. 1, page 12). The philtrum is derived from a midline **anlage** (embryological precursor), whereas the rest of the lip is derived from two lateral anlagen, one on either side. Thus the vestiges of an embryological process of fusion are visible in these structures.

With the aid of a magnifying glass, study the surface of the upper lip, on a male and on a female model. What differences are visible between the hair types seen in males and females?

Red (Prolabial) Portion of the Lip

Study the red (prolabial) portion of the lip using a magnifying glass where appropriate, and note whether any hair is present and whether any **sebaceous glands,** seen as small, white, submerged dots under the hand lens, are visible. Sebum (fatty substance) is normally secreted (put out) onto the skin surface by small glands such as these, situated on the face and elsewhere. Note the situation of the **mucocutaneous junction** between the dry red portion of the lip and the moist mucous membrane (muco meaning of the mucous membrane; cutaneous meaning of the skin) (see fig. 3, page 12). Observe whether or not this junction is visible with the mouth closed. Note that the major portion of the tubercle of the upper lip is situated in the red portion of the lip.

Mucous Membrane Lined Portion of the Lip

Using your magnifying glass, study the surface of the lip. Note the pattern created by the small blood vessels. Examine the **labial glands,** which present as elevated, whitish structures close to the surface, and are responsible for production of some of the **saliva** which keeps the mucous membrane moist. Gently palpate the surface of the lip noting the granulated consistency which these glands give it on palpation (fig. 3, page 12). These glands can also be palpated on yourself by gently running the tip of the tongue across the posterior surface of the lip. Review now the meaning of the term mucous membrane, as described and defined on page 9.

Toward the **angle of the mouth** (page 10) an arterial pulse can be palpated on either side in both upper and lower lips. Palpate this pulse bidigitally (between two fingers) and bimanually (using both hands). Use your index fingers. Attempt to determine whether the pulse lies nearer to the mucous membrane or skin aspect of the lip. These pulses are due to **labial branches** of the **facial artery.** The latter passes lateral to the angle of the mouth on each side (fig. 8, page 16). After palpating the **labial arteries,** observe them briefly on an anatomical specimen and in an atlas.

Note in both the superior and inferior labioalveolar sulcus, a mucosal fold in the midline, extending from the inner surface of the lip, across the sulcus, toward the inner boundary of the vestibule. Observe in detail the attachments of these folds, called the **superior** and **inferior labial frenula.** (Frenum indicates bridle. Frenulum is the diminutive of frenum and describes a small fold which limits movement of an organ.)

Palpate the upper and lower lips bidigitally, between index finger and thumb, while the subject is asked to purse his or her lips. Perform this palpation around the entire circumference of the mouth opening, feeling the contraction of a deep-lying muscle known as the **orbicularis oris** (os means mouth; oris means of the mouth; and orbicularis means ring-shaped). This muscle is a **sphincter muscle,** (i.e., a muscle that surrounds an opening and, by its contraction, narrows or closes that opening). Study this function of the orbicularis oris muscle. Note that at the angles of the mouth, a thick muscular pad is palpable, evidently within the orbicularis oris muscle. This is known as the **labial commissure** (commissure indicating a site of crossing over of the fibers). In the labial commissure, muscle fibers originating from the cheek pass toward the lips, with fibers from the upper portion of the cheek passing to the lower lip and vice versa. The commissure is the site where these fibers cross. These fibers proceed in to merge with the fibers of orbicularis oris.

The **oral region,** defined and outlined previously (page 10), is divided into the regions of the upper and lower lips by the **oral fissure** (the space between the lips, opening into the mouth) and an imaginary, horizontal line continuing laterally in the plane of the fissure. Study the relationship between the oral fissure and the teeth by gently inserting a dental probe into the oral fissure between the closed lips of your subject. What is the position of the fissure in relation to the cutting edge of the upper incisors?

TONGUE (LINGUA)

As mentioned previously, the tongue arises developmentally from the floor of the mouth. As seen in the adult, a portion of it is situated in the floor of the mouth, and a posterior portion of it is situated in the pharynx. The tongue is accordingly divided into an **oral portion** and a **pharyngeal portion.** As described on pages 9–10, the oral cavity may be thought of as having an anterior boundary (the lips) and a posterior communication into the pharynx. Thus the oral cavity, as well as the portion of the pharynx into which it opens (the **oral part of the pharynx),** can be thought of as lying in a horizontal plane, and having an anteroposterior long axis.

12 YOUR PATIENT'S ANATOMY: A CLINICAL VIEW OF HUMAN MORPHOLOGY

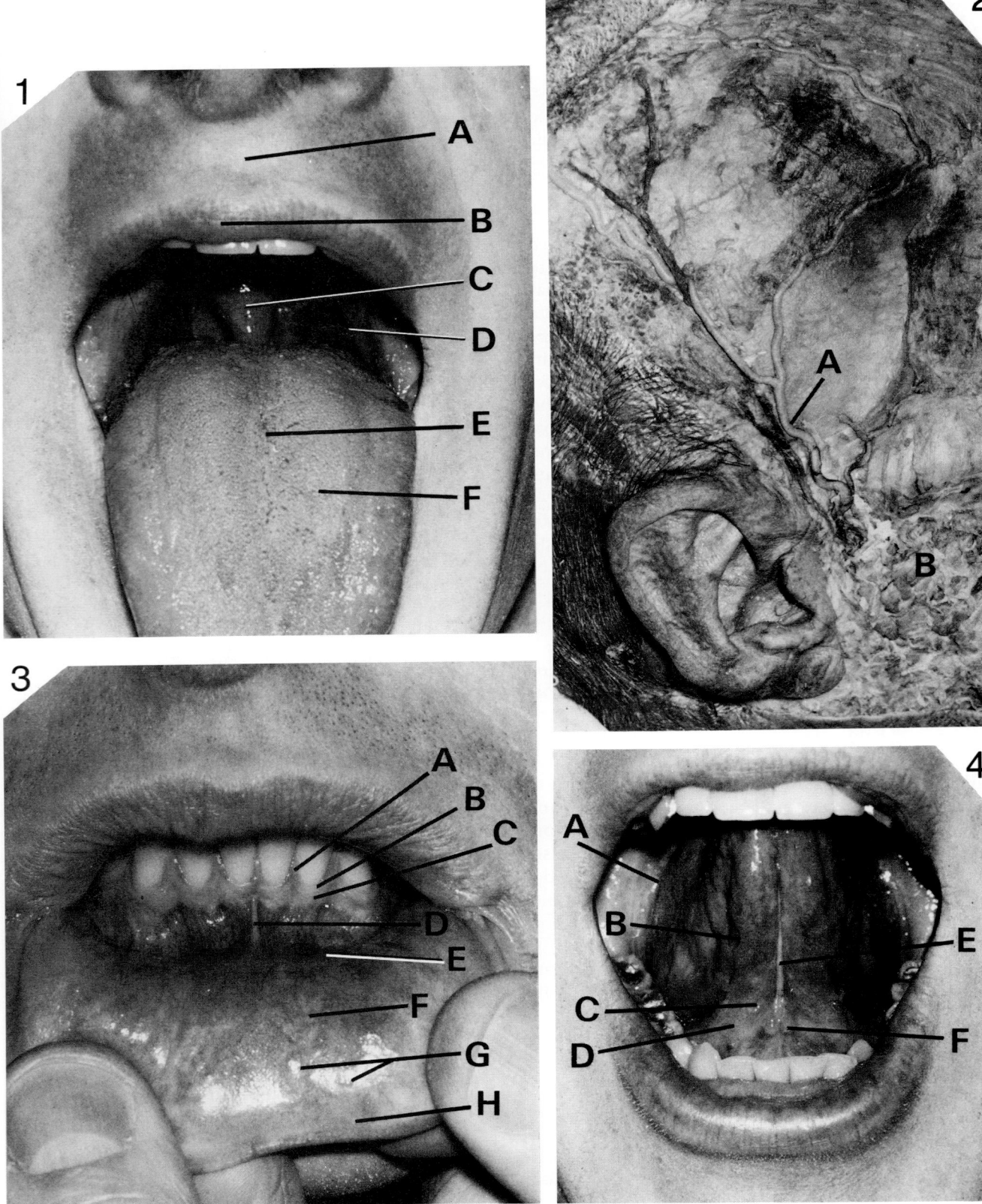

Figure 1. Structures of the oral cavity and oral region.
A. Philtrum
B. Tubercle of upper lip
C. Uvula
D. Palatoglossal arch
E. Median sulcus of tongue
F. Filiform papillae

Figure 2. Superficial dissection anterior to ear.
A. Superficial temporal artery
B. Parotid gland

Figure 3. Structures of the oral vestibule.
A. Interdental papilla
B. Free marginal gingiva
C. Alveolar gingiva
D. Inferior labial frenulum
E. Inferior labioalveolar sulcus
F. Mucous membrane lining of lower lip
G. Labial glands
H. Mucocutaneous junction of lower lip

Figure 4. Structures of the oral cavity.
A. Pterygomandibular fold
B. Plica fimbriata
C. Deep lingual vein
D. Sublingual plica (on crest of sublingual eminence)
E. Frenulum linguae
F. Sublingual caruncle (papilla)

Morphological Note
As is common knowledge, food, on being swallowed, proceeds from the mouth, and the part of the pharynx in communication with it, in a downwards (inferior) direction. (The pharynx has three parts: an upper region behind the nasal cavity; a middle region called the oral part of the pharynx and located behind the mouth; and a lower region behind the **larynx** or voice box. This is shown in Figure 9, page 22). Thus the digestive canal changes direction at a right angle bend, the first portion (oral cavity and the oral part of the pharynx) lying in a horizontal plane, and the subsequent portions lying in a vertical plane. The horizontal portion is encased within the skull, and delimits a "visceral" part of the cranium, or **viscerocranium**. Thus the viscerocranium is that part of the skull that surrounds the first part of the digestive tube. The use of the term visceral in this context corresponds to its usage as defined on page 5. Review this point now, and recall that the prototype visceral structure of the primitive organism is the digestive canal. At this stage, you should also observe an anatomical specimen in sagittal section in order to view the "right-angle" relationships of the initial and subsequent portions of the digestive canal, as well as to identify the visible parts of the viscerocranium. In an evolutionary sense, the primitive organism may be regarded as a tube (see page 5), and this right-angle bend may be regarded as a feature of the tube and thus of the organism as such. Note, for example, in the sagittal section, that the central nervous system makes a similar right-angle bend. Also, the horizontally-lying, foremost portion of the central nervous system, or **brain,** is housed within the skull and delineates a special portion thereof, known as the **neurocranium,** which thus may be defined as the part of the skull that surrounds the brain. The vertical continuation of the central nervous system, or **spinal cord,** is housed within the canal formed by successive vertebrae. Several important morphological consequences emerge from this, two of which will be referred to in the following discussion. First, note that the skull is considered to consist of two major portions—the **neurocranium,** which houses the brain (and related structures, including organs of special sense), and the **viscerocranium,** which houses the first portion of the digestive canal and associated structures. A second consequence of the right-angle bend, which is acquired during embryological development of the organism, is that the superior surface of the brain, and the posterior surface of the spinal cord, are, developmentally speaking (i.e., morphologically) part of one and the same surface. That one portion of this surface faces **superiorly** and the other portion comes, because of a fold during development, to face **posteriorly** does not alter the fundamental continuity of the surfaces. Similarly, it is also clear that the surface that is **posterior** in the human (the "back"), corresponds morphologically to the surface that is **superior** in lower animals, such as dogs and cats. To describe this surface, irrespective of the stage in development of the individual organism **(ontogeny)** or of the stage in evolutionary development of the species **(phylogeny),** the morphological term **dorsal** is used (fig. 38, page 88). The opposite surface (i.e., the anterior surface of the spinal cord and inferior surface of the brain in adult humans and the inferior surface of the spinal cord in cats and dogs) is morphologically known as **ventral.**

Returning to the tongue, observe in a sagittal section that this organ, like the central nervous system, has a dorsal surface that in part faces superiorly (in the oral portion of the tongue) and in part faces posteriorly (in the pharyngeal portion of the tongue).

Now study as much as possible of the dorsal surface of the tongue. For this purpose instruct your model to stick his or her tongue out as far as possible and, when necessary, grasp the tip of the tongue in a piece of gauze and pull it further toward you. For this examination you will also need a pocket torch (flashlight) and a dental mirror.

1. Find a midline (median or sagittal) groove called the **median lingual sulcus** (sulcus indicates groove).

2. Determine whether the mucous membrane on the dorsal surface of the tongue is loosely or firmly bound to underlying tissues.

3. Observe carefully, with the aid of a magnifying glass, that the mucous membrane on the dorsal surface of the tongue is irregular because of the presence of minute elevations called papillae (papilla means projection or nipple). There are four types of papillae of various shapes and sizes (fig. 6, page 16).
a. **Filiform papillae** (filiform means thread-like) cover most of the dorsum of the tongue and are very fine and hair-like structures. Obtain an impression of the consistency that these papillae give to the surface of the tongue by gently stroking your finger over the surface. Observe which color these papillae impart to the surface of the tongue.
b. **Fungiform papillae** (fungiform means mushroom-shaped) are distributed much less frequently among the filiform papillae and are found mainly in the anterior portion of the tongue. The fungiform papillae are found isolated, are much larger than the filiform, and are reddish in color. Unlike the filiform papillae, the fungiform papillae carry micro-

scopic, cellular **taste buds** that are the sensory organ of taste.

c. Far posteriorly on the dorsum of the tongue, visible with the aid of a dental mirror, is a row of very large (2 to 3 mm in diameter) papillae known as **vallate papillae** (vallate means surrounded by a moat). Six to ten vallate papillae may be observed arranged in an inverted **V** on the dorsum of the tongue, the apex of the **V** pointing posteriorly in the midline, and the limbs of the **V** extending out anterolaterally (i.e., somewhat anteriorly and somewhat laterally) toward the sides of the tongue. Consult an anatomical specimen and compare it with what you have observed on the model's tongue. Observe that these papillae, as with the previous two types, are appropriately described by the names they have been given. Vallate papillae contain taste buds. In addition, small salivary glands embedded in the tongue **(middle lingual glands)** secrete their product into the "moats" of these papillae. (There are also anterior and posterior lingual glands. These will be studied later in the text).

d. The fourth type of papillae observed in humans are known as the **foliate papillae** (foliate means leaf-shaped). These are situated at the lateral edge of the tongue on either side, at the site where the most lateral vallate papillae almost reach the lateral edge. These may be observed by asking the model to place the tip of his or her tongue far over to one side. Then, with the aid of a spatula, push the tongue further to the same side and observe the foliate papillae on the opposite side. Foliate papillae also contain taste buds.

4. Posterior to the row of vallate papillae is a shallow groove that follows the inverted *V* of the vallate papillae just behind them. This groove, known as the **sulcus terminalis,** may be viewed using a dental mirror. (Take care not to touch the walls of the oral pharynx with the instrument, as this may induce a choke or a regurgitation reflex.) This sulcus is regarded as the boundary between the oral and pharyngeal portions of the tongue. It also, accordingly, demarcates the posterior limit of the floor of the mouth, and thus of the oral cavity proper, inferiorly. Note the relationship of the sulcus terminalis to the posterior demarcation of the oral cavity proper which you previously studied, the palatoglossal arch, (page 10, and fig. 1, page 12). Just posterior to the apex of the sulcus terminalis a depression or pit may be observed: the **foramen cecum** (foramen indicating hole; cecum indicating blind sac. This is the remnant of an embryological downgrowth from the ventral aspect (floor) of the primitive mouth, to form the **thyroglossal duct** (thyro means of the **thyroid gland,** an endocrine gland situated in the neck; glossal refers to the tongue). As the name implies, this structure extends in embryonic life from the tongue to the developing thyroid gland. The inferior portion of the duct contributes to the formation of the thyroid gland; the remainder of the duct obliterates and is not normally present in adults. Thus the foramen cecum merely marks a remnant of an embryological structure.

The dorsal surface of the pharyngeal part of the tongue, posteroinferior to the sulcus terminalis, can also be visualized using a dental mirror. The surface will be seen to be uneven, due to the presence of irregular elevations. These contain lymphatic tissue, and are accordingly called the **lingual tonsil** (lingual meaning of the tongue; tonsil is a collection of lymphatic tissue) (fig. 6, page 16). Interspersed among the lymphatic accumulations, embedded in the tongue, are the **posterior lingual glands.**

Study the inferior surface of the tongue in your model (fig. 4, page 12) establishing first whether the mucous membrane covering this surface is loosely or firmly bound to underlying tissue. In the midline, on the inferior surface of the tongue, observe a median fold of mucous membrane, known as the **frenulum linguae** (frenulum meaning bridle) see page 11. At the apex of the tongue, just lateral to the frenulum linguae, an additional fold can be observed on either side: the **plica fimbriata** (plica means fold; fimbriate means frilly). These folds diverge from the midline when followed posteriorly. Between the frenulum linguae and the plica fimbriata of each side, a darkly colored blood vessel may be observed. This is the **deep lingual vein.** By gentle palpation medial to the vein you may be able to feel the pulsation of the terminal portion of the **lingual artery** (also called the **deep lingual artery).** (Closely associated with the blood vessels but neither visible nor palpable are a terminal branch of the **lingual nerve** and the **anterior lingual glands.)**

The muscles of the tongue should now be studied. These are divided into two major groups. The **extrinsic muscles** of the tongue (extrinsic meaning outside of) have a muscle attachment outside of the tongue, to a skeletal structure, and a second attachment within the tongue itself. The **intrinsic muscles** of the tongue (intrinsic indicating within) are entirely within the tongue itself and have no outside skeletal attachment. Indeed, the structure of the tongue is almost entirely muscular made up by these two groups of muscles. All the tongue muscles are supplied by the **hypoglossal (twelfth cranial) nerve,** which is a peripheral nerve (see page 5).

Recall that you previously have studied (page 10) the **palatoglossus muscle,** which extends from the soft palate to the tongue. This muscle does not

16 YOUR PATIENT'S ANATOMY: A CLINICAL VIEW OF HUMAN MORPHOLOGY

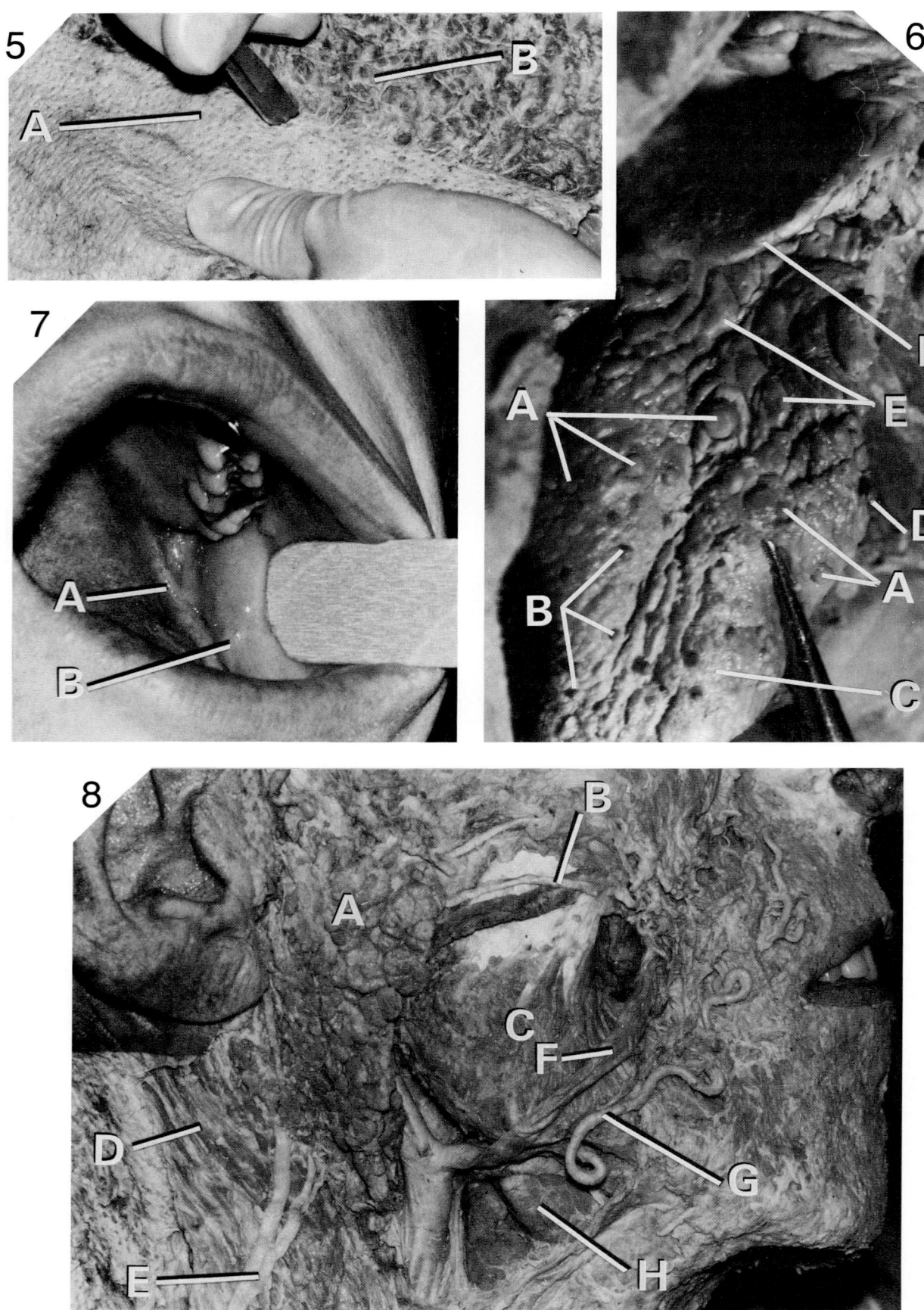

Figure 5. Dissection technique for skinning.
A. Deep surface of skin
B. Subcutaneous tissue

Figure 6. Dorsal surface of tongue.
A. Vallate papillae
B. Fungiform papillae
C. Filiform papillae
D. Foliate papillae
E. Lingual tonsil
F. Superior edge of epiglottis

Figure 7. Vestibule.
A. Pterygomandibular fold
B. Parotid papilla

Figure 8. Superficial dissection of face and neck.
A. Parotid gland
B. Parotid duct
C. Masseter muscle
D. Sternocleidomastoid muscle
E. Great auricular nerve
F. Facial vein
G. Facial artery
H. Submandibular gland (superficial part)

take its origin from a skeletal structure, and is usually excluded from the extrinsic muscles of the tongue for this reason, as well as for the fact that it is closely associated with the muscles of the palate in function, nerve supply, and developmental origin.

Extrinsic Muscles of the Tongue

There are three extrinsic muscles of the tongue.

1. Hyoglossus. Review palpation of the **hyoid bone** (page 4). While palpating the greater horn of the hyoid bone, and just above it, through the skin of the neck, ask your model to depress his or her tongue into the oral cavity. Your palpating finger will detect tension of a muscle just above the greater horn of the hyoid bone, and some slight movement of the bone itself. This muscle passes upwards from the greater horn of the hyoid bone to the side of the tongue. Compare your findings on palpation with an examination of an anatomical specimen and illustrations in an atlas.

2. Genioglossus. Palpate bimanually and bidigitally, placing the index finger of one hand within the model's mouth far anteriorly below the tongue, and the index finger of the other hand outside of the mouth just below the chin, while your model projects his or her tongue forward and pulls it back in. Compare your finding with observations on anatomical specimens and atlas illustrations. You may also be able to palpate some small projections on the inner surface of the **mandible** (lower jaw), where the two halves of the mandible meet at an angle. Observe these projections on an isolated mandible. These are called the genial tubercles (genial means of the chin; tubercle means small projection). The genial tubercles may be irregular, but often there are clearly seen to be two upper and two lower bony projections. From the **superior genial tubercles** the genioglossus muscles pass backwards and upwards into the tongue. (From the **inferior genial tubercles** the **geniohyoid** muscle pass backwards and downwards to the hyoid bone. Geniohyoid is **not** a tongue muscle.)

3. Styloglossus. Identify the **styloid processes** on the inferior surface of an isolated skull (styloid means resembling a pointed or tapering pillar). These sharp processes project inferiorly and slightly medially and anteriorly from just below the **external auditory meatus** (outer ear canal, see page 3) on each side. (In some cases the styloid process may be broken off, leaving only a short stub.)

The styloglossus muscle passes from the styloid process to the side of the tongue, which it enters posteriorly. Place an index finger on the lateral aspect of your model's tongue, as far posteriorly as possible, and ask the model to withdraw the tongue back and upwards into the oral cavity. You will be able to feel tension of the styloglossus muscle in this action.

Intrinsic Muscles of the Tongue

The intrinsic muscles of the tongue comprise muscle fibers that run very regularly in the vertical (superoinferior), longitudinal (anteroposterior), and transverse (lateromedial) planes, at right angles to each other. Four muscle groups are distinguished.

1. The vertical muscles of the tongue.
2. The superior longitudinal muscles of the tongue.
3. The inferior longitudinal muscles of the tongue.
4. The transverse muscles of the tongue.

Like the extrinsic muscles, these muscles play very important roles in swallowing. Whereas the extrinsic muscles are capable of changing the position of the tongue in relationship to its surroundings, the intrinsic muscles are those that alter the actual form of the tongue itself. Observe the tongue and changes in its form, as your model swallows with his or her mouth open. What is the initial movement that the tongue executes in commencing the swallowing process? What movements follow this initial act? What other important function, specific to the human species, are intrinsic movements of the tongue involved in?

Your model should attempt to carry out the following movements, during which you should observe the tongue, and attempt to determine which of the intrinsic muscles are involved in each movement.

1. Roll the tongue by making the dorsum of the tongue concave (i.e., creating an exaggerated groove on the dorsal surface).
2. Create a dorsal convexity of the tongue.
3. Flatten and broaden the tongue.
4. Extend the tongue, making it long, narrow, and high.

The ability to carry out these movements appears to be genetically determined: some models may be unable to perform these movements, while others will execute them with ease. If your model is unable to perform them, a neighboring subject might oblige.

FACIAL MUSCLES

The muscles of facial expression, unlike most other muscles of the body, lie in a very superficial plane, immediately deep to the skin of the face. In most other parts of the body a tough, fibrous layer of connective tissue known as **deep fascia** separates the **skin** and **subcutaneous tissue** (sub meaning deep to; cutaneous meaning of the skin) from the muscles themselves. The facial muscles, however, are situated in the subcutaneous tissue. These muscles play an extremely important role in human communication. They are all supplied by the **seventh cranial nerve** (a peripheral nerve: see page 5). Accordingly, the action of individual muscles of facial expression may be used as indicators of whether the seventh cranial nerve, also known as the **facial nerve,** is intact. Thus in testing for disease of this nerve, the physician will ask the patient to perform the various movements of facial expression described below.

Because these muscles are superficial, they can be palpated in movement as individual functions are performed. However, because they are extremely thin, and, in some situations, not sharply delineated from each other, the actual borders of each individual muscle are not usually palpable. In the following exercises, therefore, you should merely attempt to ascertain the general position of individual muscles by feeling movement or slight tension, deep to the skin, as your model performs the particular change of facial expression instructed. You should consult an anatomical specimen and work with an atlas illustration at hand to identify the general position of the muscles described. You should also have a disarticulated skull available as you perform these exercises and note the position on the skull to which each muscle attaches.

One of the most important muscles of facial expression, the orbicularis oris, has been studied previously (page 11). Several of the muscles mentioned in the table insert their fibers into the orbicularis oris, and indeed contribute to formation of this muscle. In addition, one remaining very important muscle of facial expression constitutes an important layer of the cheek, and some of its fibers enter orbicularis oris, decussating to form the **commissure** (page 11). This remaining facial muscle of the cheek, called **buccinator,** will be studied in the next session. A muscle of lesser importance but mentioned here for completeness is one that combines the functions of two of those described in the list below. The levator labii superioris alequae nasi muscle has a slip which assists in dilating the nostril, and a slip which assists in everting the lip.

A test of the integrity of the facial (seventh cranial) nerve that physicians often use is to ask the subject to show his or her front teeth. This exercise employs several of the above muscles as you should confirm by palpation on your model.

Actions Which Model Should Perform	Situation Where Examiner Should Palpate	Name of Muscle Being Palpated
Raise eyebrows in expression of surprise	Forehead and front of scalp	Frontal belly of occipitofrontalis
Raise eyebrows in expression of surprise	Back of scalp	Occipital belly of occipitofrontalis
Close eyes tightly	Upper and lower eyelids, skin around eyes	Orbicularis occuli
Wrinkle the skin over the nose	Bridge of the nose	Procerus
Narrow the nostril	Side of nostril	Compressor naris
Dilate nostril	Side of nostril	Dilator naris
Evert (turn inner surface outwards) upper lip	Upper lip	Levator labii superioris
Raise the angle (corner) of the mouth as in smiling	Above angle of mouth	Levator anguli oris
Elevate angle of mouth as in laughing	Superolateral to angle of mouth	Zygomaticus major
Elevate the corner of the mouth as in a sneer	Above angle of mouth	Zygomaticus minor
Protrude the lower lip and pucker skin of the chin as in expressing doubt	Below lower lip and anterior surface of chin	Mentalis
Draw lower lip downwards	Below lateral extremity of oral fissure	Depressor labii inferioris
Draw angle of mouth downwards and laterally	Inferolateral to angle of mouth	Depressor anguli oris
Leering grimace	From in front of the ear to the angle of mouth	Risorius

OTHER SUPERFICIAL STRUCTURES OF THE FACE

Some other superficial structures of the face include the branches of the **facial nerve** (seventh cranial nerve) that supply the individual muscles mentioned in the table above, and the sensory branches of the fifth cranial or **trigeminal nerve** (tri means three; gemini means twins; trigemini means triplets). This nerve is so named because it has three **divisions:** the **ophthalmic** (eye region) **nerve,** supplying the area around and above the eye including the forehead; the **maxillary** (upper jaw region) **nerve,** supplying the area about the upper jaw; and the **mandibular** (lower jaw region) **nerve,** supplying the area about the lower jaw. With the aid of Figure 105 (page 202) and a skull, identify and then palpate on your model the **supraorbital, infraorbital,** and **mental foramina** (holes) through which terminal branches of these three nerves respectively emerge to supply portions of the three areas indicated above. (The supraorbital **foramen** may be incomplete, that is, in the form of a **notch.**)

Certain arteries can now be studied on your model. By palpating along the inferior border of the **mandible** (lower jaw), pulsation of the **facial artery** can be felt in a groove in the posterior half of this border (compare with a skull and fig. 8, page 16). From this point the artery passes up into the face. The **labial arteries** (page 11) are branches of the facial artery. Also, palpate the **superficial temporal artery** about 1 cm in front of the outer ear, (fig. 2, page 12). The artery is situated at the posterior extremity of the bony **zygomatic arch,** which should also be palpated, and observed on a skull, projecting back from the cheekbone. Finally, instruct your model to clench his or her teeth, and palpate 1 to 2 cm below the zygomatic arch moving your palpating finger vertically up and down. Below and parallel to the arch (i.e., in a horizontal plane) you should palpate the **parotid duct,** a cord-like structure which slips away from the palpating finger (Fig. 8, page 16). This duct conveys saliva to the mouth, from the **parotid gland** (par means next to; otid means of the ear) (fig. 2, page 12 and fig. 8, page 16) that can be palpated as a vague mass in front of the ear.

To complete this learning session, you should now review the emphasized key terms of the entire session.

Laboratory Session 2

GENERAL ORIENTATION

The cadaver will have been prosectioned. The head and part of the neck will be sagittally sectioned, and the neck transversely sectioned on one side, resulting in a detached half of the head and neck and a half which remains attached to the rest of the body. In your study of this region you will need to examine both right and left halves of the specimen. However, for the dissection that ensues you should use the detached half.

Start by identifying the oral cavity, using the lips and the tongue as landmarks to find your whereabouts. Review (page 14 and fig. 9, page 22) the right angle bend that the digestive tract makes, the oral cavity and oral part of the pharynx lying in a horizontal plane, and subsequent portions of the upper digestive tract proceeding down vertically. Reconfirm that the central nervous system similarly has a first horizontal portion and a subsequent vertical portion. The horizontal portions of these two systems are encased in the skull. Review the names of the respective parts of the skull that the two systems define.

REVIEW OF BOUNDARIES AND SUBDIVISIONS OF THE ORAL CAVITY

Examine the oral cavity identifying, with the aid of the descriptions given in Clinical Session 2 (page 9), the following structures: dental arcade, gingiva, alveolar process of the maxilla, and alveolar part of the mandible. Review the division of the oral cavity into two components: the oral cavity proper and the vestibule. Study the inner lining of the cheek that bounds the vestibule laterally. Identify the buccoalveolar sulcus superiorly and inferiorly.

Anteriorly, study the lips, identifying the superior and inferior labioalveolar sulci and the upper and lower labial frenulae (page 11).

Within the oral cavity proper, study the roof. The palate is clearly seen in a sagittal section to comprise an anterior, horizontal portion known as the hard palate within which the bony palate can be seen and palpated, and the posterior soft palate, which hangs down like a veil. In the floor of the mouth, identify the tongue.

Posteriorly, identify the landmarks that define the imaginary posterior wall of the oral cavity; the palatoglossal arch (page 10 and fig. 1, page 12) and the sulcus terminalis of the tongue (page 15). Very carefully incise the mucous membrane of the palatoglossal arch with a scalpel and expose the muscle fibers of the palatoglossus extending the incision upwards and downwards to expose the entire muscle.

Having reidentified the boundaries of the oral cavity and its subdivisions into the vestibule and the proper portion of the cavity, turn now to a more detailed examination of the following structures.

Lips

Identify the three portions of the lips (page 10). Reexamine the various structures you previously identified in Clinical Session 2 including all the emphasized key terms. Study carefully the orbicularis oris muscle, the fibers of which can be clearly identified in the sagittal section. Using a pair of sharp-pointed scissors, as well as a scalpel and a pair of forceps, dissect these fibers carefully following them from the sagittal plane around onto the anterior surface of the face as they encircle the opening of the mouth.

Tongue

Examine the tongue in more detail. Identify on its dorsal surface the median sulcus and the four types of papillae (pages 14–15). Reidentify the sulcus terminalis and locate the foramen caecum. (Recall that the portion of the dorsum of the tongue posterior to the sulcus terminalis is pharyngeal. This will be studied further with the oropharynx.)

Elevate the tip of the tongue and examine the inferior surface of the tongue, identifying the structures you have previously seen on a living subject (page 15). These structures should include the **frenulum linguae,** the **plica fimbriata**, and the **deep lingual vein.**

Identify, on the sagittally sectioned surface of the tongue the **genioglossus** muscle (page 18 and fig. 10, page 22). Identify its **origin** (immobile attachment) from the superior genial tubercles and note that the fibers radiate up into the tongue, with the most anterior fibers passing upwards and then forwards toward the tip of the tongue, the posteroinferior fibers passing horizontally backwards, and intermediate fibers spreading out fan-wise, extending toward the entire length of the dorsum of the tongue. The inferior horizontal fibers should be traced posteriorly where they have their **insertion** (mobile attachment) on the body of the hyoid bone. Reidentify (page 4) the hyoid bone by palpation through the skin of the neck, inferior to the tip of the chin, and then identify the bone also in the sagittally sectioned surface of the specimen. Immediately inferior to the most inferior fibers of genioglossus, a second mus-

22 YOUR PATIENT'S ANATOMY: A CLINICAL VIEW OF HUMAN MORPHOLOGY

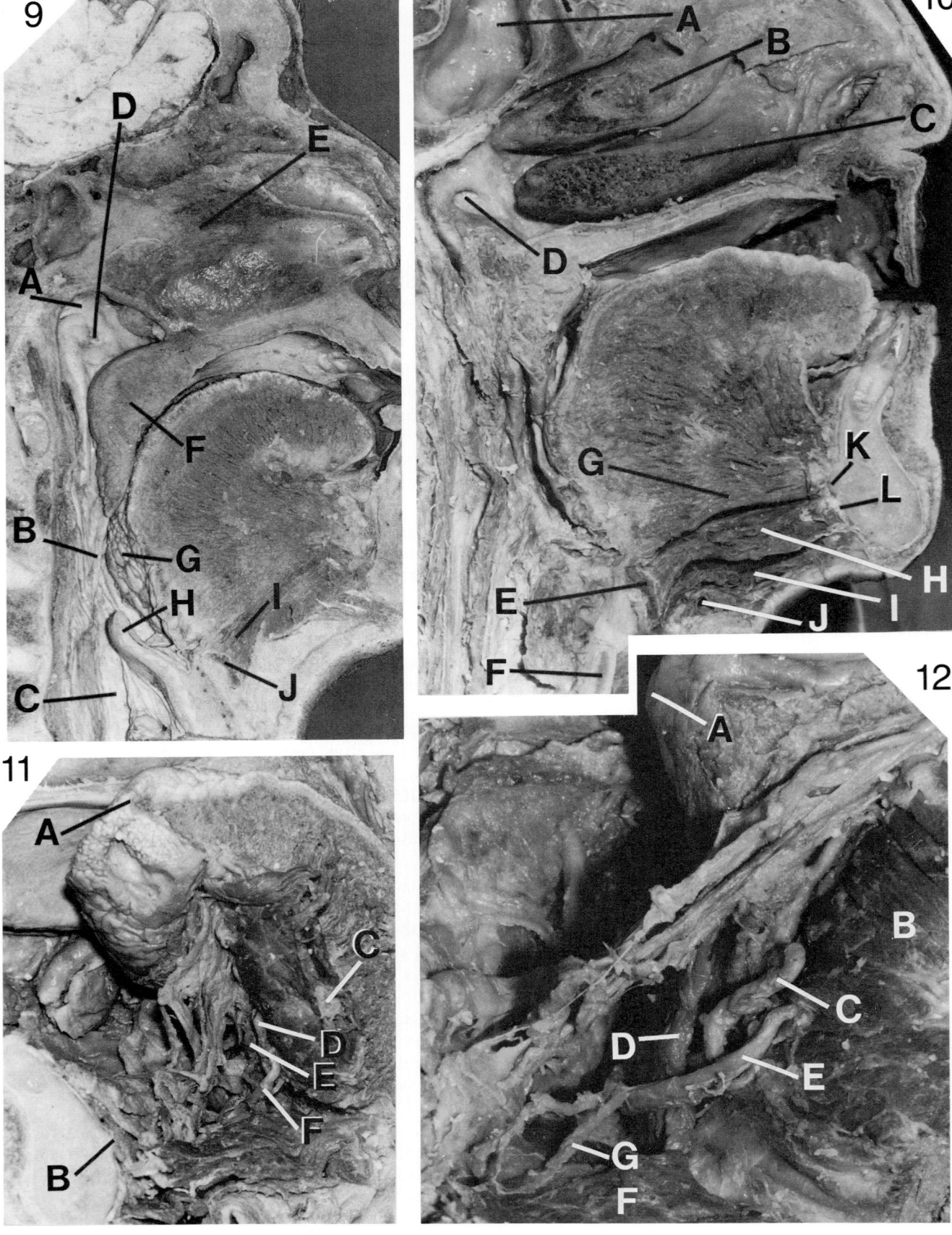

Figure 9. Sagittal section: pharynx
A. Nasopharynx
B. Oropharynx
C. Laryngopharynx
D. Opening of auditory tube
E. Nasal septum
F. Soft palate
G. Lingual tonsil (on pharyngeal part of the dorsum of tongue)
H. Epiglottis
I. Geniohyoid
J. Body of hyoid bone

Figure 10. Sagittal section: nasal cavity, nasopharynx, and floor of mouth.
A. Sphenoid sinus
B. Middle nasal concha
C. Inferior nasal concha
D. Opening of auditory tube
E. Body of hyoid bone
F. Thyroid cartilage
G. Genioglossus
H. Geniohyoid
I. Mylohyoid
J. Anterior jugular vein
K. Superior genial tubercle
L. Inferior genial tubercle

Figure 11. Dissection of floor of mouth (from sagittal surface).
A. Tip of tongue
B. Superior genial tubercle
C. Origin of genioglossus (detached from B and reflected back)
D. Lingual artery
E. Anterior edge of hyoglossus
F. Hypoglossal nerve (curving from lateral surface of hyoglossus, in front of anterior edge of hyoglossus, to reach genioglossus)

Figure 12. Dissection of right side of tongue (seen from the front).
A. Edge of tongue
B. Genioglossus
C. Lingual artery
D. Anterior edge of hyoglossus
E. Hypoglossal nerve (see F, fig. 11)
F. Geniohyoid
G. C1 fibers passing from hypoglossal nerve to geniohyoid

cle, **geniohyoid,** extends from the inferior genial tubercles also to the hyoid bone. Find the plane of cleavage between the lowermost fibers of genioglossus and the uppermost fibers of geniohyoid, and gently separate the two muscles from the midline out laterally. Inferiorly, a second plane of cleavage can be established between the geniohyoid and a thin, sheet-like muscle, **mylohyoid** which has been studied on the living body previously and should now be reviewed (page 4). Mylohyoid is also attached to the anterior surface of the body of the hyoid, and its sagittally sectioned edge is seen to extend toward the front of the mandible. The fibers of the mylohyoid can also be followed out laterally by gently separating mylohyoid from geniohyoid in the appropriate cleavage plane. (Inferior to mylohyoid, yet another muscle, the **anterior belly of the digastric** can be identified close to the midline anteriorly. It passes posterolaterally and need not be traced further at this stage.)

Returning to the cleavage plane between the lowermost fibers of genioglossus and the uppermost fibers of geniohyoid, after gently separating the two muscles the origin of genioglossus should be carefully dissected from the superior genial tubercles. By grasping the detached origin of the muscle firmly in a pair of forceps, reflect the muscle posteromedially, dissecting carefully in the plane of the lateral surface of the muscle fibers that you are now exposing. Gently elevate the fibers of genioglossus from geniohyoid proceeding with care as you expose the full extent of the attachment of geniohyoid to the hyoid bone. As you proceed with this dissection, and reflect genioglossus progressively further backwards, look for the **hypoglossal nerve** (page 15) and the **hyoglossus muscle** (page 18 and fig. 12, page 22).

Recall, from your previous palpations of the hyoid bone (page 4) that this structure is horseshoe-shaped. Having repalpated the body of the hyoid bone (page 21), now palpate the greater horn of the bone curving posteriorly from the body. This palpation should be carried out externally (i.e., through the skin) and internally by placing a palpating finger below the lateral edge of the tongue. The **hyoglossus** muscle, you will recall (page 18) arises from the greater horn of the hyoid bone, and its fibers pass upwards to the lateral edge of the tongue. The muscle has an anterior free edge, which you will encounter in this dissection as you reflect genioglossus backwards and dissect on its lateral surface. The hypoglossal nerve lies on the **lateral** aspect of hyoglossus parallel to and just above the greater horn of the hyoid, and the nerve, as it passes anteriorly beyond the anterior free edge of hyoglossus, sends branches to the genioglossus muscle. You will encounter these branches at this stage of your dissection. Furthermore, you will also encounter the terminal portion of the **lingual artery** that lies on the deep or medial surface of hyoglossus and that ascends just in front of the anterior free border of the muscle. Thus the hyoglossus separates the hypoglossal nerve, lying on its lateral surface, from the lingual artery, lying on its medial surface (fig. 12, page 22). The lingual artery also runs parallel to and just above the greater horn of the hyoid just before emerging medial to the anterior border of hyoglossus. After passing upwards, medial to the free edge of the muscle, it proceeds anteriorly toward the tip of the tongue, medial to the deep lingual vein, where you previously palpated pulsation in the artery on a living subject (page 15). Relationships of the nerve, muscle, and artery should be clearly demonstrated (see fig. 12, page 22).

You might also observe a branch of the hypoglossal nerve that passes to the geniohyoid muscle. Geniohyoid is not, however, an extrinsic muscle of the tongue and does not receive innervation from the hypoglossal nerve as such. The nerve fibers you observe originate from C1 (i.e., the first cervical spinal nerve), and the fibers merely "hitchhike" passage along the hypoglossal nerve to reach the geniohyoid muscle.

In the dissection just performed, you will have first separated genioglossus, and, in identifying the hypoglossal nerve, the hyoglossus from the overlying mucous membrane on the lateral surface of the tongue. You should proceed with this, to the lateral edge of the tongue, along which you may now cut the mucous membrane of the side of the tongue from that of the dorsal surface of the tongue, reflecting that of the side of the tongue downwards. This will further expose the course of the hypoglossal nerve, as it passes along the lateral surface of the hyoglossus muscle. Superior to the hypoglossal nerve, other nerve fibers will be encountered, and should be preserved. These belong to the **lingual nerve** (page 15), which will be dissected subsequently. At this time, however, attempt to expose hyoglossus in its full extent posteriorly. Observe the direction of the fibers of hyoglossus, running upwards and slightly anteriorly from the greater horn of the hyoid bone to the side of the tongue. At the posterior edge of hyoglossus near its superior border, you will find muscle fibers running in a posteroanterior direction, meeting the upper ends of the hyoglossus fibers and merging with them to insert into the lateral edge of the tongue. These fibers belong to the **styloglossus** muscle (page 18). You should now dissect styloglossus as far posteriorly as you are able to, reflecting the mucous membrane of the lateral surface of the tongue further inferiorly and posteriorly as necessary. Reidentify the **styloid** process on a skull and gently running your finger along the lateral surface of styloglossus as you expose it, attempt to follow the muscle to the styloid process in your cadaver dissection.

Intrinsic Muscles of the Tongue
On the opposite half of your cadaver specimen (i.e., on the attached half of the head and neck), cut a coronal section through the tongue, about halfway between the tip of the tongue and the sulcus terminalis, and identify, consulting an atlas illustration to aid you, the intrinsic muscles of the tongue as listed and described previously on page 18. On the sagittally sectioned surface of that half of the tongue, identify the anterior lingual gland (page 15), which can be seen embedded in the most anterior fibers of the genioglossus close to the tip of the tongue and close to the terminal portion of the deep lingual vein and artery.

Facial Muscles
Turning to the exterior surface of the head, dissect in the subcutaneous tissue that has been exposed by prosectional removal of the skin and identify the facial muscles you have previously studied (pages 18–19). Recall (page 19) that these muscles are virtually unique in lying immediately deep to the skin without a covering layer of deep fascia. Connective tissue and superficial fat should be carefully dissected away to identify the boundaries of the individual facial muscles. You should make frequent use of atlas illustrations, as well as prosections provided for this purpose, to identify the individual muscles and review their names and functions. You should attempt to preserve the small nerve fibers, that you will encounter in this dissection. Motor branches of the facial (seventh cranial) nerve should be traced from the **parotid gland** where they branch out of the main trunk of the nerve to the individual muscles that they supply. Identify with the aid of an atlas illustration the major named branches of the nerve as they radiate from the parotid gland.

Other Superficial Structures of the Face
Having identified the facial muscles and their motor nerves, identify and dissect the **parotid duct**, the **superficial temporal** and **facial arteries**, and the **sensory nerves** of the face mentioned previously (page 20, and fig. 2, page 12 and fig. 8, page 16). The **sensory nerves** should be identified with the aid of an atlas illustration. Consult an illustration of the distribution of the three main subdivisions of the fifth cranial (trigeminal) nerve, i.e., the eye region component or **ophthalmic nerve**, the upper jaw component or **maxillary nerve**, and the lower jaw component or **mandibular nerve** as you study the following branches. Dissect the **infraorbital nerve** first. This represents the termination of the upper jaw component. Note that a branch (the **mental nerve**) of the lower jaw component and a branch (the **supraorbital nerve**) of the eye region component each emerge in the same vertical plane as the **infraorbital nerve**. Finally, dissect the **auriculotemporal** nerve, posterior to the superficial temporal artery. (The remaining branches should be studied if encountered but may be difficult to find.)

Clinical Session 3

■ **TOPICS**
Vestibule
Buccinator
Bony Structures of Vestibule
Bony Structures of Oral Cavity Proper
Gingiva
Floor of Mouth
Roof of Mouth

Review the boundaries of the **oral cavity** and its subdivision into the **vestibule,** and the **oral cavity proper** (pages 9–10). In this session you will examine in greater detail the vestibule and its boundaries, and further study the roof, floor, and lateral wall of the oral cavity proper. (The posterior and anterior boundaries of the oral cavity, and part of the floor [i.e., the tongue] have been studied in detail previously.)

VESTIBULE

Study the mucous membrane of the inner surface of the cheek. Palpate the surface and compare it with that of the inner surface of the lips. Determine the presence of similar, granular structures in both surfaces. What would the structures be called in the cheek?

Using a sterilized spatula, gently hold the cheek away from the superior dental arch and find a small projection, about 2 to 3 mm in length and about 1 to 2 mm in diameter, at a site opposite the upper second molar. This is the **superior salivary papilla,** also called the **parotid papilla** (papilla means a small, nipple-shaped projection or elevation). Study this papilla as follows. Dry the mucous membrane of the cheek over and around the papilla, using sterile gauze, and then, while observing the papilla carefully, allow a few drops of lemon juice, from the container provided, to drop onto the subject's tongue.

The reaction you could observe is due to the fact that the secretion of the large salivary gland, the **parotid gland,** empties into the mouth through an opening on the tip of this papilla. The parotid gland is situated near the external ear, and its secretion reaches this papilla through the **parotid duct** (page 20).

In the reaction you have just studied, **sensory cells** (in **taste buds** in the tongue), in response to the stimulus, initiated an impulse that was carried from the tongue centripetally in **sensory nerves** of the **peripheral nervous system (PNS)** to the **central nervous system (CNS).** From there an impulse was again carried in the PNS, this time centrifugally, in **secretomotor nerves** supplying the parotid gland (see pages 5–6). The gland was thus stimulated through what is called a **simple reflex arc** (reflex indicates reflection). By analogy with reflection of light, the impulse going in toward the CNS (the **sensory,** or **afferent** [centripetal] impulse) gets "reflected" back from the CNS and "returns" to the effector organ, in this case the salivary gland in **(secreto)motor** or **efferent** (centrifugal) fibers. The connections of the afferent and efferent PNS fibers with the CNS are, anatomically, in the form of an **arc.** To understand this further, you should observe an anatomical specimen and figure 51 and lower half of figure 52, page 104.

Review on your model the position of the parotid gland (page 20), making use of anatomical specimens and atlas illustrations, as well as a disarticulated skull, to guide you. Begin by palpating the inferior border of the mandible on your model. Trace this posteriorly, to the point where the inferior border becomes continuous with the posterior border of that part of the mandible which passes up vertically toward the ear. The vertical portion of the mandible is known as the **ramus of the mandible** (ramus meaning branch), and the posterior border of the ramus meets the inferior border of the mandible at an obtuse **angle of the mandible.**

Now palpate on your model the cheekbone or **zygomatic bone,** and its posterior continuation into the **zygomatic arch.** The arch can be palpated as far back as the ear. Confirm on a skull that the zygomatic arch is partly made up of a process from the zygomatic bone, and partly by a process coming from the ear, and belonging to the **temporal bone.** Ask your model to clench his or her teeth firmly together, and palpate the powerful **masseter muscle,** extending from the zygomatic arch downwards and inferiorly to the outer surface of the ramus of the mandible and the angle of the mandible. About 1 to 2 cm inferior to the zygomatic arch, and parallel with it (i.e., running horizontally), palpate, while your model tenses his or her masseter muscle by clenching the teeth, the cord-like structure which slips away from under your fingers as you palpate it (i.e., the **duct of the parotid gland,** page 20). The **parotid gland** itself lies mainly on the lateral surface of the masseter muscle, anteroinferior to the ear, where it may be palpated as a vague and indefinite mass. Its borders cannot, however, be clearly felt, because its consistency is not nearly as firm as, for example, that of the tensed masseter muscle. (The parotid gland also extends medially, posterior to the ramus of the mandible, and curves round onto the medial aspect of the ramus of the mandible.) The **parotid duct** passes anteriorly from the anterior edge of the parotid gland, crossing the lateral surface of the masseter muscle. At the anterior border of the masseter muscle it leaves this relationship, proceeds further anteriorly, and then curves inward (medially) to pierce the tissues of the cheek, including the contained muscle (described later in this session) to end on the **parotid papilla.**

You have now studied examples of the **small salivary glands** (labial glands in the lips and buccal glands in the cheeks) and **large salivary glands,** of which the parotid is one of three paired (bilateral) glands.

BUCCINATOR

Palpate the cheek bidigitally between the index finger and thumb and study the action of **buccinator** (trumpeter) the flat muscle which lies within the substance of the cheek. The muscle's contraction can be felt when the model sucks forcibly upon the examining index finger within the mouth. Remove your index finger and ask the model to blow his or her cheeks out against a closed mouth, as in blowing the trumpet. Push on the blown out cheek with your index finger, while the model attempts to resist this pressure. This exercise constitutes a test for the integrity of the nerve supply of the buccinator. This and other facial muscles receive their innervation (supply of nerve-carried impulses) from the seventh cranial nerve (the facial nerve). In paralysis of this nerve, the subject would demonstrate reduced ability to resist pressure from the examining finger on the affected side.

Examine now, on the buccal wall of the vestibule, the posterior attachment of buccinator. This attachment is to a vertical, ligamentous band that is easily visible and palpable through the **mucosa** (mucous membrane) of the cheek. The band attaches just behind the posterior molars of upper and lower jaws. You can ascertain that you have identified it correctly by the following.

Place a palpating index finger about 1 cm posterior, and slightly medial, to the posterior upper molar tooth of your subject and, through the tissues of the soft palate, feel a small, bony projection known as the **hamulus** (little hook). You can feel your own hamulus with the tip of the tongue. The ligament to which the buccinator attaches is itself attached superiorly to the hamulus, and passes inferolaterally to a point just behind the posterior molar of the mandible. Observe by inspection a fold of mucous membrane which covers this ligament, the **pterygomandibular fold** (fig. 4, page 12 and fig. 7, page 16). The fold is clearly visible when your subject opens his or her mouth. Palpate the ligament from the hamulus to its attachment to the mandible behind the posterior molar. Identify the hamulus on the isolated skull, posteromedial to the third upper molar and just behind the hard palate. Note that it constitutes a hook-like projection downwards, from a vertical plate of bone that is continuous with the lateral bony wall of the **nasal cavity** (the cavity of the nose). This plate is called the **medial lamina** of the **pterygoid process** (lamina means plate; pterygoid means wing-shaped). Also identify the **lateral lamina** of the pterygoid process, which flares out posterolaterally. Study the site on the isolated mandible, just posterior to the last molar, where the ligament you have palpated on your model attaches.

Ligamentous and other structures are often named according to the bony attachments between which they extend. This ligament is accordingly known as the **pterygomandibular ligament** (or **raphe).** The mucosal fold which it raises, and which you can observe in your model when he or she opens the mouth is the pterygomandibular fold mentioned previously.

Note the proximity of the **palatoglossal arch** (the boundary between the mouth and the pharynx) and the **pterygomandibular raphe** (the posterior limit of buccinator, the muscle of the lateral wall of the mouth). The pterygomandibular raphe separates the mouth muscle, buccinator, attached to the anterior edge of the raphe, from the upper muscle of the wall of the pharynx, the **superior pharyngeal constrictor,** which is attached to its posterior edge. (Raphe means a seam; here a seam between the two muscles.) The origin of the latter muscle extends to the posterior border of the medial pterygoid plate. The right and left muscles meet in the posterior midline of the pharynx in another seam known as the **pharyngeal raphe.**

Palpate again the posterior attachment of buccinator to the pterygomandibular ligament. In addition, palpate the upper and lower attachments of buccinator, which are to the alveolar process and alveolar part of the upper and lower jaws respectively. Finally, complete your examination of buccinator by palpating the anterior extent of the muscle. The upper and lower fibers of buccinator decussate (cross over each other) in passing, respectively, to the lower and upper lips, to mingle with fibers of orbicularis oris. Palpate once again the site of this decussation, known as the **labial commissure** (page 11).

Palpate the boundaries of the vestibule, noting differences in the consistency of the mucous membrane and its deep-lying tissues at different sites. Establishing where attachment of mucous membrane to deep-lying structures is loose and where it is firm. Compare in this way the gingiva to the buccoalveolar and labioalveolar sulci. Review (page 11) the attachments of the superior and inferior labial frenulae.

Note the color differences in different parts of the mucous membrane lining the vestibule. How can these be accounted for?

BONY STRUCTURES OF VESTIBULE

Making frequent reference to an isolated skull, study the following bony structures which are accessible to examination within the vestibule, by palpation through mucous membrane-lined soft tissues. In the vestibule lateral to the upper jaw, palpate the **zygomatic process of the maxilla,** extending upwards to articulate (join with) the **zygomatic bone.** You have previously studied two other **processes** of the maxilla: the **alveolar process** (page 9) and the **palatine process** (page 10). Review these on the skull now, and, in addition, find the fourth and remaining process of this bone: the **frontal process,** which passes up, medial to the **orbit,** to articulate with the **frontal bone** (forehead).

In the vestibule surrounding the upper jaw you can also palpate the bony elevations in the alveolar process which the roots of the teeth make. In particular, the canine teeth produce noticeable elevations. Palpate also the **anterior nasal spine,** which is found high in the superior labioalveolar sulcus (page 10), in the midline, projecting up from the maxilla.

In the vestibule of the lower jaw, elevations for the roots of the teeth can similarly be palpated. In addition, palpate the anterior margin of the **ramus of the mandible** by placing an index finger posterolateral to the last molar, and running it backwards and upwards. Follow the anterior border to its upper extremity, into the **coronoid process.** Ask your model to clench his or her teeth while palpating the coronoid process, and note that fibers of insertion of a second powerful chewing muscle, the **temporalis muscle,** can be felt here. You have previously palpated another chewing, or **masticatory muscle,** the **masseter** (page 26). These two, and two other major masticatory muscles, as well as several accessory masticatory muscles, will be studied in greater detail later.

LATERAL BOUNDARY OF THE ORAL CAVITY PROPER

Palpate the medial aspect of the **body of the mandible** (the curved, horseshoe-shaped portion that meets the **ramus of the mandible** posteriorly on each side at the **angle of the mandible**). By palpating inferior to the medial aspect of the last molar, a bony ridge, the **mylohyoid line,** can be felt, and, just superior to it, the **lingual nerve** (pages 15 and 24). Identify the mylohyoid line on an isolated mandible. The mylohyoid muscle (pages 4 and 24) attaches to this line (fig. 13, page 34). By palpating carefully (because the structures are sensitive, and because they may not be easy to palpate), attempt to follow the lingual nerve as it passes anteroinferiorly and medially, to reach the side of the tongue (figs. 14 and 15, page 34). (Lingual means of the tongue. The lingual nerve is a branch of the mandibular division of the **fifth cranial nerve,** and it mainly contains sensory nerve fibers from the tongue.) The relationship of the lingual nerve to the mandible, inferior to the last molar, is of major clinical importance. It is a useful site for administration of a local anesthetic injection. Furthermore, as you will see by observing the medial surface of the ramus of an isolated mandible, it is close to the **mandibular foramen,** through which the **inferior alveolar nerve** enters the **mandibular canal** to subsequently supply sensory nerves to the teeth. In dentistry, local anesthetic injections are frequently administered to the inferior alveolar nerve at this site. Such injections may also anesthetize the lingual nerve, explaining why the tongue may become numb during local anesthesia of teeth of the lower jaw. (The inferior alveolar nerve is also a branch of the mandibular division of the fifth cranial nerve.)

GINGIVA (GUMS)

Review the form of the maxillary and the mandibular dental arches. On an isolated skull, study the anatomical parts of the teeth: the expanded **crown,** the narrow **roots** (lodged in the alveolus), and the constricted **neck** between crown and roots. Compare the **anatomical crown** with the **clinical crown** (the portion of the tooth projecting beyond the gingiva [gums]). In the following, examine the gums both in moist and dry conditions. An absorbent gauze roll should be placed in the alveolobuccal (buccoalveolar) and alveololabial (labioalveolar) sulci, and the mucous membrane dried with gauze and with an air stream. In each of the regions of gingiva to be studied in the following paragraphs, observe whether the surface is rough or smooth, what the color is and whether it is even, what the consistency is to palpation, and whether the gingiva is loosely or firmly bound to underlying hard tissues (teeth or alveolar bone). Make constant use of comparisons with the skull, and study the gums both on the **lingual aspect** (i.e., the aspect facing in toward the tongue) as well as the **vestibular** (or **facial) aspect** (both buccal and labial surfaces). Study the following three named parts of the gingiva (fig. 3, page 12).

Free Marginal Gingiva

This term refers to the part of the gingiva which surrounds the individual teeth on the oral and vestibular aspects. It is not tightly adherent to the tooth, but can, on the contrary, be separated from the surface of the tooth, to a varying extent of up to several millimeters. Thus a **periodontal sulcus** exists, at the bottom of which (in the lower jaw) or at the top of which (in the upper jaw) the gingiva does indeed become closely adherent to the tooth, and, beyond that, to alveolar bone. Gently insert the point of a **measuring probe** into the periodontal sulci of your model's gums and measure the depth of the sulci on several teeth. It is important that you appreciate that the surface epithelium of the free marginal gingiva, as seen on the buccal or labial aspect of the gum, or on the oral aspect, is continued over the free edge of the gingiva, into the sulcus, extending to the bottom of the sulcus. It is also important to understand that the existence of the sulcus is a normal phenomenon; abnormal increase in size of the sulcus is a pathological state.

The extent of the sulcus is usually indicated, on the vestibular surface of the free marginal gingiva, by a corresponding groove known as the **free gingival groove.** Observe this groove and its relationship to the anatomical crown of the tooth and the alveolar bony margin.

Interdental Papillae

This term applies to the parts of the gingiva which extend between the teeth. (Papilla means a small nipple-like process.) Observe the form of these papillae, and their extent.

Alveolar Gingiva

This term indicates that portion of the gingiva which overlies the alveolar bone. Study the continuation of this portion of the gingiva with the ordinary mucous membrane of the mouth. Note color differences, as well as differences in consistency and binding to deep-lying hard tissues.

FLOOR OF THE MOUTH; SUBLINGUAL REGION

The floor of the mouth is comprised of the superior and inferior surfaces of the tongue, as well as the structures lying inferior to the tongue, visible when the tongue is lifted: the latter region is known as the **sublingual region** (sub meaning under; lingua meaning tongue). That the inferior surface of the tongue also may be regarded as part of the floor of the mouth is explained by the fact that the tongue, developmentally, arises as a swelling which grows up from the primitive floor. The superior and inferior surfaces of the tongue have previously been studied (pages 14–15). Ask your model to place the tip of his or her tongue on the roof of the mouth, and study the structures visible:

1. Review (page 15, and fig. 4, page 12) the midline fold and the two laterally diverging folds on the inferior surface of the tongue, as well as the lingual artery and vein.

2. Observe the sublingual mucous membrane, while the model keeps the tip of his or her tongue fairly well forward (thus avoiding tension of the sublingual mucous membrane). Just lateral to the **lingual frenulum,** on either side, observe a small papilla, the **sublingual papilla** or **sublingual caruncula** (caruncula indicates papilla). This papilla is the site of secretion of saliva from the **submandibular gland.** Study this secretion process, using the lemon juice test, as previously performed for study of the **parotid papilla.** See page 26 for details of the technique.

From the sublingual papilla, observe a fold which passes backwards and laterally in the sublingual region. This fold is called the **sublingual plica** (plica meaning fold). The fold contains the **submandibular duct,** which ends at the sublingual papilla.

3. Ask the model to move the tip of his or her tongue further backwards along the roof of the mouth, thus placing some tension on the sublingual mucous membrane. The sublingual plica will now be seen to be situated at the crest of a broad (0.5 cm) swelling in the sublingual region, the **sublingual eminence.** This eminence, or swelling, is due to the underlying **sublingual gland.** Thus the **submandibular duct** runs along the superior surface of the **sublingual gland.** (Note that some textbooks use the term sublingual plica or fold to include what is here described as the sublingual eminence: in that usage, the term sublingual fold would include both the narrow fold within which the submandibular duct lies, as well as the broader swelling caused by the sublingual gland.)

Palpate gently over the sublingual eminence, and feel the extent of the sublingual gland on either side. Review the situation of the mylohyoid muscle (pages 4 and 24) and visualize that the sublingual gland and the submandibular duct lie superior to the muscle. Also palpate the **submandibular gland** which lies mainly **inferior** to the muscle (see fig. 8, page 16), and curves over its posterior edge, where the gland continues into the duct. This palpation may be facilitated by bimanual examination (i.e., by placing the other hand below the skin of the lower jaw to give upward counter pressure).

4. Note that the sublingual gland does not normally empty its secretion into the submandibular duct, nor secrete on the sublingual papilla, (though it may, unusually, do so). The normal anatomical site of sublingual secretion is through small ducts running directly from the gland to empty at various sites along the course of the sublingual fold.

ROOF OF THE MOUTH

The roof of the mouth is constituted by the palate. The anterior two thirds of the palate are known as the **hard palate,** since this part is comprised of a thin layer of bone covered by soft tissues. The posterior third of the palate is known as the **soft palate,** as it contains no bony component.

Review the bony structure of the hard palate on an isolated skull: the anterior portion of it is formed by the **palatine process of the maxilla** (page 10), whereas the posterior portion is formed by a separate bony sheet known as the **horizontal plate of the palatine bone.** Note that the soft palate does not continue back horizontally, but slopes downwards and backwards. Posteriorly, it ends in a free edge, in the midline of which a rounded enlargement, the **uvula,** is seen. Observe the inferior surface of the palate, noting color differences between the soft and hard palate. Also note the small openings of **palatine glands,** the small salivary glands comparable to the buccal and labial glands, the largest openings of which can be seen with the naked eye. In particular, two very large such glandular openings may be present, close to the midline, at the boundary between the hard and soft palates. If present, these are known as the **palatine foveolae** (foveola means depression). In the hard palate, a midline division is seen; the **palatine raphe (seam).** Close to the anterior extremity of the palatine raphe, a small swelling is often visible. This is the **incisive papilla** (incisive meaning related to the **incisor** or cutting **teeth;** papilla meaning nipple-like projection). This papilla is very sensitive to touch. Irradiating from the incisive papilla, and from the anterior part of the palatine raphe, several transverse folds of mucous membrane are seen; these are the **transverse palatine rugae (folds).**

On palpation, confirm the presence of palatine glands, giving a consistency similar to that previously noted for the labial and buccal glands. Palpate the consistency of the hard palate, noting that it differs as you proceed from the median plane out laterally. Close to the palatine raphe, the consistency is firm, due to deep-lying glands and, far anteriorly, fatty tissue in the submucosal layer. Laterally, as the palpating finger approaches the angle between the palatine and the alveolar processes of the maxilla (compare with a skull), the consistency becomes softer to the palpating finger, due to the presence of loose connective tissue containing blood vessels and nerves. Finally, as one proceeds further laterally to the continuity between the hard palate and the gingiva, the consistency becomes firm again, as is characteristic for the gums.

While observing the soft palate, ask your model to say "ah." Note that the soft palate moves. Similarly, ask your model to attempt to swallow, keeping the mouth open, and observe the movements of the soft palate in this action. You can conclude from your observations that the soft palate contains muscle. You have previously studied one of these muscles, the **palatoglossus muscle** (page 10). Other muscles of the palate will be studied elsewhere. Recall, however, (pages 15–18), that palatoglossus is to be regarded as a palatine rather than a tongue muscle, for reasons of its function as well as its nerve supply. The tongue muscles are supplied by the twelfth cranial nerve (page 15). The nerve supply to the palatine muscles is complex. Most of the muscles receive motor fibers which originate from the eleventh cranial nerve, but which leave this nerve to join the tenth cranial nerve (the vagus nerve) and reach the palate in a branch of the latter.

Complete your study of the soft palate by again palpating the **hamulus** of the **pterygoid process** (page 27). The hamulus is palpable, through the soft palate, about 1.5 cm posteromedial to the last molar. Review, on a skull, the hamulus, projecting down as the lower free tip of the posterior border of the **medial pterygoid plate.** Review also the **lateral pterygoid plate.** These two plates both constitute parts of the **pterygoid process** (of the **sphenoid bone of the skull**). Note again, that the medial pterygoid plate forms the posterior part of the lateral wall of the **nasal cavity.**

Now review the material studied in this session, by reidentifying all the structures and reviewing the meanings of the terms which are **emphasized** in the text.

Laboratory Session 3

Begin this laboratory session by identifying on your cadaver the vestibule of the oral cavity. You should review its extent and its boundaries at this time (page 9), identifying on your specimen each of the emphasized structures you previously studied on a living model.

Pull the tongue downwards and medially, thus stretching the palatoglossal arch, which then becomes easily identifiable. Just anterior to it, the **pterygomandibular raphe** (page 27) is palpable. Start superiorly, by palpating the **hamulus** through the soft palate, posteromedial to the last molar (page 27). Your objective in the dissection that follows is to remove the mucous membrane lining of the cheek, to display the medial surface of the buccinator muscle. The muscle should be traced posteriorly to its origin from the pterygomandibular raphe (page 27). Thereafter, the origin of the **superior pharyngeal constrictor,** from the posterior aspect of the pterygomandibular raphe (page 27) and the adjacent pterygoid and mandibular structures (pages 27 and 28), will be studied.

In the following section, you will dissect the medial aspect of the buccinator muscle. In order to gain access to this from the vestibule of the oral cavity, it is necessary to open the jaws of your cadaver. Access may be difficult in subjects who are not edentulous. Fixation of the muscles with the jaw in the closed position at the time of embalming of the cadaver may make it very difficult to depress the mandible in your specimen. If necessary, you should make use of metal instruments, which should be wedged between the teeth to increase mobility at the temporomandibular (jaw) joint.

Make an incision in the mucous membrane overlying the pterygomandibular raphe (i.e., from the site in the soft palate where the hamulus can be palpated, to the inferior attachment of the raphe, posterior to the posterior extremity of the **mylohyoid line** [page 28] and close to the last mandibular molar tooth). Reflect the mucous membrane anterior and posterior to this incision **very carefully.** The underlying muscle fibers of the buccinator muscle anteriorly, and, in particular those of the **superior constrictor of the pharynx** posteriorly, are extremely thin and delicate, and very easily damaged. With care, the raphe and its muscular origins both anteriorly and posteriorly can be very clearly demonstrated by this approach. Dissect the mucous membrane posteriorly, to meet your previous incision over the palatoglossus, thus exposing the superior constrictor (fig. 13, page 34), and then anteriorly, gently elevating it from the medial surface of the buccinator.

The buccinator is a very thin, flat, rectangular muscle. The posterior border of the rectangle is that edge which attaches (page 27) to the anterior edge of the pterygomandibular raphe. Anteriorly, the superior fibers of the muscle pass inferiorly, and the inferior fibers superiorly, to decussate in the **labial commissure,** which you previously palpated in a living subject (page 11). After passing through the commissure, the buccinator fibers insert into the lips, blending with the **orbicularis oris** muscle. Because of the decussation, the superior fibers of the buccinator insert into the inferior lip, and vice versa. The rectangular muscle also has a superior and an inferior attachment. The superior border attaches to the **alveolar process of the maxilla,** and the inferior border attaches to the **alveolar part of the mandible.** You should attempt now to remove as much as possible of the mucous membrane of the cheek, starting posteriorly at the pterygomandibular raphe and proceeding anteriorly. Identify the **parotid papilla** (page 26) as you proceed anteriorly. As you reflect the mucous membrane, preserve the papilla for later study.

Turn now to the external aspect of your specimen. Identify the **parotid duct.** Follow the duct from its origin at the anterior border of the **parotid gland** to its termination on the **parotid papilla** by dissecting it out anteriorly and then medially. This will involve removing piecemeal the **buccal pad of fat,** through which the duct passes as it bends medially. You will also follow the duct through the buccinator muscle, and thus remove a small portion of the muscle to expose the terminal portion of the duct as it approaches the papilla. In the course of this dissection you will work through the plane of the lateral surface of the buccinator muscle, which you should demonstrate by removing fascial coverings and fat.

Study the **gingiva,** identifying the various portions of it, as previously described (pages 28–29).

THE SUBLINGUAL REGION

Dissect the sublingual region as follows.

1. You have previously separated the mucous membrane of the side of the tongue from its underlying structures and studied the relationships of the hyoglossus muscle, the hypoglossal nerve, and the lingual artery. Now complete the elevation of this layer of mucous membrane from the sublingual region

by stripping it out laterally, all the way to its continuity with the mandibular gingiva. As you separate the mucous membrane from the underlying structures, branches of the lingual nerve will come into view. This nerve, you will recall, crosses the floor of the mouth, from the inner surface of the mandible toward the side of the tongue (page 28). In its path to the side of the tongue, the lingual nerve comes in relationship to the lateral surface of the hyoglossus muscle, a little above the hypoglossal nerve. In this dissection, you will also encounter the submandibular duct (page 29). Carefully dissect in the region where the submandibular duct and the lingual nerve approach each other: the lingual nerve is initially lateral to the duct, as the nerve leaves the inner surface of the mandible and crosses the floor of the mouth. On the lateral surface of hyoglossus, the submandibular duct crosses the lingual nerve, the duct lying superior to the nerve. After passing inferior to the duct, the nerve recurves up, to lie first medial to and then above the duct. Thus the nerve starts off lateral to the duct and ends up medial to it, having looped below it (figs. 14 and 15, page 34). Inferior to the submandibular duct, you should also dissect out the sublingual gland (page 29). The gland lies immediately superior to the mylohyoid muscle (pages 28–29). You should dissect the relationship of the gland to the muscle (page 30), and follow the muscle out laterally to its attachment to the **mylohyoid line,** on the inner surface of the mandible (page 28). Reidentify the mylohyoid line on a disarticulated mandible. Frequently, a small extension of the sublingual gland will appear to penetrate through the mylohyoid muscle, to its inferior surface.

2. In relation to that part of the lingual nerve which lies on the lateral surface of the hyoglossus, look for the **submandibular ganglion,** a round structure about 1 to 2 mm in diameter, suspended from the lingual nerve by thin nerve fibers, about 0.5 cm below the nerve. This ganglion is a component of the autonomic nervous system (pages 5–6), and is the site of connections **(synapses)** between neurons that mediate the autonomic function of salivation (page 26). It is characteristic of centrifugal pathways of the autonomic nervous system (page 6), in contrast to those of the somatic nervous system, that synapses occur between the **nerve cell process** of one neuron and the **nerve cell body** of the next, in **ganglia,** outside of the CNS (i.e., in the PNS). (The nerve cell body of a neuron is the central part of the nerve cell, including its nucleus. From it the nerve cell processes [page 5] extend. A **ganglion** is a cluster of nerve cell bodies outside the CNS. In some ganglia, such as the submandibular ganglion, synapses occur on the nerve cell bodies within the ganglion.)

3. Trace the branches of the lingual nerve to their destinations, in the mucous membrane of the tongue, and branches from the submandibular ganglion to their destinations in the sublingual (and submandibular) glands. Follow the submandibular duct posteriorly, where, with careful dissection, you should be able to establish its continuity with the **superior (deep) portion of the submandibular gland** (page 30). You should also carefully dissect the mylohyoid muscle posteriorly to establish that the deep portion of the submandibular gland that you have exposed by following the duct posteriorly is situated at the posterior, free edge of the mylohyoid muscle. A second, superficial, portion of the submandibular gland is situated inferior to the mylohyoid muscle (page 30). Thus the submandibular gland "hooks around" the posterior edge of the mylohyoid muscle. The superficial portion of the submandibular gland will be dissected at a later stage. However, at this time, you can identify, on the skinned, lateral surface of the head, inferior to the base (lower border of the body) of the mandible, the **superficial portion of the submandibular gland** (fig. 8, page 16).

4. By carefully replacing the structures you have just dissected, review the relationships, in the floor of the mouth, of the mucous membrane, the lingual nerve, the submandibular duct, the deep portion of the submandibular gland, the sublingual gland, the hyoglossus muscle, the geniohyoid muscle, the mylohyoid muscle, and the superficial portion of the submandibular gland. Inferior to the mylohyoid muscle, you will subsequently study in greater detail the anterior belly of the digastric muscle (page 24 and fig. 16, page 34). To complete this reconstruction of the structures dissected in this region, note that inferior to the muscles of the floor of the mouth lie subcutaneous tissue, in which the **anterior jugular vein** can be seen (fig. 10, page 23), and, finally, skin.

ROOF OF THE MOUTH

Study the roof of the mouth, identifying the features previously observed in the living model (pages 30–31). Carefully dissect a short distance into the cleavage plane between the soft tissues inferior to the bony palate, and the bone of the palate itself. You should establish the close adherence of the submucosa to the **periosteum** (the thin layer of connective tissue which covers bony surfaces). This fusion of mucosa and periosteum is known as **mucoperios-**

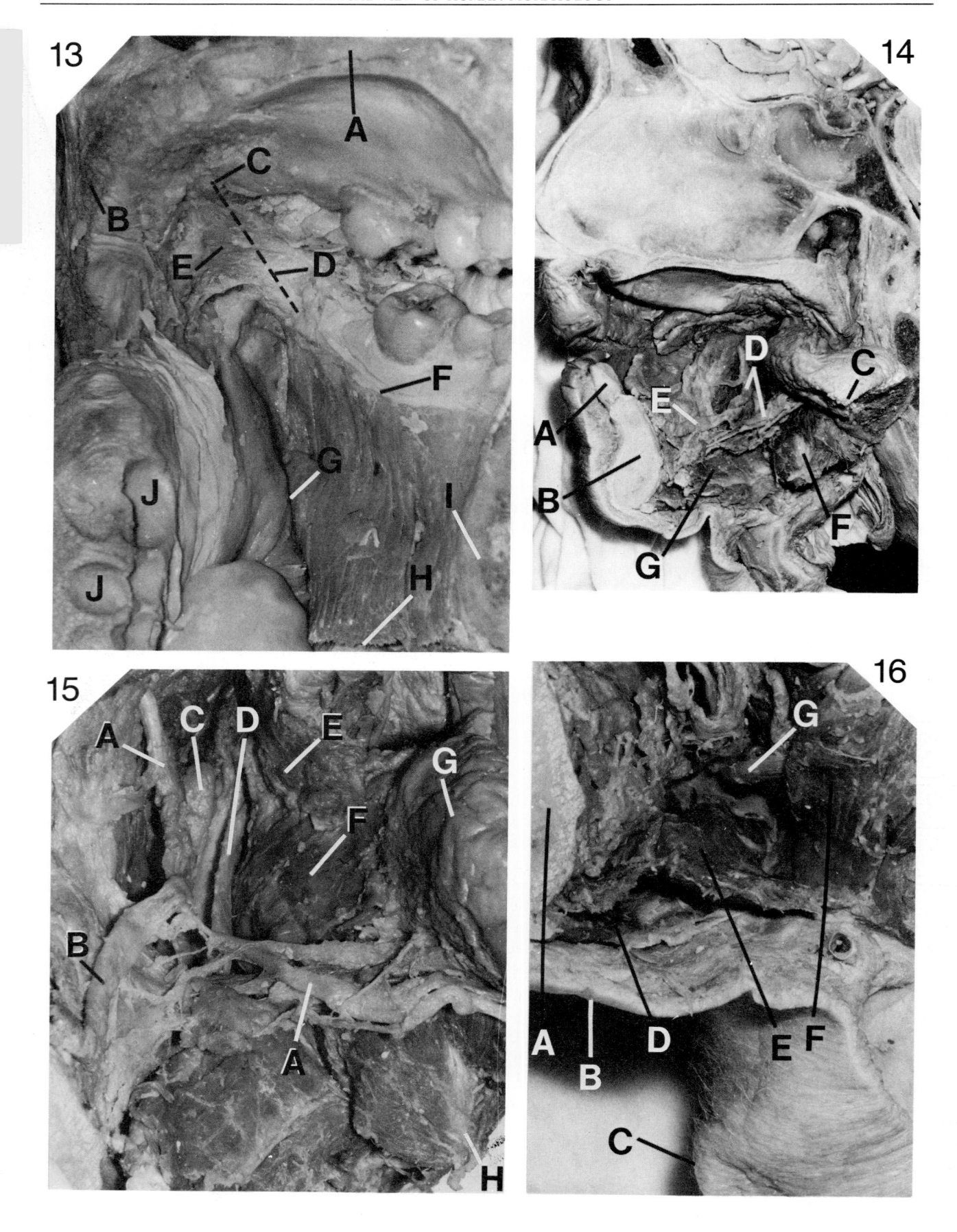

34 YOUR PATIENT'S ANATOMY: A CLINICAL VIEW OF HUMAN MORPHOLOGY

Figure 13. Sagittal section, left side of oral cavity.
A. Sagittal section of hard palate
B. Sagittal section of soft palate
C. Hamulus
D. Site of pterygomandibular raphe **(broken line)**
E. Superior pharyngeal constrictor
F. Mylohyoid line
G. Posterior, free edge of mylohyoid
H. Sagittally cut edge of mylohyoid
I. Sagittal section through mandible
J. Tooth marks (from embalming), in lateral aspect of tongue, (being held to expose mylohyoid)

Figure 14. Sagittal section, right side, sublingual region.
A. Incisor tooth
B. Mandible
C. Tip of tongue, pulled back
D. Lingual nerve, looped under submandibular duct (E)
E. Submandibular duct
F. Genioglossus
G. Geniohyoid

Figure 15. Right sublingual region, seen from front and above.
A. Lingual nerve, looped under submandibular duct (B)
B. Submandibular duct
C. Deep part of submandibular gland
D. Hypoglossal nerve
E. Styloglossus
F. Hyoglossus
G. Lateral edge of tongue
H. Reflected origin of genioglossus

Figure 16. Sagittal section, right side, chin region and floor of mouth.
A. Mandible in sagittal section
B. Inferior surface of chin
C. Front of neck (thyroid cartilage)
D. Anterior belly of digastric
E. Mylohyoid
F. Geniohyoid (anterior end detached and turned upwards)
G. Hypoglossal nerve

teum. Within the mucosa, distinguish, by fine dissection, the palatine glands (page 30).

Posterior to the hard palate, you should identify, in the sagittal section, the muscular layer of the soft palate. Preserve this layer for future dissection.

BONY AND RELATED STRUCTURES PALPABLE IN THE VESTIBULE AND LATERAL BOUNDARY OF THE ORAL CAVITY PROPER

Reidentify, in your cadaver, the structures you previously palpated within the vestibule, and in relation to the lateral wall of the oral cavity proper, in the living model (page 28). These should include the zygomatic process of the maxilla and the ramus of the mandible, including the coronoid process (page 28). By careful dissection, follow the lingual nerve posteriorly to its situation on the medial aspect of the mandible. Palpate the posterior extremity of the dental arcade, and, posterosuperior to the last molar, the coronoid process, to get your bearings in relationship to the path of the lingual nerve. In the portion of its course which you have been dissecting, the nerve is related mainly to the floor of the mouth (remaining, of course, deep to the mucous membrane when the latter is intact). As you trace the nerve posterosuperiorly, observe that it leaves the vicinity of the oral cavity by passing inferior to the lowest fibers of the superior pharyngeal constrictor.

Clinical Session 4

■ **TOPICS**

The Masticatory (Chewing) Apparatus
Maxilla
Mandible
Masseter Muscle
Temporal Muscle
Medial Pterygoid Muscle
Lateral Pterygoid Muscle
Temporomandibular Articulation (Joint)

The masticatory apparatus includes the upper and lower jaws (maxilla and mandible), the muscles of mastication and the joint between the upper and lower jaw, called the temporomandibular joint. In addition to the jawbones themselves, the actual articulation and the four primary muscles of mastication, certain additional bony structures, ligaments, and secondary (accessory) muscles play a role in mastication. Some of these structures will be dealt with in later sessions.

In this session you should identify and examine the structures described on your model, making free use of the anatomical specimens provided, as well as atlas illustrations, to assist you. The structures to be examined in this session are palpable both externally, through the skin, and internally, within the oral cavity. Remember to wash your hands anew, each time you examine your model within the oral cavity, after having handled unsterile objects such as skulls and books.

1. Identify by external palpation (through the skin) the following features of the maxilla.

The inferior border of the **orbit,** or **infraorbital margin** (infra meaning below), is formed in its medial half by the **maxilla.** Where this articulates laterally with the **zygomatic bone,** a notch is palpable in the infraorbital margin. This notch marks the upper end of the **zygomaticomaxillary suture.** The inferior extremity of the zygomaticomaxillary suture can also be palpated, as can the entire extent of the suture in many subjects. Examine the course of the suture on the skull, and attempt to palpate it on your model. After having identified this suture, palpate the maxilla as it extends up onto the medial aspect of the orbit, projecting upwards as the **frontal process of the maxilla,** which articulates with the **frontal bone.** Now palpate the **zygomatic process of the maxilla** as it curves upwards and laterally to articulate with the **zygoma** (zygomatic bone, page 28).

On the anterior surface of the maxilla, just inferior to the middle of the infraorbital margin, palpate the **infraorbital foramen,** after having identified it on the skull. Care should be exercised, as an important sensory nerve emerges through this foramen, and it may be sensitive to deep pressure. Remember (page 25) the name of this nerve, its cranial nerve origin, and the extent of its distribution in the sensory supply of the face.

The mandible should now be palpated through the skin. Start at the chin, where a midline groove marks the fusion of right and left halves of the mandible at the **symphysis menti** (sym means together; phyein means to grow; mentum means the chin). Proceed posterolaterally, palpating the inferior border of the **body of the mandible.** Before you reach the **angle of the mandible,** you will encounter a notch in the inferior border, visible on the skull, and palpable on your model, just anterior to the anterior border of the **masseter muscle** (which your model should tense by clenching his or her teeth). The **facial artery** curves around the inferior border of the body of the mandible in this notch, passing from the deep surface to the superficial surface, just in front of the masseter. Palpate the pulsation of this artery at this site (fig. 8, page 16). (Note that this artery then passes up in the superficial tissues of the face, giving off, among other branches, the **labial arteries** which you previously palpated in the lips (page 11.) Proceed with palpation of the mandible to the **angle,** and then the posterior border of the **ramus.** As can be seen on the skull, the posterior border of the ramus continues up into the **condyloid process,** so called because at its upper extremity the articulating **condyle** is perched. A condyle indicates a knuckle; this descriptive term is frequently used for the rounded extremities of bones which participate in articulation (joining) with other bones. With the jaw open, the anterior border of the ramus of the mandible can be palpated through the skin, and observed to proceed upwards toward a second process known as the **coronoid process** (page 28) (coronoid indicates hooked, or shaped like a crow's beak).

Within the oral cavity, palpate the **zygomatic process of the maxilla,** the **alveolar process of the maxilla,** and the **palatine process of the maxilla** (review from page 28). By following the alveolar process posteriorly, the finger will encounter the **tubercle of the maxilla,** a small swelling on the posterior surface, just above the last molar. By palpating medially from the tubercle, you can, in some cases, just reach the **lateral pterygoid plate,** which you have studied previously (page 27).

Palpate the mandible from within the oral cavity, reviewing the features previously identified (page 28). In particular review the **mylohyoid line,** and the course of the **lingual nerve** in relationship to this. In the vestibule, passing your finger posteriorly, palpate the anterior border of the ramus of the mandible, as it ascends towards the coronoid process.

2. PRIMARY MUSCLES OF MASTICATION

Masseter Muscle
While your model intermittently clenches his or her teeth, palpate the extent and direction of the masseter muscle (page 26), from the **zygomatic arch** to the lateral surface of the **ramus of the mandible** and the **angle.** Consult your anatomical specimens and atlas illustrations as needed. Within the mouth, palpate the anterior border of the masseter through the mucous membrane of the cheek with one index finger. Then palpate the anterior border bidigitally, with one index finger within the cheek and the other feeling the anterior edge of the muscle through the skin. Follow the muscle superiorly to the zygomatic arch, both internally and externally. Note that motor innervation of the masseter and other muscles of mastication is through the **mandibular division of the fifth cranial nerve.** Compare this with the motor source of supply to the muscles of facial expression. From the direction of the muscle fibers of masseter, what actions could one predict this muscle may have?

Temporalis Muscle
This muscle attaches extensively to the lateral side of the skull, or **temporal region** (colloquially called the temple). Observe its position and extent in specimens and illustrations, and then palpate the muscle through the skin and hair, as your model chews or clenches the teeth. Attempt to palpate the fibers of the muscle as they converge on the **coronoid process,** noting that the anteriormost fibers are vertical, whereas the posteriormost fibers are horizontal, and the remaining fibers spread out as a fan between these two extremes. Palpate first the anterior, vertical fibers, and then the posterior, horizontal fibers, first while the model clenches his or her teeth, and thereafter while the model **retracts** the mandible (draws it backwards) from a **protracted** (pulled forward) position. What do you conclude about the functions of the different fibers of this muscle? Now palpate the temporalis muscle from within the oral cavity. This can be done by palpating the coronoid process, high up in the vestibule, as described previously (pages 28 and 37). Since the fibers of the muscle insert into the coronoid process, tension in the muscle can be felt as the model bites while the examiner palpates just above the coronoid process. Be aware of the proximity of the masseter muscle and attempt to appreciate the separate tension within the inserting temporalis fibers.

Medial Pterygoid Muscle
Palpate on the medial aspect of the angle of the mandible by indenting the skin just behind and below the angle. Next, push inwards and forwards, and palpate on the medial aspect of the angle to feel as your model clenches his or her teeth, the insertion of the medial pterygoid muscle. The muscle reaches this insertion from its origin on the medial aspect of the **lateral pterygoid plate.** The fibers pass inferolaterally and posteriorly, about parallel to those of masseter. The two muscles can be thought of as forming a sling for the angle of the mandible. Visualize the position of the medial pterygoid muscle on a skull, as well as in anatomical specimens and atlas illustrations.

The medial pterygoid muscle can also be palpated, though with difficulty, from within the oral cavity. To do this, reidentify the **pterygomandibular fold** (page 27). By placing an index finger posterior to this raphe, and palpating laterally toward the medial aspect of the ramus of the mandible, the muscle can be felt in some subjects. However, the mucous membrane at this site is very sensitive in many subjects, and this may render this particular exercise impossible. On the other hand, the following exercise should be possible in most subjects. Refer to the exercise described on pages 37–38, in which you palpated the **maxillary tubercle** and lateral aspect of the lateral pterygoid plate. Although most of the fibers of the medial pterygoid muscle arise from the **medial** aspect of the lateral pterygoid plate, a small, second "head" of the muscle takes it origin from the **lateral** aspect of the lower part of the lateral pterygoid plate and from the adjacent maxillary tubercle. These fibers can be palpated by placing a finger in the **superior alveolobuccal sulcus,** on the medial aspect, far back behind the last molar, as described previously. The muscle will be activated when your model clenches his or her teeth.

Lateral Pterygoid Muscle
This muscle can be palpated by repeating the exercise you have just performed, but attempting to reach even higher on the lateral aspect of the lateral pterygoid plate. Also, the exercise should be performed while your model attempts to **protract** the mandible against the resistance of his or her hand. The muscle fibers you are attempting to palpate extend from the lateral surface of the lateral pterygoid plate pos-

terolaterally toward the **mandibular condyle.** Thus they are functional in pulling the mandible forwards. Visualize the position of these muscle fibers on an isolated skull, as well as in anatomical specimens and illustrations. Note that the lateral pterygoid muscle also has two heads: one that takes its origin, as described previously, from the lateral surface of the lateral pterygoid plate (the **inferior head**); and a **superior** head that takes its origin from the horizontal bony surface which lies at right angles to the lateral surface of the lateral pterygoid plate, i.e., the **infratemporal surface** of the **greater wing** of the **sphenoid bone.** (This surface is the superior boundary, or roof, of the **infratemporal region,** which the muscle is therefore situated in.) The fibers of the two heads converge to insert mainly in a pit on the anterior surface of the condyle. The right and left lateral pterygoid muscles, acting together, function in concert to protract the mandible, as mentioned earlier. What function would the lateral pterygoid muscle of one side have, acting on its own?

3. THE TEMPOROMANDIBULAR JOINT

Palpate the lateral extremity of the **condyle of the mandible,** just anterior to the **external auditory meatus.** By referring to an isolated skull you will appreciate the close proximity of these structures, which allows one to use the meatus as a landmark in finding the condyle. Just superior to the condyle, a slight depression can be palpated, especially easily when the mouth is opened, and, above the depression, another bony structure is palpable. Examination of the skull will reveal that the depression is situated between the condyle and the **zygomatic arch,** or, more specifically, the posterior component of this arch, known as the **posterior root** of the **zygomatic process of the temporal bone.** What you have examined here are the **articulating skeletal elements,** and the **joint space** between them, of the temporomandibular joint.

A **joint** or **articulation** is a connection between two skeletal structures. Joints are usually classified into three main types: **synovial, cartilaginous,** and **fibrous,** according to the nature of the tissues intervening between the two skeletal elements. The temporomandibular joint is of the **synovial** type. This group of joints have, between the two skeletal elements, an actual **joint space,** surrounded by a more or less stable **capsule** which attaches the free ends of the two skeletal structures. Because of the joint space, a relatively large degree of free movement is possible between the two articulating bones. In contrast, in cartilaginous joints, the space between the articulating bone ends is filled with cartilage, and in fibrous joints with tough fibrous tissue. Both of these tissues allow less movement than the lumen within the capsule of a synovial joint. The capsule of a synovial joint is made of fibrous tissue and is reinforced on the outside by tough fibrous bands, known as **ligaments,** which extend between the articulating bones. The inner surface of the fibrous capsule of a synovial joint is lined by a soft membrane which secretes a cloudy fluid lubricant, which gives this type of joint its name (syn means with; oon means egg; the appearance of the **synovial fluid** being reminiscent of egg white). Examples of cartilaginous and fibrous joints are also found in the skull. Anteroposterior growth of the skull takes place until the age of 25, at the site of a **cartilaginous joint** or **synchondrosis** (syn indicating joint; chondros indicating cartilage) on the inferior surface of the skull between the **sphenoid** and **occipital bones** (just anterior to the largest of all foramens, the **foramen magnum**). Observe this **spheno-occipital synchondrosis** in an appropriate immature skull specimen. The various **sutures** which are seen between most other skull bones are examples of **fibrous** joints.

On your model palpate, on both sides simultaneously, the **posterior roots** of the **zygomatic processes of the temporal bones** mentioned previously and the joint spaces (the **slight depressions** between these roots and the condyles). Then place a palpating index finger on the lateral extremity of each condyle while your model performs the following exercises.

1. The model repeatedly opens and closes his or her mouth slightly, the incisor teeth of the upper and lower jaws separated by not more than 1 to 2 cm. During this movement, the condyle may be felt to move slightly. The movement that can be detected is a **rotation** about a **transverse** axis. However, the basic position of the condyle does not change in this movement.

2. The model opens his or her jaw as wide as possible. In this movement, the condyle performs the rotation as previously, but thereafter, in the lower portion of the opening movement (i.e., as the incisor teeth move further away from each other than 2 cm), the condyles also move **anteroinferiorly.** Thus the basic position of the condyle actually changes, a type of movement which, in contrast to rotation, is called **translation.** Stated in another way, in rotation, the posterior part of the condyle moves up while the anterior part of the condyle moves down; in translation, all parts of the condyle move in the same direction simultaneously.

3. The model opens the jaw slightly and then moves it strongly to the left side. In this movement, the left condyle performs a rotation about a **vertical** axis; the right condyle performs a translation as in the movement described in #2, as well as a translation medially (i.e., toward the left side).

From your study of the muscles of mastication, determine which muscles are involved in the movements outlined in the preceding list.

Examine the shape of the condyle on an isolated mandible. Note the direction of the long axis of the condyle, and that an imaginary line projected through the long axis of the right condyle would meet a similar line through the long axis of the left condyle at a point just in front of the large foramen in the base of the skull, the **foramen magnum,** through which the brain and spinal cord are continuous. Having observed the articulating surface on the condyle of the mandible, examine the corresponding surface on the inferior surface of the temporal bone. Note that in the resting position of the mandible (i.e., jaws almost closed) the condyle fits into an elongated depression which corresponds in shape to the condyle. This is the **mandibular fossa** of the temporal bone. Laterally it is limited by the posterior root of the zygomatic process of the temporal bone which you palpated on your model above the joint space (page 39). Anteriorly, the smooth surface of the mandibular fossa continues onto a bony protruberance, called the **articular tubercle.** This also constitutes part of the articulating surface of this joint, since the condyle of the mandible moves forward, from its position in the mandibular fossa, to a position inferior to this tubercle, when the jaw is open wide, as you were able to palpate in the exercise (#2 above) performed previously. (Note that the articular tubercle also constitutes the **anterior root** of the **zygomatic process of the temporal bone.** Review now the posterior root, studied previously [page 39].)

Correlating your findings on palpation of the mandibular condyle during various movements, with your inspection of the articulating surfaces of the temporomandibular joint performed here, it will be clear to you that the movements performed under #1 in the previous list take place with the condyle within the situation of the mandibular fossa. The movement in #2 involves movement of the condyle from a position within the mandibular fossa to a position inferior to the articulating tubercle. The movement in #3 involves vertical rotation of the left condyle within the mandibular fossa, and movement of the right condyle from the mandibular fossa anteroinferiorly and medially to a position inferior to the articulating condyle.

For completeness it will be mentioned here, although the following facts cannot be confirmed by palpation in the normal subject, that the synovial cavity of the temporomandibular joint is subdivided into a superior and an inferior compartment by a tough, fibrous **disc.** Study an anatomical specimen and illustrations to visualize the **intraarticular disc** of this joint. The disc is so shaped that its superior surface follows the contours of the temporal surfaces of the joint: thus the disc has a posterior portion which follows the concavity of the mandibular fossa, and an anterior portion which covers the convexity of the articular tubercle. As a consequence of the compartmentalization of the joint which the disc creates, the movement which occurs in exercise #1 in the previous list takes place entirely within the lower compartment of the joint. In this movement, the disc remains in the position described earlier. In the movement which occurs in exercise #2, the disc itself slides anteroinferiorly, carrying the condyle with it, to a position inferior to the articulating tubercle. In the movement in exercise #3, the disc on the left side more or less retains its position, while the disc on the right side moves anteroinferiorly and medially. Also not palpable on the normal living subject, but to be mentioned here for the sake of completeness, are the ligamentous reinforcements of the articular capsule. A **lateral temporomandibular ligament** (see specimens and illustrations) extends from the zygomatic arch at the site of the articulating tubercle, posteroinferiorly to the lateral tip of the condyle. By the nature of the direction of these fibers, which movement would this ligament restrict?

On the medial side of the joint, no true joint ligaments (in the sense of capsular reinforcements) exist. Instead, two ligaments, placed some distance from the joint itself, provide slight additional stability. The first is the **stylomandibular ligament,** which extends from the styloid process (page 18) to the posterior border of the ramus of the mandible. The second ligament is the **sphenomandibular ligament.** The sphenomandibular ligament is attached superiorly to the **sphenoid spine.** This small, bony projection may be observed immediately medial to the mandibular fossa of the temporal bone. Note the suture between the temporal bone and this spine, confirming that the spine belongs to a separate bone, the **sphenoid bone.** Confirm that the parts of the sphenoid bone you previously have studied (i.e., the pterygoid process [page 27] and the inferior surface of the greater wing [page 39]) are parts of the same bone that you are now observing.

The sphenomandibular ligament is inferiorly attached to a small tongue of bone situated just anterior to the **mandibular foramen** (page 28) on the medial surface of the ramus of the mandible. The small tongue-like process is called the **lingula** (lingua meaning tongue). Visualize the position of this ligament in a skull with the mandible in place and observe it also in anatomical specimens and illustrations.

Complete this session by reviewing the concepts related to each emphasized key word.

Laboratory Session 4

The objective for this laboratory session is to study, by dissection, the masticatory apparatus which you previously examined in a living subject (page 37). Structures will be initially palpated from the external aspect of a sagitally sectioned head and neck specimen, then palpated from the oral aspect, and then dissected by a combined internal and external approach.

Palpation of deep structures through the skin, in a living subject, is easier to perform than similar palpation on an embalmed specimen because of hardening of the soft tissues in the latter. However, identification of deep-lying, bony landmarks is considerably facilitated by removal of the skin, prior to palpation. You should make free use of a disarticulated skull and an atlas in the ensuing dissection.

Begin by confirming, by palpation from the external aspect of the specimen, the bony features of the maxilla and mandible you previously studied on a living model (pages 37–38). Identify, on the disarticulated skull and on your specimen, the infraorbital foramen, which is easily palpable inferior to the orbit. In your previous dissection of the superficial tissues of the face (page 25), you will have dissected the large sensory branch of the fifth cranial nerve, the **infraorbital nerve.** Trace the various branches of this nerve centripetally (i.e., inwards) to the stem, emerging from the infraorbital foramen. In the lower jaw, you should similarly dissect the branches of the **mental nerve** to the **mental foramen,** which you should also identify on a mandible. Other sensory branches of the fifth cranial nerve should also be traced to their foramina, but none of the other foramina are as large, and as easy to identify as the above two.

Palpate in the oral cavity to identify those features of the maxilla and the mandible which you previously (pages 37–38) identified from within the mouth in the living subject. Some of these landmarks will be more easily accessible in the cadaver than in the living subject. In particular, the maxillary tubercle and the lateral pterygoid plate will be more readily identified.

TEMPORALIS

Prior to dissecting the temporalis muscle reidentify and review overlying and neighboring structures that might get damaged or removed in the ensuing dissection. Structures that you should review at this time are the fronto-occipital aponeurosis, the superficial temporal vessels, and superficial sensory nerves of the side of the face (pages 19–20 and 25). Identify, on a skull, the **temporal line,** arching back from the **zygomatic process of the frontal bone.** (The latter process extends from the frontal bone, to articulate with the frontal process of the zygomatic bone, thus forming the lateral border of the orbit.) The temporal line branches into superior and inferior lines.

Having identified the superior temporal line on a skull, palpate it on your dissection specimen. Inferior to this line, dissect through the superficial tissues of the face, and identify the tough, white layer of deep fascia (page 19), known as the **temporal fascia.** Incise into the temporal fascia, with a curved incision parallel to and 1 cm inferior to the temporal line, and elevate the fascia by inserting, deep to your incision, the back end of a pair of forceps. Elevate the temporal fascia above the cut by blunt (finger) dissection, and, where necessary, scissors dissection, up to the temporal line, from which you should detach the fascia superiorly. Inferior to the incision, reflect the fascia downwards. Observe as you expose the muscle the direction of the muscle fibers, correlating your observations with your previous findings on the actions of the different fibers in the different parts of this muscle, as studied on the living subject (page 38 and fig. 17, page 44). Continue reflection of the fascia inferiorly, as far down as its attachment to the zygomatic arch, where attachment should be clearly demonstrated. In so doing, you will observe that muscle fibers of the temporalis muscle actually, in part, take their origin from the temporal fascia. The muscle fibers pass deep to (i.e., medial to) the zygomatic arch, as the fascia attaches to the arch. Separate the attachment of the temporal fascia from the frontal process of the zygomatic bone. These muscle fibers should be gently eased away from the zygomatic bone, in order to follow them inferiorly, toward their attachment to the coronoid process of the mandible (page 38). From this anterior site, work your way posteriorly, elevating the temporalis muscle and detaching it from the temporal line and from the deep-lying bony surfaces of the **temporal fossa** (fig. 18, page 44) by scalpel dissection. Identify on your specimen, and on a skull, the component bony parts of this surface (i.e., **zygomatic bone, frontal bone, greater wing of the sphenoid bone, parietal bone,** and **squamous portion of the temporal bone).** With care, the periostial layer covering the bone can be elevated with the muscle insertion into it, thus separating the muscle completely from the bone, while clearing the latter of soft tissues. On the deep surface of the muscle, observe the

deep temporal nerves and vessels that supply the muscle.

SUPERFICIAL DISSECTION OF MASSETER AND DEEP DISSECTION OF TEMPORALIS

Review the area inferior to the zygomatic arch, to identify, as you did prior to dissection of the temporalis muscle, neighboring structures to the masseter muscle, which might be damaged in the following dissection. You should at this time, reidentify the parotid gland and duct; branches of the facial nerve radiating from the parotid gland and passing to individual facial muscles; the more important of these muscles, including the zygomaticus major; sensory branches of the fifth cranial nerve; and branches of the facial artery and vein. Identify at this time a tough deep fascia, almost comparable in thickness to the temporal fascia, which covers the parotid gland and tightly binds it to the masseter muscle. This fascia is known as the **parotideomasseteric fascia.** (Note that the temporal fascia and the parotideomasseteric fascia are the only two sites in the face where a true **deep fascia** exists. Elsewhere in the face, as indicated previously [pages 19 and 25], deep fascia is absent, and the facial muscles and related structures lie in the plane of superficial, subcutaneous tissue.)

Carefully dissect the parotideomasseteric fascia from the parotid gland and the masseter muscle studying the direction of fibers of the masseter muscle as you do so. Correlate your observations on the muscle fibers with your previous deductions concerning the action of the muscle in a living subject (page 38). Observe that the parotid gland extends medially, behind the ramus of the mandible, to a deep portion that lies medial to the ramus.

Having displayed the direction of the superficial fibers of the masseter muscle, define the anterior free edge of the muscle, sloping inferoposteriorly from the zygomatic arch toward the angle of the mandible. Using blunt dissection, elevate this anterior border of the muscle, separating fibers carefully from the lateral surface of the ramus of the mandible. Proceed with this, as you previously did with the temporalis muscle, displaying the lateral surface of the bone, and elevating the masseter muscle as far back as the posterior border of the ramus. When you have clearly separated the muscle fibers from the ramus, at the lower portion of the ramus (close to the angle and base of the mandible), insert a pair of forceps in the space between the muscle and the bone, and work your way upwards with the forceps toward the zygomatic arch. In so doing, you will be separating the **superficial fibers** of the masseter, which constitute a **superficial layer,** from deeper fibers. You should be able to elevate the anterior edge of the masseter to visualize these deeper fibers. Having done so, use scalpel dissection to separate the superficial fibers from the zygomatic arch. This will enable you to reflect the superficial portion of the masseter muscle, and the overlying parotid gland, inferoposteriorly. You should cut the parotid duct, leaving the anterior portion visible, as it passes medially through the buccal pad of fat to pierce the buccinator muscle, and leaving the posterior portion of the duct connected to the gland. This will allow more free reflection of the masseter.

At this point, you will have exposed the **deeper layer of masseter.** Carefully observe the direction in which the fibers pass. Having studied the deeper portion of masseter, make use of an electric saw, under supervision, to cut through the zygomatic arch at two sites: far anteriorly, by a vertical cut extending downwards from the posterior border of the frontal process of the zygomatic bone; and posteriorly, by a vertical incision just anterior to the articular tubercle (page 40). Now, retaining the attachments of the deep fibers of masseter, dissect the separated portion of the zygomatic arch away from its deep-lying tissues. You should proceed with this dissection with utmost care, as it will enable you to visualize relationships you may not be able to study in any other way. You should observe that the deep fibers of masseter take origin both from the lateral and the medial aspects of the arch, and that the deeper (medial) fibers have to be carefully dissected away from fibers of temporalis muscle. Thus temporalis fibers also arise, in part, from the zygomatic arch, and the two muscles merge with each other at this site. This is not surprising, in view of the fact, which this dissection will underscore, that the two muscles (and the other muscles of mastication) develop from the same embryological anlage, the **first branchial arch** of mesoderm. Your dissection will also allow you to observe the insertion of temporalis in detail. Note that the anterior, vertical fibers insert into, and anterior to, the coronoid process, and the posterior, horizontal fibers insert into the posterior aspect of that process. Some intermediate fibers of the temporalis muscle insert further back, on the ramus of the mandible.

Use careful scissor dissection (with the closed points of small, sharp pointed scissors) to gently dissect on the deep surface of the deep fibers of the masseter muscle. In this manner you will be able to display the **masseteric nerve and vessels,** which pass through the **mandibular notch** (the notch between the coronoid and condyloid processes of the mandible) to reach the deep surface of masseter. Preserve the nerve and vessels carefully. Follow

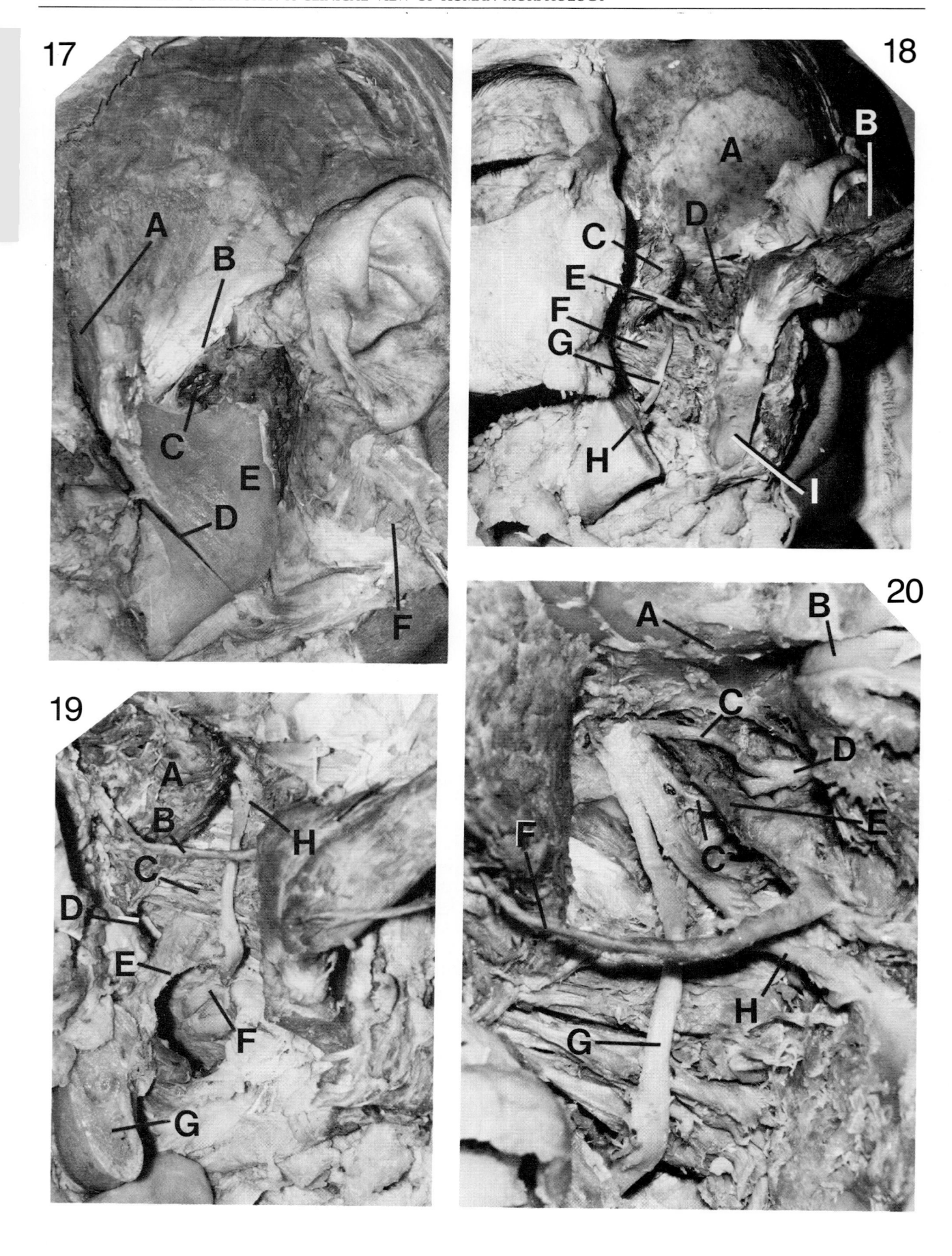

44 YOUR PATIENT'S ANATOMY: A CLINICAL VIEW OF HUMAN MORPHOLOGY

Figure 17. Elevation of temporalis and masseter muscles.
A. Vertical fibers of temporalis
B. Horizontal fibers of temporalis
C. Masseteric vessels
D. Saw cut through mandible
E. Lateral surface of ramus of mandible
F. Masseter (reflected from lateral surface of ramus of mandible)

Figure 18. Infratemporal region, exposed by rotating ramus of mandible at site of saw cut (see D, Fig. 17).
A. Temporal fossa
B. Temporalis muscle (reflected from surface of temporal fossa)
C. Anterior part of lateral pterygoid muscle
D. Posterior part of lateral pterygoid muscle (rotated with ramus of mandible)
E. Maxillary artery
F. Medial pterygoid muscle
G. Lingual nerve
H. Saw-cut surface of mandible
I. Saw-cut surface (rotated from surface indicated by H)

Figure 19. Detail of region shown in figure 18 (anterior part of mandible rotated).
A. Anterior part of lateral pterygoid muscle
B. Maxillary artery
C. Medial pterygoid muscle
D. Window into oral cavity, dissected in buccinator
E. Inferior fibers of superior pharyngeal constrictor
F. Lingual nerve
G. Saw-cut surface of mandible (rotated)
H. Chorda tympani nerve

Figure 20. Detail of region shown in figure 18.
A. Infratemporal surface
B. Mandibular condyle (displaced by rotation of ramus)
C. Roots of auriculotemporal nerve, encircling middle meningeal artery
D. Auriculotemporal nerve
E. Middle meningeal artery (unusually large in this specimen)
F. Maxillary artery
G. Lingual nerve
H. Inferior alveolar nerve

them to their insertion into the deep surface of the masseter, and dissect away from the bone a small portion of the muscle, at that site of insertion, as a tag or marker for the nerve and vessels. Having done this, reflect the isolated portion of the zygomatic arch and the attached deep fibers of masseter, downwards, leaving the inferior attachment of the masseter fibers to the lateral surface of the ramus of the mandible intact (fig. 17, page 44). In performing this reflection, you should take care to avoid damage to deep-lying structures. Note, for example, that you are very close to the temporomandibular articulation, and immediately deep to the area in which you are now dissecting you will soon encounter the insertion of the lateral pterygoid muscle (pages 38–39).

Having reflected the deep fibers of masseter downwards, complete any residual dissection that may be necessary to free fibers of temporalis from their origin in the temporal fossa. Such residual dissection may include separating merged temporalis fibers from those of the deep-lying lateral pterygoid muscle. The significance of such fusion of masticatory muscles was discussed previously. You should also, at this time, study the deep surface of the reflected temporalis muscle and reidentify the **deep temporal nerves and vessels.** Preserve the connections of these with their stems of origin, emerging from the infratemporal region (page 39).

LATERAL PTERYGOID

To facilitate access to the infratemporal region (page 39), make an electric saw-cut through the condylar process of the mandible. This will allow a degree of lateral displacement of the body of the mandible.

Trace fibers of the temporalis muscle onto the medial aspect of the coronoid process, where you previously palpated them in the living subject (pages 38–39). Using scissor dissection, proceed with separation of the temporalis muscle from the lateral surface of the lateral pterygoid muscle. In the ensuing dissection, you should first attempt to visualize the position and shape of the lateral pterygoid muscle, making use of atlas illustrations, anatomical specimens, and the brief description from which you previously studied the muscle in a living subject (pages 38–39 and fig. 18, page 44). To achieve this visualization, first identify the uppermost fibers of the lateral pterygoid muscle, which run more or less at right angles to the vertical anterior fibers of temporalis from which you have been separating them. With a disarticulated skull in front of you as you proceed with the following dissection, localize the cleavage plane at the upper border of the lateral pterygoid muscle, between its uppermost fibers and the infratemporal surface (page 39). Follow the superior fibers of the muscle posteriorly as they converge with inferior fibers (passing superoposteriorly from the lateral surface of the lateral pterygoid plate) in a tendon. This tendon inserts into the anterior surface of the **neck of the mandible** (the narrow communication between the **condyle,** or **head of the mandible,** and the condyloid process). In addition, fibers of the muscle insert directly into the intraarticular disc, as well as into the joint capsule (page 39). Identify the superoposteriorly-sloping lowermost fibers of the muscle, as they enter the tendon.

Having visualized the muscle, proceed with the dissection instructions below, taking due care to avoid damage to the **maxillary artery** and its branches, lying lateral to the muscle; the termination of this artery medial to the anterior part of the muscle; and the **mandibular nerve** medial to (deep to) the posterior part of the muscle (figs. 18–20, page 44).

At this time, identify on a skull the vertical fissure (crack or narrow opening) between the anterior edge of the lateral pterygoid plate and the posterior surface of the maxilla (the **pterygomaxillary fissure).** The pterygomaxillary fissure opens into a space, the **pterygopalatine fossa,** which is situated between the pterygoid process and the back of the maxilla.

On the lateral aspect of the lateral pterygoid muscle, between it and the temporalis muscle anteriorly, and between the muscle and the medial surface of the ramus of the mandible more posteriorly, you will encounter the **maxillary artery.** This artery passes from behind the condylar process of the mandible, toward the pterygomaxillary fissure. Posteriorly, behind the condylar process, it can be traced to meet the **superficial temporal artery,** which you have previously studied (page 20 and 25). The small masseteric artery which you previously identified sinking into the deep surface of the masseter, and marked with a small piece of muscle, is a branch of the maxillary, and following this artery centripetally will facilitate your finding the maxillary artery. The maxillary artery also gives off the deep temporal branches you will have found on the deep surface of the temporalis muscle, as well as other branches you will study and dissect later in this session. The artery and its branches are accompanied by the **maxillary vein** and appropriately named tributaries. The maxillary and superficial temporal veins unite to form the **retromandibular vein.**

Medial to the lateral pterygoid muscle, the mandibular division of the fifth cranial nerve (trigeminal nerve) emerges from the skull through the **foramen ovale,** which should now be identified on the skull. Because of this close proximity of this very important nerve, and its important related structures, you should proceed slowly and with great care in mobilizing the lateral pterygoid muscle.

Separate the superior fibers of the lateral pterygoid from the infratemporal surface in the cleavage plane you previously identified. Proceed inwards, by blunt dissection with the back of a pair of forceps, toward the top of the lateral surface of the lateral pterygoid plate. Then work your way posteriorly, toward the tendon. In this dissection you will partly separate the **superior head** of the muscle (page 39) from its site of origin. The origin of the **inferior head** (page 39) from the lateral surface of the lateral pterygoid plate should be left intact at this time. However, the inferior sloping border of the muscle should be carefully mobilized from its surroundings and similarly traced back to the tendon. Your objective now is to separate the entire circumference of the lateral pterygoid tendon, both on lateral and medial surfaces, from surrounding tissues. When you have so isolated the tendon, cut through it, separating the distal portion of the tendon, and its insertion into the capsule posteriorly from the muscular bellies of both heads which will remain attached anteriorly.

You will now be able to gently rotate the mandible on a vertical axis. Displace the anterior extremity of the mandible laterally. Further rotation of the mandible will be restrained by a tough white band which will make its appearance, extending from the vicinity of the pterygoid process, to the inner surface of the ramus of the mandible. Confirm, by reviewing your previous dissection of the medial aspect of this region (page 32), that the band you are now identifying is the **pterygomandibular raphe.** Careful dissection just anterior to the band will reveal fibers of buccinator muscle. Posterior to the band, fibers of the superior pharyngeal constrictor muscle will come into view.

MEDIAL PTERYGOID

Posterolateral to the constrictor muscle, you will now see fibers of the medial pterygoid muscle, passing posteroinferiorly from their origin (page 38) to the medial aspect of the angle of the mandible. Use blunt dissection, and, where necessary, scissor and scalpel dissection, to fully display the medial pterygoid muscle.

Define the anterior edge of the temporalis muscle, dissecting the anteriormost fibers, at their insertion into the anterior surface of the coronoid process, free of surrounding connective tissue. Similarly, dissect the pterygomandibular raphe clear of connective tissue, and separate a space between the raphe and the anterior temporalis fibers. Using the electric saw under guidance, cut through the mandible posterior to the raphe and anterior to the coronoid process and temporalis muscle insertion, separating the ramus from the rest of the body of the mandible. Your cut should be made in such a way that the temporalis muscle and the medial pterygoid muscle are carried with the posterior portion of the mandible, whereas the pterygomandibular raphe, with the buccinator and pharyngeal constrictor intact, remain attached to the anterior part of the mandible (fig. 17, page 44). This cut through the mandible will enable you to rotate the posterior portion of the mandible further laterally, thus exposing the deep surface of the ramus (fig. 18, page 44). Gently separate the medial pterygoid muscle from the inner surface of the ramus of the mandible, dissecting in the plane between the muscle and the mandible, to demonstrate the **lingual nerve.** Establish continuity between the lingual nerve, as demonstrated in this plane, and its continuation in the sublingual region, where you have dissected it previously (page 33 and fig. 19, page 44).

Using scissor dissection, follow the lingual nerve superiorly, tracing it to its origin from the mandibular division of the trigeminal nerve. As you trace it upwards, you will encounter other branches of the nerve, including the **inferior alveolar nerve,** passing down toward the mandibular foramen (page 28). Visualize the course of the latter nerve from the intraoral aspect of your specimen, in relation to the site for administration of local anesthesia (page 28). In relationship to the latter branch of the mandibular nerve, you will also identify the tough **sphenomandibular ligament** (page 40). Carefully dissect the ligament free of surrounding connective tissue and demonstrate its inferior attachment to the lingula. By reflecting the anterior portion of the lateral pterygoid muscle downwards, at the site where you previously cut through the tendon of this muscle, you will be able to follow the lingual and inferior alveolar nerves, and other branches of the mandibular division of the trigeminal, right up to the foramen ovale. Similarly, the sphenomandibular ligament can be followed to the sphenoid spine.

Referring freely to atlas illustrations and demonstrated anatomical specimens, dissect out the branches of the **mandibular nerve** and of the **maxillary artery** in the infratemporal fossa. By replacing the temporalis muscle, identify the three parts of the maxillary artery: the first part, posterior to the posterior edge of the muscle; the second part, deep to temporalis; and the third part, anterior to the muscle (passing into the pterygopalatine fossa, where you will study it again later).

Carefully reflect the lateral pterygoid muscle anteriorly, tracing its attachments to the infratemporal surface (page 39) and the infratemporal crest (the lateral edge of the infratemporal surface). In addition, similarly trace the medial pterygoid muscle anteriorly, establishing the separate origins of its lateral head from the lateral aspect of the lateral pterygoid plate, as well as the maxillary tubercle, and its medial head from the medial aspect of the lateral pterygoid plate. Anterior to the lingual nerve, the **buccal branch of the mandibular nerve** should be traced, curving to the lateral aspect of the buccinator muscle, giving branches which supply both the skin of the cheek and the mucous membrane within the vestibule.

On the medial aspect of the mandibular nerve, just after its emergence from the foramen ovale, identify a swelling, about 0.5 cm in diameter known as the **otic ganglion.** This ganglion, which is entirely comparable to the submandibular ganglion (page 33), mediates nerve impulses that stimulate secretion from the parotid gland.

TEMPOROMANDIBULAR JOINT

You have now mobilized the surrounding tissues of the joint sufficiently to be able to open the joint capsule and study the joint itself. Before disrupting the capsule, study the **lateral temporomandibular ligament** of the joint (page 40), by scalpel dissection. Then gently displace the coronoid process downwards and forwards, thus levering the condyle of the mandible out of its relationship in the mandibular fossa (fig. 20, page 44). This places the articular disc on stretch, and allows you to study its attachment to the anterior surface of the condyle. Carefully detach the disc from this anterior attachment by scalpel dissection. Posteriorly, a thick fibrous band, the **stylomandibular ligament,** attaches the mandible to the styloid process (page 40). Trace the ligament to the styloid process by scalpel dissection. Detach the inferior attachment of the medial pterygoid muscle from its insertion on the medial aspect of the angle of the mandible, after having established the "sling" relationship of the medial pterygoid and the masseter on either side of, and inferior to, the angle of the mandible (page 38). The head of the mandible and ramus can now be rotated even further laterally, allowing further scissor dissection in the infratemporal region. Identify the **middle meningeal branch** of the maxillary artery, ascending to the **foramen spinosum** (so named because of its relationship to the **sphenoid spine**), and the small roots of the **auriculotemporal nerve** that embrace the middle meningeal artery (fig. 20, page 44). Establish continuity of the auriculotemporal nerve with the mandibular division of the trigeminal nerve, emerging through the foramen ovale. Then trace the auriculotemporal nerve posteriorly, passing lateral to (or through) the sphenomandibular ligament, to gain the anterior surface of the **tympanic plate** of the temporal bone (the anterior wall of the external auditory meatus). Here the nerve turns laterally, passing behind the temporomandibular joint (i.e., posterior to the joint capsule), and then ascends, in front of the ear, where it should be traced. Establish continuity here, with the superficial branches of the nerve, which you previously traced posterior to the superficial temporal artery (page 25).

Secretomotor nerve impulses which originate from the ninth cranial nerve are carried, after relay through synapses in the otic ganglion, through the auriculotemporal nerve, to the parotid gland. With the increased access to the medial aspect of the ramus of the mandible which you now have, complete your study of the deep portion of the parotid gland, with the aid of atlas illustrations. Observe the relationship of the gland to the styloid process; the **external carotid artery** (the artery which gives off, as its terminal branches, the maxillary and superficial temporal arteries); and the facial and auriculotemporal nerves; and attempt to trace the secretomotor branches of the latter to the gland.

Returning to the lingual nerve, trace it up to its origin from the mandibular nerve and observe the **chorda tympani** branch joining the lingual nerve from behind. Trace the chorda tympani nerve upwards and posteriorly, where it disappears medial to the sphenoid spine. (The origin of the chorda tympani nerve is from the seventh cranial nerve.)

Complete your laboratory study of this region by reviewing the anatomy of the muscles of mastication, and of the temporomandibular joint, identifying all the emphasized structures mentioned in the description given on pages 37 to 41. You should also ensure that you have identified the branches of the mandibular division of the trigeminal nerve and of the maxillary artery.

Clinical Session 5

■ **TOPICS**

Anterior Triangles of the Neck
  Digastric Triangle
  Submental Triangle
  Carotid Triangle
  Muscular Triangle

Submandibular Gland

As stated in the first session, the division of the body into regions is arbitrary, and regions often overlap with each other. In respect of the particular objectives of study for the present session, it should be emphasized that it is impossible to demarcate a boundary between the head and the neck. In this session, structures such as the submandibular gland, which are related to the floor of the mouth and in that sense constitute part of the head, are to be studied. Clinical examination of these structures is made, in part, from the oral cavity. However, it is also made in part through a region of the skin of the neck. As will be seen in the following section, the region defined anatomically as the anterior triangle of the neck in fact extends from the neck into the head.

1. Observe and palpate the margins of the right and left **anterior triangles of the neck** as follows.

The base of each triangle faces superiorly (and the apex, correspondingly, inferiorly). The base, which you should palpate now on the right side, is formed by the inferior border of the body of the mandible, from the **symphysis menti** (page 37) to the **angle of the mandible,** and extends from there by an imaginary line to the **mastoid process** (mastos means breast; oid means resembling in form), a bony projection which can be palpated immediately posteroinferior to the external ear. Identify this process on your model, making reference to the isolated skull. The **medial border** of the anterior triangle of the neck corresponds to the median plane, from the symphysis menti to the **suprasternal notch** (the notch at the superior end of the sternum). To establish the **lateral border** of the triangle, ask your model to rotate his or her head to the left against resistance (i.e., without moving the shoulders, turn the head to the left, as if looking over the left shoulder, but against resistance applied to the left temporal region). This exercise will produce tension in a long muscle which extends from the right mastoid process to the region of the suprasternal notch and the right **clavicle** (collarbone) lateral to the notch. The muscle is called the **sternocleidomastoid muscle** (sterno means of the sternum or breastbone; cleido means of the clavicle or collarbone; mastoid refers to the mastoid process). Palpate the medial edge of the sternocleidomastoid muscle (fig. 21, page 50). This edge forms the lateral side of the anterior triangle of the neck. Note, in passing, the **external jugular vein** (fig. 33, page 82), which passes inferiorly and slightly posteriorly across the anterior surface of the muscle. This vein drains blood from the face and ear regions, and passes deep posterolateral to the muscle.

2. The **digastric triangle,** also known as the **submandibular triangle,** will now be examined. This triangle, and the others to be studied in the following paragraphs, are subdivisions of the anterior triangle.

The base of this triangle is the same as the base of the anterior triangle of the neck. To establish the other two sides of the smaller triangle, palpate the **hyoid bone,** as described previously (pages 4 and 18). The **body of the hyoid bone** can be identified by running a finger down the front of the neck from your model's symphysis menti, in the midline (i.e., along the medial border of the anterior triangle of the neck). The first hard tissue structure the finger encounters is the hyoid bone; the body of the bone can be palpated to either side of the midline. Observe an anatomical specimen and atlas illustration, and palpate out laterally to feel the **greater horn of the hyoid bone,** as it curves out on either side into the lateral structures of the neck. Now visualize, making use of anatomical specimens and atlas illustrations, the **digastric muscle**. This muscle has two **bellies:** di indicates two; gaster indicates belly. The **posterior belly of the digastric** has its origin on the medial aspect of the mastoid process. It extends from there to a point situated at the junction between the body of the hyoid bone and the greater horn of the bone. The **anterior belly of the digastric** has its origin on the inner surface of the mandible, close to the symphysis menti (see fig. 16, page 34). It extends from this attachment to the same point of junction between the body and the greater horn of the hyoid bone. At this site, the tendons of the bellies unite, in an **intermediate tendon** (fig. 22, page 50), which is tethered down to the hyoid bone by a connective tissue sling. Thus you will appreciate that both bellies acting simultaneously would be capable of elevating the hyoid bone. Attempt to palpate the bellies

49

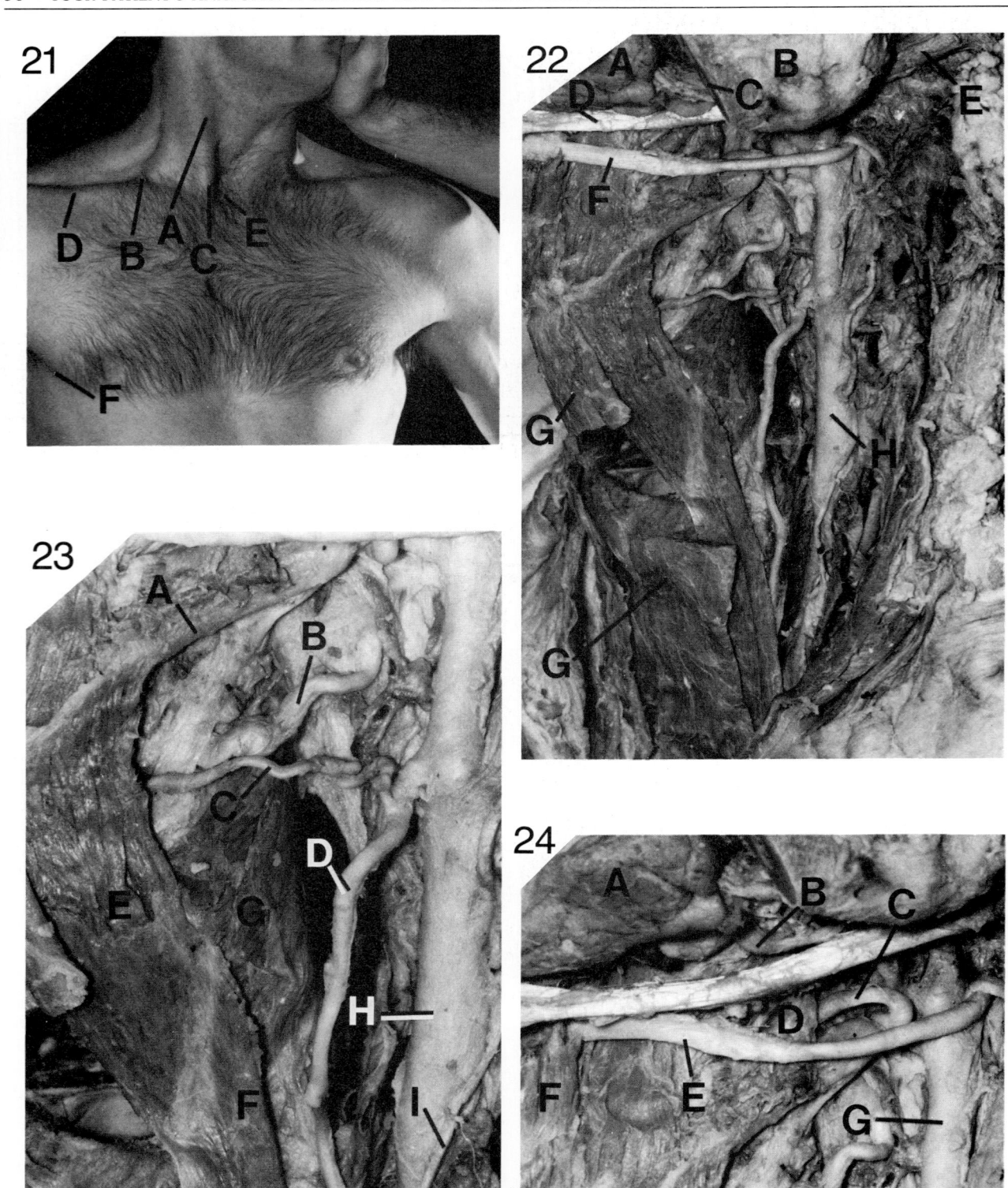

50 YOUR PATIENT'S ANATOMY: A CLINICAL VIEW OF HUMAN MORPHOLOGY

Figure 21. Anterior triangle of neck and related structures.
A. Sternocleidomastoid muscle
B. Clavicular head of sternocleidomastoid
C. Sternal head of sternocleidomastoid
D. Clavicle
E. Suprasternal notch
F. Inferolateral border of pectoralis major muscle

Figure 22. Left anterior triangle of neck.
A. Submandibular gland (exposed by elevating anterior part of mandible)
B. Angle of mandible
C. Saw-cut surface of mandible
D. Intermediate tendon of digastric
E. Posterior belly of digastric
F. Hypoglossal nerve
G. Sternohyoid muscle (middle portion removed to expose larynx)
H. Common carotid artery

Figure 23. Detail of anterior triangle of neck.
A. Greater horn of hyoid bone
B. Internal laryngeal nerve
C. Superior laryngeal branch of superior thyroid artery, passing deep to thyrohyoid muscle
D. Superior thyroid artery
E. Thyrohyoid muscle
F. Sternothyroid muscle
G. Inferior pharyngeal constrictor
H. Common carotid artery
I. Descending branch of hypoglossal nerve

Figure 24. Detail of anterior triangle of neck.
A. Submandibular gland
B. Facial branch of external carotid artery, passing up to come in relation to the submandibular gland
C. Loop of lingual branch of external carotid artery, passing medial to hyoglossus muscle
D. Hyoglossus muscle
E. Hypoglossal nerve
F. Mylohyoid muscle
G. External carotid artery

of the digastric muscle, as your model swallows: the hyoid bone is elevated, in part due to the action of the digastric muscle. The posterior belly of digastric forms the posterior side and the anterior belly of digastric forms the anterior side of the digastric triangle.

By careful palpation at the site of junction between the greater horn and the body of the hyoid bone, it may be possible to palpate the **lesser horn of the hyoid bone,** a small projection 1 to 3 mm in height from this junction on the anterolateral aspect. At or just behind this site, a second muscle attaches, and reinforces the posterior belly of the digastric as part of the posterior boundary of the digastric triangle. This is the **stylohyoid muscle,** which lies very slightly anterosuperior to the posterior belly of the digastric (and therefore mainly within the digastric triangle). Its tendon of insertion into the hyoid bone splits to enclose the intermediate tendon of the digastric muscle, which thus has two slings that tether it to the hyoid bone: the stylohyoid muscle and the connective tissue loop mentioned previously. Note that the stylohyoid muscle has its origin from the back of the styloid process (which you should now reidentify on the skull) and its insertion into the hyoid bone. It therefore also participates in elevation of the hyoid bone during swallowing. Ask your model to swallow again, and attempt to palpate both the stylohyoid and posterior belly of digastric in the posterior border of the digastric triangle.

The main contents of the digastric triangle are the **submandibular salivary gland** and the structures related to it. This, of course, is the reason for the triangle also being known as the submandibular triangle.

Palpate the submandibular gland by bimanual examination, with one index finger within the oral cavity proper and a second palpating hand in the digastric triangle (page 30). Gentle pressure should be exerted from within the oral cavity toward the inferior border of the body of the mandible, while counterpressure is gently applied from below. The gland will be felt as a more or less diffuse structure; sharp borders will not be discernible. In pathological states, for example the presence of salivary calculi (calculus means stone), enlargement of the gland or actual palpation of the stone is possible. It should be appreciated, in performing this palpation of the submandibular gland, that the tissues intervening between the intraoral finger and the major part of the gland being palpated include not only the mucous membrane of the oral cavity, but also the **mylohyoid muscle,** (page 24), which extends from the hyoid bone to the inner (medial) surface of the mandible (figs. 13 and 16, page 34).

Reidentify on an isolated mandible the bony ridge on the medial surface, immediately superior to which you previously studied the lingual nerve (page 28). This ridge slopes down in an anteroinferior direction from its origin behind the last molar, fading away toward the inferior border of the mandible in the direction of the symphysis menti. The ridge, known as the **mylohyoid line,** is the attachment of the mylohyoid muscle onto the mandible (mylo means mill, pertaining to the molar teeth). Inferior to the mylohyoid line, on the medial surface of the body of the mandible, is a slight depression into which the major portion of the submandibular gland is lodged. Thus in the palpation performed, the intraoral finger palpated the gland through the oral mucous membrane and through the mylohyoid muscle. From the posterior extremity of the mylohyoid line, the posterior free edge of the mylohyoid muscle extends downwards and anteromedially to the body of the hyoid bone. The submandibular gland extends back on the inferior surface of the muscle to this posterior free edge and then recurves up and over the free edge (pages 30 and 39). The gland is therefore described as having a major, **superficial portion** (inferolateral to the mylohyoid muscle, and lying mainly within the digastric or submandibular triangle), and a smaller, **deep portion,** lying superior to the posterior edge of the mylohyoid muscle. The deep portion of the gland lies, of course, in the plane of the **sublingual region** of the oral cavity (page 32), where you have previously dissected it (page 33). The submandibular duct takes the same course (of recurving over the posterior free border of the mylohyoid muscle) to enter the sublingual region. This explains how the submandibular duct comes to lie in this region, superior to the sublingual gland, to form the sublingual fold or plica, and to end in the sublingual papilla.

Several lymph glands are found in intimate relationships to the submandibular gland. These may be bimanually palpated as small, more or less firm, granular structures.

Study the three-dimensional arrangement of the submandibular gland on an anatomical specimen and in atlas illustrations and repeat the palpation as described previously, bearing in mind the relationships to the mylohyoid muscle as you perform the examination. It will be clear to you, as you visualize the position which the mylohyoid muscle has in relation to the hyoid bone and the mandible, that this muscle forms a muscular, diaphragmatic reinforcement of the floor of the mouth, and that contraction of this muscle, too, will lead to elevation of the hyoid bone. With your intraoral palpating finger in the region of the posterior border of the muscle, ask your model to swallow, and appreciate the tension in this muscle during swallowing and the resulting upward movement of the entire floor of the mouth.

To complete this clinical examination of the submandibular triangle, review (page 20) palpation of the facial artery. It is an intimate posterior relation of the submandibular gland in the digastric triangle, where the gland recurves over the mylohyoid muscle (pages 30 and 33). From this point the artery curves over the lower border of the mandible to enter the face, where you have dissected it (pages 20 and 25).

Having now studied the third of the large, paired salivary glands, it is appropriate here to summarize the salivary glands. These are grouped into

1. The large, paired glands
    a. parotid glands (pages 20 and 26 to 27)
    b. sublingual glands (pages 29 and 30)
    c. submandibular glands (page 30)
2. The small salivary glands
    a. labial glands (page 11)
    b. buccal glands (pages 26 and 27)
    c. palatine glands (page 30)
    d. lingual glands
        1) posterior lingual glands (page 15)
        2) middle lingual glands (page 15)
        3) anterior lingual glands (page 15)

3. The **carotid triangle** is so named because of its major contents, the carotid arteries (common carotid artery, internal carotid artery, and external carotid artery). Using anatomical specimens and illustrations visualize, on your model, the boundaries of this triangle. The superomedial boundary is the posterior belly of the digastric muscle. This boundary therefore separates the carotid triangle from the digastric or submandibular triangle. The posterior boundary of the carotid triangle is the anterior border of the sternocleidomastoid muscle. The inferior boundary is formed by the superior belly of another two-bellied muscle, the **omohyoid muscle.** The **superior belly of omohyoid** is attached superiorly to the inferior border of the hyoid bone, close to where the greater horn joins onto the body. This belly extends inferolaterally from that attachment to pass behind (posterior to, or deep to) the sternocleidomastoid muscle, about one quarter of the way up the length of the muscle (i.e., a few centimeters above the medial end of the clavicle). Thus the lateral border of the superior belly of the omohyoid forms the inferomedial border of the carotid triangle. (Deep to sternocleidomastoid, this belly of omohyoid ends in an intermediate tendon, which links it to the inferior belly of the omohyoid. The inferior belly passes posterolaterally toward the shoulder, and will be studied later (omos means shoulder). The intermediate tendon of the omohyoid muscle is slung to the medial end of the clavicle, by a fibrous loop lying deep to sternocleidomastoid.)

Palpate the carotid triangle of your model and feel the powerful pulse of the carotid arteries. In the lower portion of the carotid triangle, it will be the **common carotid artery** that you are palpating (fig. 22, page 50). This is the major arterial trunk that carries blood from the outflow of the heart to the head. In the upper portion of the carotid triangle, the common carotid artery divides into its two branches, the **internal carotid artery,** which lies lateral, and which is destined to pass into the skull (hence its name); and the **external carotid artery,** which initially lies medial, and, again as implied by its name, is destined to supply the structures exterior to the skull. To more accurately determine the site of bifurcation (bi meaning two; furca meaning fork), palpate the greater horn of the hyoid bone again, and, thereafter, the region immediately inferior to the greater horn. Here you will feel the superior edge of the **thyroid cartilage** ("Adam's apple," page 4). A palpating finger can be depressed into a groove between the greater horn of the hyoid bone, and the upper edge of the thyroid cartilage. The superior extremity of the thyroid cartilage is situated at the level of bifurcation of the common carotid artery into its internal and external branches. This position corresponds to a horizontal plane passing through C3 (the third cervical vertebra). For completeness, it should be mentioned that the common and internal carotid arteries are accompanied by the internal jugular vein, lateral to the arteries, and the vagus (tenth cranial) nerve, posterior to the vessels.

Two branches of the external carotid artery should be further studied here, with the help of anatomical specimens and illustrations. The first, and more **proximal** (closer to, in this case, closer to the heart) is the **lingual artery.** You previously palpated the pulsation in this artery on the inferior surface of the tongue (page 15) and have dissected that portion of it, medial to the hyoglossus (page 24, and fig. 12, page 22). The artery arises from the medial aspect of the external carotid artery, at about the level of, or just above, the tip of the greater horn of the hyoid bone. After making a slight loop upwards (fig. 24, page 50) the artery comes to pass medially, just above the greater horn of the hyoid bone, to gain the deep surface of the hyoglossus muscle which, as you previously studied (pages 18 and 24), passes upwards from the greater horn of the hyoid bone to the side of the tongue. In this manner the artery enters the tongue, passing anteriorly, deep to hyoglossus, to attain its final position, where it can be palpated on the inferior surface. The **facial artery** leaves the external carotid a little **distal** (further) to the lingual, and passes upwards and medially to gain the posterior extremity of the submandibular gland (fig. 24, page 50), where this gland recurves from the inferior surface of the mylohyoid muscle to its superior surface. Having reached the gland, the facial artery passes onto the lateral surface of the superficial part of the gland (pages 50–53), thus intervening between the gland and the inner surface of the mandible. From this point, it curves around the inferior border of the mandible (where you have previously palpated it just anterior to the masseter muscle (page 20)), to enter the superficial layer of the face. Here it ascends toward the medial angle of the eye, giving off, among other branches, the labial arteries (fig. 8, page 16).

Superiorly, the external carotid artery ends in the parotid gland (page 48) by dividing into the superficial temporal artery (pages 20 and 25) and the maxillary artery (page 46).

4. Unlike the other subdivisions of the anterior triangles studied in this session, the **submental triangle** is not a bilateral region but a single, centrally placed space, the right and left halves of which belong to the respective anterior triangles. Thus the submental triangle is bounded above by right and left anterior bellies of the digastric muscles, and inferiorly the base of the triangle is formed by the entire body of the hyoid bone. Deep to skin, this triangle contains, as its most important component, several small lymph nodes, which drain lymph from the tip of the tongue and the anterior portion of the floor of the mouth. These are of considerable clinical importance. These **submental lymph nodes** are easily palpable, and should be examined in your model by careful bimanual examination. In addition to the submental lymph nodes, there are, as elsewhere in the superficial tissue of the face and neck, blood vessels, including the anterior jugular vein (fig. 10, page 22) and nerves. More deeply in this triangle (i.e. superiorly) the mylohyoid muscle is found, in the floor of the mouth. Superior to this, as previously studied, (pages 21–22), the **geniohyoid** and **genioglossus muscles** are situated.

5. The **muscular triangle** is the triangle enclosed superolaterally by the medial border of the superior belly of the omohyoid muscle; inferolaterally by the medial border of the sternocleidomastoid muscle; and medially by the median plane. Its name comes from the fact that its main contents are muscles extending between the hyoid bone, the superior end of the sternum, and intervening skeletal structures between these two points, including the thyroid cartilage. At this time you should attempt to visualize the situation of these muscles, by palpating the front of the neck, between the hyoid bone above and the suprasternal notch (page 49) below as your model swallows.

Begin by reviewing the situation of the hyoid bone and of the thyroid cartilage (pages 4 and 53). Palpate the midline ridge of the thyroid cartilage (the **angle**, formed where the two **laminae** (plates) of this cartilage meet in the midline). Below it, palpate a horizontal, curved bar of cartilage. This is the anterior portion of a horizontal ring, part of the **cricoid cartilage.** Below this, extending down to the suprasternal notch, are a further series of 6 to 10 palpable cartilaginous bars which belong to the **trachea** (wind-pipe).

Consult an atlas illustration, anatomical specimens, and figures 22 and 23, page 50, to aid you in localizing the **sternohyoid muscle**. This muscle can be felt as a narrow strap on either side of the midline in the initial stages of swallowing. Deep to sternohyoid, but extending further laterally, and therefore also palpable, are the **sternothyroid** and **thyrohyoid muscles.** Because of their flat, narrow shape, these muscles and the **omohyoid** are known as the **strap muscles.** Due to their position, they are also known as the **infrahyoid muscles.** This triangle, these muscles, and the related structures and organs will be studied in further detail in a subsequent session.

To complete this session, review the emphasized topics and concepts.

Laboratory Session 5

Your objective for this laboratory session is to study the anterior triangle of the neck, identifying on your cadaver structures which you previously studied in the living subject.

Begin by removing the skin, according to the prescribed technique (page 10, and fig. 5, page 16). In outlining the triangular area to be skinned, make your initial scalpel incisions according to the descriptions (page 49) previously given for the base and the **median border** of the triangle, using the half head and neck specimen which is still attached to the rest of your cadaver. For the **third** side of the triangle, make your incision slightly lateral to the actual border of the triangle, so that the incision includes the sternocleidomastoid muscle within the area to be skinned. This will allow a more detailed study of the actual lateral border of the triangle (i.e., the medial border of this muscle).

On the now skinned specimen, identify fibers of the **platysma muscle.** This muscle belongs to the facial muscles (pages 18–19), and as such, is situated in the subcutaneous tissue, rather than deep to deep fascia (page 19). Consult an atlas illustration to guide you in identifying the platysma. In addition, also in the subcutaneous layer, dissect out the **external jugular vein** (page 49), crossing more or less vertically from the anterior border of the sternocleidomastoid, superficial to the muscle, to the region posterior to the posterior border of the muscle. Note that superiorly, this vein is formed by a posterior branch of the **retromandibular vein** (page 46) uniting with the **posterior auricular vein.** Thus, the external jugular vein drains blood from in front of and behind the ear. Other superficial structures which you may encounter in removing the superficial connective tissue and fatty layer include a cervical branch of the facial nerve crossing down to innervate the platysma and branches of the **cervical plexus**—a network of nerves derived from cervical spinal, peripheral nerves (plexus means network). This plexus lies deep to sternocleidomastoid and its branches emerge at the muscle's posterior border. Identify, with the aid of an atlas, the **supraclavicular, transverse cervical, great auricular,** and **lesser occipital** branches. Also in the subcutaneous plane, running vertically, close to the midline, identify and dissect out the anterior jugular vein (page 54, and fig. 10, page 22).

As you dissect through and remove the fatty, subcutaneous connective tissue, you will encounter a tough layer of **deep fascia** which ensheaths the entire neck (pages 18 and 19). Expose and identify this **investing layer** of the **deep cervical fascia,** and then incise it anterior to the sternocleidomastoid muscle, and elevate it both laterally and medially. In so doing, you should confirm that a second layer of the fascia is found posterior to the sternocleidomastoid muscle, and that the two layers fuse at the lateral and medial edges of the muscle. Thus, the fascia splits to enclose the sternocleidomastoid muscle.

Medial to sternocleidomastoid, the split layers refuse, as stated, and when traced medially, may be seen to split again, to enclose muscles of the **muscular triangle** (page 54). These muscles will be further studied at a later stage. At this time, identify, but do not dissect, the **sternohyoid, sternothyroid,** and **thyrohyoid muscles** (page 54, and figs. 22 and 23, page 50).

CAROTID TRIANGLE

Proceed now to carefully separate the anterior layer of the deep fascia covering sternocleidomastoid, from the muscle throughout the exposed length of the muscle. Thereafter, separate the muscle from its posterior layer of fascia and clearly define its medial border, taking care to avoid damage to the deep-lying structures, including the **carotid arteries** (page 53) and the accompanying **internal jugular vein** which lies lateral to the arteries.

As you proceed inferiorly, carefully identify and dissect out the **superior belly of the omohyoid muscle** (page 53). Your search for this structure will be assisted by repalpating the hyoid bone and attempting to identify the hyoid attachment of the muscle at the site previously described (page 53). Trace the omohyoid tendon posterior to the sternocleidomastoid and identify its fascial attachments. However, do not attempt to trace its sling downwards to the clavicle at this stage. Note that to expose the superior belly of the **omohyoid,** you will have to incise the fascial sheath of deep cervical fascia which encloses this and other muscles of the muscular triangle, as well as the sternocleidomastoid. To facilitate this and subsequent dissection, carefully transsect (cut across) the belly of the sternocleidomastoid to allow the cut ends to be reflected.

When you have cleared the posterosuperior border of the anterior belly of the omohyoid of fascia, two of the three borders of the carotid triangle (page 53) will have been defined. Now attempt to expose the third side of this triangle, the inferior border of the posterior belly of the digastric muscle. To do this, proceed from the site on the hyoid bone, where the superior belly of omohyoid attaches, and, working your way posteriorly, remove adjacent connective tissue. Since the submandibular gland quite often exceeds the boundaries of its region (the submandibular or digastric triangle [page 49]), you might encounter the gland overlapping the posterior belly of the digastric muscle in this dissection. If so, gently elevate the gland, to display the inferior border of the muscle, and thus the remaining boundary of the carotid triangle. As you elevate the submandibular gland, and dissect by blunt finger dissection into the cleavage plane deep to the gland, grasp the hyoid bone and place tension on it by pulling it medially. This will bring the intermediate tendon of the digastric muscle into view (page 49 and fig. 22, page 50). Take care, as you dissect the tendon out, to avoid damage to the fibers of the closely related stylohyoid muscle (page 52), the insertion of which, into the hyoid bone, splits to enclose the digastric tendon. Demonstration of the latter calls for careful scissor dissection. Similarly, delicate structures just inferior to the digastric, in the superior part of the carotid triangle, also should be exposed by careful scissor dissection.

When the superior border of the carotid triangle has been clearly identified and defined, to facilitate further dissection and demonstration of structures in the upper reaches of the triangle (and, subsequently, structures in the digastric triangle), you should make an incision, using an electric saw under guidance, through the body of the mandible, as you previously did (page 47) on the opposite half of your cadaver. Ensure that you make this incision posterior to both the submandibular gland and to the most posterior mandibular molar, but anterior to the insertion of the masseter muscle (fig. 22, page 50). Having made this incision, rotate the inferior border of the anterior portion of the mandible superolaterally, thus exposing the digastric tendon and related structures.

Inferior to the digastric tendon, the large **hypoglossal nerve** will come into view and should be carefully followed by scissor dissection. You have previously dissected the terminal portions of this nerve (page 24). You will recall that in the portion which you previously dissected, the important relationships of this nerve, the hyoglossus muscle and the deep lingual artery were demonstrated (page 24, and fig. 12, page 22). In the present dissection, you will demonstrate how the nerve gains the relationship you previously studied. By carefully removing the connective tissue around the nerve, you can demonstrate that it is in its situation just inferior to the digastric tendon, in relation to a delicate muscle lying medial to the nerve (fig. 24, page 50). This is the **hyoglossus,** and the relationship of the nerve to the muscle is as you previously have observed it (i.e., the nerve is lateral to the muscle). Confirm, also, that the fibers of the muscle run vertically and are seen to take their origin inferiorly from the greater horn of the hyoid bone, which can easily be palpated at this time. By following the nerve anteriorly, you will see that it disappears by passing medial to a free muscle edge. By putting traction between the hyoid bone and the free, anterior portion of the mandible, it will be seen that the free muscle edge is that of mylohyoid. Turn to the median (intraoral) surface of your sagittally sectioned specimen. From that approach, gently dissect in the cleavage plane between mylohyoid and geniohyoid, until you reach the posterior free edge of mylohyoid, and can establish continuity between the two parts of the hypoglossal nerve which you have dissected, from lateral and median approaches respectively.

Posteriorly, the hypoglossal nerve should be traced across the lateral aspect of the external and internal carotid arteries, enclosed (with the internal jugular vein) in their own sheath of deep fascia, known as the **carotid sheath.** The carotid sheath encloses, in addition to these vessels, the **tenth cranial (vagus) nerve** (page 53). To demonstrate this relationship of the hypoglossal nerve to the carotid sheath and its contents, remove the loose connective tissues and other structures surrounding the sheath. In so doing, you may remove some lymph glands, and in addition, you will encounter the **common facial vein,** as it approaches and enters the internal jugular vein. The common facial vein is constituted by the junction of the facial vein (fig. 8, page 16) and the anterior branch of the retromandibular. Recall (page 55) the fact that the posterior branch of the retromandibular vein unites with the posterior auricular vein to form the external jugular vein. As the hypoglossal nerve crosses the carotid sheath, search for a small branch, which leaves the hypoglossal to descend, in the carotid sheath, on the anterior aspect of the carotid vessels. Although evidently a branch of the hypoglossal nerve, this **descending branch** (fig. 23, page 50) is in fact comprised of cervical spinal nerve fibers (from the first cervical nerve, C1), "hitchhiking" transit in the hypoglossal nerve. You have previously encountered similar cervical fibers which run in the hypoglossal nerve in a previous dissection of the distal portion of the nerve (page 24).

Carefully dissect the carotid sheath, to expose the internal jugular vein (laterally) and the internal and external carotid arteries lying medial to the vein. A second descending nerve fiber, parallel to the descending branch of the hypoglossal nerve, but lying lateral to it, anterior to the internal jugular vein, will also be found in the carotid sheath. This is similarly formed by cervical fibers and is called the **descending cervical nerve.** The two descending nerves join to form the **ansa hypoglossi** (ansa means loop). The ansa hypoglossi has branches which supply muscles of the muscular triangle.

In clearing the fascia of the carotid sheath from the enclosed vessels, take care to avoid damage to the arterial branches which you will encounter. In particular, preserve a branch which runs downwards and medially, to pass deep to the superior belly of the omohyoid muscle, and subsequently deep to the lateral edge of the **sternothyroid muscle** (page 54). This vessel, the **superior thyroid branch** of the external carotid artery (fig. 23, page 50), will be followed to its destination in a subsequent dissection. At this time, palpate the superior border of the thyroid cartilage to establish the level (C3) at which the common carotid artery bifurcates into the internal and external carotids (page 53). Display this bifurcation of the artery in your specimen by careful dissection. Rotate the anterior mandibular fragment as previously discussed and perform careful scissor dissection of the external carotid artery (the medial of the two branches of the common carotid), following its course deep to the hypoglossal nerve. Just superior to the hypoglossal nerve, display by scissor dissection the origin of the lingual artery from the external carotid artery (page 53). Observe the upward loop of the artery (which the hypoglossal nerve usually crosses) (fig. 24, page 50), confirming by palpation the relationship of the loop to the tip of the greater horn of the hyoid bone. Very carefully follow the artery anteriorly, until it disappears medial to the posterior free edge of the hyoglossus muscle, thus gaining the relationship of artery to muscle which you have previously observed (page 24). Just slightly superior to the origin of the lingual artery from the external carotid, display the origin of the facial artery, which passes superiorly. Carefully trace this artery to its relationship with the submandibular gland, described on page 54. Complete this dissection by tracing the artery around the lower edge of the mandible, to its position anterior to the masseter muscle, where you have palpated it in the living subject (page 20). In dissecting these branches of the external carotid artery, take great care to avoid damage of the related structures, in particular the delicate muscles of the pharynx. Confirm, by examining the sagittally sectioned median plane of your specimen, that you are in the vicinity of the pharynx.

DIGASTRIC TRIANGLE

Dissect in the plane between the inferior border of the mandible, and the submandibular gland, to free the gland of surrounding connective tissue, and to display the relationship of the gland to the mandible and to the anterior belly of the digastric muscle. In this dissection, take care to preserve the **nerve to the anterior belly of digastric,** a branch of the **mylohyoid nerve.** Examine a disarticulated mandible to reidentify the mandibular foramen on the medial aspect of the ramus of the mandible, as well as the mylohyoid line on the medial aspect of the body of the mandible. As the inferior alveolar nerve enters the mandibular foramen (page 28), it gives off the mylohyoid nerve, which passes inferior to the mylohyoid muscle's attachment to the mylohyoid line, thus coming to run on the inferior surface of the mylohyoid muscle. In your present dissection, you should be able to demonstrate this nerve on the inferior surface of the mylohyoid muscle, by elevating the submandibular gland. The mylohyoid nerve is accompanied by mylohyoid branches of the inferior alveolar vessels, as well as by a small branch of the facial artery. Carefully dissect the superficial portion of the submandibular gland free of the mylohyoid muscle, and of its relationship to the mandible, reexamining these relationships (page 52). Study any lymph glands you encounter, and display in detail the relationship of the facial artery to the medial and posterior surfaces of the gland (page 54). Display from this approach, as you previously have displayed internally on the other side of your cadaver specimen, the relationship of the **deep portion of the submandibular gland** to the posterior free edge of the mylohyoid muscle. Return to the sagittally sectioned aspect of your specimen, and the cleavage plane between the mylohyoid muscle and the geniohyoid, to complete this dissection by establishing continuity between the submandibular gland as dissected externally, and the gland as seen from the superior surface of the mylohyoid. Utilizing this approach, you should also be able to demonstrate the submandibular duct and its relationship to the lingual nerve in this dissection, as previously done on the other side. Finally, make an incision through the mucous membrane of the sublingual region, superior to the sublingual eminence (page 29) and demonstrate, from the oral cavity, the structures which you have just dissected from the external and median approaches.

Trace the sternocleidomastoid and digastric muscles superiorly. To do so, expose the region by making a deep incision through skin and soft tissues, at the level of the zygomatic arch. Extend superiorly the incision which you previously made when cutting through the mandible with the electric saw (page 56), to meet the incision over the zygomatic arch. Reflect posteriorly the flap of skin and soft tissues thus created. This exposure will allow you to trace both the sternocleidomastoid and the posterior belly of digastric toward the mastoid process. Grasp the angle of the mandible firmly, and retract it laterally. You will thus place tension on the stylomandibular ligament (pages 40 and 48), which will enable you to identify this ligament and dissect it free of surrounding connective tissue. By following the ligament posteromedially, a palpating finger will thus be able to identify the styloid process. From the styloid process, trace the stylohyoid muscle fibers downwards to meet the inferior insertion of this muscle, where you previously dissected it at the hyoid bone (page 56). Finally, further dissect and demonstrate the external carotid artery as it passes toward its termination, within the parotid gland (page 48), where it branches into the superficial temporal artery (pages 20 and 25) and the maxillary artery (page 46). Identify the remaining branches of the external carotid artery, using an atlas illustration to aid you. The **occipital artery** is a useful landmark, passing just above the inferior border of the posterior belly of the digastric, just deep to the mandible.

Clinical Session 6

■ **TOPICS**
   Nose
   Nasal Cavity
   Paranasal Sinuses
   Nasal Portion of the Pharynx (Nasopharynx)
   Muscles of the Soft Palate

As described previously (page 10), the nose may be regarded as being pyramidal in shape, the base of the pyramid being flush with the anterior surface of the face, and the apex of the pyramid corresponding to the tip of the nose. In this concept, the inferior surface of the pyramid is that surface of the nose in which the two nostrils are situated, and the remaining two surfaces are the right and left sides of the nose.

1. The nostrils may be elongated, having a long axis and a short axis, or they may tend to be round. The direction of the long axis of the nostril may vary with racial type. The nostrils are separated from one another by the inferior edge of the **nasal septum,** a partition which divides the nose and the nasal cavity into two approximately equal halves. Gently palpate the separation between the nostrils, as well as the **nasal alae,** the flanges which form the lower parts of the lateral surfaces of the nose. The lower edge of each ala forms the periphery of each nostril. Ascertain that the tissue has a firm, though not hard, consistency. This consistency is typical of **cartilaginous tissue,** of which this part of the nose is made.

2. Gently palpate the sides of the nose. Confirm, by the consistency of the tissue forming these surfaces, that only the lower portion of each side of the nose is composed of **cartilage.** The uppermost portion of the side of the nose is bony. By gentle palpation, and comparison with the disarticulated skull, identify the border between the bony and cartilaginous portions of the nose. Similarly, identify on the skull the extent of the **nasal bones,** and the sutures which unite them to the **frontal processes of the maxillae** (page 28) on either side and to the **frontal bone** superiorly.

3. Examine the interior of the nose, using a nasal **speculum** (an instrument which aids inspection) and making use of anatomical specimens, illustrations, and a skull. As the speculum passes through a nostril, it enters one half of the **nasal cavity,** passing first into the **vestibule** (bounded laterally by the cartilaginous ala of the external nose) and then into the **respiratory region** of the cavity. Examine the **lateral wall** of the cavity. It will be seen that this wall is irregular, the nature of the irregularities being in the form of longitudinal (anteroposterior) grooves. Between these grooves, folds of tissue project inwards toward the lumen (space) of the cavity. Careful observation will reveal that the grooves extend not only laterally but also superiorly; the projections between the grooves are thus seen to constitute curved shelves, which overhang the space of the groove. Two such overhanging curved shelves are visible: these are known as the **inferior** and **middle nasal conchae** (concha means shell; the shell-like shape of the curved, paper-thin sheets of bone that form the skeletons of these structures can be observed in the disarticulated skull) (see fig. 10, page 22). As implied by the names of the two visible conchae, a third, the **superior nasal concha,** is situated higher up on the lateral wall, and cannot be seen by this method of inspection. The spaces inferior to each concha, which open into the nasal cavity through the visible grooves, are called the **inferior, middle,** and **superior nasal meatuses** (meatus indicates space or canal: compare with external auditory meatus).

Not visible by this method of inspection are openings into the meatuses, from adjacent **air spaces,** the so-called **paranasal sinuses** (para means next to; nasal means of the nose; sinus means enclosed space). These are lined by mucous membrane which is in continuity with that of the nasal cavity itself. Visualize, with the aid of anatomical specimens, atlas illustrations, and a skull, the positions of the **maxillary sinus,** the **frontal sinus** (visible but not marked, in fig. 9, page 22, and fig. 14, page 34), the **ethmoid sinus** and the **sphenoid sinus** (fig. 10, page 22), each of which are situated in the cranial bone of the same name. The particular site, in the respective meatuses, at which each sinus opens into the nasal cavity, should be ascertained. In addition visualize, on a skull, the passage of the **nasolachrymal duct** from the orbit to the inferior meatus. These communications will be studied in greater detail subsequently.

4. By directing a speculum medially, it is possible to examine the **nasal septum** (fig. 9, page 22), which separates the right from the left half of the nasal cavity. Comparison with the disarticulated skull will reveal that only a portion of the septum is formed by bone. The anteroinferior portion of the septum is cartilaginous, and is, in fact, continuous anteriorly with part of the cartilage of the lower parts of the sides of the nose that were previously palpated. Confirm by palpation, by grasping the lower part of the nasal septum of your subject between index finger and thumb, that this portion of the septum is freely mobile and not bony in nature.

5. Superiorly, the narrow **roof** of the nasal cavity may be seen. The roof and adjacent superior portion of the septum, and the superior concha, are lined by the **olfactory epithelium,** containing the **olfactory cells,** the sense organs of smell (olfactore means to smell). This constitutes the **olfactory region** of the nasal cavity. Review the **respiratory region** and the **vestibule,** studied in point 3 above.

6. Finally, the floor of the nasal cavity can also be observed by speculum examination through the nostril. Comparison with the skull should confirm that the floor of the nasal cavity corresponds to the roof of the oral cavity, or **palate.** Review the bony components of the hard palate (page 10).

7. Posterior examination of the nasal cavity is possible, in suitable subjects, with the help of a postnasal mirror (or, when available, by fiber-optic illumination). This instrument can be inserted through the mouth, passed into the oral part of the pharynx (oropharynx), and directed upwards and forwards, so that the posterior aspect of the nasal cavity, and its communication with the nasal portion of the pharynx (nasopharynx), can be studied. Use this technique, under guidance, to study the **posterior nasal apertures** or **choanae** (choana means opening). The choanae should be simultaneously studied on a skull, noting that they are bounded laterally by the **medial plate of the pterygoid process of the sphenoid bone,** a structure which you have studied previously when identifying the **hamulus** (pages 27 and 30–31). Medially, the choanae are bounded by the nasal septum, which at this site is mainly formed by another paper-thin cranial bone, the **vomer.**

It will now be evident that the choanae mark the boundary between the nasal cavity and the **nasopharynx.** The nasal mirror technique of examination also allows observation of structures in the nasopharynx. Identify the following.

8. The lateral wall presents, posterior to the inferior nasal concha, the **pharyngeal opening of the auditory tube** (figs. 9 and 10, page 22). From this opening, the auditory tube extends posterolaterally, to enter the middle ear. The proximal portion of the auditory tube (i.e., the portion nearest the pharynx), is cartilaginous, lying in a groove between the **petrous part** of the **temporal bone** and the **greater wing** of the **sphenoid bone.** Identify these bony structures on the skull now. The groove in which the cartilaginous portion lies ends proximally near the root of the **medial pterygoid plate.** The bony portion of the auditory tube may be traced, on the inferior surface of the skull, as a posterolateral continuation of the groove described previously (i.e., the groove lodging the cartilaginous portion of the canal): the junction between the two being situated just medial to the **sphenoid spine,** which you previously have identified, medial to the mandibular fossa (page 40). Visualize, on the skull and with the aid of specimens and atlas illustrations, the continuity of the auditory canal from the nasopharynx to the **middle ear** (the portion of the ear situated medial to the **external auditory meatus**). This visualization can be facilitated by the following exercise: Gently insert a probe (e.g., a piece of wire) into the bony portion of the auditory tube and pass it posterolaterally. By looking in through the external auditory meatus the probe can be seen to enter the middle ear. Note that in the **living** state the middle ear is **separated** from the external auditory meatus by the **tympanic membrane.** In the dried skull this membrane is absent.

9. The pharyngeal opening of the auditory tube is bounded above and behind by the **tubal elevation** (also called the **taurus tubarius:** taurus meaning elevation) This is shown, but not marked, in fig. 10, page 22). This prominence is merely the anterior end of the medial wall of the cartilage of the auditory tube, lying deep to the pharyngeal mucous membrane and raising a fold in it.

10. A vertical fold of mucous membrane extends from the lower part of the tubal elevation downwards to disappear in the wall of the pharynx. This fold (shown, but not marked, in fig. 9, page 22), contains the salpingopharyngeus muscle (salpinx means tube) which, as the name implies, extends from the auditory tube to the pharyngeal wall.

11. Immediately posterior to the tubal elevation a depression or recess in the mucous membrane is seen. This is known as the **pharyngeal recess.** This may extend somewhat laterally and represents the most lateral extent of the pharynx at this level.

12. Immediately inferior to the pharyngeal opening of the auditory tube a small swelling may be observed due to the **levator veli palatini muscle** (levator means elevator; velum means veil: the soft palate is called velum palatinum since it hangs down like a veil from the back of the hard palate, thus partly separating the oral cavity from the pharynx). This muscle extends from the auditory tube and the tem-

poral bone just medial to the tube, down to the soft palate which, as its name implies, it elevates (for example, in swallowing).

13. The roof of the nasopharynx is related to the inferior surface of the skull, more specifically the **basal** portions of the **sphenoid** and **occipital bones** (i.e., the site of the **spheno-occipital synchondrosis,** the cartilaginous joint where, as previously mentioned [page 39], longitudinal growth of the skull continues until the age of about 25 years). Identify this region on the base of the skull, anterior to the **foramen magnum.** Observe, about 0.5 to 1 cm anterior to the anterior edge of the foramen magnum, a small tubercle, the **pharyngeal tubercle.** This is the superior site of attachment of the **pharyngeal raphe,** the posterior, midline seam in which the two halves of the pharyngeal musculature meet (page 27). As will be evident on inspection of the nasopharynx of your model, using the nasal mirror, the roof of the nasopharynx slopes down to continue into the posterior wall. Confirm this point by examination of anatomical specimens, and of an articulated skeleton in which the skull is in its natural relationship to the **first cervical vertebra** (the **atlas**). Thus the anterior surface of the body of the atlas is in direct relationship to the posterior wall of the pharynx, as the **basisphenoid** and **basi-occipital** bones are in relationship to the sloping roof.

14. Situated deep to the mucous membrane of the roof and posterior wall of the nasopharynx is a collection of lymphatic tissue, the **pharyngeal tonsil.** This tissue is much more prominent in children than in adults, but may be visible in the nasal mirror, of your model. When enlarged, by infection (usually in children), these structures are called "adenoids." As will be seen later, this region of lymphatic tissue, with the lymphatic tissue previously mentioned in the pharyngeal portion of the dorsum of the tongue (the **lingual tonsil,** page 15), and other lymphatic tissue in the oral part of the pharynx, constitutes a ring of such tissue which surrounds the site of entry, into the digestive system, of potentially harmful foreign particles from outside the organism.

15. Note that the pharynx is a muscular tube which is open anteriorly to the nasal cavity above, the oral cavity in the middle, and the **larynx** (voice box) inferiorly. The nature of the muscular wall and the attachments of the three muscles which form it, the **superior, middle,** and **inferior pharyngeal constrictors** (constrictor means sphincter, a circular or ring-muscle, capable of constricting a lumen), will be studied in detail elsewhere. Also to be studied later, but mentioned here for the sake of completeness, are the **longitudinal muscles** of the pharyngeal muscular coat: the **stylopharyngeus,** extending from the **styloid process,** which you should reidentify on the base of the skull now, to the pharyngeal wall; the **salpingopharyngeus,** mentioned in point number 10 above; and the **palatopharyngeus.** The latter muscle raises a fold of mucous membrane, which should be observed now in your model, extending from the soft palate posterolaterally to merge into the wall of the pharynx. Note that this fold of mucous membrane, known as the **palatopharyngeal** arch, originates close to the **palatoglossal arch,** which you previously studied (page 10 and fig. 1, page 12), but diverges from it posteriorly as the two arches and their contained palatal muscles pass inferiorly to enter the pharynx and the tongue respectively. The three longitudinal muscles of the pharynx mentioned here, and especially the palatopharyngeus, lie mainly internal to the circular or sphincter muscle fibers. The longitudinally arranged muscle layer of the pharynx is thus an exception to the rule which pertains in the rest of the digestive tract (i.e., an inner circular and an outer longitudinal layer of muscle).

16. In addition to being the third of the longitudinal pharyngeal muscles mentioned here, palatopharyngeus is also the third muscle of the palate that you have encountered (the previous two having been the levator veli palatini, mentioned above in point 11, and the palatoglossus, reviewed in the previous paragraph point 15, and previously studied on pages 10 and 15 through 18).

For completeness, a fourth muscle of the palate will be mentioned here, although it cannot be observed in the living subject. It should, however, be visualized, with the aid of a skull, anatomical specimens, and illustrations. This is the **tensor veli palatini,** which has its origin lateral to the auditory tube (compare with the levator veli palatini, medial to the auditory tube). The tendon of the tensor veli palatini passes inferomedially to curve around in front of the hamulus, and then passes medially into the soft palate (figs. 25 and 26, page 64). The relationships of this and the other palatine muscles will be studied in greater detail in the dissection laboratory. Notwithstanding this and the fact that further detailed study of the pharyngeal musculature similarly is deferred (point 15 above), the **actions** of the palatal muscles, and of a portion of the **superior pharyngeal constrictor,** should be studied now as follows.

17. Instruct your model to swallow, keeping his or her mouth as wide open as possible. Observe with the aid of a focused beam of light and a dental or nasal mirror, the soft palate and the posterior wall of the pharynx during the process. The muscles of

the palate, and in particular the **levator** and **tensor** muscles, act in consort to tense and lift the soft palate upwards and posteriorly, to come in contact with the posterior wall of the pharynx. Simultaneously, that wall of the pharynx is tensed and brought forwards to meet the soft palate. This action is performed by a specialized **portion** of the **superior pharyngeal constrictor,** which curves around from the lateral part of the soft palate into the pharyngeal wall. This band of muscle is known as the **palatopharyngeal sphincter,** and the rounded ridge on the posterior pharyngeal wall, which swells forwards to meet the elevated soft palate in swallowing, is known as the **ridge of Passavant,** named after an early observer of this phenomenon. The function of this total closure of the passageway between the **oral** and **nasal** parts of the **pharynx** during swallowing is obvious. The structures described here as creating this partition, are also arbitrarily designated as the **boundaries** between these two parts of the pharynx.

In conclusion, to complete this session of study you should review the emphasized key words and topics.

Laboratory Session 6

Use the half head and neck specimen that remains attached to the rest of the cadaver for the following laboratory study of the nose, nasal cavity, nasopharynx, and soft palate.

Begin by examining the **nasal septum.** (If the sagittal section cut through your cadaver has left the nasal septum on the detached half head and neck specimen, make use of that specimen for study of the septum, and then return to the attached half for subsequent sections.) Using scalpel and forceps dissection, elevate the **mucoperiosteum** (pages 33 to 36), or **mucoperichondrium,** in the anterior portion of the nasal septum, comprised of cartilage rather than bone. As you elevate the mucosal lining layer identify, with the aid of an atlas illustration, the various components of the nasal septum: the **vomer** (page 60), the **perpendicular plate of the ethmoid bone,** and the **septal cartilage,** anteriorly. Inferiorly, note that the **palatine process of the maxilla** and the horizontal plate of the palatine bone (page 30) each make a small, perpendicular contribution to the nasal septum.

Having completed your study of the septum, proceed to remove it, if necessary, for study of the lateral wall of the nasal cavity. Identify the **vestibule, the respiratory region,** and the **olfactory region** of the nasal cavity (pages 59 and 60). Refer to an atlas illustration and anatomical specimens where necessary to observe and study the various structures of the nasal cavity previously examined in the living subject. Ensure that you identify each of the emphasized structures mentioned in pages 59 to 62, of Clinical Session 6.

When you have completed your study of the conchae and meatuses of the lateral wall of the nasal cavity, use a pair of bone scissors, as well as a scalpel and a pair of forceps, to dissect the inferior, middle, and superior conchae away from their attachments to the lateral wall, to expose the more lateral structures, including the paranasal sinuses (page 59).

In the inferior meatus, find the opening of the **nasolacrimal duct** under cover of the anterior part of the inferior concha. In the middle meatus, inferior to the middle concha, identify the **bulla ethmoidalis,** on the apex of which the opening of the **middle ethmoidal sinus** will be seen. Inferior to the bulla ethmoidalis, observe a groove, the **semilunar hiatus.** At the anterior extremity of this, observe the opening of the **infundibulum,** into which the **frontal sinus** and the **anterior ethmoidal sinus** open. Identify the opening into the **maxillary sinus,** also in the middle meatus, immediately inferior to the semilunar hiatus. Underlying the superior concha, observe the openings of the **posterior ethmoidal sinus** into the superior meatus. Further openings of the posterior ethmoidal sinus might be seen above the superior conchae. Finally, an opening from the space above the superior conchae into the **sphenoidal sinus** (fig. 10, page 22) might be observed.

Use a blunt, soft probe to examine the various openings, passing the probe toward the various sinuses that communicate through these openings into the nasal cavity. Identify the frontal sinus in sagittal section in the frontal bone, above the nasal cavity and anterior to it.

Having identified all of the openings, and having proved the communication with the respective sinuses, perform further dissection with bone scissors and forceps to explore the frontal, maxillary, ethmoidal, and sphenoidal sinuses. Ensure that you obtain thorough understanding of the three-dimensional relationships of the sinuses to the nasal and the oral cavities. In particular, study the relationship of the maxillary sinus to the maxillary teeth. To do so, you should, if necessary, dissect extensively into the bony wall of the maxillary sinus, inferior to the opening of this sinus into the nasal cavity.

Identify the orifice of the auditory tube (see fig. 10, page 22) and the related taurus tubarius, pharyngeal recess, and the salpingopharyngeal fold (pages 60 to 61).

At the posterior end of the middle concha, slightly anterosuperior to the opening of the auditory tube, gently elevate the mucoperiosteum by scalpel dissection to demonstrate the **sphenopalatine foramen,** which you should also identify on a skull. This foramen establishes communication between the **pterygopalatine fossa** (page 46) and the nasal cavity. With careful use of bone-cutters, you should expose the pterygopalatine fossa, inferior and superior to the sphenopalatine foramen, to demonstrate its contents, the third part of the maxillary artery and its branches (pages 46 to 48), and the **pterygopalatine ganglion (sphenopalatine ganglion),** a component of the visceral or autonomic nervous system (pages 5 and 6) entirely comparable to the submandibular (page 33) and otic (page 48) ganglia you previously studied. Nasal and **nasopalatine nerves** (branches from the ganglion), and the **sphenopalatine branch of the maxillary artery,** pass from the pterygopalatine fossa through the sphenopalatine foramen to reach the mucosal lining of the nasal cavity. With the aid of an atlas, identify as many branches of the ganglion, and of the third part of the maxillary artery as possible. In particular, the **greater and lesser palatine nerves** should be traced

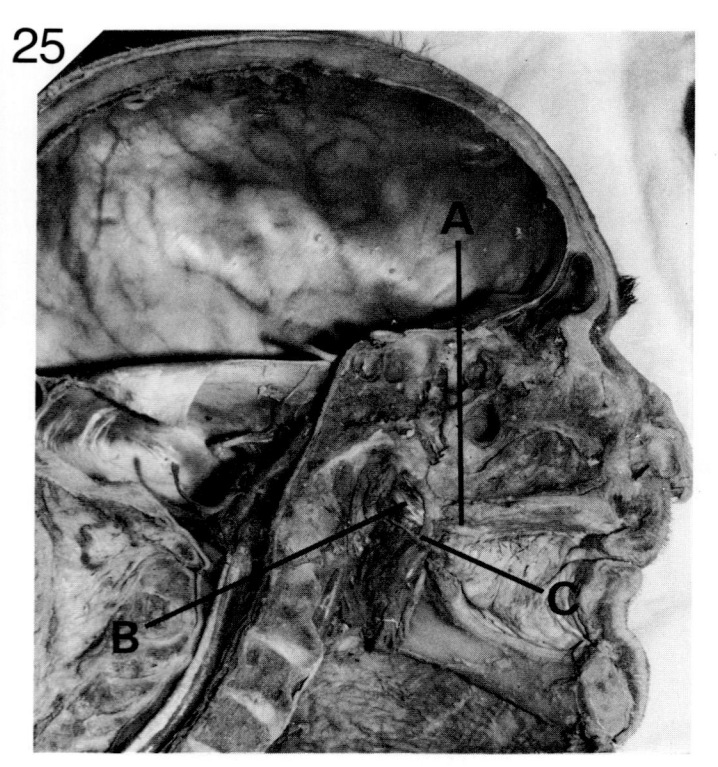

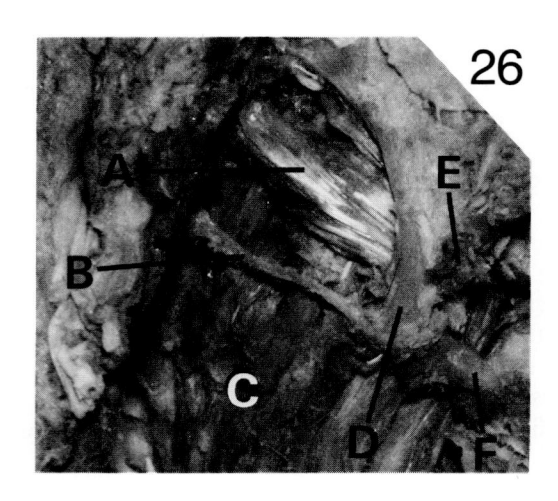

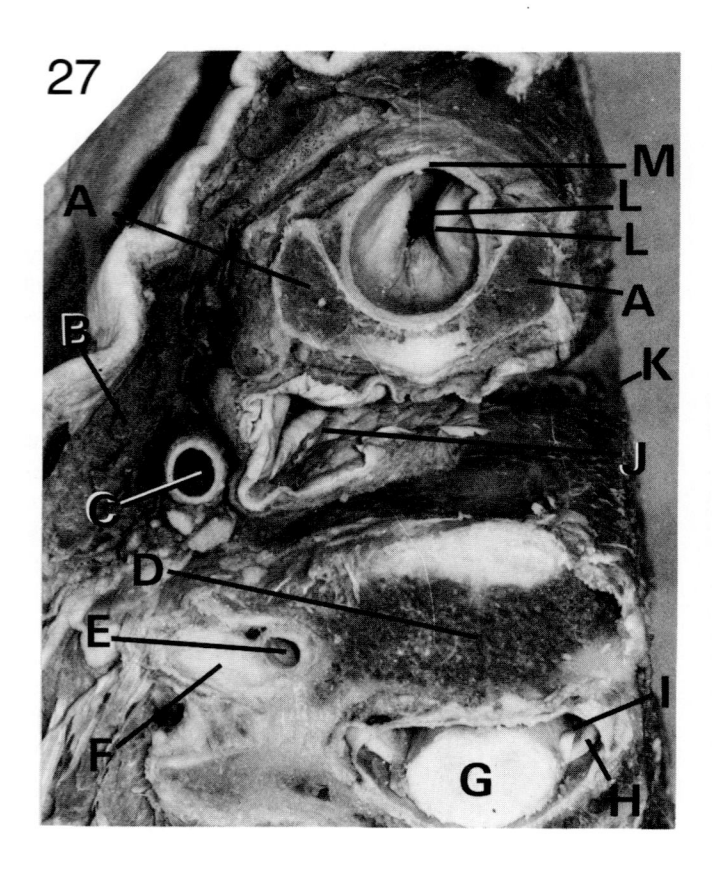

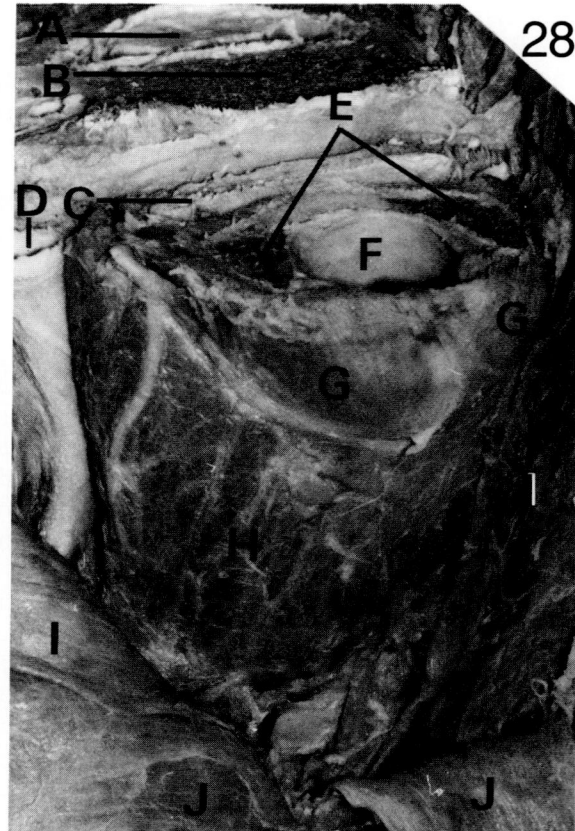

64 YOUR PATIENT'S ANATOMY: A CLINICAL VIEW OF HUMAN MORPHOLOGY

Figure 25. Sagittal section, overview, indicating site of tensor veli palatini muscle.
A. Soft palate
B. Tensor veli palatini muscle
C. Hamulus

Figure 26. Detail of sagittal section indicating site of tensor veli palatini muscle.
A. Tensor veli palatini muscle
B. Fiber of superior pharyngeal constrictor, attached anteriorly to the hamulus. Remaining fibers of the constrictor and the pharyngobasilar fascia have been dissected away
C. Medial pterygoid muscle
D. Hamulus
E. Tendon of tensor veli palatini, dissected from the palatine aponeurosis
F. Remnant of pterygomandibular ligament

Figure 27. Transverse section through larynx, viewed from below.
A. Cricoid cartilage
B. Sternocleidomastoid muscle
C. Common carotid artery
D. Vertebral body of C5
E. Vertebral artery
F. Spinal nerve
G. Spinal cord
H. Dorsal root
I. Denticulate ligament
J. Lumen of pharynx
K. Sagitally cut surface
L. Vocal folds
M. Conus elasticus

Figure 28. Transverse section of larynx, viewed from above and anteriorly.
A. Spinal cord
B. Vertebral body of C5
C. Pharynx (lumen collapsed)
D. Common carotid artery
E. Cricoid cartilage
F. Lumen of lower part of larynx
G. Cricothyroid muscles
H. Thyroid gland
I. Sternothyroid muscles (right reflected)
J. Sternohyoid muscles (right and left reflected)

to the palatine foramina. Demonstrate the connections of the ganglion to the maxillary nerve (page 20), which passes superior to the ganglion. Use the bone scissors as necessary to expose the maxillary nerve as it passes to the infraorbital canal (pages 25 and 37). Dissect into the maxillary sinus to expose the **superior alveolar branches** of the maxillary nerve.

In the following dissection, use a sharp scalpel, fine-pointed scissors, and a pair of forceps to carefully elevate the mucous membrane of the nasopharynx from the deep-lying muscle layers of the pharyngeal wall. Begin posteriorly, at the median sagittal section through the posterior wall of the pharynx (which will more or less have passed through the pharyngeal raphe [pages 61 and 27]). Carefully work your way forwards to the superior surface of the soft palate, preserving the deep-lying muscle layers. Since the mucous membrane is very tightly adherent to the subjacent muscles, and these muscles are relatively delicate, this dissection should be performed with great care. Adhere strictly to the dissection technique described on page 10, and illustrated in figure 5, page 16.

The **salpingopharyngeus muscle** is relatively easy to identify and define because of its origin from the auditory tube and because it raises a clear fold in the mucous membrane. Trace the muscle, from its origin on the tube to the site where the fibers merge with **palatopharyngeus,** forming an inner, longitudinal muscle layer of the pharynx (item 15, page 61). Having defined the salpingopharyngeus, gently elevate it from the circular muscular fibers of the **superior pharyngeal constrictor,** and then cut across the belly of the salpingopharyngeus. Reflect the cut ends of the muscle superiorly and inferiorly, and proceed to expose the superior constrictor fully. Trace the fibers of the superior constrictor anteriorly. The uppermost fibers of this muscle can be traced to the lower end of the posterior edge of the medial pterygoid plate, from which they take origin. Review (pages 27 and 32) the fact that the superior constrictor arises mainly from the posterior edge of the pterygomandibular raphe. Some additional fibers of the muscle take origin from bone, adjacent to where the raphe attaches inferiorly and superiorly: on the mandible inferiorly; and from the hamulus and adjacent portion of the posterior edge of the medial pterygoid plate superiorly. Identify the superiormost fibers of the muscle, and trace them as they curve posterosuperiorly to attain the pharyngeal tubercle (page 61). By visualizing on an isolated skull the course of these uppermost fibers, from the middle of the posterior edge of the medial pterygoid plate to the pharyngeal tubercle of the skull, it will be clear to you that a gap in the muscular wall of the pharynx will exist between the superior edge of this muscle and the base of the skull. This gap is filled by a layer of fascia, mainly formed by an extension of connective tissue of the submucosal layer of the pharynx. This fascia is called the **pharyngobasilar fascia** (pharyngobasilar indicates from pharynx to base of skull).

Dissect carefully above the superior edge of the superior constrictor, to demonstrate the pharyngobasilar fascia. The **pharyngeal opening of the auditory tube** is situated in this fascia. A second structure which pierces the fascia is the **levator veli palatini muscle** (item 12, page 60), which you now should dissect out to its insertion in the palate. Finally, you will encounter a third structure piercing the pharyngobasilar fascia, the **ascending pharyngeal artery,** a branch of the external carotid artery.

The **tensor veli palatini muscle** (point 16, page 61) will not be visible as long as the pharyngobasilar fascia remains intact. This muscle, unlike the levator, does **not** pierce the fascia-closed gap between the superior constrictor and the base of the skull to reach the palate. Rather, it passes anterior to the hamulus, and thus remains entirely outside of (anterior to) the pharynx. It will be studied in the following paragraphs.

Define the lowermost fibers of the superior pharyngeal constrictor muscle inferiorly, and identify the **palatopharyngeal sphincter** (ridge of Passavant) (page 62).

Gently elevate levator veli palatini and cut it across its belly, reflecting the superior portion to the auditory tube and the inferior part toward the palate. After studying the pharyngobasilar fascia, dissect through it to expose the tensor veli palatini muscle (figs. 25 and 26, page 64). Reidentify the hamulus by palpation through the soft palate (page 27). Trace the fibers of the tensor downwards to their convergence in the tendon and define the latter as it hooks around the hamulus anteriorly to pass medially into the soft palate.

With the hamulus as your landmark, carefully remove the mucous membrane lining the inferior surface of the soft palate, anterior to the coronal plane through the hamulus. Deep to the mucosa, you will encounter a tough, fibrous aponeurosis (flattened tendon), known as the **palatine aponeurosis.** This aponeurosis is derived from the tendon of the tensor veli palatini muscle. It can be traced anteriorly to an attachment to the posterior edge of the hard palate, and, posterolaterally, its fibers can be dissected out to converge to the round portion of the tendon of tensor veli palatini, as the latter hooks around the hamulus. The tendon passes, as previously seen, anterior to the hamulus, from lateral to medial. Visualize on a disarticulated skull the position of the muscle, from its origin lateral to the auditory tube, to the passage of the tendon anteromedially, in front of the hamulus, to insert, as the aponeurosis, into the posterior edge of the horizontal plate of the palatine bone. Having defined the tensor veli palatini tendon in this way, attempt to follow it superiorly to the muscle itself.

Note that the **palatoglossus muscle** (pages 10 and 30) takes its origin from the inferior surface of the palate, inferior to the palatine aponeurosis. Trace fibers of this muscle, in the palatoglossal arch, down to the side of the tongue. The **palatopharyngeus muscle** (page 61) should be similarly identified, in the palatopharyngeal arch. Note that its fibers originate superior to the palatine aponeurosis. The insertion of the levator veli palatini muscle into the palate separates the palatine origin of palatopharyngeus into two layers, one superior to and one inferior to the levator insertion.

To complete this session, review your previous dissections of the styloglossus muscle (page 24), and the lingual nerve (page 36), noting that these two structures pass **below** the **inferior edge** of the **superior pharyngeal constrictor** to "enter" the lumen of the digestive tract. Two other structures, one of which, the stylopharyngeus muscle, has been mentioned previously (page 61), also enter by passing through this gap. Review also the structures that enter by passing **above** the **superior** edge of this constrictor (i.e., by piercing the pharyngobasilar fascia) (page 66).

Clinical Session 7

■ **TOPICS**

Oropharynx
  Palatine Tonsil
  Waldeyer's Lymphatic Ring

Laryngopharynx
Larynx
Muscular Triangle
Strap Muscles
Thyroid Gland

The **oropharynx** (**oral part of the pharynx,** or the part of the pharynx that opens into the oral cavity) has been delineated from the nasopharynx previously. Review this boundary, as defined on page 62. The anterior extent of the oropharynx has previously been studied: the **palatoglossal arches** demarcate the posterior limit of the **oral cavity,** and thus the anterior limit of the oropharynx (page 10). In the floor of the oral cavity, the sulcus terminalis constitutes an additional landmark for the posterior extent of the oral cavity (page 15). An actual posterior wall of the oral cavity is, however, lacking, leaving free communication between the oral cavity and the oropharynx. In the following, the lateral wall of the oropharynx, the further extent of this cavity and its communication inferiorly, will be studied.

1. The lateral wall of the oropharynx should be examined, by observing the structures posterior to the palatoglossal arch. You have previously (page 61) observed the **palatopharyngeal arch** diverging posteriorly from the palatoglossal arch (and studied its contained muscle, the **palatopharyngeus,** [pages 61 and 67]). Between the two arches (palatoglossal and palatopharyngeal) the clinically important **palatine tonsil** (commonly referred to as the **tonsil**) is situated. Study this structure in the situation described.

The extent of the palatine tonsil varies greatly between individuals. Among the factors which may influence its size are age and previous bouts of infection of the tonsil (tonsillitis). The tonsils are usually larger in younger people, and may be considerably regressed in older persons. When fully developed, the tonsils may be seen to extend from the site mentioned, upwards into the soft palate. Even when they are not obviously expanded to this extent, deep-lying lymphatic tissue does continue into this site, thus accounting for the name of this organ. Another major cause for the varying size of the palatine tonsil is the surgical removal of it. If your subject has had a tonsillectomy, there may be no trace of the organ to be observed. However, some remnants of the organ may be present despite previous surgical removal. Small remnants left at surgery can subsequently redevelop and expand.

Observe the surface of the tonsil noting its irregularity. The surface is deeply pitted by numerous depressions or recesses, called **tonsillar crypts.** One particular large depression may be observed in the upper part of the organ. This is known as the **intratonsillar cleft.** The irregularities in the surface of this organ facilitate contact between foreign particles, such as bacteria, and the **immunocompetent lymphocytes** of the **lymphatic tissue** of the tonsil. Immunocompetent indicates capable of performing an immune function. Review (page 6) the two types of immune reaction, **cellular** and **humoral,** and that both are executed by lymphocytes.

2. It was previously mentioned in the discussion of the **nasopharynx** that the **pharyngeal tonsil** constitutes part of a lymphatic ring, to which the lymphatic tissue of the tongue (the **lingual tonsil**) also contributes (page 61). You should have no difficulty in visualizing that the **palatine tonsil** is in intimate relationship with the lingual tonsil and that the palatine and pharyngeal tonsils form, with the lingual tonsil, a ring of lymphatic tissue which surrounds the oropharyngeal "portal of entry" to the organism. This ring of lymphatic tissue is known as **Waldeyer's lymphatic ring.**

3. The posterior wall of the oropharynx is formed by mucous membrane and deep-lying pharyngeal musculature. As indicated, the pharyngeal tonsil may extend down from the nasopharynx into the oropharyngeal posterior wall to a varying degree.

4. Inferiorly, the oropharynx is continuous with the third and last portion of the pharynx, the **laryngopharynx,** so named because it is situated behind the larynx or voice box (of which the thyroid cartilage is a part). Using a laryngeal mirror and a focused beam of light, determine the arbitrary border between these two continuous spaces (i.e., the oropharynx and the laryngopharynx) by identifying, anteriorly, the upper edge of the **epiglottis** just posteroinferior to the tongue [fig. 6, page 16]). The epiglottis is a cartilaginous, "leaf-shaped" structure. Consult an anatomical specimen and atlas illustra-

tions, and with the aid of a laryngeal mirror, visualize an imaginary horizontal plane passing through the superior edge of the epiglottis, thus dividing the oropharynx from the laryngopharynx.

5. Using a laryngeal mirror, examine the lumen of the pharyngeal tube inferior to the arbitrary border you have just established. Your immediate objective now should be to visualize the division of the tube into two separate passageways, or lumens. The anterior of the two enters the larynx. This lumen is guarded in front by the epiglottis, and posteriorly and laterally by a ridge or fold continuous with the upper edge of the epiglottis. Posterior to this fold is a second lumen, which enters the laryngopharynx. Compare what you are able to observe with an anatomical specimen and atlas illustrations to appreciate that the anterior opening (i.e., that of the larynx) belongs to the respiratory tract, and the posterior (i.e., that of the laryngopharynx) belongs to the digestive tract.

6. In the adult, the **laryngopharynx** marks the site at which, in embryological life, the respiratory system develops from the digestive canal. At approximately 3 weeks of fetal life, a tube buds off and grows out on the ventral surface of the digestive tube, then grows **caudally** (toward the tail) to create a second tube lying ventral to the digestive tube. This ventral tube gives rise to the larynx and other respiratory organs. The embryological course of events described here accounts for why the **thyroid cartilage** or Adam's apple, a part of the larynx (point 4), lies ventrally and can be palpated immediately deep to the skin (page 4).

The morphological terms ventral and dorsal were previously (page 14) introduced, and the relationship of these terms to the anatomical terms anterior and posterior was explained. In a similar fashion, the term **caudal** (tail) corresponds to inferior with respect to the trunk in the adult human in the normal anatomical position. However, during development, the human embryo is curled up into a position which is not comparable to the normal anatomical position, and the usefulness of a term to indicate direction toward the tail end of the organism, which applies to structural relationships throughout **ontogeny**, will be obvious. Also, the tail end of lower animals (e.g., dogs and cats) is not inferior but posterior; thus morphological terms can also be used to compare different stages in **phylogeny**. The morphological directional opposite of caudal is **cranial** (cranium means skull), and, again this corresponds to superior in the normal anatomical position of an adult human (fig. 38, page 88).

Since the third part of the pharynx, or laryngopharynx, is the site at which the developing respiratory tube grows out ventrally from the digestive tube in the adult, it is this site at which the common lumen (space) of the pharynx divides into two channels: a ventral canal for air passage, palpable anteriorly, and a dorsal canal for food passage. The first part of the respiratory outgrowth from the digestive tube is called the **larynx** (voice box), which extends downwards to the level of the sixth cervical vertebra (C6), marked anteriorly by the level of the ring of the cricoid cartilage (page 54). The lower part of the pharynx similarly extends down to C6, and therefore has the larynx lying immediately anterior to it, for which reason the third part of the pharynx is called the laryngopharynx.

7. Examine the portion of the respiratory tube known as the **larynx**. Unlike the digestive tube, the walls of the larynx (and the subsequent portions of the respiratory tract) are composed of a semi-rigid material, called cartilage. In light of the respective functions of the two tubes, the selective advantage of this difference should be apparent. The anterior, cartilaginous portions of the larynx are easily palpable through the skin of the neck. You have previously palpated the superior edge of the **thyroid cartilage** (page 53) and noted the vertebral level which this landmark corresponds to, and which you should now review. In palpating this superior edge again now, note that in the midline there is a notch, concave upwards, from which a right and a left superior edge diverge upwards and laterally. These are respectively the upper borders of the right and left **laminae** (plates) of the thyroid cartilage. Each upper border ends laterally in a **superior horn**. From this you should palpate the lateral border of each lamina down to the **inferior horn**. Then palpate the **inferior border**, which curves gently upwards to the midline. The right and left laminae of the thyroid cartilage are set at an angle to each other (the **thyroid angle**), such that they meet in the midline in the **laryngeal prominence**, which is more marked in males than in females. The reason for this is that the angle the laminae make with each other is more acute in males than in females. Because the laminae meet at an angle, they form the anterolateral walls of the larynx. (Thyroid means shield-shaped.) The remainder of the walls of the larynx at this level are formed by structures to be studied later.

8. Immediately inferior to the inferior edge of the thyroid cartilage, repalpate (page 54) the second major cartilage of the larynx, the **cricoid cartilage**. It is palpable, across the midline, as a curved horizontal cartilaginous bar, immediately inferior to the inferior extremity of the thyroid prominence. The shape of

the cricoid cartilage, as can be seen on an anatomical specimen, resembles that of a signet ring, the flat or signet portion being situated posteriorly in the larynx. The portion palpable anteriorly, immediately inferior to the thyroid cartilage, is the anterior curvature of the ring.

9. Inferior to the cricoid ring, palpate the uppermost six to ten **tracheal rings** (page 54) which continue in series inferior to the cricoid ring. These are cartilaginous components of the wall of the **trachea,** the portion of the respiratory tract which follows caudally after the larynx. Note, however, (although this cannot be confirmed by palpation), that the tracheal rings, unlike the cricoid ring, are not true rings but are incomplete posteriorly and are in fact horseshoe-shaped. Although only six to ten tracheal rings are palpable, above the **suprasternal notch,** the trachea continues inferiorly, posterior to the sternum, to the level of a horizontal plane passing between the fourth and fifth thoracic vertebrae. This will be studied in detail later.

10. At this time, examine an important lateral relation of the trachea by palpating the very distinct arterial pulsation of the **common carotid artery,** immediately lateral to the trachea on either side. As mentioned previously, the larynx and the laryngopharynx are coextensive inferiorly: both end at the level of C6, which is also the vertebral level of the **cricoid ring.** You have established that inferior to the larynx (i.e., below the cricoid cartilage), the respiratory tract continues as the trachea. Below the level of C6, the digestive tract continues as the **esophagus,** a more or less vertical tube with a muscular wall, continuous with the pharynx superiorly (and the **stomach** inferiorly). Visualize, with the aid of anatomical specimens, that since the esophagus lies posterior to the trachea, the common carotid artery, being posterolateral to the trachea, lies in the groove between the trachea and the esophagus. Palpate the common carotid artery, extending from the site where you have just identified it, lateral to the trachea, upwards to where you previously palpated it, close to its bifurcation, in the **carotid triangle** (page 53).

11. In the following exercise, you should palpate anterior structures in the neck to appreciate movements of the larynx during swallowing and the relationships of these movements to others, in the overall **mechanism of deglutition** (swallowing). Palpate with your right index finger the body of your model's hyoid bone. Place your left index finger over your model's laryngeal prominence (page 69). Then instruct your model to swallow, simulating the normal process as nearly as possible (i.e., picturing to him- or herself that there actually is food in the mouth at the commencement of the process). You will detect movements upwards of both the hyoid bone and the thyroid cartilage. The movement of the hyoid bone precedes that of the thyroid cartilage, indicative of the fact that deglutition takes place in stages (three in all). In the first stage, the tip of the tongue is pressed up against the palate and a wave of contraction passes through the tongue anteroposteriorly, due to the **intrinsic muscles** of the **tongue** (page 18). In this **first stage of deglutition,** the hyoid bone moves upwards and anteriorly. Attempt to visualize, from your previous knowledge of muscular attachments to the hyoid bone (pages 21 to 25, 49 to 52, 53, 54) which muscles are involved in these movements. As the food bolus (mass) passes backwards, the posterior portion of the tongue is elevated and the communication between the oral cavity and the oropharynx becomes constricted. Which muscles may be responsible for these movements?

As these functions occur and the bolus moves beyond the back of the tongue, the soft palate becomes elevated and opposed to the posterior wall of the pharynx (pages 61 to 62), (with development of Passavant's ridge, the causation of which you should review now [page 62]); and the back of the tongue gets elevated to bulge into the oropharynx. These movements all initiate the **second stage of deglutition** in which the bolus passes through the oropharynx, assisted by contraction of the superior constrictor muscle of the pharynx and gravity. In this stage of deglutition, the larynx is pulled upwards and you should therefore detect, by palpation of the thyroid cartilage, this movement in this stage. Thus movement of the larynx comes after the previously mentioned movements of the hyoid bone have been performed.

In the **third stage of deglutition,** the bolus is moved on downwards, toward the esophagus, by action of the inferior pharyngeal constrictor muscle fibers.

12. The muscles involved in movement of the larynx during swallowing are, among others, fairly superficial muscles situated in the **muscular triangle,** the boundaries of which you should now review (page 54). You will recall that because the muscles of this region are narrow and flat or thin, they have been given the descriptive name of **strap muscles** (page 54). These include small muscles extending between the hyoid bone and the thyroid cartilage (the **thyrohyoid muscles**); strap muscles between the thyroid cartilage and the back of the upper part of the sternum (the **sternothyroid muscles**); as well as strap

muscles from the sternum to the hyoid (**sternohyoid muscles**). Palpate the strap muscles during deglutition and particularly the thyrohyoid as the thyroid cartilage moves upwards, and the sternothyroid as it moves down again.

13. For completeness, you should attempt to palpate the **thyroid gland** (an important **endocrine** gland [page 7]). Place yourself **behind** your model and pass a hand round each side of the neck to palpate on the anterior surface of the neck. The normal thyroid gland is generally not palpable, but slight enlargement, which is quite common, makes it so. It comprises **right and left lobes,** which are largely covered by the respective sternothyroid muscles and an **isthmus,** which connects the two lobes by passing anterior to the second and third tracheal rings (and which derives from the thyroglossal duct [page 15]). An enlarged gland will thus be felt (and seen) as a swelling which raises the sternothyroid muscles and overlying tissues. This gland may extend laterally and medially beyond the muscle, and follow the movements of the muscle and the thyroid cartilage in swallowing.

14. Making reference to anatomical specimens and atlas illustrations, use a laryngeal mirror to visualize the laryngopharynx, and the posterior aspect of the larynx facing into the laryngopharynx. The **epiglottis** is much foreshortened when viewed from this aspect but its superior edge and the swelling on its posterior surface, called the **tubercle,** which faces posteriorly into the laryngopharynx, are usually easily visible. From the margins of the epiglottis a fold can be followed around on either side, to meet in the midline posteriorly. The entire edge of this fold and the epiglottis constitute the anterior or respiratory opening previously referred to (point 5, page 69). The fold is known as the **aryepiglottic fold,** because it extends from the epiglottis, anteriorly, around on each side to enclose the **arytenoid cartilages** posteriorly. Through the opening within these folds, two parallel anteroposterior white folds are visible close to the median plane. These are the **vocal folds.** Through the space between them, called the **rima glottidis** (rima means cleft or slit; **glottis** means vocal folds), the rings of the trachea may be seen. The vocal folds move during production of voice sounds and this may be observed. Lateral to and above the white vocal folds, a pink **vestibular fold** may be visible on either side. Anteriorly, in the median plane, a median and two lateral folds can be observed between the anterior surface of the epiglottis and the back of the tongue; these are the **glossoepiglottic folds.** On either side of the median fold, between the epiglottis and the back of the tongue, is a small space the **vallecula,** which has the clinical significance of being a space in which foreign bodies sometimes lodge. Posteriorly, between the posterior pharyngeal wall and the posterior extent of the aryepiglottic fold, the lumen of the pharynx will be seen as a mere slit lying in a frontal (lateral to lateral) plane.

During swallowing, access to the lumen of the larynx is restricted to allow the bolus to pass through the pharyngeal lumen downwards toward the esophagus. This occurs partly as a result of the upward movement of the larynx which you have palpated, and partly because of actual closure of the rim of the aryepiglottic folds.

To complete this session of study, review the emphasized key concepts.

Laboratory Session 7

In the following dissection of the pharynx and larynx, you should work primarily with the attached half of the head and neck specimen, in order to obtain an impression of continuity of the digestive and respiratory tracts. However, when necessary, the detached half of the specimen may be used for study of a specific structure.

OROPHARYNX

Review (page 68) the palatoglossal and palatopharyngeal arches and their contained palatoglossal and palatopharyngeal muscles, dissecting the delicate muscle fibers if this has not adequately been done previously. Between the arches, identify and study the palatine tonsil, observing the features described on page 68. Using scissor and scalpel dissection to elevate the covering mucous membrane, follow the extent of the tonsillar material toward the soft palate, establishing thereby the reason for the name of this organ.

On the posterior wall of the pharynx, observe the inferior extent of the pharyngeal tonsil by similarly elevating the mucous membrane at that site.

At the anterior wall of the oropharynx, reexamine the sulcus terminalis of the tongue and study the pharyngeal portion of the dorsum of the tongue. Observe the foramen caecum, just behind the apex of the sulcus terminalis as well as the elevations of the lingual tonsil (fig. 6, page 16). On the posterior aspect of the tongue, observe the median **glossoepiglottic fold** and, on either side of it, the **valleculae** (page 71).

Now turn to the external surface of your specimen where you have previously dissected the anterior triangle of the neck. Reidentify the hyoid bone, the digastric and stylohyoid muscles, the lingual artery, the hypoglossal nerve, and the hyoglossus muscle. Review the relationships of the latter three structures, the artery passing medial to and the nerve lateral to the muscle. Recall that the artery, just before passing medial to the posterior border of the hyoglossus muscle, takes a small loop upwards and that the nerve crosses this loop of the artery before the nerve passes onto the lateral aspect of the muscle (page 57, and fig. 24, page 50). At this site, posterior to the posterior edge of the hyoglossus muscle, the artery and nerve lie lateral to the **middle pharyngeal constrictor**. You should now carefully dissect connective tissue away to expose this muscle. It takes its origin from the upper border of the greater horn of the hyoid bone. Deep to (medial to) hyoglossus, a prominent ligament, the **stylohyoid ligament**, extends from the styloid process to the lesser horn (page 52) of the hyoid bone. This ligament makes an acute angle with the greater horn of the hyoid bone and the middle pharyngeal constrictor's origin from the greater horn extends to the stylohyoid ligament. The muscle also takes origin from the lowest portion of the ligament itself; thus the origin of the muscle occupies the acute angle between the ligament and the greater horn of the hyoid bone. From this origin, the fibers of the middle constrictor sweep posteriorly to meet fibers of the opposite side in the pharyngeal raphe (page 61). The superior fibers of the middle constrictor curve upwards and overlap (i.e., pass lateral to) the lowermost fibers of the superior constrictor.

By following the superior edge of the middle constrictor posteriorly, identify the **stylopharyngeus muscle** passing down from the styloid process to enter the gap between the superior and middle constrictors, and thus attain the inner, longitudinal layer of pharyngeal musculature (page 61). In intimate relationship with the muscle, identify also the **glossopharyngeal (ninth cranial) nerve.** This nerve can be readily observed posterior to the muscle from where it curves around the muscle laterally. It enters the gap between the superior and middle constrictor muscles together with the stylopharyngeus. Dissect the nerve forwards, tracing its sensory branches to their destinations in the palatine tonsil, the mucous membrane of adjacent parts of the pharynx and mouth, and the posterior part of the tongue including the taste buds of the vallate papillae (page 15). The nerve also supplies motor fibers to the stylopharyngeus.

With the stylopharyngeus muscle and the glossopharyngeal nerve, you have now observed all the structures which "enter" the lumen of the digestive tract by passing between the superior and middle constrictors. Review the structures you previously have observed doing so (page 67), as well as the structures which pass above the superior border of the superior constrictor (page 66).

LARYNGOPHARYNX

Review the extent of the laryngopharynx, from the upper border of the epiglottis to the level of the cricoid cartilage. What vertebral level does the latter structure correspond to?

Study the walls of the laryngopharynx, noting that the anterior wall is comprised, inferiorly, of the posterior wall of the larynx and, superiorly, the opening into the larynx, the aditus. Observe the limits of this opening, formed by the upper edge of the epiglottis, anteriorly; the arytenoid cartilages, which should be palpated through their covering mucous membrane, posteriorly; and the **aryepiglottic folds** of mucous membrane, extending from the lateral edge of the epiglottis to the corresponding arytenoid cartilage. On either side of the aditus, the anterior wall of the laryngopharynx is formed by a shallow depression; the **pyriform fossa** (pyriform means pear-shaped). Confirm by palpation that the pyriform fossa is related, anteriorly, to the space between the hyoid bone and the upper edge of the thyroid cartilage (in which space the **thyrohyoid membrane,** as well as the thyrohyoid muscles, extend between the bone and cartilage). Below the aditus, the anterior wall of the laryngopharynx is formed by the posterior surfaces of the arytenoid cartilages and the "signet" surface of the signet-ring shaped cricoid cartilage, covered by muscles and mucous membrane.

LARYNX

Observe in sagittal section the extent of the larynx (fig. 29, page 74). Reidentify the **aditus** and observe the epiglottis in sagittal section. Identify the **vestibular fold** (superiorly) and the **vocal fold** (inferiorly) (page 71).

Identify, using detached and attached half sections as required, the following parts of the larynx. The **vestibule of the larynx** is the part between the laryngeal opening and the vestibular folds. Between the vestibular and vocal folds is a region known as the **middle part of the larynx.** On either side, it opens through the slit between the vestibular and vocal folds into the **sinus** of the larynx. Explore the sinus with a blunt probe, passing the latter anteriorly to enter the **saccule** of the larynx, a blind extension from the sinus. The **rima glottidis** (page 71) is the space between the right and left vocal folds. These mucous membrane folds overlie the so-called **vocal cords,** the ligamentous bands which, with their mucosal covering, comprise the anatomical basis for the production of voice sounds. The **lower part of the larynx** is the portion extending below the level of the vocal folds, to the lower border of the cricoid cartilage.

The **vocal cords** extend from an anterior attachment on the inner (posterior) surface of the thyroid angle (page 69) to a posterior attachment to the **vocal process** of the arytenoid cartilage on each side. The arytenoid cartilages are small, pyramidal structures, the apex of each pyramid pointing superiorly, medially, and posteriorly, and the base resting inferiorly on the superior edge of the **cricoid lamina** (the "signet" part of the cricoid cartilage). The three sides of the pyramid face medially, posteriorly, and anterolaterally. Small muscles attached to the arytenoid cartilages and to the vocal cords themselves, alter the tension within the vocal cords and are capable of modulating the pitch of the sounds created by the cords. The vocal process and a **muscular** process project from the anteromedial and posterolateral angles of the base, respectively.

Using anatomical specimens and atlas illustrations to guide you, identify and dissect (preferentially in the detached half specimen) the **intrinsic muscles of the larynx.** To study the posterior cricoarytenoid, transverse and oblique arytenoid, and aryepiglottic muscles, you should elevate the mucous membrane on the posterior surface of the larynx (i.e., the anterior surface of the laryngopharynx) (fig. 32, page 74). For study of the lateral cricoarytenoid, thyroepiglottic, and thyroarytenoid muscles, including vocalis, which constitutes the lowermost fibers of the latter muscle, a lateral approach should be used (figs. 29, 30, and 31, page 74). The "sagittal" section through your specimen has been cut in such a way that it should be only slightly displaced from the median plane (i.e., should in fact be a parasagittal section passing through the thyroid lamina of one side). Gentle reflection of the thyroid lamina allows the access illustrated in Figure 31, page 74. If the parasagittal section is lateral to the lamina or in the case of a true sagittal section, the posterior part of the lamina should be dissected away to simulate the section shown in Figure 31.

The remaining intrinsic muscle of the larynx, the **cricothyroid,** should be studied on the anterior (external) surface of your specimen (fig. 28, page 64).

After identifying the muscles, your remaining project, necessary for the understanding of the anatomy of the larynx, is to expose and visualize in three dimensions the **conus elasticus** (shaped like a truncated [cut off] cone and consisting of elastic tissue). The horizontal plane of section which separates the detached half head and neck specimen from the remainder of your cadaver has been placed so that it should allow you a view from below of the cone with the rima glottidis (pages 71 and 73) forming the truncated apex of the cone. This is seen in Figure 27 on page 64. The truncated cone shape arises as follows: the conus elasticus attaches inferiorly to the circumference of the upper edge of the **ring** of the cricoid cartilage. Posteriorly, this attachment, as it approaches the **lamina** of the cricoid cartilage, continues onto the inferior surface and then the tip of the **vocal process** of the arytenoid

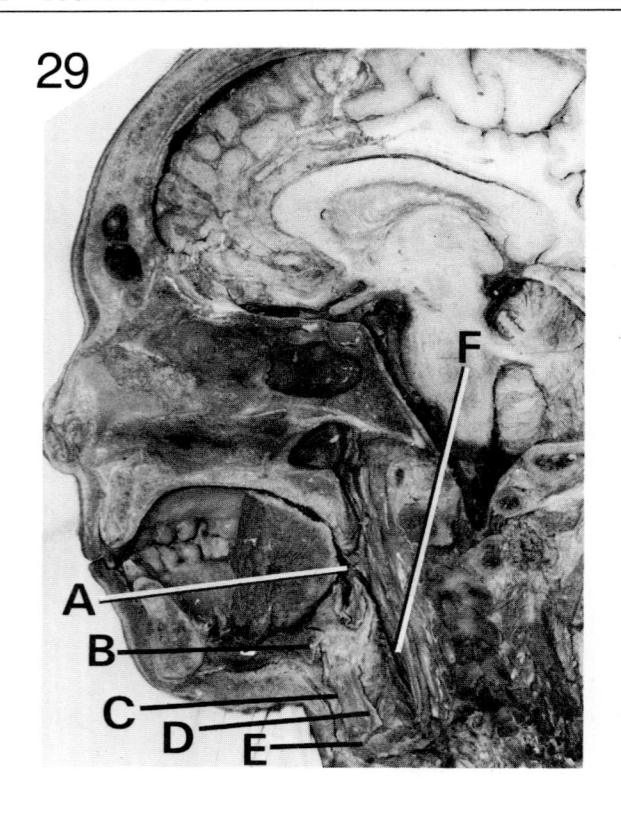

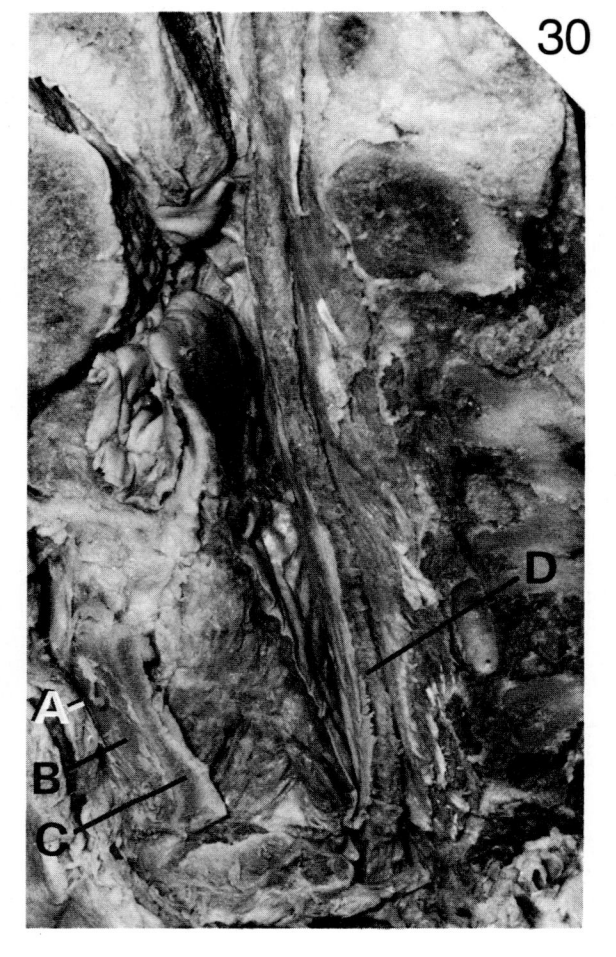

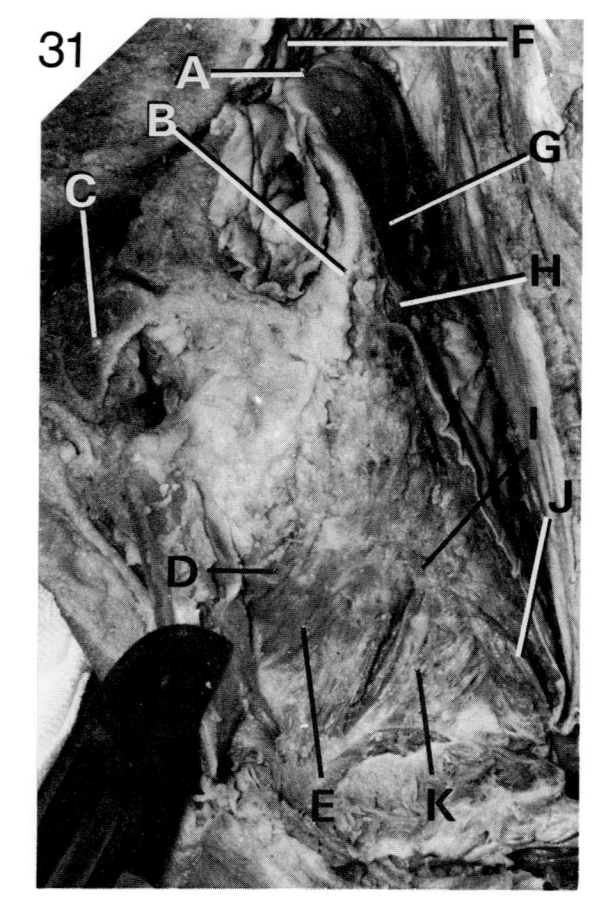

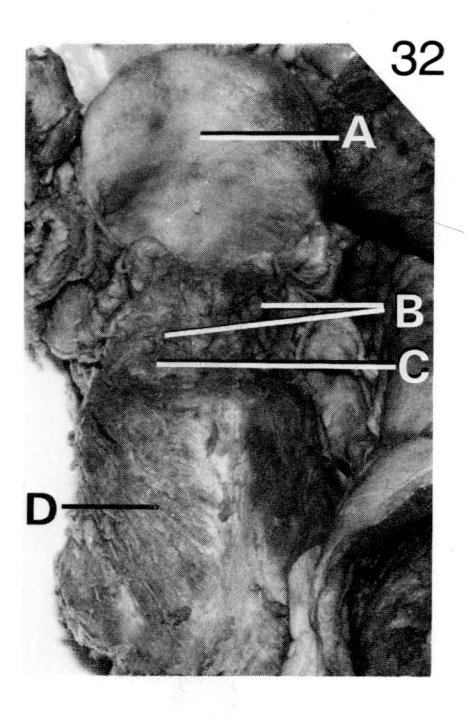

74 YOUR PATIENT'S ANATOMY: A CLINICAL VIEW OF HUMAN MORPHOLOGY

Figure 29. Overview: extent of larynx.
A. Superior edge of epiglottis
B. Body of hyoid bone
C. Thyroid angle
D. Parasagittal section through thyroid lamina
E. Anterior part of cricoid ring
F. Lumen of laryngopharynx

Figure 30. Detail of larynx.
A. Thyroid angle
B. Thyroid lamina
C. Parasagittal section through thyroid lamina
D. (Para)sagittal section through posterior pharyngeal wall

Figure 31. Detail of larynx showing thyroid lamina reflected anteriorly, for lateral access to larynx.
A. Superior edge of epiglottis
B. Lateral edge of epiglottis
C. Body of hyoid bone
D. Thyroepiglottic muscle fibers
E. Thyroarytenoid muscle fibers
F. Pharyngeal part of dorsum of tongue
G. Aditus
H. Aryepiglottic muscle fibers
I. Position of muscular process of arytenoid cartilage
J. Lateral edge of posterior cricoarytenoid muscle
K. Lateral cricoarytenoid muscle

Figure 32. Larynx from behind (mucous membrane removed).
A. Posterior surface of epiglottis
B. Fibers of oblique arytenoid muscle
C. Fibers of transverse arytenoid muscle
D. Posterior cricoarytenoid muscle

cartilage. At this level, the conus elasticus has a superior free edge extending anteriorly from the vocal process of the arytenoid cartilage to a site of attachment on the deep surface of the thyroid angle. It will be clear to you that this description of the attachments of the superior free edge of the conus elasticus corresponds precisely with the description given previously for the attachments of the vocal cords (page 73). Stated in another way, the **vocal cord** or **ligament is** the thickened upper free edge of the conus elasticus. (The conus elasticus is also known as the **cricovocal membrane** because it extends from the cricoid cartilage to the vocal cord.) Anteriorly, a thickening of the cricovocal membrane extends in the anterior midline between the lower border of the thyroid cartilage and upper border of the cricoid cartilage: this condensation of fibers is known as the **cricothyroid ligament.**

MUSCULAR TRIANGLE, THE STRAP MUSCLES, AND THE THYROID GLAND

Using preferentially the attached half specimen, you should now identify and dissect from ensheathing deep fascia, the infrahyoid strap muscles of the muscular triangle (pages 54 and 55). In dissecting the muscles free of deep fascia, note that the fascia constitutes part of the same layer of **deep cervical fascia,** the **investing layer,** which splits to enclose the sternocleidomastoid muscle (page 55). The strap muscles are embedded on the posterior surface of the posterior lamina of the fascia enclosing the sternocleidomastoid.

The **carotid sheath** (page 56) constitutes a separate and independent component of the deep cervical fascia. You have previously opened the carotid sheath and identified the contained common carotid artery, its bifurcation, and its branches—the internal and external carotid artery. You have also studied some of the branches of the external carotid artery (page 57). Of these, you should now reidentify and follow to its destination, the superior thyroid artery (page 57 and fig. 23, page 50). Following the main trunk of this branch of the external carotid artery will lead you to the superior pole of the thyroid gland (page 71).

Reidentify the sternothyroid and thyrohyoid muscles, reviewing their common site of attachment: the **oblique line of the thyroid lamina,** into which the sternothyroid muscle inserts superiorly and to which the thyrohyoid muscle has its inferior attachment.

The sternohyoid muscle should be divided across the middle of its belly and the superior and inferior portions reflected upwards and downwards to their respective attachments. Similarly, the sternothyroid muscle should be divided and reflected in the same way. This will facilitate exposure of the thyroid gland and the superior thyroid artery.

The thyroid gland is firmly bound down to the larynx and the trachea by another layer of **deep cervical fascia,** the **pretracheal fascia.** It will be necessary for you to incise the pretracheal fascia, in order for you to expose the thyroid gland. Do this, to demonstrate the **right and left lobes** of the gland, and the **isthmus,** situated as described on page 71. An asymmetrical structure, the **pyramidal lobe,** may be present on the right or left side of the isthmus, passing up in continuity with a cord-like remnant of the **thyroglossal duct** (page 15). On the **posterior** aspect of the thyroid gland, look for the four small (about 1 cm diameter) **parathyroid glands** embedded in the thyroid gland. The hormone produced by these small endocrine glands (page 7) influences calcium metabolism.

Demonstrate the relationships of the thyroid gland to the superior thyroid artery, the **inferior thyroid artery** (which curves in from a lateral approach to attain the posterior surface of the inferior portion of each lateral lobe), and the **recurrent laryngeal nerve.** The latter, a branch of the **vagus (tenth cranial) nerve,** ascends in the groove between the trachea and esophagus to similarly reach a position posterior to the inferior extremity of each lateral lobe of the gland.

Returning to the superior thyroid artery, identify, in addition to branches to the gland itself, other branches to the strap muscles; an infrahyoid branch running inferior to the hyoid bone; a sternomastoid branch (close to the origin of the superior thyroid artery, which, in passing to the sternomastoid muscle, often crosses the hypoglossal nerve); and the important **superior laryngeal branch.** Together with the **internal laryngeal branch** of the **superior laryngeal nerve** (a branch of the vagus), the superior laryngeal artery (fig. 23, page 50) pierces the **thyrohyoid membrane** (page 73) to enter the larynx. Its branches can be dissected out deep to the mucous membrane of the pyriform fossa: this dissection should be performed now. The internal laryngeal nerve supplies sensory fibers to the larynx down to and including the vocal folds.

Note that, in piercing the thyrohyoid membrane, the superior laryngeal artery and the internal laryngeal nerve pass below the inferior border of the middle pharyngeal constrictor muscle. Review (pages 67 and 72) the structures which pass through the gap between the middle and superior constrictors, and the structures which pass above the superior border of the superior constrictor (page 66).

INFERIOR PHARYNGEAL CONSTRICTOR AND RELATED STRUCTURES

The dissection you have just completed will allow exposure of another muscle originating from the oblique line of the thyroid lamina, in this case from the posterosuperior aspect of this line. This is the **inferior pharyngeal constrictor** (fig. 23, page 50). Observe that this muscle, in addition to taking origin from the thyroid cartilage, also is attached inferiorly to the ring of the cricoid cartilage. A fibrous band connects these two origins, arching across the lateral surface of the **cricothyroid muscle** (page 73). Fibers of the lowest constrictor also originate from this fibrous arch. Carefully define the entire extent of the inferior pharyngeal constrictor muscle. Following the superior fibers of the muscle as they pass back toward the pharyngeal raphe (page 61), confirm that, as in the case of the middle constrictor (page 72), the muscle overlaps the lowest fibers of the constrictor above it. Clear the lower border of the inferior pharyngeal constrictor muscle of connective tissue to demonstrate the two structures that "enter" the pharynx by passing inferior to the lowest constrictor: the **recurrent laryngeal nerve** and the **inferior laryngeal artery** (a branch of the inferior thyroid artery). The recurrent laryngeal nerve is responsible for the sensory supply of the larynx below the level of the vocal folds, and for motor supply of all the intrinsic laryngeal muscles except the cricothyroid, which is supplied by the **external laryngeal** branch of the superior laryngeal nerve.

Review at this time the structures which traverse the various pharyngeal gaps above the superior constrictor (page 66); between the superior and middle constrictors (pages 67 and 72); between the middle and inferior constrictors (page 76); and those entering below the inferior constrictor.

Recall (page 70) that the respiratory and digestive tracts continue below the C6 vertebral level as the trachea and esophagus respectively. Although these structures thus commence in the neck region (the arbitrary boundaries of which you should review [page 4]), they are more conveniently studied in relation to the thorax. With this proviso, you have thus completed your study of the **viscera** of the **head and neck region**. In the process of learning about these you also studied related **somatic** or **body wall** components of this region, such as the strap muscles. To consummate your understanding of this you should review (pages 5 and 14) the concepts of visceral and somatic components of the living organism.

Contents of Section 2

Thorax and Abdomen

8. **Chest Wall; Segmentation**
   Clinical Session 8 81
   Laboratory Session 8 86

9. **Lungs; Pleura; Respiration; Respiratory Mechanism**
   Clinical Session 9 93
   Laboratory Session 9 96

10. **Mediastinum; Great Vessels; Nervous System**
    Clinical Session 10 98
    Laboratory Session 10 101

11. **Heart; Ascending Aorta; Pulmonary Trunk**
    Clinical Session 11 108
    Laboratory Session 11 111

12. **Diaphragm; Esophagus; Descending Thoracic Aorta**
    Clinical Session 12 120
    Laboratory Session 12 122

13. **Abdominal Regions ; Abdominal Wall**
    Clinical Session 13 125
    Laboratory Session 13 131

14. **Upper Abdominal Structures; Upper Abdominal Digestive Tract; Esophagus, Stomach, and Duodenum; Spleen; Pancreas; Liver; Gallbladder**
    Clinical Session 14 135
    Laboratory Session 14 138

15. **Abdominal Structures; Abdominal Descending Aorta; Cecum; Jejunum; Ileum; Colon: Ascending; Transverse; Descending; Sigmoid**
    Clinical Session 15 149
    Laboratory Session 15 151

16. **Posterior Abdominal Structures; Kidneys; Suprarenal Glands; Abdominal Aorta**
    Clinical Session 16 158
    Laboratory Session 16 163

17. **Pelvis**
    Clinical Session 17 168
    Laboratory Session 17 171

18. **Perineum; Urogenital Triangle; Anal Triangle**
    Clinical Session 18 175
    Laboratory Session 18 179

Clinical Session 8

■ **TOPICS**
   Chest Wall
   Segmentation

Your overall objective for this session is to familiarize yourself with some of those features of the anatomy of the anterior **chest wall** which can be identified and examined in the living body.

Many diagnoses depend on information, concerning internal organs, which has to be obtained by examining organs **through** overlying tissues such as the **body wall** (page 5), of which the chest wall is a part. Therefore it is necessary to acquire expertise in the techniques of such examination. The student should not be discouraged by initial difficulty, for example in palpating deep-lying structures, but should realize that facility is quickly developed by practice.

1. Commence your physical examination by reviewing the anatomical position (page 3) and placing your model in this position. The subject should be standing erect, with the feet together and the toes pointing forward, the palms of the hands facing forward and the head in the so-called **Frankfurt plane** (page 3). Review the meanings of the terms **sagittal, median, transverse, medial, lateral, anterior, posterior, superior,** and **inferior**.

2. Review the division of the human body into regions (pages 3 to 5). The major regions are, by convention, the **head and neck,** the **upper limbs,** the **lower limbs,** the **abdomen** including **pelvis** and **perineum,** and the **thorax**. The thorax is the region extending between the head and neck and the abdomen, and is in communication laterally with the upper limbs. To perform the following examination, your model should strip to the waist, in a well heated room. When reviewing the region on yourself, you should preferably stand before a full length mirror.

3. Delimit the extent of the **thorax**. Its junction with the neck is at the collar bones, or **clavicles**. Palpate these, and note that they are relatively superficial. Laterally, the upper limbs are attached to the thorax, and inferiorly the thorax is continuous with the abdomen. Anteriorly you can locate the level of this continuity at the inferior end of the breastbone or **sternum**. Palpate the sharp pointed lower end of the sternum in the midline, where the firm bony cage of the thorax gives way to the soft anterior wall of the abdomen.

4. On inspection, the thoracic region demonstrates more obviously than other parts of the body the fundamental **morphological** characteristic of **segmentation**. Segmentation refers to the division of the body along its long axis into repeating units or **segments**. In the anatomical position the long axis of the body extends from superior to inferior, and the repeating units you can see are the **ribs,** running in a more or less transverse plane. These are the externally visible features of a basic general pattern of the organism which has persisted through evolution, from the segments of the centipede to the vertebrae, ribs, spinal nerves, arteries, and muscles of the human body. Segmentation is thus, as stated, a morphological trait (page 3), and illustrates the meaning of **morphology** (the study of the general rules of form and its correlates) and its relationship to **anatomy** (the study of the structure of a particular organism). Morphology is concerned with form in general, and is thus concerned with such correlates as: what determines form (including developmental, evolutionary, and genetical factors), and what the consequences of form are (including the way form influences function). Anatomy is, in the strictest sense, concerned only with the actual structure of a particular organism (or its parts). Clearly, an understanding of morphological aspects of anatomy is a prerequisite for insightful application of knowledge to the context of medical and paramedical practice.

Observe the slight elevations on the surface of the chest wall which the ribs make, and the slight depressions (**intercostal spaces [inter means between, costa means rib]**) between the ribs. Palpate the ribs and the intercostal spaces. The latter are soft to the touch. They are filled by muscles (**intercostal muscles**) which extend from the lower border of one rib to the upper border of the next. Thus the chest wall is comprised of the ribs, which together form the **rib cage,** and the intercostal muscles.

There are three layers of intercostal muscles: an outer, middle, and inner (fig. 36, page 82). This, too, is a morphological trait, in that the body wall, in general, is comprised of three muscle layers.

5. The intercostal muscles play a role in breathing (**respiration**). Observe your model's rib cage as he or she takes a deep breath. The ribs move upwards as well as swinging outwards (laterally and anteriorly), thus increasing the transverse and anteropos-

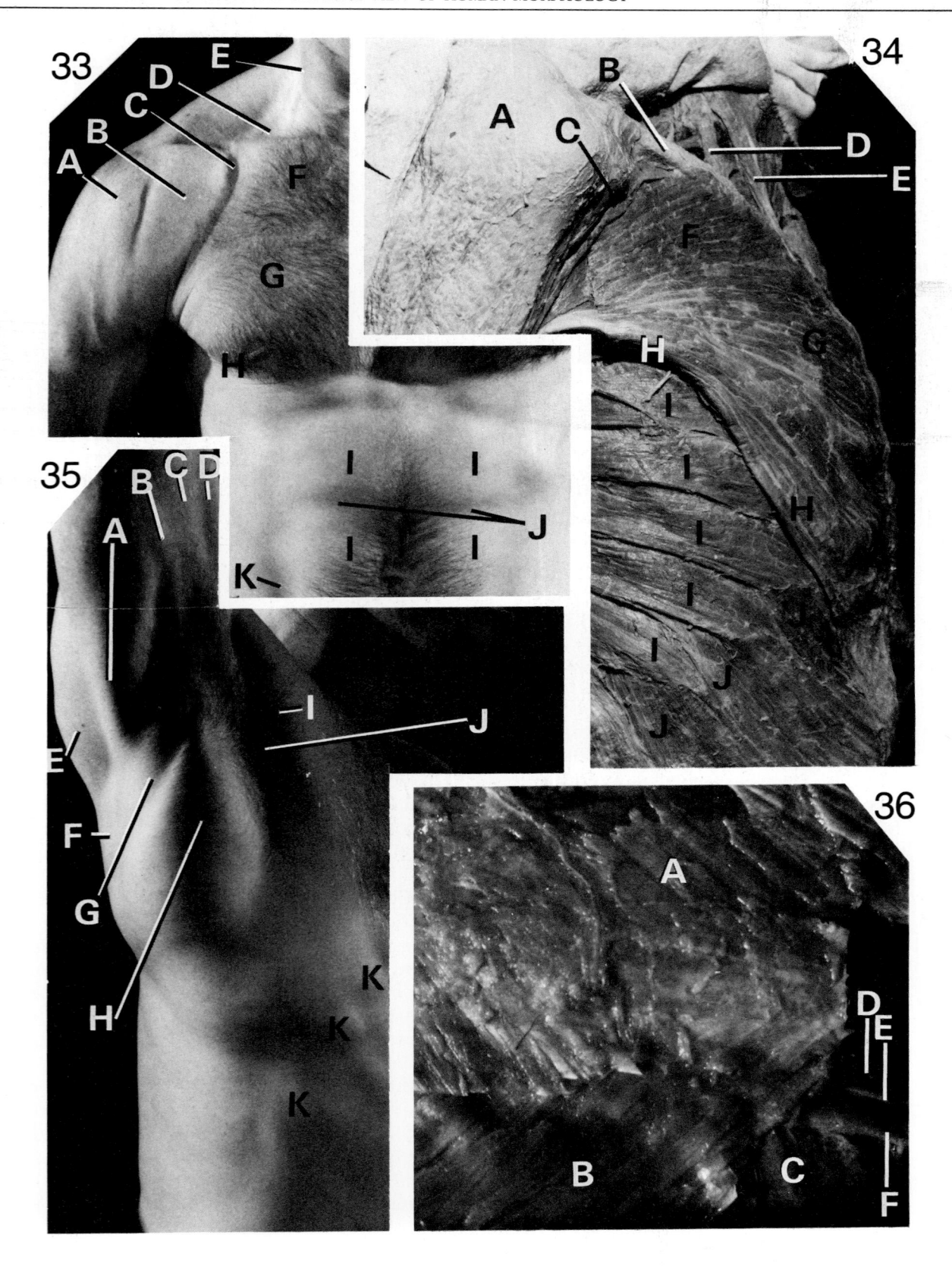

82 YOUR PATIENT'S ANATOMY: A CLINICAL VIEW OF HUMAN MORPHOLOGY

Figure 33. **Neck, shoulder, and upper portion of trunk, anterior aspect.**
A. Deltoid muscle (lateral, acromial portion)
B. Deltoid muscle (anterior, clavicular portion)
C. Deltopectoral triangle
D. Supraclavicular fossa
E. External jugular vein
F. Pectoralis major muscle (clavicular portion)
G. Pectoralis major (sternocostal portion)
H. Pectoralis major (abdominal portion)
I. Rectus abdominis muscle
J. Tendinous intersection of rectus abdominis muscle
K. Linea semilunaris

Figure 34. **Neck, shoulder, and upper portion of trunk, right lateral aspect, skin reflected.**
A. Deltoid muscle (covered by fascia)
B. Clavicle
C. Cephalic vein, passing deeply, in deltopectoral triangle
D. Clavicular head of sternocleidomastoid muscle
E. Sternal head of sternocleidomastoid muscle
F. Clavicular portion of pectoralis major muscle
G. Sternocostal portion of pectoralis major muscle
H. Abdominal portion of pectoralis major muscle
I. Digitations of serratus anterior muscle
J. Digitations of external oblique muscle of the abdomen

Figure 35. **Axilla and upper portions of arm and trunk, anterolateral aspect.**
A. Long head of triceps muscle
B. Coracobrachialis muscle
C. Brachialis muscle
D. Biceps muscle
E. Deltoid muscle
F. Infraspinatus muscle
G. Teres major muscle
H. Latissimus dorsi muscle (in posterior axillary fold)
I. Pectoralis major muscle (in anterior axillary fold)
J. Axilla
K. Digitations of serratus anterior muscle

Figure 36. **Dissection of medial part of a right anterior intercostal space.**
A. External intercostal muscle
B. Internal intercostal muscle
C. Sternocostalis muscle (belonging to innermost layer of intercostal musculature)
D. Intercostal vein
E. Intercostal artery
F. Intercostal nerve

terior diameters of the chest. These movements are most obvious in the lower ribs. Part of this movement is due to contraction of the intercostal muscles, which, in each intercostal space, elevate the rib below.

6. Palpate the clavicles to ascertain their shape, noting that they are gently curved, the left forming a flattened "S" (starting at the left shoulder, curving slightly upwards as it extends medially, and then curving downwards to approach the upper end of the sternum) and the right clavicle correspondingly forming a reversed "S." Where the clavicles meet the sternum, an expanded end can be felt on each at its medial end.

7. Now palpate the approximately square upper portion of the sternum. Start by palpating its upper end, between the two expanded medial ends of the clavicles. Crossing the midline, you can feel a curved upper border (concave upwards), called the **suprasternal notch** (page 49). Run your palpating finger down the anterior surface of the sternum from the suprasternal notch, in the midline, until it encounters (about 4 cm down) a transverse horizontal ridge. This demarcates the upper portion of the sternum, called the **manubrium** (manus means hand; manubrium means handle. The sternum is said to resemble a dagger and the manubrium is the handle). At this ridge, the manubrium forms a slight angle with the rest of the sternum, and the junction is known as the sternomanubrial joint **(sternal angle)**. A horizontal (transverse) plane through this angle passes through the space between the fourth and fifth thoracic vertebrae, and this plane, the **sternomanubrial plane (transthoracic plane)**, is an extremely important landmark for structures within the chest, which will be studied later. Use a skin marker to indicate the anterior part of this plane.

Continue your palpation of the sternum downwards in the midline. The lower end of it consists of a pointed section, the **xiphisternum** (xiphos means sword, here indicating the point of the dagger), the lowest extremity of which you have already palpated in the soft abdominal musculature (point 3, page 81). Like the manubrium, the xiphisternum is a separate portion of the sternum. The manubrium and the large middle portion of the sternum (or **body of the sternum**) are usually fully fused (at the sternal angle) in the adult, but the joint between the xiphisternum and body may not be fused and some mobility of the **xiphoid process** (xiphisternum) on the body may be possible in some adults. This latter joint, like the sternomanubrial, marks an important horizontal plane. The **xiphisternal plane** lies at the level of the space between the eighth and ninth thoracic vertebrae. Mark this plane parallel with the sternomanubrial.

8. Palpate the individual ribs, identifying them by number. It is easiest to start by identifying the second rib, because this attaches to the sternum at the sternal angle. To find the second rib, identify the angle as a ridge on the anterior surface of the sternum. Then run your finger out laterally to each side, where it will encounter, at the medial, anterior end of each second rib, a cartilaginous process belonging to the rib, called the **costal cartilage.** The costal cartilages extend a few centimeters from the sternum and are then replaced by the bony ribs proper. The first rib's costal cartilage attaches to the manubrium above the second, and below the manubrial attachment of the clavicle. Try to palpate it, and the first rib as it curves sharply backwards to disappear behind the clavicle. Then identify the third and subsequent ribs and intercostal spaces. Only the upper seven ribs attach by their costal cartilages to the sternum. The eighth, ninth, and tenth attach to the costal cartilage of the rib above. The eleventh and twelfth ribs do not attach anteriorly, and are called **free** or **floating** ribs. Palpate these and their free anterior ends in the soft musculature of the lateral abdominal wall. The tip of the eleventh rib lies further anteriorly in the lateral wall than that of the twelfth. The intercostal spaces are also numbered, the space between the first and second rib being the **first intercostal space,** and so forth.

9. Return to an inspection of the anterior thoracic or chest wall. Identify the **nipple,** which in the male can be ascertained by palpation to lie at approximately the level of the fourth intercostal space. The position of the female nipple is more variable. The nipple is surrounded by a pigmented region of skin called the **areola.** In the male, a curved prominence below the nipple, caused by the lower border of the underlying **pectoralis major** muscle (pectoral meaning of the chest, major means large) should be identified by inspection and palpation (figs. 33, 34, and 35, page 82). To confirm that you have identified the muscle, ask the model to press his or her elbow (or press your own elbow in self-examination) against the side of the body. The pectoralis major muscle attaches medially on the anterior chest wall and laterally to the bone of the arm (the **humerus**), and is therefore capable of pulling the arm in toward the thorax. Therefore it will be felt to contract when performing this exercise.

More precisely, the medial attachment, or **origin** (page 21), of the pectoralis major muscle has three portions: a clavicular, from the medial part of the clavicle; a sternocostal, from the sternum and upper seven costal cartilages; and an abdominal, arising from the surface of the outer muscle layer of the abdominal body wall. Palpate the three portions, referring to Figures 33, 34, and 35 (page 82).

Palpate the muscle out toward the arm, and note that it becomes part of the anterior wall of the **axilla** (armpit) (fig. 35, page 82). In the anatomical position, the axilla is bounded **anteriorly** by this muscle; **posteriorly** by a second muscle mass, extending from the shoulder blade (**scapula**) to the arm, and comprising largely the **latissimus dorsi muscle; laterally** by the arm; and **medially** by the lateral thoracic wall. The anterior and posterior "walls" of the axilla can be grasped between a palpating thumb and index finger, and the folds thus identified are known respectively as the **anterior and posterior axillary folds.** Where these folds meet the thoracic wall, imaginary vertical lines called the **anterior** and **posterior axillary lines** are described, extending the length of the lateral surface of the body. The **midaxillary line** is the line between the anterior and posterior. These imaginary lines are often used as landmarks, and should be drawn in on your model.

Return to palpation of the free ribs (point 8, page 84) and confirm that the tip of the eleventh rib is approximately in the anterior axillary line and that of the twelfth rib is approximately in the posterior axillary line.

In the female, the pectoralis major is partly covered by the **breast (mammary gland),** a specialized **cutaneous gland** situated in the **subcutaneous** tissue (cutaneous means of the skin; subcutaneous means deep to the skin). The breast shows considerable mobility, in part dependent on movement of the upper limb (fig. 39, page 88) due to its relationship to the deep-lying pectoralis major. Confirm that the breast extends from the second to the sixth rib, and from the lateral border of the sternum to the anterior axillary line (see fig. 39, page 88). In pubertal and young adult females it is approximately hemispherical in shape. Projecting from its anterior aspect is the **nipple,** surrounded by a pigmented area, the **areola.** The breast is the **lactiferous** (milk-secreting) gland (lactis means milk). During lactation, the **lactiferous ducts** open on the surface of the nipple. The areola is lubricated by sebaceous **areolar glands** (see fig. 40, page 88).

Below and lateral to the curved margin of pectoralis major, find the digitations (digit means finger; digitations means finger-like processes) of **serratus anterior** muscle (figs. 35 and 34, page 82). The digitations attach to the lateral surfaces of the ribs where they can be seen and palpated as your model pushes, with a flat hand, against a wall. At this site they interdigitate (fit between other digitations) with the **external oblique muscle of the abdomen,** the outermost of the three muscle layers of the abdominal wall (fig. 34, page 82).

10. Just inferior to the lateral third of the clavicle a slight depression can be seen in most subjects. This corresponds to a deep-lying triangular space, the **deltopectoral triangle** (figs. 33 and 34, page 82), which separates the upper (clavicular) fibers of the pectoralis major muscle from the **deltoid** muscle of the shoulder. Palpate deeply into this depression (or just inferior to the lateral third of the clavicle, if no depression is obvious) to identify the **coracoid process** (coracoid means crow-like, referring to a curved, beak-like structure), a bony projection extending forwards and downwards from the scapula. About 5 cm lateral to the coracoid process, a second bony protuberance may be palpated; this is the **lesser tubercle of the humerus.** To distinguish between these two structures, ask the model to **rotate** the upper limb about its vertical axis. This is done as follows. Start in the anatomical position with the palm of the hand facing anteriorly. Keeping the limb in a straight line, turn the palm to face posteriorly (i.e., swing the thumb from a lateral position to a medial position). This movement brings about **medial rotation** of the arm (as well as an additional movement of the forearm, to be studied later). In returning the palm to face anteriorly, the subject performs **lateral rotation** of the arm. On rotating the limb medially and laterally several times, any structure on the humerus would rotate with the limb. Because it is part of the scapula and not of the humerus, the coracoid process would **not** rotate. Your test of having palpated this structure is, therefore, that it does not move during rotation of the upper limb (though you may feel some movement of the adjacent muscles as you palpate the process during rotation). This process is an important landmark, and is the attachment site of muscles you will be dissecting.

Refer back to the emphasized terms and test your recall of the anatomical structures and your understanding of the concepts. Review (page xii) the guidelines on how to ensure that you have achieved the objectives of the session.

Laboratory Session 8

Begin your dissection of the thoracic wall by making the following skin incisions. Starting in the suprasternal notch, make a vertical incision down to the tip of the xiphisternum (page 84). Cut laterally on either side, following the **costal margin.** The latter term refers to the palpable, sloping, inferior edge of the rib cage (page 81), which passes inferolaterally on each side from the xiphisternal joint. Because the eighth, ninth, and tenth costal cartilages each attach to the preceding costal cartilage, the costal margin is formed by the seventh, eighth, ninth, and tenth costal cartilages (page 84, point 8).

Carry the skin incision from the costal margin inferolaterally to the midaxillary line (page 85). Make a vertical incision in this line, continuing upwards to the apex of the axilla. Thereafter, carry the incision forwards and then upwards to meet the lateral extremity of the clavicle. Finally, make an incision over the clavicle, to meet your starting point in the suprasternal notch.

Following the dissection instructions given on page 10, and illustrated in Figure 5, page 16, reflect the skin from both sides of the anterior thoracic wall, preserving superficial structures, including the breast, on one side and performing the rapid skinning procedure on the other.

In the subcutaneous tissue, identify, in the intercostal spaces, superficial branches of the **intercostal** segmental (page 81) **veins, arteries,** and **nerves.**

As a morphological generalization, the veins, arteries, and nerves of the body wall lie in a **neurovascular plane** situated between the innermost and middle of the three muscular layers which constitute this wall (page 81, and fig. 36, page 82). In each intercostal space, the intercostal nerve which runs in this plane represents a **ventral ramus** of a **mixed spinal nerve** (ramus means branch). Recall (pages 5 to 6) that the peripheral nervous system (PNS) comprises peripheral nerves connected to the brain, known as cranial nerves, and other peripheral nerves connected to the spinal cord, known as spinal nerves. The term **mixed spinal nerve** implies that the nerve contains both afferent fibers, carrying impulses in toward the central nervous system (CNS), and efferent fibers, carrying impulses out from the CNS. The (mixed) spinal nerve is situated in an **intervertebral foramen,** which you should now identify on the skeleton. Within the CNS, the afferent and efferent neurons are linked by a **connector neuron,** to complete the **simple reflex arc** (page 26) as follows. From the mixed spinal nerve, in the intervertebral foramen, the afferent fibers pass toward the spinal cord, situated in the **vertebral canal** of the vertebral column (page 5). Because they enter the dorsal aspect of the spinal cord, these afferent fibers constitute the **dorsal root.** A **ventral root** carries efferent fibers from the spinal cord to the mixed spinal nerve. The afferent neurons' **nerve cell bodies** (the part of the nerve cell containing the nucleus, and from which the nerve cell **processes** extend) form a cluster, called the **dorsal root ganglion,** situated on the dorsal root. Afferent neurons **synapse** with (make contact with) the nerve cell bodies of connector neurons, which are clustered in the **dorsal horn of gray matter** in the spinal cord. The connector neurons, in turn, synapse with the nerve cell bodies of efferent neurons clustered in the **ventral horn of gray matter,** from where, as mentioned previously, the efferent nerve fibers emerge in the ventral root to reach the mixed spinal nerve. (Note that a cluster of nerve cell bodies within the CNS is called **gray matter,** and such a cluster outside the CNS [i.e., in the PNS] is called a **ganglion** [pages 6 and 33].) The intercostal nerve is a **ventral ramus** of the spinal nerve because it supplies the region ventral to a coronal plane through the intervertebral foramina. **Dorsal rami** pass back to supply the region dorsal to this plane. The plane is accordingly called the **coronal morphological plane.** Dorsal and ventral rami contain both afferent and efferent fibers. The entire pattern of the simple reflex arc (page 26) is repeated segmentally throughout the spinal cord. The **spinal segment** is part of the morphological segmentalization of the entire organism (page 81) and is the basis of the intercostal nerves (fig. 51, page 104).

The intercostal nerves, after coursing in the neurovascular plane of the body wall, from the intervertebral foramen around to the anterior end of an intercostal space, terminate by passing superficially, where they can be dissected in the subcutaneous tissue, close to the sternum. Also, each intercostal nerve gives off a **lateral cutaneous branch,** which similarly passes through the outer two muscle layers of the chest wall, to become superficial close to the midaxillary line. Branches extend anteriorly and posteriorly, and can be dissected out in each intercostal space.

The extent of the female **breast,** or **mammary gland,** should be confirmed (page 85), and its lobular structure studied by dissection. The gland consists of about 15 lobes, separated by connective tissue septa, and is mainly situated superficial to pectoralis major and serratus anterior muscles (page 85). Having originated in embryological development as an ingrowth from skin, the lobes of glandular tissue retain their contact with the skin through their **lactiferous ducts,** which open on, and, during lactation, convey milk to, the surface of the **nipple** (fig. 40, page 88). An extension of the gland passes superolaterally, at the lower edge of pectoralis major, as the **axillary tail.** The breast receives blood supply from intercostal arteries, and from branches of arteries of the upper limb. The lymphatic drainage, of clinical importance in disease of the breast, follows both these sources of blood supply. Thus lymph glands of importance in relation to the breast are found near the main upper limb artery, in the axilla (page 85) and near the intercostal vessels, close to the sternum. These structures will be further studied subsequently.

The pectoralis major muscle, and the serratus anterior muscle (pages 84 and 85, and figs. 33, 34, and 35, page 82) should be identified and defined, and, in the female, their relationships to the breast studied. After identifying the three portions of the pectoralis major muscle (pages 84 and 85), incise the muscle, starting superiorly in the deltopectoral triangle (fig. 34, page 82), and proceeding downwards toward the inferolateral free edge of the muscle. The two portions of the muscle thus separated should be reflected laterally and medially respectively. Care should be exercised in performing these reflections, in order to study and preserve the pectoral nerves, which may be seen supplying the muscle from its deep surface. The **lateral pectoral nerve** may be seen superiorly, and the **medial pectoral nerve** inferiorly. The lateral pectoral nerve will be seen to pass superior to, and the medial pectoral nerve through, the **pectoralis minor muscle,** which will be exposed to view as the pectoralis major is reflected.

The superior attachment of pectoralis minor is to the coracoid process which you previously palpated in the living body in the deltopectoral triangle (page 85). Study this process now on a skeleton, and then identify the muscle extending from this superior attachment on the process, to its inferior attachments on the third, fourth, and fifth ribs, close to their costal cartilages. Review (page 21) the terms **origin** and **insertion** of a muscle, and determine under which conditions and in which muscle actions the superior attachment would be the origin, and in which actions it would be the insertion, of pectoralis minor.

The muscle is enclosed between two layers of the **clavipectoral fascia,** which attaches superiorly to the clavicle. Inferiorly the layers fuse to form the **suspensory ligament of the axilla,** which attaches to the fascial roof of the axilla. Identify also, along the lateral border of pectoralis minor, the **lateral thoracic artery,** important because it supplies the female breast.

Expose the interdigitations of **serratus anterior** and the **external oblique muscle of the abdomen** (fig. 34, page 82). Identify also the vertical fibers of another muscle of the abdominal wall, the **rectus abdominis** (fig. 33, page 82), the upper attachment of which should be defined on the xiphoid process, and the fifth, sixth, and seventh costal cartilages.

Dissect in an intercostal space, close the sternum, to demonstrate the three muscle layers of the thoracic wall, and the directions of their fibers (fig. 36, page 82). Note that the fiber directions of the outermost muscle layer (**external intercostal muscle:** downwards and toward the anterior midline) and the middle muscle layer (**internal intercostal muscle:** downwards and away from the anterior midline) are constant all the way around the chest wall. On the other hand, the innermost muscle layer has the direction shown in Figure 36, that is parallel to the external intercostal, only in the anterior portion of this layer, known as the **sternocostalis muscle** (also called the **transversus thoracis**). The sternocostalis passes from the deep surface of the lower portion of the sternum, upwards and laterally to attach to the deep surface of costal cartilages and ribs (fig. 43, page 90). Elsewhere in the chest wall the innermost muscle layer may be thin or deficient, but where present it has the opposite fiber direction to that of sternocostalis: in the lateral region of the intercostal spaces the muscle fibers of this layer, here called the **intimal intercostal muscle,** run parallel to those of the internal intercostal muscle. Posteriorly, where they span two intercostal spaces at a time, they are called **subcostalis** and also have the latter direction.

Carefully dissect the external and internal intercostal muscles away, in one intercostal space, to demonstrate the neurovascular structures (fig. 36, page 82). Review the morphology of intercostal nerves (page 86).

Bisect pectoralis minor muscle by cutting across its belly, and reflect proximal and distal portions to their respective bony attachments.

Using an electric saw, under guidance, cut through the second to eighth ribs, close to the anterior axillary line on either side of the chest wall. Cut through the sternomanubrial and xiphisternal joints (page 84), and gently elevate the anterior chest wall, by finger dissection, from the membrane which lines it internally: the **parietal pleura.** This membrane can be gently peeled off the inner surface of the ribs and intercostal muscles, allowing elevation of the separated portion of the anterior chest wall from the membrane. In certain places the separation may be difficult to perform, in which case

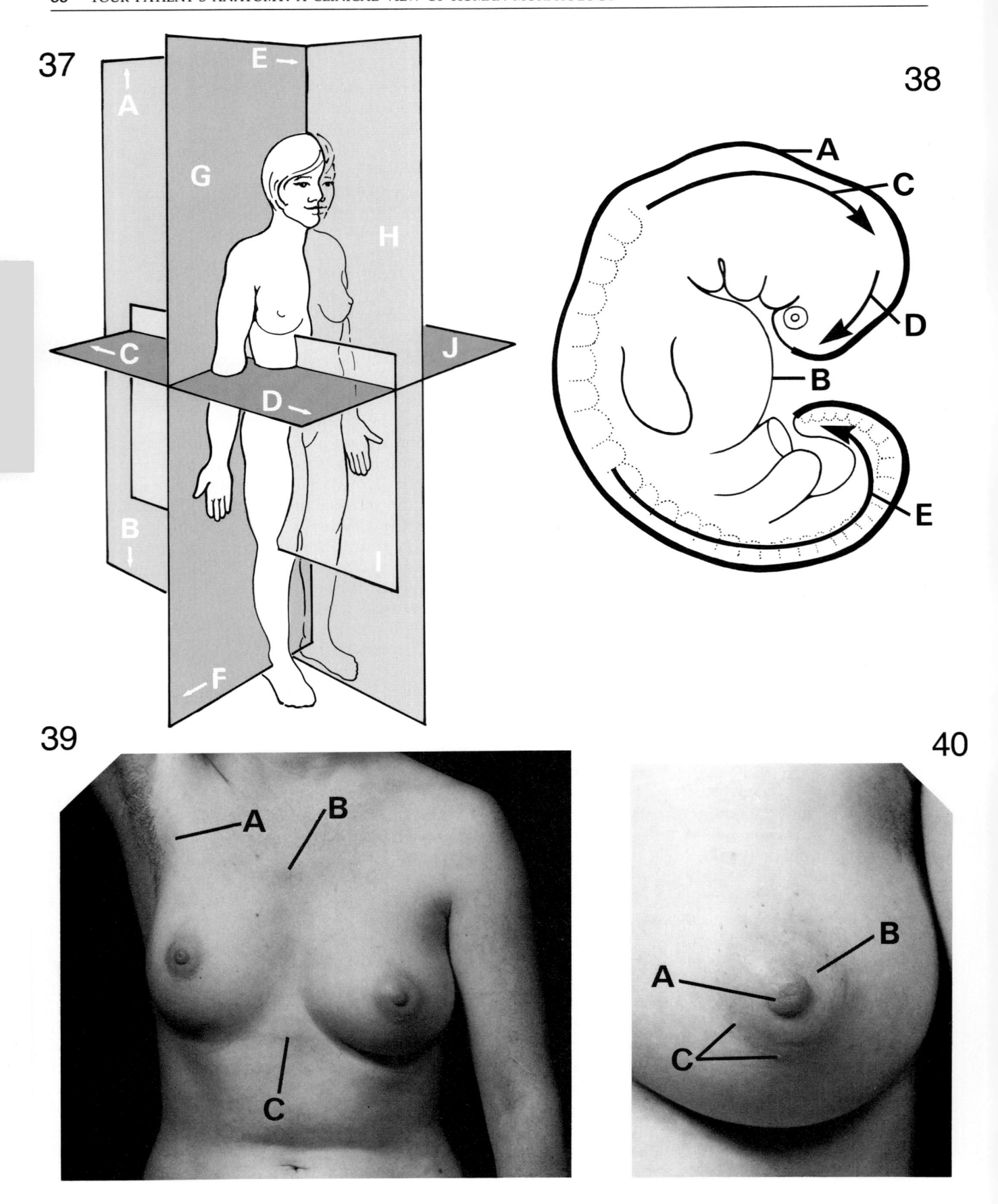

88 YOUR PATIENT'S ANATOMY: A CLINICAL VIEW OF HUMAN MORPHOLOGY

37

38

39

40

Figure 37. Planes and directional terms, used in reference to normal anatomical position.
A. Superior
B. Inferior
C. Posterior
D. Anterior
E. Medial
F. Lateral
G. Coronal or frontal plane
H. Median or sagittal plane
I. Parasagittal plane
J. Transverse or horizontal plane

Figure 38. Morphological terms denoting surfaces and directions.
A. Dorsal surface (heavy outline)
B. Ventral surface (light outline)
C. Cranial (toward head end)
D. Rostral (within head, toward rostrum [i.e., frontmost part of brain])
E. Caudal (toward tail)

Figure 39. Female breasts, showing mobility of breast with movement of arm.
A. Right anterior axillary fold
B. Sternal angle
C. Xiphisternal joint

Figure 40. Detail of female breast.
A. Nipple
B. Areola
C. Areolar glands

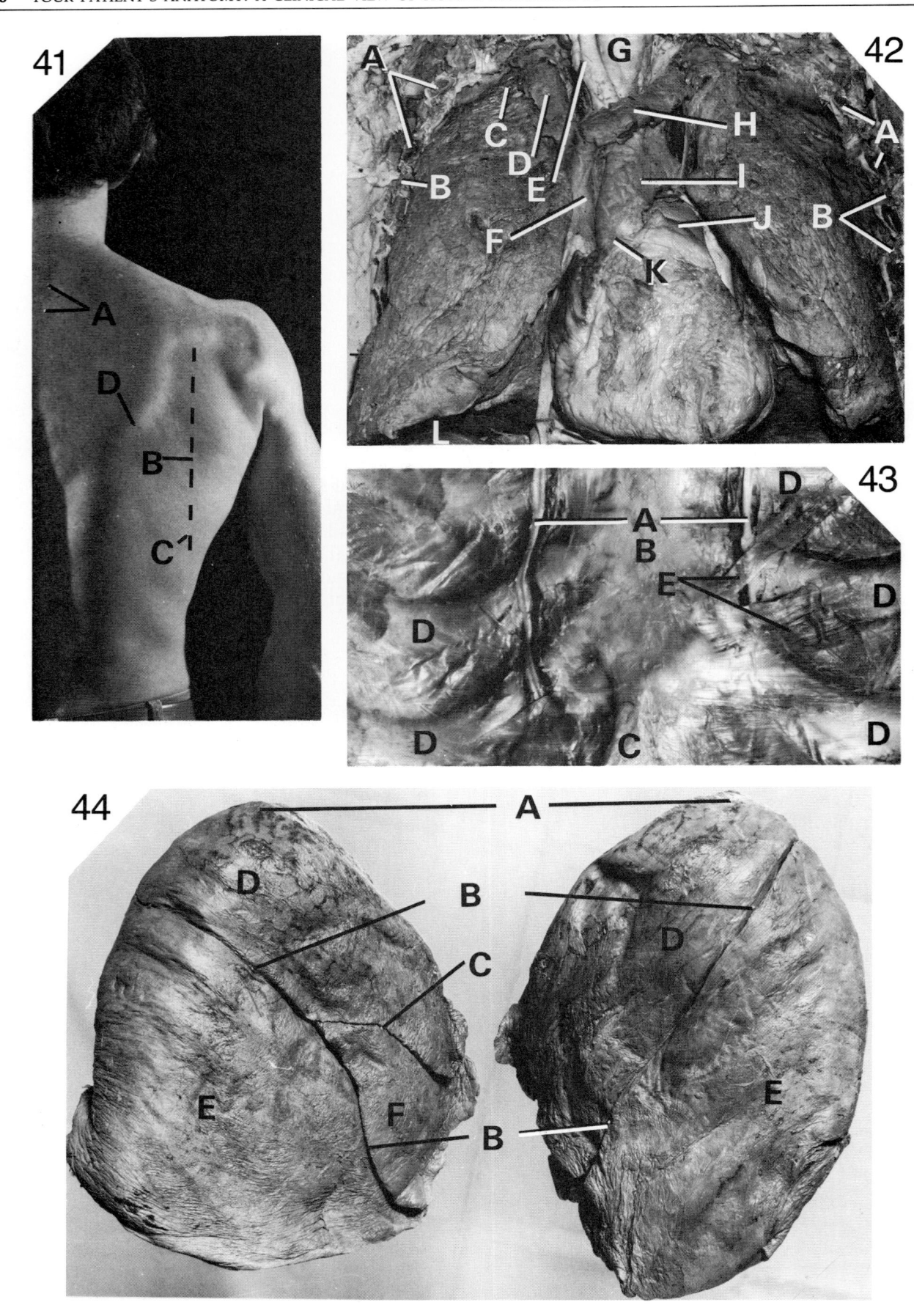

90 YOUR PATIENT'S ANATOMY: A CLINICAL VIEW OF HUMAN MORPHOLOGY

Figure 41. Dorsal view of upper back.
A. Vertebral spines
B. Inferior angle of scapula
C. Broken line: scapular line
D. Edge of trapezius muscle

Figure 42. Ventral view of thoracic contents, anterior chest wall removed.
A. Cut ends of ribs (in thoracic body wall)
B. Sectioned intercostal muscles (in thoracic body wall)
C. Apex of right lung
D. Right brachiocephalic vein
E. Bifurcation of brachiocephalic trunk (into right subclavian and right common carotid arteries)
F. Superior vena cava
G. Trachea
H. Left brachiocephalic vein
I. Arch of aorta
J. Pulmonary trunk
K. Ascending aorta
L. Right dome of diaphragm

Figure 43. Inner (posterior) surface of anterior chest wall, lower end.
A. Internal thoracic arteries
B. Body of sternum
C. Xiphisternum
D. Costal cartilages and ribs
E. Sternocostalis in characteristic relation to neurovascular plane

Figure 44. Lateral aspects of right and left lungs.
A. Apex
B. Oblique fissure
C. Horizontal fissure
D. Superior lobe
E. Inferior lobe
F. Middle lobe

the parietal pleura may be removed with the chest wall, and subsequently dissected away from it. As will be seen in due course, the parietal pleura is, morphologically, related to the chest wall (parietes means wall). Removal of the **chest wall** with the parietal pleura exposes the **visceral** structures of the thorax, including the lungs.

Study the deep surface of the anterior thoracic wall, identifying the sternocostalis muscle and the **internal thoracic artery** which runs vertically, close to the lateral edge of the sternum (see fig. 43, page 90). Note that this artery conforms to the morphological rule concerning the neurovascular plane, in lying between the sternocostalis and internal intercostal muscles. In the intercostal spaces, it gives off small, segmentally arranged anterior intercostal branches which, just lateral to the sternum, **anastomose** (communicate) with the posterior intercostal arteries, which sweep around the chest wall to this site (page 86). These arteries also supply the breast, and the lymph glands associated with them receive some of the breast's lymph drainage (page 87). In breast cancer it may be necessary to remove these glands because of spread of the disease to them. Inferiorly, the internal thoracic artery ends by dividing into a medial branch, the **superior epigastric artery,** and a lateral branch, the **musculophrenic artery.** Superiorly, the internal thoracic artery arises as a branch of the subclavian artery, to be studied later. Inferiorly, the superior epigastric artery passes downwards, posterior to the rectus abdominis muscle (page 87 and fig. 33, page 82) to anastomose with the **inferior epigastric artery,** which passes superiorly from below, in the same plane.

Study, with the aid of an atlas, the bones of the rib cage, as seen in a skeleton, and in a disarticulated state. Finally, observe the displayed prosections of the lungs (figs. 42 and 44, page 90).

Clinical Session 9

■ **TOPICS**

Lungs
Pleura
Respiration
Respiratory Mechanism

Your objective for this session is to study the lungs and the respiratory mechanism in the living state. In so doing you will explore the potentials and the limitations of the techniques of physical examination, for obtaining information concerning internal organs by examination through the body wall. Landmarks such as bony prominences, imaginary "planes" or "lines" and surface markings of deep structures such as the lungs should be drawn on the skin surface with a wax marking pencil.

The lungs perform the function of respiration (page 6) by taking in air and transferring oxygen from this air to the bloodstream. To understand the mechanism whereby air is taken in, the lungs may be compared to two elastic balloons each filled with spongy tissue. From the **air spaces** of the sponge (called **alveoli** in the lung) oxygen is transferred to small blood vessels in the walls of each **alveolus**. The sponge-filled balloons may be thought of as being enclosed in an air-tight box (the chest). However, the balloons have access to the exterior through a common tube, to which the necks of the two balloons are joined. If the volume of the box increases, for example by one or more of the walls of the box moving outwards, a suction or negative pressure will be created between the walls of the box and the balloons. This negative pressure leads to air being sucked in through the common tube.

1. Reidentify the "common tube" of the above analogy, the "windpipe" or **trachea** (page 70), by palpating in the midline above the suprasternal notch. By running your finger lightly up and down 1 or 2 cm above the suprasternal notch, you will palpate firm ridges, each 2 or 3 mm in height and separated from each other by spaces of about the same distance. The ridges you can feel are the anterior portions of cartilaginous "rings." These so-called **tracheal "rings"** are in fact horseshoe-shaped rather than ring-shaped because they are incomplete posteriorly. They constitute part of the wall of the trachea or air tube which conveys air down toward the lungs. In the sternomanubrial plane (page 84, point 7) the trachea divides into a right and left **main bronchus**, each of which slopes inferolaterally toward the corresponding lung.

2. The rigid box of the analogy is the **thoracic cavity,** the walls of which are made up of the rib cage, the intercostal muscles, and an inferior partition between the thorax and the abdomen, called the **diaphragm.** The latter is a large muscle which is dome-shaped, the peak of the dome facing upwards toward the contents of the thorax.

For completeness, it should be mentioned that the lung of each side is **invaginated** into a **pleural sac.** The term invaginated derives from the word vagina, which means a closely fitting sheath or covering. The process of invagination can be understood by picturing a balloon, into which a solid object, such as a finger, is pushed: the object acquires a sheath or covering. In the embryo, after development of the respiratory system is initiated as a ventral outgrowth from the digestive tube (page 69, point 6), this outgrowth divides into right and left **lung buds,** each of which grows out laterally toward a pleural sac. The developing lung (the solid object) then invaginates the pleural sac (the balloon) from its medial aspect. In so doing, the lung acquires a closely fitting covering, which, because it is intimately applied to a **viscus** (i.e., one of the **viscera,** or internal organs, [page 5]) is called the **visceral layer of the pleura,** or simply the **visceral pleura.** The visceral, invaginated layer remains in continuity with an outer layer of pleura, which lines the inner surface of the body wall and is therefore called the **parietal layer of pleura,** or **parietal pleura** (parietes meaning wall), which you have already dissected (page 87). In continuation of the previous analogy then, the invagination of the lungs into the pleural sacs constitutes the invagination of the sponge-filled balloons into outer, unfilled balloons. Because of the adherence of the visceral layer of pleura to the lung, and of the parietal layer to the thoracic wall, the negative pressure created between the walls of the "box," and the lungs, is in fact created **within the pleural sacs.** Puncturing of the sacs, or accumulation of fluid in them, will therefore interfere with respiration.

Review and reidentify the following features of the thoracic cage (pages 84 and 86): the sternal angle, the xiphisternum, the second costal cartilage, and the inferior margin of the rib cage (the **costal margin**).

3. Identify the **supraclavicular fossa** (fig. 33, page

93

82), a depression immediately superior to the middle third of the clavicle.

4. Identify the following landmarks on the posterior surface of your model's thoracic wall. (In reviewing this session by self-examination, the use of two mirrors will be helpful. Most of the structures to be palpated are readily accessible to self-examination.)

In the midline, the **segmental sequence** of the vertebrae is evident in the form of the **vertebral spinous processes** (fig. 41, page 90). These are projections from each vertebra and they point mainly posteriorly and somewhat inferiorly. They are usually easily visible and palpable. Try to identify individually the seven cervical, the twelve thoracic, and the five lumbar vertebral spines. A useful landmark is that the spine of the seventh cervical vertebra is usually especially prominent (page 4), and indeed is known for this reason as **vertebra prominens.** As a further landmark, palpate the lowest point of the shoulder blade (the **inferior angle** of the **scapula;** see fig. 41, page 90). This point on the scapula usually overlies the seventh intercostal space, or the eighth rib as the latter slopes downwards and laterally at this point. A horizontal line carried medially from the inferior angle of the scapula would meet the vertebral column at approximately the eighth thoracic **spinous process** (spine of vertebra).

5. The lungs can neither be inspected nor palpated in the living body. However, the respiratory mechanisms which are responsible for lung function can be. Previously, in studying the intercostal muscles (page 81, point 5) you observed the movements of the chest wall during inspiration. The volume of the rigid **thoracic cavity** (the box of the analogy) is increased by movement of the ribs (which, in part, is due to function of the intercostal muscles), as well as by a movement downwards of the diaphragm. The latter movement involves compensatory movements of the abdominal muscles. Thus two mechanisms of respiration co-exist: **thoracic** and **abdominal respiration.** Observe the thoracic cage and the abdominal musculature of your model on deep inspiration and determine which of the two types of movements predominates. Individuals vary in the degree to which the one or other mechanism is used. Infants usually rely largely on abdominal breathing. In adults, men tend to use abdominal respiration more, while women tend to use thoracic respiration.

In addition, certain so-called **accessory muscles of respiration** come into use in clinical situations of **respiratory distress,** such as breathing against resistance or excessively deep breathing. Such muscles include the **sternocleidomastoid** (page 49) and other muscles of the neck which attach to the thoracic cage. Observe and palpate the sternocleidomastoids and the **platysma** (page 55) muscles of the neck, for example, during extremely deep inspiration. Also review in this context, pectoralis minor and the question previously posed about its origin and insertion in various functions (page 87).

6. Palpation of respiratory movement is carried out by the physician in order to determine whether **respiratory excursion** (the degree of movement) is satisfactory, and whether it is symmetrical. Place the palms of both hands on your model's chest wall anteriorly, so that your thumbs are parallel and meet in the midline. Ask your model to breathe in deeply. Your palms and your thumbs should separate and move out to their respective sides as the subject breathes in. The distance that your thumbs separate from each other is a measure of respiratory excursion. This may vary from 1 cm in quiet respiration to several (up to 5 cm or more in physically fit individuals) in deep respiration.

7. An extremely useful method of examination of the lungs is percussion. Percussion is performed by placing the fingers of one hand flat on the surface to be percussed, and one or more of those fingers are tapped by the index finger of the other hand. Start by percussing your model's chest on the right side, 3 or 4 cm below the clavicle. In this region, the typical **resonance** of normal underlying lung tissue should be perceived. By contrast, percuss on the side of the neck, 3 or 4 cm below the angle of the mandible. Here you should feel the typical "**dull**" note of solid underlying tissue.

Now attempt to determine the extent of the lungs by percussing on each side of the chest anteriorly and posteriorly. On the anterior surface use as a landmark the **midclavicular line** (an imaginary vertical line through the midpoint of the clavicle) and on the posterior surface use the **scapular line** (an imaginary vertical line through the inferior angle of the scapula [see fig. 41, page 90]). Mark these lines with a wax pencil. Thereafter, attempt to confirm the following facts concerning the extent of the lungs, making reference to Figures 42 and 44, page 90, and an atlas.

8. On both sides, the highest point of the lung (**apex of the lung:** see figs. 42 and 44, page 90) extends up to, and provides the characteristic resonant percussive sound in, the **supraclavicular fossae.**

9. On the posterior aspect, the apex of the lung reaches the level of the first thoracic spinous process, about 5 cm from the midline.

10. Anteriorly, the border of each lung passes from the supraclavicular fossa obliquely through the **sternoclavicular joint** (the joint between the medial end of the clavicle and the manubrium of the sternum), and then passes vertically downwards behind the upper part of the body of the sternum, where the two sides meet in the midline, and proceeds down as far as the level of the fourth costal cartilage. On the right side, this border then slopes laterally to the junction of the sixth costal cartilage with the sternum, and then curves out laterally, following the sixth intercostal space, meeting the midclavicular line at the seventh rib. On the left, the lung has a **cardiac notch** to make a space for the heart. The outline is thus represented by a concave line drawn from the junction of the fourth costal cartilage and the sternum, passing laterally to a distance 4 cm from the sternum and then bending back to the 6th costal cartilage, 2.5 cm from the sternum. Thereafter, the left lung follows a similar contour to that of the right, passing through the sixth intercostal space and crossing the seventh rib in the midclavicular line. Posteriorly, on both sides, the medial border of the lung passes 2 cm from the **median** line, extending from the level of the first thoracic spine to that of the tenth or eleventh thoracic spine. To complete an outline of the **lung contour,** a slightly curved line, convex downwards, should connect this inferior extremity of the posterior border to points in the scapular line, at the level of the tenth rib; in the midaxillary line at the level of the eighth rib; and in the midclavicular line at the level of the seventh rib. Having confirmed the extent of the lungs by percussion, draw the surface markings described above on the skin surface, using a wax marking pencil.

11. The positions just described correspond to quiet respiration. In deep inspiration, the inferior border can be shown to move a full 5 cm downwards. Confirm this by percussion in the midaxillary line.

12. Conclude your examination of the lungs by auscultation, using a stethoscope. Commence your auscultatory examination by listening on the right side, 3 to 4 cm below the scapula, in the scapula line. Ask the model to breathe in and out deeply. By contrast, auscultate on the side of the neck, 3 to 4 cm below the angle of the mandible. In the latter situation you should hear no breathing sounds whereas what you heard in the former situation should be those of normal lung movement. As a further contrast, auscultate over the trachea, 2 to 3 cm above the suprasternal notch.

13. Now confirm by auscultation the extent of the lungs, and the borders which you established previously by percussion, and which are described above. Confirm that **"lung sounds"** (sounds of normal respiration) can be heard in the supraclavicular fossae, due to the apex of the lung (page 94, point 8, and figs. 42 and 44, page 90), and both anteriorly and posteriorly within, but not outside of, the lung borders described.

14. Each lung is divided into **lobes,** separated by **fissures.** There are normally three lobes on the right side and two on the left. Because infection or other disease processes may be limited to one lobe at a time, it is of clinical importance to be able to examine each lobe separately, through the chest wall. On the left side, examine by percussion and auscultation first the **superior lobe** and then the **inferior lobe,** respectively above and below the following oblique surface demarcation. Starting posteriorly at the third thoracic spinous process, the line curves laterally and downwards to the fifth intercostal space, then anteriorly around the lateral thoracic wall following the sixth rib. The line meets the inferior border of the lung at about the sixth costal cartilage, 4 or 5 cm lateral to the lateral edge of the sternum. On the right side, a similar line delineates superior and inferior lobes, and in addition, a **middle lobe** is demarcated from the superior lobe by a horizontal line on the anterior chest wall. This line extends from the fourth costal cartilage, to meet the oblique fissure described previously, in the posterior axillary line. Examine all three lobes on the right side.

To complete this session, review the key topics.

Laboratory Session 9

Having removed the anterior thoracic wall and exposed the thoracic viscera, at the completion of your last laboratory session, you can now proceed with examination of the structures studied on the living body in Clinical Session 9 (page 93 et seq.).

Because the parietal pleura is closely adherent to the inner surface of the chest wall (page 93), you will, as previously described (page 92), probably have removed it with the rest of the anterior chest wall. On observing the lungs **in situ** (fig. 42, page 90), the exposed surface of each lung is therefore only covered by the visceral layer of pleura.

The lungs should now be mobilized from their surroundings by hand dissection. Using the right hand for the left lung of the cadaver, and vice versa, the palm of the hand should be placed on the anterolateral surface of the lung (the dorsal surface of hand thus facing the inner surface of the rib cage). You should then gently work your fingers around each lung, posteriorly, following the curvature of the rib cage, toward the vertebral column; superiorly, toward the apex of the lung; and inferiorly, toward the diaphragmatic surface (that surface of the lung in contact with the diaphragm). Note that because of the adherences of the parietal and visceral layers of pleura, to the chest wall and the lung respectively, your dissecting hand is actually within the pleural sac.

When the lung is thoroughly mobilized on the aspects mentioned above, you should then proceed similarly to mobilize the medial surface, identifying the structures which form the **root of the lung,** and which enter the medial surface of the lung at a site called the **hilus.** Because of the attachment of the root of the lung to the lung's medial surface, the medial aspect can be mobilized from related structures adjacent to, and peripheral to, the hilus, but not at the hilus itself.

With mobilization of the lung completed, the root of the lung should be cut with a scalpel, relatively close to the hilus. When this is completed, each lung should be carefully removed from the thoracic cage.

Study the external form of the right and left lungs, correlating your observations with the description that you followed (page 95, point 10) in visualizing the extent of the lungs in the living body. You should identify the fissures and lobes (fig. 44, page 90), making use of atlas illustrations to guide you. Observe the differences between the right and left lungs, including the numbers of lobes and placement of fissures, the cardiac notch (page 95), and other features of size and general form.

Study the hilus, with the aid of atlas illustrations and anatomical specimens. Identify the components of the hilus by sight and touch. Recall (page 93) that the trachea terminates, in the sternomanubrial plane, by dividing into a right and a left **main bronchus,** each of which slopes downwards toward its own lung. The bronchus can be distinguished from the blood vessels at the hilus, by the fact that the rigid consistency of its wall, due to cartilaginous structures similar to the rings of the trachea, maintain the lumen widely patent. This rigid consistency can also be easily distinguished on palpation. Finally, the bronchus characteristically occupies a posterior position within the hilus.

The **pulmonary arteries** and **veins** (pulmonary indicates of the lungs) can also be distinguished by patency, consistency of the wall, and position. The arterial walls are firmer than those of veins (though not as firm as those of bronchi). As a result, the arteries have a more patent lumen than the veins. Furthermore, the characteristic position for the pulmonary artery is superior. The somewhat collapsed lumens of the relatively thin-walled pulmonary veins can be observed in their characteristic positions, which are anterior and inferior in the hilus (there being two pulmonary veins from each lung).

In the nature of the description of the origin of the pleural sac (page 93), it will be clear that the hilus is the site where the visceral pleura is in continuity with the parietal. Thus a cut edge of visceral pleura will be seen, circumscribing the structures of the root of the lung, at the hilus. Note, however, that this border does not merely surround the structures, but that it is also extended somewhat inferiorly, as if pulled downwards from within the sheath of pleura which surrounded the root on ingination. This downward extension of the pleural sheath is known as the **pulmonary ligament.** (The term ligament is frequently used in this context to imply a fold of a membrane such as the pleura, derived from the embryological **coelom.** This usage thus implies a completely different meaning to that of the ligaments of articulations of the skeleton.)

The region between the lungs is known as the **mediastinum,** which means standing in the middle. The major viscus (internal organ) situated in this region is the heart (fig. 42, page 90). Examine the right aspect of the mediastinal content (i.e., that aspect which faces the medial surface of the right lung). Compare that aspect to the lung surface, replacing, when necessary, the lung in its natural position within the thorax, and identify the important mediastinal structures which are related to the lung, and which imprint a groove or impression on the lung. Similarly, study the left aspect of the mediastinum and its relationship to the medial surface of the left lung.

On each side, you should start by identifying a large blood vessel which has a vertical part posterior to the hilus of the lung, and a curved portion which arches superior to the hilus. On the right side, the vessel is the **azygos vein** (azygos means unpaired), and on the left side the impression is due to the **arch** and **descending thoracic** part of the **aorta**. Identify in each case the arching vessel, and its impression behind and superior to the hilus of the respective lung. Communicating with the arch of the azygos vein, find, anteriorly, the **superior vena cava** (fig. 42, page 90) and its marking on the lung. The azygos vein drains into the superior vena cava, which then terminates by entering the right side of the heart. On the left side, two large branches of the **arch of the aorta,** and their markings, should be identified both in the mediastinum and on the lung. These are the **left common carotid artery** anteriorly, and the **left subclavian artery** posterolateral to it. On the right side, find the common **brachiocephalic trunk,** which passes superolaterally before dividing into the **right common carotid** and **right subclavian arteries.** On both sides, identify the large impression made by the heart, anterior to the hilus and the pulmonary ligament. Posterior to the vertical portions of the aorta and azygos vein, markings of the **vertebral bodies** can be identified on each lung.

Review (page 95) the division of the lungs into **lobes.** With the help of an atlas illustration, and study of anatomical specimens, explore the branching of the main bronchus of each lung into its **lobar** and **segmental** subdivisions. You can obtain an impression of this branching by dissecting the spongy lung tissue (which represents the smaller subdivisions and their terminations in the alveoli [page 93]), from the firm wall of the main bronchus. You should proceed with this dissection far enough to observe the first branchings of the bronchus, but do not attempt to dissect the entire bronchial tree. While a detailed knowledge of the pattern of branching is not essential, the student should have an understanding of the morphology of the lung in **lobes,** and of each lobe in **bronchopulmonary segments.** Since each lung has its own primary bronchus, the secondary bronchi are those which supply the lobes and the tertiary bronchi are those which supply a bronchopulmonary segment.

In addition to the lateral aspect of the mediastinum on each side, examine the other boundaries of the thoracic space, within which each lung is confined in the living body. Explore the relations of the apex of the lung, confirming by palpation that it is related, anteriorly, to the supraclavicular fossa (page 94). Inferiorly, observe the diaphragm, which bulges upwards, to form a separate dome on each side, each impinging on, and indenting the inferior surface of, the respective lung. Note that the crevices created by the curvature of the diaphragm, as it approaches the body wall, are filled by **recesses** of the **pleural sac** (the **costodiaphragmatic recesses**). Anteriorly, the **costomediastinal recess** of each pleural sac is situated posterior to the sternum.

Clinical Session 10

■ **TOPICS**

Mediastinum
Great Vessels
Nervous System

Your goal for this session is to study the surface projections of the space, within the thoracic cavity, which lies between the right and left **pleural sacs** (and therefore between the right and left lungs, each of which is invaginated into the corresponding pleural sac). This space is called the **mediastinum,** which means situated in the middle.

1. The mediastinum is situated posterior to the sternum (but extends further laterally than the lateral borders of the sternum). You should start by drawing the surface markings of the sternum. This is done by first palpating the sternal borders and related structures, and then marking what you have palpated, as follows. Commence by outlining the suprasternal notch. Proceed to palpate and mark the sternal ends of the clavicles and the notches in the manubrium into which the clavicles **articulate** (join). Palpate and mark the sternal angle, the second sternocostal **articulation,** and then, between the latter and the sternoclavicular articulations, the first costal cartilages, curving sharply upwards to disappear deep to the clavicles. By palpating deeply in the first to sixth intercostal spaces, delimit the medial extent of these, and thus the lateral borders of the sternum. Palpate and draw the third to seventh costal cartilages and their articulations with the sternum. Palpate and draw the xiphisternum, and draw in the xiphisternal and sternomanubrial planes (page 84). Through which vertebral levels do these planes pass?

Some structures situated within the mediastinum extend to regions outside the mediastinum (e.g., the neck or abdomen), and are more easily accessible to examination in the latter situations. By examining them in these more easily accessible situations, and extrapolating their continuity to within the mediastinum, a conceptual understanding of the relationships of **mediastinal structures** in the living human body can be obtained.

2. Some structures of the **superior portion of the mediastinum** (the part above the sternomanubrial plane) extend up into the neck. Start by palpating the **trachea,** superior to the **suprasternal** notch. The trachea descends in an almost vertical, median plane, and passes posterior to the manubrium. It then deviates slightly from the median plane to the right and then **bifurcates** (divides into two) at the level of the sternomanubrial plane. Now draw in a surface marking of the lower portion of the trachea starting about 2 cm above the sternal notch. To do this, gently (in order not to displace the trachea laterally) palpate, and then mark, first the right border of the trachea, and then the left, down to the suprasternal notch. Then extend both borders down to the sternomanubrial plane, inclining slightly to the right. At the sternal angle, indicate the **bifurcation of the trachea** by drawing right and left branches **(right and left main bronchi).** The right main bronchus should slope down at slightly more than a 45 degree angle with the sternomanubrial plane, should be about half the diameter of the trachea, and should be drawn about 2.5 cm in length. The left main bronchus should be drawn at a slightly smaller angle to the sternomanubrial plane (i.e., slightly less vertical than the right main bronchus), should be slightly narrower in diameter, and should be about 5 cm in length.

3. Using a stethoscope, auscultate the suprasternal part of the trachea while your model breathes deeply. Thereafter, continue your auscultation inferiorly down to the sternomanubrial plane, and then attempt to continue tracing first the right and then the left main bronchus by means of auscultation. By contrast, auscultate over the lungs (about 3 cm inferior to the clavicle, in the midclavicular line) and note the differences in the nature of the breath sounds audible over the trachea, main bronchi, and lung.

4. Palpate deeply (but gently) first to the right of the suprasternal portion of the trachea and then to the left of it. On each side of the trachea, a pulse should be easily palpable. You are palpating the **right and left common carotid arteries** on each side (page 70, point 10; page 55). Mark these arteries on your model's skin surface, from about 3 cm above the suprasternal notch to the sternoclavicular joint on either side. On the right side, the common carotid artery arises behind the sternoclavicular joint, by the division of the **brachiocephalic trunk** (brachio indicates of the arm; cephalic means of the head: this is a common trunk for supply of both regions, the common carotid being the cephalic component). Draw in the brachiocephalic trunk (fig. 42, page 90), a little more than half a centimeter in thickness,

from the center of the manubrium to the right sternoclavicular joint where the trunk divides. The right common carotid artery ascends, as indicated, more or less vertically, lateral to the trachea. The other branch of the trunk (the upper limb component) is the **right subclavian artery** (sub means under; clavian refers to the clavicle). From the sternomanubrial joint this artery passes superolaterally, where you should now palpate the subclavian pulse in the **supraclavicular fossa** (page 94 and fig. 33, page 82). The artery then curves laterally and downwards, deep to the clavicle. It crosses the first rib as it passes toward the upper limb.

5. The brachiocephalic trunk arises, behind the central point of the manubrium, from the **arch of the aorta** (page 97 and fig. 53, page 112). This can be indicated on your model by drawing a semicircular vessel about 2 cm in diameter, commencing at the right border of the manubrium in the sternomanubrial plane, curving up to the middle point of the manubrium, and then curving down to the left border of the manubrium in the sternomanubrial plane. At the right extremity of this **arch** blood enters from the **ascending aorta** (fig. 42, page 90), which comes from the **left ventricle** of the **heart.** At the left extremity, the arch of the aorta is continuous with the **descending aorta** (page 97). Although the curvature of the arch can only be indicated by the surface marking as curving from right to left, it should be noted at this stage (and correlated with your previous observations, page 97), that the arch also curves from an anterior position on the right to a posterior position on the left. Thus the end of the arch at the left border of the manubrium is in a posterior position, and the descending aorta likewise.

After giving off the brachiocephalic trunk, the **arch of the aorta** gives off two further branches. The **left common carotid artery** (page 97) comes off just to the left of, and posterior to, the brachiocephalic trunk, passes behind the left sternoclavicular joint, and then ascends lateral to the trachea, where you have already palpated it. The **left subclavian artery** (page 97) comes off just beyond the left common carotid, and curves up laterally, deep to the sternoclavicular joint, to then follow a course similar to that on the right side (see above). Palpate the left subclavian pulse in the supraclavicular fossa.

6. In order to get an impression of the contents of the **inferior portion of the mediastinum** (the portion inferior to the sternomanubrial plane), the situation of the main component of this part of the mediastinum, the heart (see fig. 42, page 90), will now be ascertained. Examine your model by percussion.

Percuss across the anterior thoracic wall, starting in the anterior axillary line on one side and moving progressively across to the anterior axillary line on the opposite side, (repeating the process at various transverse levels (from the clavicles downwards). Confirm by this process, that the heart lies approximately behind the following marking. From the lower edge of the left second costal cartilage, about 2 cm from the edge of the sternum, draw a straight line to the upper edge of the third costal cartilage on the right, about 2 cm from the sternum. This line represents the superior limit of the heart. From the right extremity of this border, draw a line having a slight convexity to the right, to the lower edge of the sixth costal cartilage, about 2 cm from the sternum. From the inferior extremity of the right border, draw a line slightly concave downwards through the xiphisternal plane, reaching to the fifth intercostal space on the left side, about 3 cm from the border of the sternum. Finally connect this latter point (which marks the **apex of the heart** (fig. 53, page 112), by a line with slight convexity to the left, to the left end of the upper border.

7. Now reconfirm by percussion, that the area of **cardiac dullness to percussion** corresponds to the outline you have drawn. At the lateral borders of the heart, the cardiac dullness gives way to the typical **lung resonance to percussion.**

8. Using a stethoscope, auscultate the heart within the area of the superficial markings you have made, and confirm that you can hear the **heart sounds** (two sounds per pulse beat). These will be further studied in a later session.

9. Note, from the surface markings of the heart which you have drawn in on your model, that the heart is mainly situated posterior to the **body of the sternum.** This corresponds to a space situated between the sternomanubrial and xiphisternal planes, and anterior to the bodies of thoracic vertebrae 5, 6, 7, and 8. For this reason, the spinous processes of these thoracic vertebrae are landmarks, on the posterior body surface, for surface markings of the heart. Identify these four vertebral spines in the posterior midline of your model.

10. Now review the surface markings of the anterior borders of the lungs, which you studied previously (page 95). You will have noticed that these markings of the lung overlap those of the heart which you have just drawn in. These anteromedial portions of the lungs do indeed overlap the heart anteriorly, and therefore lie in a space, within the inferior mediastinum, which is anterior to the space occupied by the heart. The **inferior mediastinum** is subdivided into an **anterior region** (occupied mainly by these portions of the lungs), a **central region** (occupied mainly by the heart and its surrounding membranes), and a **posterior region** which contains, among other things, the **thoracic part** of the **descending aorta**. Refer to point 5, (page 99), where the aorta was described as continuing down in a posterior position, after completing its arch from a right-sided and anterior position to a left-sided and posterior position.

11. The arch of the aorta and its three branches in the superior mediastinum, described above in point 5, are among the vessels known as the **great vessels of the mediastinum**. In addition to those already mentioned, several large veins are also found in the superior mediastinum. These in part accompany the arteries described, or correspond to them, and in addition, there are veins which correspond to and run with the great arteries of the neck and upper limb, also described in points 4 and 5 above. In the neck, lateral to the common carotid artery on either side, the **internal jugular vein** (page 55) is found. Anterior to each subclavian artery is a **subclavian vein**. The internal jugular and the subclavian vein of each side meet and unite at a position posterior to the sternoclavicular joint, on the right side just anterolateral to the bifurcation of the brachiocephalic trunk, and on the left side anterolateral to the point at which the left common carotid artery passes posterior to this joint. The veins formed by the junction of internal jugular and subclavian veins are known as the **brachiocephalic veins** (fig. 42, page 90). Note that there are two brachiocephalic veins, but only one brachiocephalic arterial trunk. The brachiocephalic veins are contents of the superior mediastinum.

12. Mark the **right brachiocephalic vein** on your model from its position of origin as described above (at the right sternoclavicular joint) to the first right intercostal space immediately to the right of the sternum. To this same point, connect the left brachiocephalic vein from its origin (at the left sternoclavicular joint). The **left brachiocephalic vein** thus crosses anterior to the branches of the arch of the aorta (fig. 42, page 90 and fig. 53, page 112).

13. From the junction of the two brachiocephalic veins, mark the **superior vena cava** (the large venous trunk which collects all the deoxygenated venous blood from the upper part of the body). Draw the superior vena cava from its origin to the upper border of the third right sternochondral junction, at which point the superior vena cava enters the **right atrium** of the heart (see fig. 42, page 90).

14. Among other structures contained in the mediastinum, but not accessible to easy physical examination in the living body, are numerous nerves, including the autonomic nerves (page 6) which control the functions of the heart. Of these, the **vagus nerves** are associated in the neck, on each side, with the large blood vessels described earlier in the session. The vagus nerve lies in the posterior groove between the internal jugular vein and the common carotid artery (and thus within the carotid sheath: see page 56) on each side. The vagus nerves thus enter the superior mediastinum very close to the respective veins and arteries.

An aspect of **vagal** control of **cardiac function** which can be simply confirmed on your model is the following. Palpate your model's pulse (at the wrist) while asking your model to breathe in deeply and hold his or her breath momentarily. Observe that the pulse quickens on **inspiration**. As the model then **expires,** the heart rate slows. This respiratory variation in heart rate is known as **sinus arrhythmia,** so called because this arrhythmia originates through an inhibitory influence by the vagal nerve on the **sinoatrial node**. This node is a small collection of cells situated at the junction of the superior vena cava and the right atrium. The rhythmic impulse which causes the heart to contract originates in this node.

Complete your study of this session by reviewing the emphasized key words.

Laboratory Session 10

Your objective this session is to study the mediastinum, its subdivisions, and their contents.

Review the situation and relations of the mediastinum, as observed at the time you removed the lungs from the thorax (page 96 and fig. 42, page 90), and as visualized, and projected onto the anterior thoracic wall in the living subject (pages 98–100). Review the divisions of the mediastinum into superior and inferior regions (pages 98 and 100), and of the inferior mediastinum into anterior, middle, and posterior subdivisions (page 100).

Some of the major structures of the mediastinum were observed previously, when you were studying the relationships and impressions of the medial surfaces of the lungs (pages 96–97). These structures can be discerned through the parietal pleura which covers them (the mediastinal pleura), but for more detailed study of these structures, this layer of pleura should now be dissected away.

Review and study in detail, as your main landmark on each side, the vascular structure which runs vertically, posterior to the hilus of the lung, and forms an arch superior to the hilus. On the right side, this structure is the azygos vein (fig. 47, page 102). Study its course from where it pierces the diaphragm inferiorly, passing superiorly on the anterolateral aspect of thoracic vertebra, to arch forward over the root of the right lung, in the sternomanubrial plane, and end by emptying into the superior vena cava. The latter vessel should be traced from its formation, by the junction of the right and left brachiocephalic veins (page 100, and fig. 48, page 102), to its termination in the upper chamber of the right side of the heart, called the **right atrium**. Note that the right atrium constitutes the major part of the right aspect of the heart, which impinges on, and creates the cardiac impression on the right lung, anterior to the hilus. (The heart, as visualized at this stage in your dissection, is enclosed in a tough connective tissue membrane, known as the **fibrous pericardium.**)

Posterior to the root of the right lung, and anterior to the vertical portion of the azygos vein, locate the esophagus, which you previously encountered as the continuation of the digestive tract below the level of C6 (page 70). Trace the esophagus superiorly. Anterior to the esophagus, above the level of the sternomanubrial plane, the trachea (page 93) should be palpated, identifying it by the characteristic cartilaginous rings. Trace the trachea downwards to its bifurcation in the sternomanubrial plane, and follow the right main bronchus into the root of the lung.

On the lateral aspect of the trachea the **vagus nerve** (the tenth cranial nerve: page 100) will be observed sloping posteroinferiorly (fig. 45, page 102) to pass onto the lateral aspect of the esophagus, behind the root of the lung. On the esophagus, the vagus nerve branches into a rich plexus (network). Identify the vagus nerve and the esophageal plexus on both sides.

As indicated previously, the vagus nerve is, in part, an autonomic nerve, carrying visceral innervation to such internal organs as the heart (page 100), as well as a large part of the digestive tract. Anatomically, the nervous system consists of the **central nervous system** (CNS), comprising the brain and spinal cord, and the **peripheral nervous system** (PNS), comprising cranial nerves and spinal nerves. In a functional and morphological sense, it is useful to think of these anatomical constituent parts as also being subdivided into components of the CNS and PNS which innervate **somatic** (body wall) structures, and other components which innervate internal, **visceral** structures (pages 5–6). Nervous control of somatic structures is under **voluntary** control, and is represented largely in the conscious mind, whereas visceral structures are under involuntary (**autonomic**) control, which proceeds subconsciously.

Anatomically, visceral innervation differs from somatic in that, whereas somatic afferent nerve fibers are connected to somatic efferent fibers by connector neurons situated entirely **within** the central nervous system (page 86), visceral connector neurons leave the CNS to synapse (link) with efferent neurons **outside of** the CNS. Thus the structure which you dissect here as the vagus nerve comprises connector neurons. These synapse on their corresponding efferent nerve cell bodies on the surface of the organ being supplied, or actually within it. These efferent nerve cell bodies are gathered in clusters in the organs concerned (e.g., the esophagus). A cluster of nerve cell bodies outside of the CNS is known as a **ganglion** (pages 33 and 86), and the ganglia on which the vagus nerve (connector neuron) fibers synapse are minute structures situated within the muscle wall of the digestive tract, or in close association with other viscera such as the heart and lungs. These autonomic ganglia of the vagus nerve (the tenth cranial nerve), placed close to or in the organ they supply, are homologous to the submandibular ganglion (page 33) of the seventh cranial nerve, the otic ganglion (page 48) of the ninth cranial nerve, and the sphenopalatine ganglion (page 63), also of the seventh cranial nerve, which you

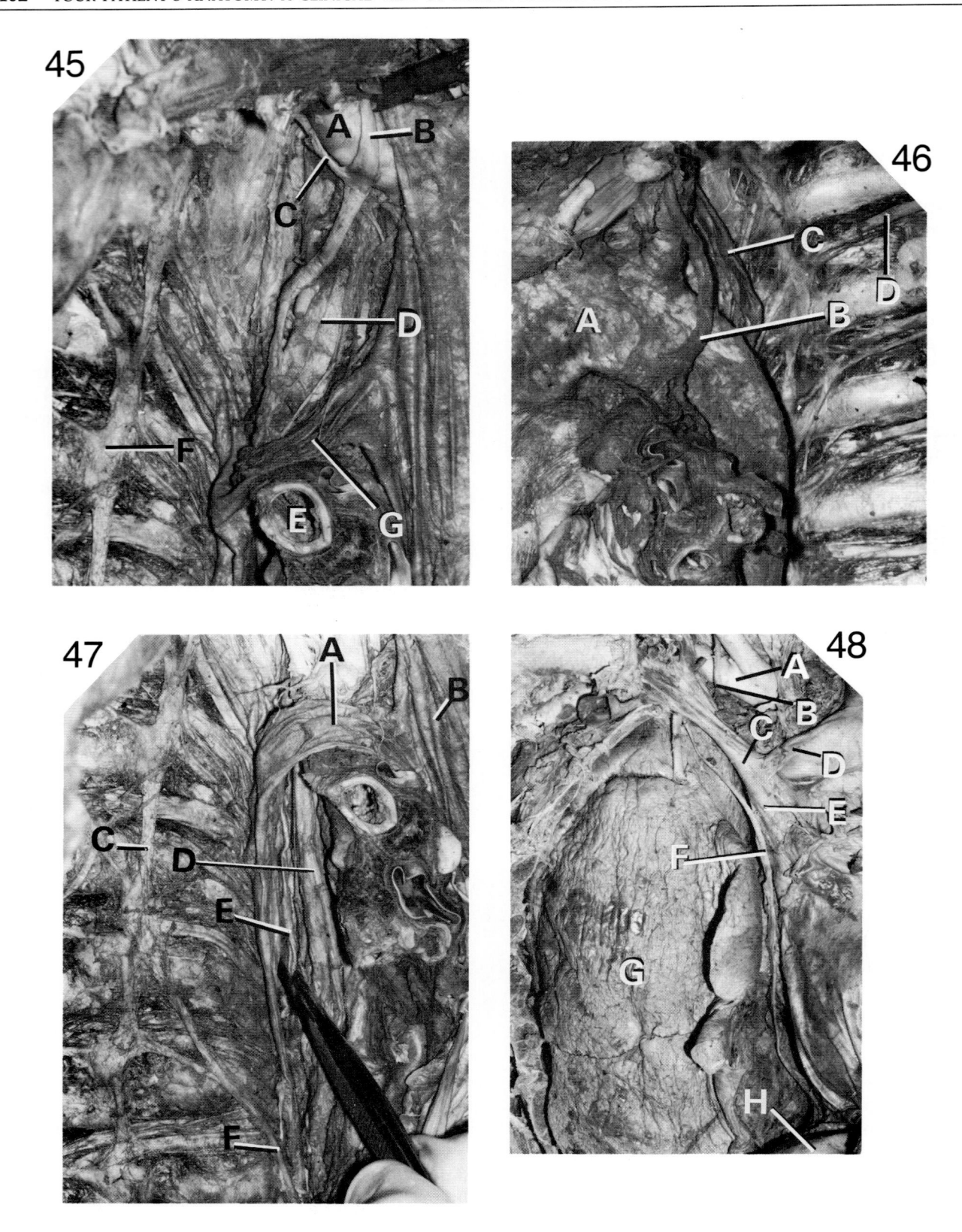

102 YOUR PATIENT'S ANATOMY: A CLINICAL VIEW OF HUMAN MORPHOLOGY

Figure 45. Right aspect of mediastinum, above root of lung.
A. Right subclavian artery
B. Right vagus nerve sloping posteroinferiorly on lateral aspect of trachea
C. Right recurrent laryngeal nerve recurving behind right subclavian artery
D. Trachea
E. Right main bronchus
F. Sympathetic ganglion

Figure 46. Left aspect of mediastinum, at and above lung root.
A. Arch of aorta
B. Left vagus nerve
C. Thoracic duct
D. Intercostal nerve

Figure 47. Right aspect of mediastinum.
A. Arch of azygos vein
B. Superior vena cava
C. Right sympathetic trunk
D. Esophagus
E. Thoracic duct (held in forceps)
F. Splanchnic nerve

Figure 48. Right side of thorax, anterior view.
A. Right subclavian artery
B. Right vagus nerve
C. Right brachiocephalic vein
D. Left brachiocephalic vein
E. Superior vena cava
F. Right phrenic nerve
G. Right lung
H. Right dome of diaphragm

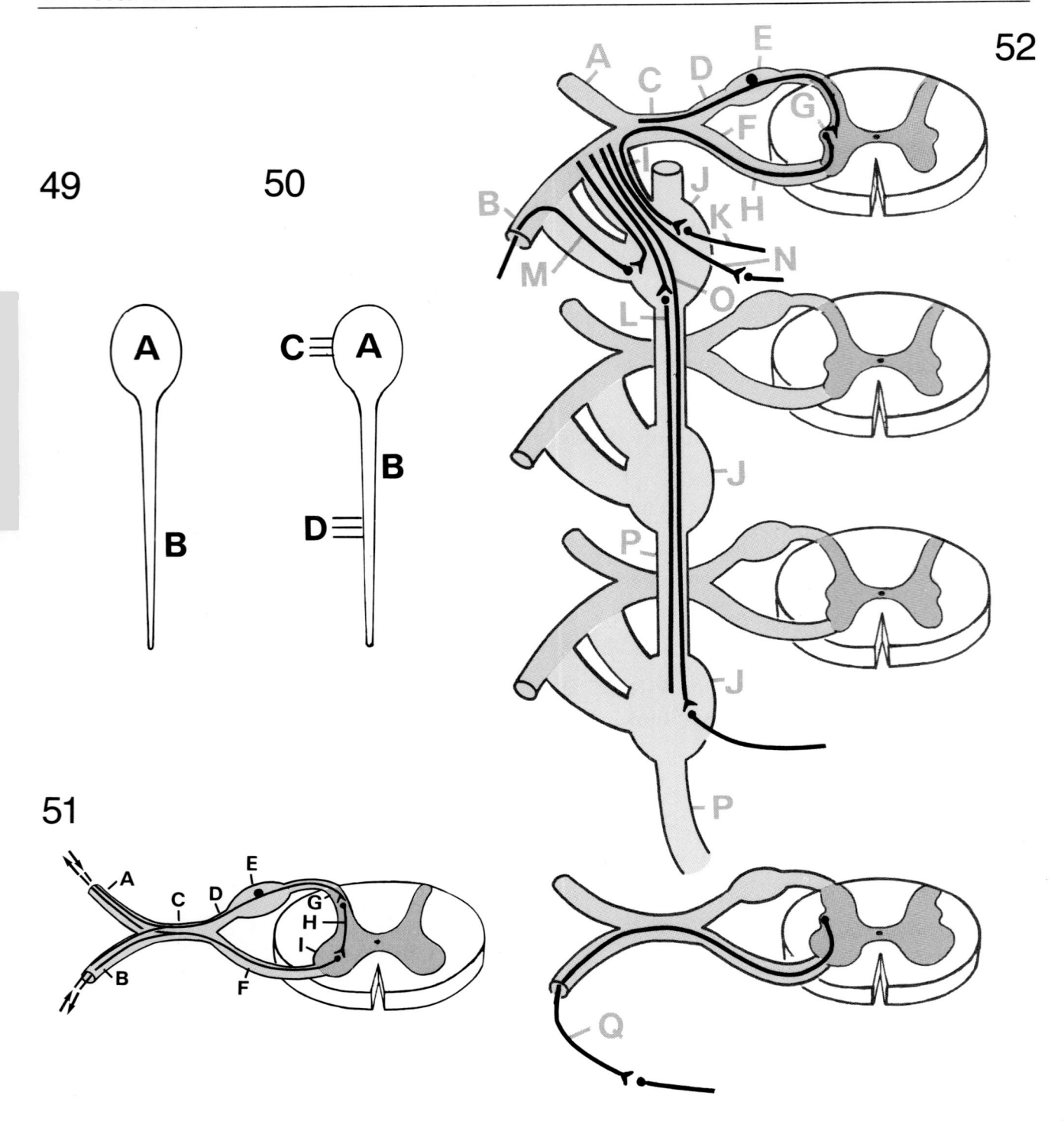

104 YOUR PATIENT'S ANATOMY: A CLINICAL VIEW OF HUMAN MORPHOLOGY

49

50

51

52

Figure 49. Central nervous system (CNS).
A. Brain
B. Spinal cord

Figure 50. CNS and Peripheral nervous system (PNS).
A. Brain
B. Spinal cord
C. Three of the 12 cranial nerves
D. Three of the 32 spinal nerves

Figure 51. Simple reflex arc. Spinal segment of the CNS and segmental nerves of the PNS, showing somatic (voluntary) nervous system (SNS) connections.
A. Dorsal ramus, containing afferent (sensory) and efferent (motor) fibers
B. Ventral ramus, containing afferent (sensory) and efferent (motor) fibers
C. Mixed spinal nerve
D. Dorsal (afferent) root
E. Dorsal root ganglion
F. Ventral (efferent) root
G. Dorsal horn of gray matter, containing nerve cell bodies of connector neurons of the SNS
H. Nerve cell processes of connector neurons of SNS, situated entirely within the CNS
I. Ventral horn of gray matter, containing nerve cell bodies of efferent neurons of the SNS

Figure 52. Segmental pattern of the nervous system, showing autonomic (visceral) nervous system (ANS) connections.
A. Dorsal ramus
B. Ventral ramus
C. Mixed spinal nerve
D. Dorsal root
E. Dorsal root ganglion
F. Ventral root
G. Lateral horn of gray matter, containing nerve cell bodies of connector neurons of the ANS (sympathetic component in segments T1–L2)
H. Nerve cell process of connector neuron of ANS, extending into PNS
I. Connector neuron processes passing in white ramus communicans to sympathetic ganglion (J), where it may synapse on efferent, motor neurons (K, L, M) or pass through the ganglion without synapse (N, O)
J. Sympathetic ganglion
K. Postganglionic efferent, motor fiber passing to viscera
L. Postganglionic efferent, motor fiber passing to another segmental level
M. Postganglionic efferent, motor fiber returning, in gray ramus communicans, to ventral ramus, for distribution to blood vessels, sweat glands, and hair follicles
N. Presynaptic fiber passing, via splanchnic nerve, to celiac or other abdominal ganglia
O. Presynaptic fiber passing to another segmental level
P. Sympathetic trunk
Q. Presynaptic parasympathetic fiber (in segmental levels S1–S3) passing, via pelvic splanchnic nerve, to ganglia situated in viscera

studied previously. Also belonging to this group, and to be studied later, with the eye, is the **ciliary ganglion** of the third cranial nerve. In contrast to a **ganglion**, which is a cluster of nerve cell bodies **outside** of the CNS, a cluster of nerve cell bodies **within** the CNS is called **gray matter** (page 86). Note that a localized mass of gray matter within the brain is often referred to as a "nucleus."

Strip the parietal pleura from the posterior thoracic wall and demonstrate another component of the autonomic nervous system, running vertically on the lateral aspect of the vertebral column, overlying the sites of articulation of the ribs with the transverse processes of the vertebra. This structure is known as the **sympathetic trunk** (fig. 47, page 102).

The autonomic nervous system, unlike the somatic nervous system, is not represented in all cranial and spinal nerves. On the contrary, it has "outflow" from the CNS in only the following peripheral nerves:

1. (a) **four cranial nerves:** (the third; the seventh [as exemplified by the secretomotor fibers which pass via the chorda tympani [page 48], and the lingual nerve [page 28] to the submandibular gland [page 33]; the ninth (giving secretomotor fibers which pass, via the auriculotemporal nerve [page 48] to the parotid gland); and, as described on page 101 the tenth cranial nerve; (b) in sacral segments **S2 and S3;**
2. in all the thoracic and the first two lumbar segments **(T1 to L2).**

The portion of the autonomic nervous system represented in spinal nerves **T1 to L2** is called the **sympathetic** part of the autonomic nervous system; the portion represented in the **third, seventh, ninth, and tenth cranial nerves** and sacral spinal nerves **S2 and S3** (1 above) is called the **parasympathetic** part of the autonomic nervous system. In addition to the differing sites of origin (thoracolumbar and craniosacral, respectively), these two portions of the autonomic nervous system also differ in their function, in that the one component usually opposes the actions of the other. For example, you have studied secretomotor, parasympathetic fibers of the seventh cranial nerve which stimulate the submandibular glands to secrete (page 29), these seventh cranial nerve fibers being conveyed to the sublingual region via the chorda tympani and the lingual nerve (pages 48 and 28). Similarly, you studied salivary secretion by the parotid gland (page 26). In this case, the parasympathetic secretomotor fibers originate from the ninth cranial nerve (the glossopharyngeal nerve), and are conveyed to the parotid gland by the auriculotemporal branch of the mandibular nerve (page 48). Sympathetic fibers to the submandibular and parotid glands counteract this parasympathetic function, and will tend to inhibit secretion of saliva. As another example, vagal (parasympathetic) stimulation tends to slow the heart beat (page 100), and sympathetic impulses tend to speed it up.

Like the parasympathetic nerves, so too with the sympathetic nerves, connector neurons emerge from the CNS before synapsing on the nerve cell bodies of their corresponding efferent neurons. Accordingly, clusters of efferent nerve cell bodies (ganglia) are again found outside the CNS. Sympathetic ganglia are, however, unlike parasympathetic ganglia, not situated within or on the surface of the organ of supply. They constitute an intermediate situation between the latter on the one hand and the situation of somatic efferent nerve cell bodies, situated within the CNS (in gray matter) on the other hand. Thus the ganglia of the sympathetic nervous system lie between the CNS and the viscera they supply, and constitute the ganglia visible as swellings on the sympathetic trunk (fig. 45, page 102) which you have exposed by reflecting the parietal pleura. The connections between the ganglia constitute connector nerve fibers which may pass through several ganglia (corresponding to several segments) before participating in a synapse on an efferent nerve cell body. Connector neurons pass from the central nervous system to the sympathetic ganglia by emerging from the CNS in mixed spinal nerves, then entering ventral rami (page 86), and then leaving ventral rami to pass to sympathetic ganglia.

Trace the sympathetic trunk inferiorly and superiorly, observing branches which it gives off to thoracic viscera, as well as its connections to ventral rami (intercostal nerves [page 86]) visible just lateral to the vertebral bodies. Large branches of the sympathetic trunk emerge medially, from some of the ganglia, to pass downwards toward the diaphragm. These constitute the **splanchnic nerves** (fig. 47, page 102), which, unlike the sympathetic fibers described earlier, represent connector neurons which do not synapse in the ganglia of the sympathetic trunk, but pass through these ganglia without establishing connections (i.e., are "preganglionic") to subsequently synapse in smaller ganglia in the abdomen. Some of the features of the autonomic nervous system described above are schematically depicted in Figure 52, page 104.

The **right phrenic nerve** (fig. 48, page 102) (phrenic means of the diaphragm) should be observed on the lateral aspect of the superior vena cava, and traced down on the pericardium covering the right side of the heart, to the diaphragm.

Posterior to the esophagus, dissect the **thoracic duct** (fig. 47, page 102), a major lymphatic vessel (page 6), that conveys lymph from the abdomen and lower limbs, and part of the thorax, superiorly.

The thoracic duct ascends posterior to the right border of the esophagus, to reach the level of the fifth thoracic vertebra, at which site it crosses to the left side. Dissect it to this level on the right side, and follow its further course when dissecting the left mediastinum (see below).

Before leaving the right side of the mediastinum, return to the vagus nerve and follow it superiorly to the subclavian artery. As the nerve passes anterior to the artery, it gives off its **recurrent laryngeal** branch. This nerve is so named because it recurves, posterior to the artery, to ascend posterior to the common carotid artery, in the groove between the esophagus and trachea, toward the larynx. Review your previous encounter with this nerve (page 76).

On the left side of the mediastinum, identify the **arch of the aorta** (fig. 46, page 102), superior to the root of the left lung, and trace the further course of this blood vessel inferiorwards, as it continues, below the level of the sternomanubrial plane, as the **descending thoracic aorta.** This large vessel should be dissected posterior to the root of the lung. Inferior to the latter, the esophagus should be demonstrated anterior to the aorta and just posterior to the heart. The chamber of the heart which is responsible for the cardiac impression on the left lung (i.e., the left ventricle) is visible here. (The first part of the aorta, (the **ascending aorta**), leaves the left ventricle to pass upwards and to the right, where it continues into the arch of the aorta in the sternomanubrial plane.) Superiorly, identify the three branches of the arch of the aorta: the **brachiocephalic trunk,** passing upwards and to the right; and the **left common carotid artery** and **left subclavian artery,** passing upwards and to the left. In the concavity of the arch of the aorta, find the **ligamentum arteriosum,** connecting the arch with the left pulmonary artery. The ligamentum is an embryological vestige (i.e., remnant of a structure that was functional in fetal life, but is not in the adult). The structure in the fetus is called the **ductus arteriosus.** Its function is to shunt blood from the pulmonary trunk to the aorta. Attempt to envisage why this should be necessary.

Reidentify the left vagus nerve and trace it proximally (superiorly) to locate the **left recurrent laryngeal nerve,** the course of which differs from that of its contralateral partner. On the left, it leaves the vagus as the latter passes to the left of the arch of the aorta. The branch then recurves inferior to the arch and posterior to the ligamentum arteriosum in the concavity of the arch to then attain a situation and further course similar to the right side (pages 76 and 107). This asymmetry has the following embryological explanation. In fetal life a series of six branchial arch arteries curve from ventral to dorsal, around the cranial portion of the developing digestive tract, connecting ventral and dorsal aortae. Most of the first and second arch arteries disappear early and the third contributes to the carotid arteries of the adult. The fourth arch artery becomes the aortic arch on the left and the first part of the subclavian artery on the right. This accounts for the manner in which these arteries, especially the arch of the aorta, curve from ventral to dorsal, adjacent to the esophagus. The fifth arch artery disappears on both sides. The ventral portion of the sixth artery contributes to right and left pulmonary arteries; the dorsal portion disappears on the right and becomes the ductus arteriosus on the left. Initially, the recurrent laryngeal nerve recurves around the sixth arch artery on each side, but as the dorsal part of the sixth artery on the right and the fifth artery degenerate, the recurrent laryngeal nerve "slips" cranialwards on the right side.

Demonstrate the **left phrenic nerve** in relationship to the left common carotid artery, crossing the arch of the aorta, and passing downwards anterior to the root of the lung, onto the lateral surface of the heart and to the diaphragm. The **vagus nerve** passes from its cervical situation posterior to the common carotid artery and internal jugular vein, to attain the lateral surface of the arch of the aorta (fig. 46, page 102), from where it passes behind the root of the lung and onto the right aspect of the esophagus, where it forms a plexus as on the left side. The sympathetic trunk should be demonstrated on the left as previously done on the right. Furthermore, the thoracic duct, which you traced to the level of T5 on the right side, should be identified on the left, as it crosses to this side at that level (fig. 46, page 104). Its further course upwards and into the neck will be dissected subsequently.

Clinical Session 11

■ **TOPICS**

Heart
Ascending Aorta
Pulmonary Trunk

The objectives of this session are to study, by the techniques of physical examination (inspection, palpation, percussion, and auscultation), the heart of a living subject. You will also study the surface projections of the heart and examine some functional correlates of this organ.

1. As previously, your model should be stripped to the waist and in the anatomical position for the following examination. **Inspection** is valuable in discerning only one feature of the normal heart; a visible pulsation in the **precordium** (the portion of the anterior aspect of the chest wall which overlies the heart). Visible pulsation is not always present in normal individuals, but when it is, it is usually seen in the lowest portion of the precordium, to the left, and is known as the **apex beat**. The position of the apex beat may vary considerably in normal individuals, but will most often be seen in the left fifth intercostal space about 3 cm from the border of the sternum and slightly medial (about 1 to 2 cm) to the midclavicular line. The apex beat is the visible impulse on the anterior thoracic wall of the **apex of the heart** (page 99 and fig. 53, page 112). This is seen each time the **ventricles** of the heart contract. If an apex beat is visible, palpate your model's pulse at the wrist while observing the apex beat. Note whether they are synchronous or not.

2. Palpation of the precordium will, in the normal heart, usually reveal a palpable apex beat. In order to palpate your model's precordium, you should stand to the right of your model and place your right hand (which you should have warmed, for example by rubbing your hands together) on the patient's chest, with the palm of the hand facing the chest wall, and the fingers stretched out so that the fingertips do not dig in to your model's intercostal spaces. In this manner, palpate the apex beat and localize its position. You should initially feel it with the palm of your hand but when identified, its exact position should be determined by using the tips of your fingers. You may note that the apex beat as determined by palpation may differ slightly from its position as determined by inspection: by palpation it may appear to be slightly more lateral than by inspection. When you have determined the apex beat by palpation, mark its position with a wax skin marker.

3. Proceed now to examine the extent of the heart by **percussion**. The left border is most easily determined by percussing in the fourth intercostal space, from the area over the left lung inwards medially toward the heart. Mark its position (approximately 1 cm medial to the midclavicular line). Repeat the exercise on the right side, this time over the right fourth rib. You will find the border about 1 to 2 cm lateral to the border of the sternum.

4. Now mark the outlines of the heart as projected onto the precordium, following the instruction previously given (page 99, point 6; refer also to figs. 53 and 56, page 112, in studying this and the following points). Briefly, these markings are a point on the lower edge of the second left costal cartilage, 2 cm from the edge of the sternum; a point on the upper edge of the third right costal cartilage, about 2 cm from the sternum; a point on the lower edge of the right sixth costal cartilage, about 2 cm from the sternum; and the point of the apex beat, usually in the left fifth intercostal space, about 3 cm from the border of the sternum. Draw the right side with a slight convexity to the right, the left side with convexity to the left, the inferior surface approximately horizontal and passing through the xiphisternal plane, and the superior border (base) with a straight line. This marking is approximate, and should be modified according to your findings on inspection, palpation, percussion, and auscultation.

5. The heart consists of four chambers. Previously (page 100, point 13) you drew the surface markings of the superior vena cava as it entered the right atrium of the heart. The superior vena cava collects all the **deoxygenated** venous blood from the upper part of the body. Similarly, the **inferior vena cava** returns the deoxygenated venous blood from the lower part of the body to the **right atrium**.

6. The **right atrium** is one of the four chambers of the heart. The deoxygenated, venous blood from the entire body passes from the right atrium downwards, forwards, and to the left, to reach the **right ventricle**. This direction of blood flow from right atrium to right ventricle is important to understand, as it aids in picturing, in three dimensions, the relationships of the four chambers of the heart to each other. Thus the atria are not only superior to the ventricles, but are also posterior to them: the upper (atrial) portion of the heart can be thus thought of as being tilted somewhat posteriorly, and the lower (ventricular) portion of the heart as being tilted somewhat anteriorly. In addition, the "right" side of the heart (right atrium and right ventricle) is not only to the right, but also somewhat anterior to the "left" side (left atrium and left ventricle): therefore, the heart can also be thought of as being rotated about a vertical axis.

7. Armed with this explanation, you will understand why the valve which guards the opening between the right atrium and the right ventricle (the **right atrioventricular valve**) has the following surface marking, which you should know and draw in. Start in the median plane (anterior to the sternum) with a point at the level of the upper edge of the fourth costal cartilage, and draw a line sloping downwards and to the right, toward the fifth sternochondral joint. (Because the right atrioventricular valve has three cusps, it is also known as the **tricuspid valve** (see fig. 58, page 116).

8. From the right ventricle, blood passes through the **pulmonary trunk** (fig. 53, page 112) upwards and to the left to be distributed to the lungs for reoxygenation. In accordance with the description, given above (point 6), because the right ventricle is anterior to the left ventricle, the pulmonary trunk passes anterior to the large blood vessel leaving the left ventricle. **The pulmonary valve,** which guards the exit of blood from the right ventricle through the pulmonary trunk, should now be marked as a horizontal line, anterior to the left half of the sternum, extending medially from the third sternochondral joint.

9. Reoxygenated blood from the lungs returns to the **left atrium** of the heart in the pulmonary veins (page 96). From there it passes to the **left ventricle** through a communication (the left atrioventricular opening) which is guarded by the **left atrioventricular valve.** Draw in the surface marking of the left atrioventricular valve as follows. From the upper median extremity of the marking of the **right atrioventricular valve,** draw a line sloping upwards and toward the third left intercostal space. The left atrioventricular valve has two cusps, and because their appearance reminded early anatomists of the appearance of a bishop's miter, it is also known as the **mitral valve.**

10. From the left ventricle, the oxygenated blood leaves to be distributed to the body through the **aorta.** Its passage from the left ventricle to the first part of the aorta (the **ascending aorta,** page 99 and fig. 42, page 90) is guarded by the **aortic valve.** Indicate the aortic valve with an oblique line, parallel to and just above, but a little shorter than the line for the mitral valve. It reaches the median plane and slopes upwards toward the marking for the pulmonary valve but does not quite reach the latter. Recall from the previous descriptions (points 6 and 8) that the aortic valve will be posterior to (i.e., at a deeper level than) the pulmonary valve.

11. You should now auscultate the **heart sounds.** In the healthy individual there are two sounds heard over most of the precordium, the first corresponding to the commencement of **ventricular systole** (the period of contraction of the ventricles), and the second corresponding to the commencement of **ventricular diastole** (the period of relaxation of the ventricles). The first sound is due mainly to the closure of the mitral and tricuspid valves. As the ventricles contract, blood is pumped out of the pulmonary trunk and the aorta respectively. To prevent leakage back into the atria, the atrioventricular valves close. The second heart sound is due to the closure of the aortic and pulmonary valves in ventricular diastole. In ventricular diastole, the atrioventricular valves are, on the other hand, open, allowing flow of blood from the atria into the ventricles.

Because the valves lie so close to each other, auscultation over the surface markings of each of them would not allow the examiner to distinguish from which valve a particular sound was originating. Accordingly, in practice, one auscultates the four valves as follows.

12. Auscultate the tricuspid valve by placing the stethoscope at the lower end of the sternum, to the right. The mitral valve is best heard at the cardiac apex. The pulmonary valve is best heard in the left second intercostal space very close to the sternal border, close to where the pulmonary trunk bifurcates into right and left pulmonary arteries, in the sternomanubrial plane. (Which other structure bifurcates in the sternomanubrial plane?) The ascending aorta passes, as mentioned, posterior to the pulmonary trunk. It passes from left to right (whereas the **pulmonary trunk passes anteriorly,** from right to left). The aortic valve is therefore best heard as the aorta reapproaches the anterior surface of the chest wall, at the sternal angle. Thus auscultation is performed at the medial end of the second right costal cartilage. This is also the site where the ascending aorta ends and passes into the arch of the aorta, which you will recall having previously marked as commencing at this point. Review this latter surface marking (page 99, point 5).

13. In auscultation at each of the areas designated in the above, you should simultaneously palpate your model's pulse, preferably at the common carotid artery, the position of which you should now review (page 70, point 10 and page 98, point 4).

14. Review also the exercise which demonstrates vagal inhibition of the cardiac impulse (page 100, point 14). Ask your model to breathe deeply and slowly, while you palpate the pulse. On deep inspiration, the pulse quickens, and on deep expiration it slows down. These changes are due to alterations in **"vagal tone."** Among possible mechanisms to account for this effect are afferent impulses from the lungs, variations in pressure on the venous side of the heart, or some central nervous system influence.

15. Auscultate (e.g., at the apex of the heart) in deep inspiration and deep expiration, and note that the heart's position does not change much with respiration. Compare this with the situation regarding the lungs (page 95, points 11 and 13). This corresponds with the fact that the central portion of the dome-shaped sheet of muscle separating thorax from abdomen (the diaphragm) moves less in respiration than the lateral portions.

To complete this session review the key concepts.

Laboratory Session 11

You have previously studied the position of the heart within the thorax (page 108). You have also noted (page 101) that the heart is surrounded by the **fibrous pericardium.** Superiorly, the fibrous pericardium merges with connective tissue surrounding the venous and arterial great vessels of the superior mediastinum.

To appreciate the nature and origin of the pericardium, begin by making a vertical scalpel incision through the tough membrane lying anterior to the heart, immediately to the left of the pulmonary trunk (fig. 42, page 90). Reflect the membrane to the right and to the left, and, inserting two fingers to the left of the pulmonary trunk, turn your fingers medially (i.e., to the right) to insinuate them posterior to the pulmonary trunk and the ascending aorta (fig. 42, page 90). You will be able to pass your fingers beyond the right edge of the ascending aorta, thus separating the two arterial trunks (pulmonary trunk and ascending aorta) anteriorly, from the superior vena cava (see fig. 42, page 90) posteriorly. In addition to this large vein, entering the right atrium (page 100), the pulmonary veins, which enter the left atrium (fig. 55, page 112), are also posterior to your fingers (although they are not visible at this time). Thus, in summary, you have separated arterial channels, anteriorly, from venous channels, posteriorly. Bearing in mind that the atria are superior and posterior to the ventricles, which are inferior and anterior (page 108, point 6), you will be able to picture the heart and its great vessels as being derived from a tubular loop, having right and left channels. Starting posterosuperiorly, the first part of the loop comprises the great veins, entering their respective atria. This posterior limb of the loop continues downwards, where it comprises the ventricles. Thereafter it bends anteriorly and upwards, this ascending limb of the loop comprising the arterial outflow channels of the ventricles (i.e., pulmonary trunk and ascending aorta).

In embryological development, the heart invaginates a balloon-like sac, comparable to the way in which the lungs do. As described previously (page 93), the sacs invaginated by the lungs are called the pleurae. It was mentioned (page 96) that the pleurae are parts of the coelom: the sac invaginated by the heart, called the **serous pericardium,** is similarly part of the coelom.

The lung buds grow out laterally on either side, invaginating their respective pleural sacs from the medial aspect (page 93). The heart, a midline structure, invaginates its sac from the dorsal aspect. Thus the two channeled loop, described above, should be pictured as invaginating the pericardial sac from the dorsocranial aspect. In looping forwards as it invaginates into the sac, the tube not only acquires for itself a visceral covering, but also carries a double layer of the sac behind it, as a loop-shaped structure would do, on invaginating a balloon. The two layers so created, behind the invaginating loop, are called the **mesocardium.** The layers fuse and then disintegrate, leaving the space between the arterial and venous ends of the loop, into which your fingers could be insinuated. This space is known as the **transverse pericardial sinus.**

It was stated earlier that the sac into which the heart invaginates is called the serous pericardium. As in the case of the pleural sacs and the lungs, invagination by the heart of the serous pericardium creates a visceral and a parietal layer of this sac. The visceral layer is very closely adherent to the muscular wall of the heart. The serous pericardium is, unlike the pleural sac, surrounded by a thick, exterior fibrous layer of connective tissue, called the fibrous pericardium, closely applied to the outer surface of the parietal layer. Thus the parietal layer of serous pericardium constitutes an internal lining of the fibrous pericardium. The fingers placed in the transverse pericardial sinus are, then, within the serous pericardial sac (i.e., between the visceral and parietal layers of it).

In describing the site of entry of the great veins into the heart as the venous end of the embryological cardiac loop (cardiac means of the heart), it was indicated that the superior vena cava enters the right channel of the loop, and the pulmonary veins, the left. For completeness, it should be stated that the **inferior vena cava,** conveying blood from the lower part of the body, also enters the venous, or atrial end of the right-sided channel (i.e., the right atrium). With the flaps of pericardium pulled laterally to their respective sides, identify the apex of the heart (page 108 and fig. 53, page 112). Now pull the apex anteriorly and insert your fingers below the diaphragmatic surface of the heart (i.e., that surface facing the diaphragm). By palpating toward the right, you will encounter the inferior vena cava which, having passed from the abdomen to the thorax by traversing the diaphragm, enters the pericardial sac, to then terminate in the right atrium. By palpating posterosuperiorly, your fingers are exploring the **oblique pericardial sinus.** To the right, the sinus is limited by the inferior vena cava, the two right pulmonary veins and the superior vena cava, in that order from below upwards. To the left, the sinus is limited by the two left pulmonary veins, and superiorly, your

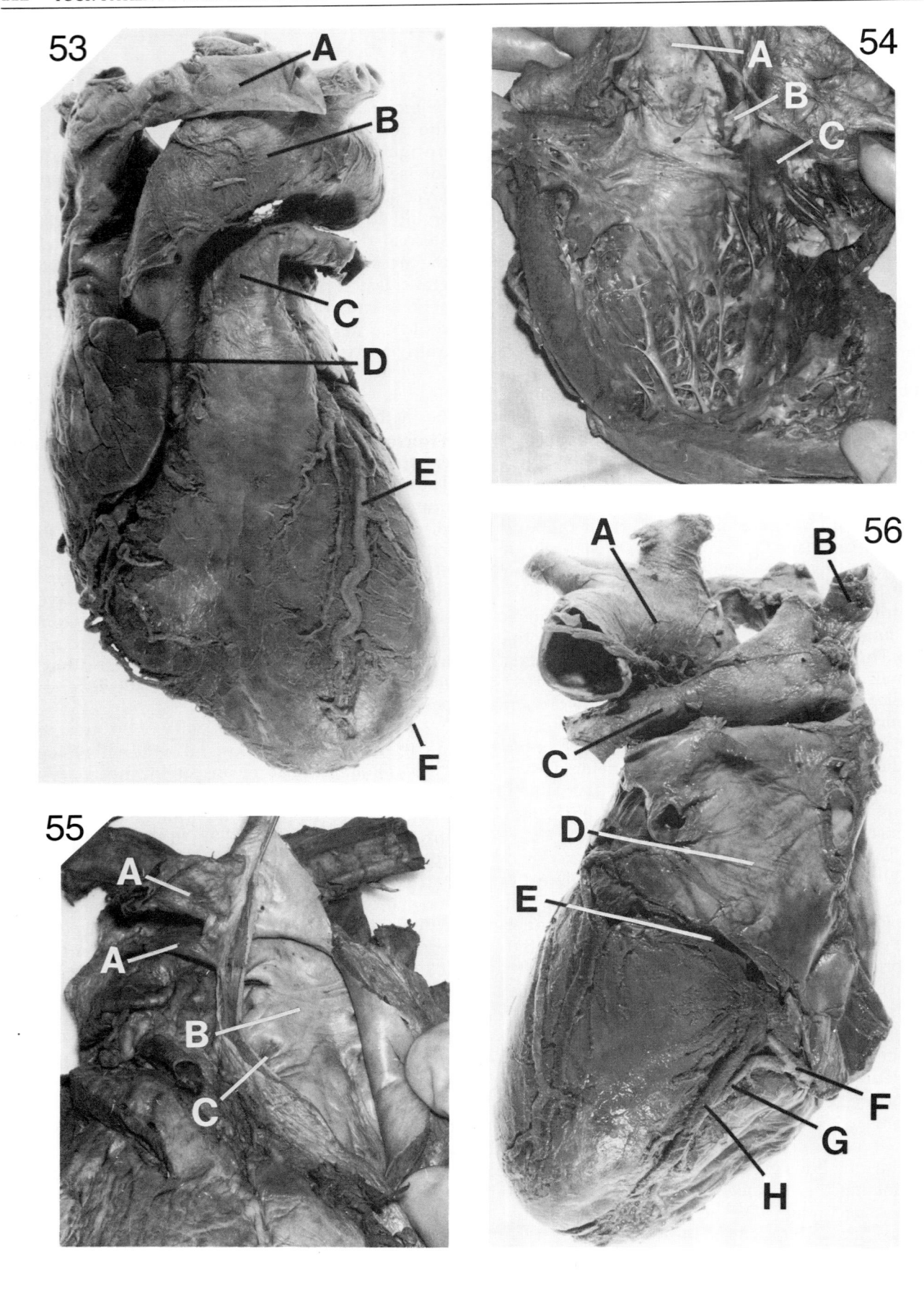

112   YOUR PATIENT'S ANATOMY: A CLINICAL VIEW OF HUMAN MORPHOLOGY

Figure 53. Sternocostal (anterior) aspect of heart.
A. Left brachiocephalic vein
B. Arch of aorta
C. Pulmonary trunk
D. Auricle of right atrium
E. Anterior interventricular artery
F. Apex

Figure 54. Interior of left ventricle, viewed from front and slightly to the left.
A. Commencement of ascending aorta
B. Left cusp of aortic valve
C. Anterior cusp of bicuspid valve

Figure 55. Left atrium opened from behind (see also fig. 60).
A. Left pulmonary veins (reflected upwards)
B. Lumen of left atrium
C. Left atrioventricular communication

Figure 56. Diaphragmatic surface and base of heart, viewed from behind and slightly to the left.
A. Arch of aorta
B. Superior vena cava
C. Left pulmonary artery
D. Left atrium
E. Coronary sinus (opened)
F. Right coronary artery
G. Posterior interventricular artery
H. Middle cardiac vein

fingers will be limited by a fold of visceral pericardium extending from the upper left pulmonary vein on the left to the superior vena cava on the right. All of these veins can be thought of as entering the venous end of the cardiac loop described previously. Thus on invagination of the serous pericardial sac, all of the veins would be tethered within the same sheath of visceral pericardium. In the adult they are therefore enclosed by this sheath, even though the veins are splayed out within the sheath in the semicircle which you have palpated as the limits of the oblique sinus.

On completion of your exploration of the oblique sinus, replace the heart in its normal position, in order to identify the situation of the four cardiac chambers, prior to removing the heart for further study. For this purpose, review the relative positions of the four chambers, referring to point 6 on page 108 and visualizing, on your cadaver specimen, the positions of each of the four chambers. In so doing, you should identify the **coronary sulcus** (corona means ring), separating the atria above and to the right, from the ventricles below and to the left. The **right coronary artery** may be seen passing downwards and to the right, in the right portion of the coronary sulcus, and the left coronary artery can be seen passing up and to the left in the left portion. Similarly, you should identify the **anterior interventricular sulcus,** separating the right from the left ventricles on the anterior surface of the heart, parallel to and about 2 to 3 cm from the left border of the heart. In the anterior interventricular sulcus, the **anterior interventricular artery** (fig. 53, page 112), a branch of the left coronary artery, passes downwards to turn onto the diaphragmatic surface of the heart, slightly to the right of the apex.

Identify the pulmonary trunk and ascending aorta (see fig. 42, page 90), noting that the former passes from the right ventricle, upwards and to the left, crossing anteriorly to the aorta, which leaves the left ventricle and continues posteriorly, upwards and to the right (page 109, point 12). Finally, observe the two small "ear-shaped" appendages, the **right and left auricles** (fig. 53, page 112), lying on either side of the pulmonary trunk, just above the coronary sulcus. (The term auricle means ear.) The auricles are small extensions, or outpouchings, from the right and left atria respectively.

To remove the heart, once again pull the apex and diaphragmatic surface forward, identify the inferior vena cava, and cut through the latter close to where it pierces the pericardium. Cut the superior vena cava about 2 cm above its site of entry into the right atrium, and similarly the pulmonary trunk and ascending aorta about the same distance from their origins, and the pulmonary veins about 1 cm from their terminations.

On removal of the heart, complete your identification of the coronary sulcus on that aspect of the heart previously invisible to you. Note that the heart is conventionally described as having an apex, previously identified, and a base (page 108, point 4) where the great vessels enter and leave superiorly. Thus, by implication, the heart is thought of as resembling an inverted pyramid, the axis of which points downwards and to the left and somewhat anteriorly, and the base of which therefore faces posterosuperiorly and to the right (page 108, point 6). Also by convention, the named surfaces of the heart are the **diaphragmatic,** the **sternocostal** (facing mainly anteriorly), and the **left pulmonary** (facing, as the name implies, the left lung).

Identify the four chambers of the heart and their boundaries with each other. This identification will be facilitated by following the right and left coronary arteries in their course in the coronary sulcus, and the anterior and posterior interventricular branches of the left and right coronary arteries in their courses in the interventricular sulci. Making reference to anatomical specimens and atlas illustrations, identify the branches of the coronary arteries, as well as their accompanying **cardiac veins** and their tributaries. Note that the veins mainly drain into the **coronary sinus** (a vessel resembling a somewhat enlarged vein situated in the posterior portion of the coronary sulcus) (fig. 56, page 112). The coronary sinus drains into the right atrium. Using careful scalpel dissection, study the cardiac veins and arteries, tracing the right and left coronary arteries to their origins from the ascending aorta, and the coronary sinus to its termination in the right atrium.

Making reference to points 8 and 10 on page 109, review the fact that the pulmonary trunk and the ascending aorta are both guarded by valves called the **pulmonary** and **aortic valves,** respectively, which allow outflow of blood during **ventricular systole,** but prevent regurgitation during **ventricular diastole** (page 109). This function, and the mechanism of action of these valves, should now be studied, in the following way.

Make incisions into the anterior walls of the right and left ventricles, parallel to and about a centimeter inferior to the coronary sulcus on either side on the anterior interventricular sulcus. Avoid cutting across the latter sulcus, and its contained anterior interventricular artery, thus retaining each incision within the confines of their respective ventricles. Now carry your heart specimen to a sink, equipped with a piece of flexible tubing attached to a cold water tap. Holding the heart under a stream of running water, wash blood clot from the right ventricle and the pulmonary trunk. Then look into the lumen of the pulmonary trunk and identify the three cusps of the pulmonary valve. A cross-section through the pulmonary trunk at its junction with the exit of the right ventricle is approximately round. The valve which closes this circular area is comprised of three folds of tissue (cusps) which constitute approximately equal segments of that circle, each of the three being capable of folding upwards to allow blood to pass superiorly. Having identified the valve, and holding the heart in such a way that you can keep the valve under constant observation, open the tap, insert the free end of the flexible tube through your insertion in the anterior wall of the right ventricle, and allow water to flow through the ventricle and out of the pulmonary trunk. This will allow you to observe the opening of the valve, as water (simulating the blood flow) gushes past the valve from below. After having observed this phenomenon, remove the flexible tube, and then, holding it above the pulmonary trunk, allow water to enter the trunk from above, under observation. This will enable you to observe the mechanism of closure of the valve, as back pressure in the trunk exceeds the diastolic pressure in the ventricle. Repeat the entire exercise you have just performed, on the left side of the heart, thus observing the similar function of the aortic valves.

Turn now to study the right atrium. The superior and inferior venae cavae enter this cardiac chamber, and identification of these great veins will assist you in delimiting the atrium itself. Recall (page 108, point 6) the situation of the right atrium in respect to the rest of the heart. On the lateral (right) aspect of the right atrium, find a thin, inconspicuous groove, connecting the anterior surfaces of the two venae cavae. This **sulcus terminalis** may be indistinct, or even unidentifiable. However, when present, it is a useful landmark for your incision into the atrium, which should be made just anterior to the sulcus. If it is not visible, make a vertical incision in the lateral wall of the right atrium, from the anterior aspect of one vena cava to the other. The sulcus terminalis is an exterior marking for a structure regularly present internally, the **crista terminalis** (fig. 57, page 116).

Starting at the superior end of your vertical incision, cut anteriorly, into the auricular appendage, following its superior border. Reflect the flap you have thus created in the atrial wall downwards, to allow you to study the interior of the atrium. Remove blood clot from the atrium and wash out the lumen of the cavity under a running stream of tap water. Note that the surface of the atrial wall is smooth posterior to the crista terminalis, but is marked by thin ridges, known as the **musculi pectinati,** anteriorly (fig. 57). Pectinate means like a comb (fig. 57, page 116). This a relic of the embryological fact that the smooth walled, posterior portion has a different origin to the anterior. Find the internal openings of the superior vena cava, and inferior vena cava, the coronary sinus, and the right ventricle. The latter three openings are all guarded by valves, though only that of the opening into the right ventricle has a true functional valve in the adult, the **right atrioventricular (tricuspid) valve** (page 109, point 7). The cusps of this valve are folded into the right ventricle (fig. 58, page 116), and will be studied later in this session. A fold of endocardium (endocardium means innermost lining layer of the heart; endo indicates internal, cardium indicates of the heart) which constitutes the **valve of the inferior vena cava** (fig. 57, page 116) can, however, be seen now. In fetal life, this fold serves the purpose of directing blood from the inferior vena cava through an opening in the wall between the right and left atria (the **interatrial septum**), but it is, as mentioned, functionless in the adult. A similar, but smaller, **valve** lies anterior to the opening **of the coronary sinus.** In the interatrial septum, an oval depression, called the **fossa ovalis,** should be identified (fig. 57, page 116). Above, it is delimited by a slight thickening in the atrial wall, the **limbus.** The fossa represents a more medial (left) layer of the septum, and the limbus represents the inferior, curved edge of a more lateral (right) layer, which, in embryological development, did not quite extend down to completely cover the first layer. The fetal opening between right and left atria was through the crevice between the fossa ovalis and the limbus. In fetal life, this opening is called the **foramen ovale.**

Returning to the incision previously made on the anterior wall of the right ventricle, cut vertically downwards from the upper extremity of that initial incision, thus creating a triangular flap which can be reflected downwards. This incision should pass just to the right of a thick band, which extends from the **interventricular septum** to the anterior wall of the right ventricle. This band is known as the **septomarginal** (or **moderator**) **band.** (However, your vertical incision might cut through the anterior insertion of this band.) Remove and wash out blood clots from the lumen of the right ventricle, and study the tricuspid valve, identifying the ventral, dorsal, and septal cusps (named according to their situations). Observe the thin **chordae tendineae** which attach to the edges of the cusps, and at their other ends are in connection with papillary muscles (fig. 59, page 116). These muscles, by contracting at the

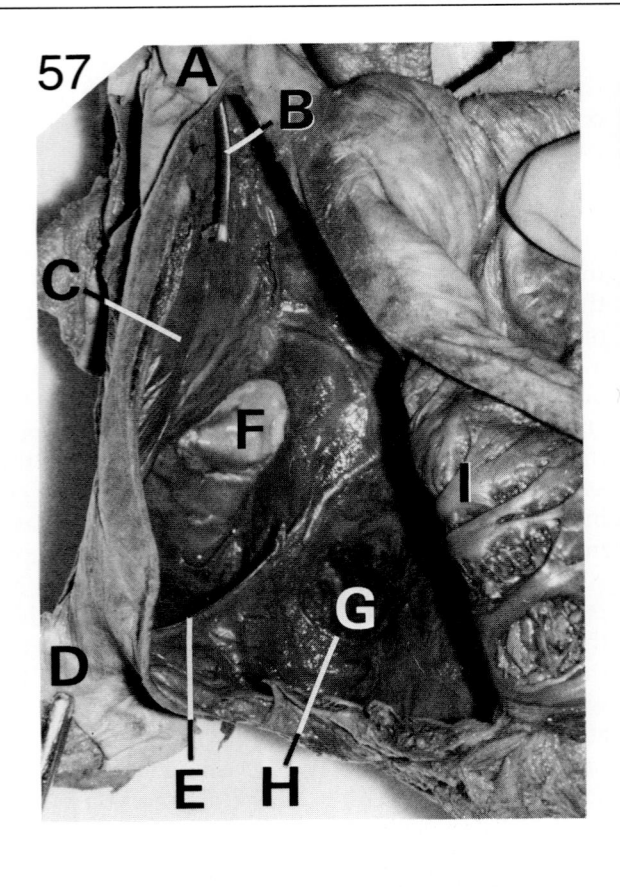

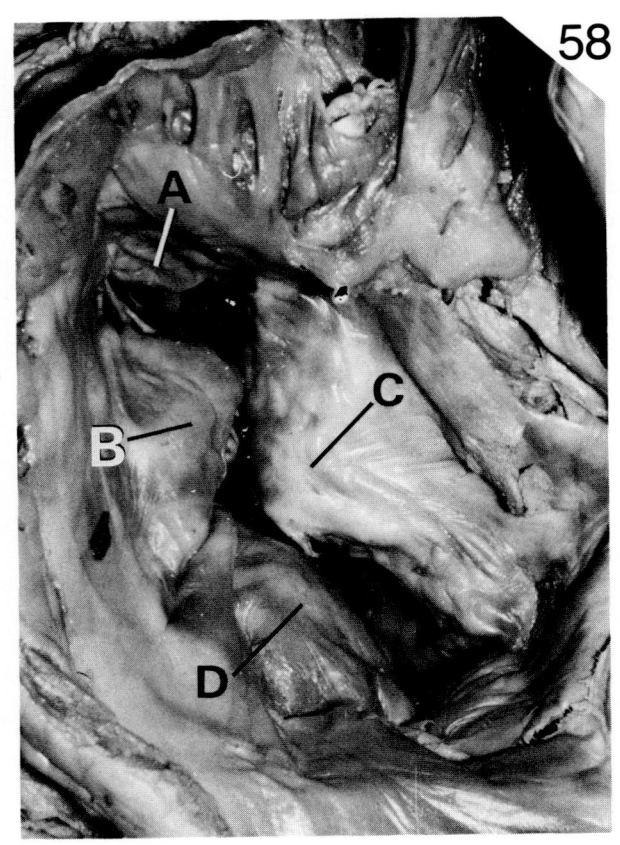

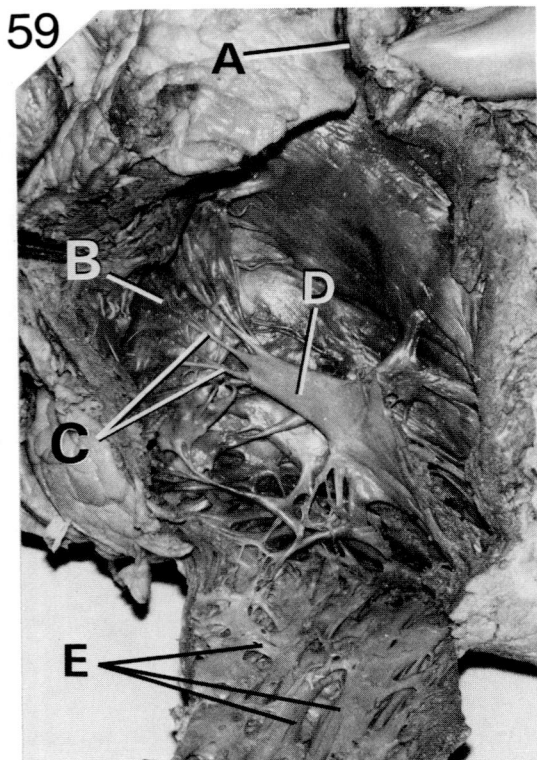

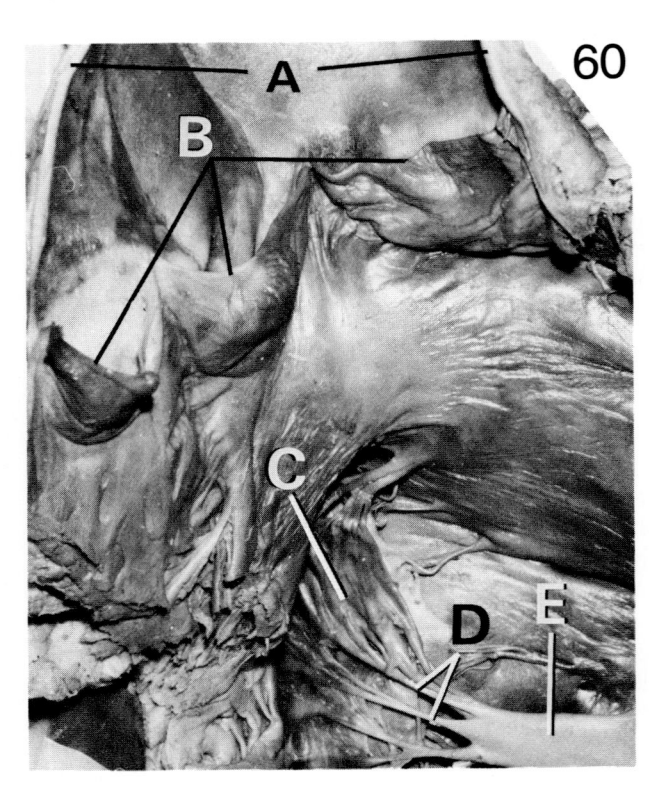

116 YOUR PATIENT'S ANATOMY: A CLINICAL VIEW OF HUMAN MORPHOLOGY

Figure 57. Right atrium, opened and viewed from right.
A. Superior vena cava
B. Thread entering superior vena cava
C. Crista terminalis
D. Inferior vena cava
E. Valve of inferior vena cava
F. Fossa ovalis
G. Coronary sinus
H. Valve of coronary sinus
I. Musculi pectinati

Figure 58. Right atrioventricular (tricuspid) valve, viewed from right atrium.
A. Intermediate (accessory) cusp segment
B. Septal cusp
C. Ventral cusp
D. Dorsal cusp

Figure 59. Right ventricle, opened and viewed from front.
A. Incision into pulmonary trunk (see fig. 60)
B. Ventral cusp of tricuspid valve
C. Chordae tendineae
D. Papillary muscle
E. Trabeculae carneae

Figure 60. Right ventricular outflow tract: region superior to that depicted in Figure 59.
A. Reflected cut edges of pulmonary trunk (see fig. 59)
B. Cusps of pulmonary valve
C. Ventral cusp of tricuspid valve
D. Chordae tendineae
E. Papillary muscle

same time as the rest of ventricular wall, are able to restrain the cusps against the back pressure created in systole, and thus prevent the cusps from being pushed into the atrium and the valve thus becoming incompetent.

Compare the thickness of the right ventricular wall with that of the atrium, and note the **trabeculae carneae** (fig. 59), elevated muscular ridges on the inner surface of the ventricular wall. Superiorly, these give way to a smooth surface which lines the outlet of the ventricle (the **infundibulum**) as this leads into the **pulmonary trunk**. Extend the vertical incision you previously made in the anterior ventricular wall, to open into the infundibulum and pulmonary trunk. This will allow you to study the cusps of the pulmonary valve in detail (fig. 60, page 116). Consult an atlas illustration and identify the individual valves and the related small outpocketings of the pulmonary trunk, called the **sinuses of the cusps.**

Make a vertical incision into the posterior wall of the left atrium, and study it as previously done in the case of the right side. Examine the left surface of the interatrial septum, the musculi pectinati, and the smooth portion of the atrial wall, surrounding the pulmonary veins. Finally, identify the left atrioventricular communication.

Open the left ventricle by making an incision parallel to the anterior interventricular sulcus, letting this incision meet the original one made inferior to the coronary sulcus. Turn the flap thus created downwards and to the left, and study the interior of the left ventricular lumen as previously done with the right ventricle, comparing features of the two ventricles as you do so. This comparison should include an estimate of the thickness of the ventricular wall, as well as a study of the **left atrioventricular (mitral) valve** (page 109), the associated chordae tendineae and papillary muscles, the trabeculae carneae, and the interventricular septum. The two cusps of the mitral valve are named ventral and dorsal respectively. The interventricular septum is mainly muscular, and bulges out toward the right side. Superiorly, the muscle of the septum is replaced by fibrous tissue, (the **membranous portion** of the septum). As on the right side, the outlet of the ventricle, here called the **vestibule** (corresponding to the infundibulum of the right ventricle), is smooth-walled, as it leads to the aortic valves which guard the entry into the ascending aorta. Note that the ventral cusp of the mitral valve is situated between the atrioventricular communication and the vestibule of the aorta (fig. 54, page 112).

Identify the site of the **sinoatrial node** (page 100), at the upper extremity of the cristae terminalis, just anterior to the opening of the superior vena cava. From this node, the rhythmic impulse responsible for contraction of the heart muscle spreads through the atrial wall, to reach a second, similar collection of highly specialized cells, the **atrioventricular node,** situated just superior to the opening of the coronary sinus. This node is in continuation with a strand of similar cellular tissue, the **atrioventricular bundle (of His)** which extends through the membranous part of the interventricular septum to reach the upper edge of the muscular portion of this septum. At this site, it divides into right and left branches, which supply their respective ventricles. On the right side, the branch of supply passes through the septomarginal or moderator band.

The sinoatrial node is the pacemaker of the heart (i.e., as mentioned on page 100, it is the site of initiation of the rhythmic impulse). The specialized cells which the nodes and the bundle and its branches consist of, are known as the **conducting tissue of the heart.**

It should be noted that although the rhythmic contraction of the heart is due to a spontaneously generated impulse, originating in the sinoatrial node, this impulse is under partial influence of the autonomic nervous system. As mentioned previously (pages 100 and 106), parasympathetic (vagal) stimuli tend to slow the rhythm, and sympathetic tend to speed it up. The small nerve branches of the vagus and sympathetic nerves destined to supply the sinoatrial node, and also the blood vessels of the heart, form a plexus, situated between the concavity of the arch of the aorta, and the bifurcation of the pulmonary trunk. This **cardiac plexus** should be identified on your cadaver, just to the right of the ligamentum arteriosum (page 107).

Complete this session by studying the parietal layer of the serous pericardium, as visualized by examination of the internal surface of that part of the fibrous pericardium which now remains in the cadaver after removal of the heart. You will recall (page 111), that in embryological development the heart, represented by a two-channeled tubular loop, invaginates the pericardial sac from dorsally and cranially. The dorsocranial end of the loop is venous, the right channel representing the right atrium and the left channel, the left atrium. The ventrocranial end of the loop is arterial. By placing your palpating fingers posterior to the pulmonary trunk and ascending aorta (page 111), you entered the space between the two ends of the loop (the transverse pericardial sinus). The venous end of the loop having subsequently been severed by your cutting through the superior and inferior vena cavae and the pulmonary veins (page 114), the sheath of serous pericardium which tethered these venous tributaries of the two channels (page 114) can now be observed at the site of that embryological invagination. Study the cut edge of this sheath, tracing it as it surrounds the superior vena cava, the superior and inferior right pulmonary veins, and the inferior vena cava on the right side, and the superior and inferior left pulmonary veins on the left side. Right and left compartments of this sheath will be connected by the double layer of serous pericardium which limited your exploration of the oblique pericardial sinus superiorly (page 114). Thus, inferior to the cut edges of the double layer, you can now study the posterior wall of the oblique sinus.

Clinical Session 12

■ **TOPICS**

Diaphragm
Esophagus
Descending Thoracic Aorta

In this session you should review and complete your study of the mediastinum, its subdivisions, and their contents.

1. Commence by drawing the surface markings of the sternum. Refer to the previous description given on page 98 (point 1). You should mark the suprasternal notch, sternoclavicular joints, sternal angle, sternomanubrial plane and second costal cartilages, xiphisternal plane, and the xiphoid process. Then draw in the manubrium of the sternum and outline the lateral borders of the sternum. The imaginary vertical line through the lateral borders of the sternum (where it is widest) is known as the **lateral sternal line** of each side. The median plane passing through the middle of the sternum is known as the **midsternal line.**

2. Draw in the surface projection of the **diaphragm.** It is a dome-shaped structure, convex upwards toward the thorax. However, this overall dome shape actually comprises three regions, as seen from the front: a right and left minor dome, and a central, flat region between these two domes. The central portion is tendinous and is called the **central tendon of the diaphragm.** Draw it in as a horizontal line passing through the xiphisternal plane and extending just lateral to the lateral sternal line on each side. The **left and right domes** or **cupolae** are muscular. Draw in the right dome as a curved line, convex upwards, commencing medially at the horizontal, central tendon and reaching up to the fifth rib in the midclavicular plane, then curving downwards and laterally to the lateral body wall. The left dome has a similar curve that rises to a point slightly (about 1 cm on average) lower than the right, in the midclavicular line. Draw in the left dome, curving down to the left lateral body wall.

The surface projection you have drawn in is an average. The position of the diaphragm varies considerably between individuals. Furthermore, its position alters constantly with respiration. In inspiration the domes descend, and in expiration they ascend to positions respectively below and above that which you have drawn. Review point 15 of Clinical Session 11, (page 110).

3. You should now mark the horizontal plane known as the **transpyloric plane** (so called because, in the recumbent (supine) position, and in the cadaver, it passes through the level of the **pylorus** of the **stomach,** which you will be studying shortly. Note that anatomical terms usually apply to the human body in the anatomical position. This exception is retained by traditional usage. The transpyloric plane passes through the intervertebral disk between L1 and L2 (i.e., the disk between the first and second lumbar vertebrae). To find this level on the anterior surface of your model, measure with a measuring tape (or, approximately, using a piece of string or even a piece of clothing) the midpoint between the suprasternal notch and the **symphysis pubis** (sym means together; phyein means to grow; see also symphysis menti, page 37). The symphysis pubis is the joint in the median plane between the right and left **pubic bones.** The pubic bones are the bony structures that you can palpate inferiorly in the anterior abdominal wall, as the palpating hand approaches the external genital area. Thus you should first palpate the junction of the pubic bones in the midline, then measure from that point to the suprasternal notch, and then find the midpoint. (Note that the right and left pubic bones comprise one of three bilateral pairs of bones which fuse to form the bony pelvis [page 4]).

The purpose of marking the transpyloric plane at this time is to be able to ascertain the approximate site of the posterior attachment of the diaphragm. Recall that the general, overall shape of the diaphragm is that of a dome, and that its curvature is not only from right to left but also from anterior to posterior. It is attached anteriorly to the inner surface of the xiphoid process and adjacent parts of the rib cage and then arches up backwards and down to attach posteriorly to the anterior surfaces of the upper lumbar vertebrae and associated structures. At the site of this attachment to the first lumbar vertebra, the descending aorta passes through the diaphragm from the thorax to the abdomen.

4. You should now draw in the surface projection of the **thoracic descending aorta.** Recall that it commences in the sternomanubrial plane, where the arch of the aorta has curved to a posterior position. The thoracic aorta should be drawn about 2 cm in diameter, extending from the sternomanubrial

plane, where it lies on the left side of the fifth thoracic vertebra (and therefore about 2 cm from the midline), passing inferiorly and veering gradually toward the midline, to reach it as it pierces the posterior attachment of the diaphragm at L1. The descending aorta is a major content of the **posterior region** of the **inferior mediastinum**. Note that because of the curvature of the diaphragm, the posterior region of the inferior mediastinum extends inferior to the xiphisternal plane. Similarly, the anterior region of the inferior mediastinum extends inferior to the xiphisternal plane anteriorly, whereas the central portion of the inferior mediastinum, which contains the heart and its surrounding membranes, is limited inferiorly by the central tendon of the diaphragm **in** the xiphisternal plane.

5. A second major component of the posterior mediastinum is the **esophagus** (page 101), the portion of the digestive canal which connects the **pharynx** (throat) with the **stomach**. The esophagus pierces the diaphragm at the level of the tenth thoracic vertebra (T10) about 1 cm to the left of the midline and about 2 cm anterolateral to the aorta. Mark this position, approximately midway between the xiphisternal plane and the transpyloric plane.

6. Now mark in at T8 (i.e., just above the xiphisternal plane), 1 to 2 cm to the right of the midline, the position at which the **inferior vena cava** pierces the diaphragm, to pass from the abdomen superiorly to the thorax, to enter the right atrium of the heart. This opening is even more anterior than the esophageal.

7. Review the contents of the **anterior region** of the **inferior mediastinum** (point 10, page 99) by marking the anterior borders of the lungs (page 95). Review also the contents of the central portion of the inferior mediastinum (pages 99 and 100, points 6 through 10).

8. Finally review the structures of the **superior mediastinum** (pages 98 and 99, points 2 through 5, and page 100, points 11 through 13). In this latter review you should redraw the arch of the aorta, its three great branches, and the great veins lying anterior to these. These are the right and left brachiocephalic veins and the superior vena cava, the latter extending down to the right atrium (refer to fig. 53, page 112).

Review the emphasized key concepts of this session before regarding it as completed.

Laboratory Session 12

In order to further study the structures of the superior mediastinum, it will now be necessary to expose these, by reflecting the manubrium of the sternum and the medial portions of the clavicles and first ribs.

The clavicle is part of the **upper limb girdle.** This term embraces the clavicle and scapula, and refers to the fact that these bones are concerned with attachment of the upper limb to the trunk. For this reason, the clavicle, although related to the thoracic wall, is more properly studied with the upper limb. Therefore, in the ensuing dissection, the sternoclavicular joint (page 95) will be preserved for further study with the upper limb.

The first ribs, together with the first thoracic vertebra posteriorly and the manubrium of the sternum anteriorly, constitute the **thoracic inlet,** the superior opening into the thoracic cage, from the neck region above, and from the upper limb laterally. The structures passing between the thorax and the neck, as well as structures passing between the thorax and the upper limb, traverse this inlet. As examples, the brachial (upper limb) and cephalic (head) blood vessels leave and enter the thorax through this opening (pages 98 to 100). Recall that the brachiocephalic trunk, for example, divides behind the right sternoclavicular joint (page 98 and 99). The cephalic component (the common carotid artery) then ascends toward the head, whereas the brachial component (the subclavian artery) crosses the superior surface of the first rib, to pass toward the upper limb. The left common carotid and subclavian arteries have similar relationships, although they come off the aortic arch independently.

With these facts in mind, you should now use the electric saw, under guidance, to split the manubrium of the sternum in the median plane. Cut through the clavicle just lateral to the attachment of the clavicular head of sternocleidomastoid muscle (page 49, and fig. 21, page 50). This will allow you to reflect the half manubrium sterni, and the medial half of the clavicle, together with the attached clavicular and sternal heads of sternocleidomastoid. However, it will first be necessary to cut through the first rib, since this structure still retains the manubrium in position. Therefore, reidentify the subclavian artery on each side (pages 97 and 99). You have previously observed the only thoracic branch of the subclavian artery (page 91). Trace the subclavian artery as it curves from its origin at the bifurcation of the brachiocephalic trunk (fig. 42, page 90), behind the right sternoclavicular joint or, on the left side, from its origin from the arch of the aorta, curving upwards and laterally to pass behind the site of the sternoclavicular joint on the left. On both sides, establish, by palpation, as well as careful dissection, the relationship of the subclavian artery to the superior surface of the first rib (page 99). Just anterior to the subclavian artery a small muscle will be found, known as the **scalenus anterior.** This muscle slopes downwards and laterally from cervical vertebrae to insert into the first rib just in front of the subclavian artery. Anterior to the muscle, the subclavian vein also crosses the first rib. Cut through the first rib just anterior to the subclavian vein on either side.

Carefully free the bisected manubrium and the medial ends of the clavicles and first ribs from surrounding tissues, and reflect these bony structures upwards and laterally on either side. This dissection will involve your cutting through the small **subclavius muscle,** extending from the inferior surface of the clavicle to the first rib. This muscle lies in a plane of fascia (the **clavipectoral** fascia), which also splits inferiorly to enclose the pectoralis minor muscle (page 87).

Immediately deep to the manubrium sterni, study the remnants of the **thymus gland,** an important component of the immunolymphatic system (page 6). This diffuse gland is large in childhood, and regresses in adulthood, which is the reason why you are likely to only see a reduced remnant of it in your cadaver. Deep to the thymus, dissect and review the great vessels of the superior mediastinum (pages 98 and 99). You should study the arch of the aorta and its branches, the brachiocephalic trunk and left common carotid and subclavian arteries, and the right subclavian and common carotid arteries (fig. 53, page 112). In addition, examine the accompanying veins of these arteries and the junction of the brachiocephalic veins to form the superior vena cava (fig. 42, page 90). In the groove posterior to the common carotid artery and internal jugular vein, identify on each side, the trunk of the **vagus nerve** (page 56). Trace its course downwards, anterior to the subclavian artery and into the thoracic cavity. On the right, it attains the lateral surface of the trachea, on which it slopes posteriorly to reach the posterior aspect of the root of the lung. This part of its course has previously been studied (page 101 and fig. 45, page 102). Note, as the nerve passes in front of the subclavian artery, the large **recurrent laryngeal** branch which the vagus nerve gives off (page 107). The recurrent laryngeal curves backwards, upwards, and medially to pass posterior to the subclavian artery (fig. 45) and ascend in the groove between the trachea and the esophagus, to attain the inferior edge of the inferior constrictor of the pharynx, below which it "enters" (page 77).

On the left, the vagus, after crossing the subclavian artery, reaches the lateral surface of the arch of the aorta (fig. 46, page 102). From here it, too, passes posteriorly to reach the posterior aspect of the root of the lung. The recurrent laryngeal branch of the left side also recurves upwards, medially, and posteriorly, as on the right. The vessel that it curves up and behind is the arch of the aorta. In so doing, the nerve passes immediately to the left of the **ligamentum arteriosum** (page 107). Its further ascent toward the larynx is similar to that of the right side.

The **phrenic nerve** originates on either side from the ventral rami of the third, fourth, and fifth cervical nerves. The nerve passes inferiorly, crossing the anterior surface of the scalenus anterior muscle (page 122), thus lying posterior to the subclavian vein as it approaches the thoracic inlet. Find the phrenic nerves at this site on each side and follow their further course into the thorax. On the right, after passing behind the subclavian vein, the phrenic nerve reaches the lateral aspect of the right brachiocephalic vein and then the superior vena cava, whereafter it passes downwards, on the lateral aspect of the pericardium, to reach the diaphragm (page 106). On the left, the nerve passes onto the arch of the aorta, crossing in front of the vagus nerve. It then proceeds downwards, anterior to the root of the lung, in a course similar to that of the right side, to reach the diaphragm.

Reidentify the **thoracic duct** on the left side (page 107 and fig. 46, page 102). Trace the duct upwards, as it passes on the left aspect of the esophagus, to attain the posteromedial (right) surface of the arch of the aorta. The duct passes upwards, to then curve forwards and laterally, and end in the left internal jugular vein, subclavian vein, or junction of these, at the formation of the brachiocephalic vein.

Review the sympathetic trunk on both sides (page 106 and figs. 45 and 47, page 102). Identify the greater, lesser, and least splanchnic nerves, and trace them down to their passages through the diaphragm. Recall (page 106) that these represent preganglionic fibers of the sympathetic nervous system, which synapse in minor ganglia in the abdomen.

The segmentally arranged **posterior intercostal arteries** (pages 92 and 86), branches of the descending aorta, should be traced out into the left and right intercostal spaces. Branches to the right side pass posterior to the esophagus. The intercostal veins of the right side should be traced to their termination in the azygos vein. Those of the left side should be traced medially, and their confluence into the **inferior hemiazygos** (also known as the hemiazygos) and the **superior hemiazygos** (also known as the accessory hemiazygos) veins should be observed, making use of atlas illustrations to guide you in your identification of these latter two equivalents of the azygos. The further passage of these two veins across the midline, posterior to the aorta and the esophagus, to drain into the azygos vein on the right side of the body, should be studied. Drainage of intercostal veins from second and third interspaces into the left superior intercostal vein, which crosses the arch of the aorta to reach the left brachiocephalic vein, should be observed.

Study the esophagus, reviewing (pages 77 and 121) its superior and inferior connections, and the vertebral levels at which these occur. Reexamine the important relationships of the neighboring organs, replacing the lungs and heart to do so. Examine the clinically important points of narrowing, at which foreign bodies may obstruct the lumen, at its commencement, at the site where the left bronchus crosses it and at its passage through the diaphragm. The relationship to the heart is also of clinical significance: enlargement of the latter organ can be diagnosed on barium contrast x-ray examination of the esophagus. Which chamber of the heart would be most easily examined by this technique?

Observe the nerve and blood supply of the esophagus. The esophageal plexus of nerves is formed mainly by the vagus nerve, with contributions from sympathetic branches. Arterial supply is mainly from the descending aorta, though the lower portion receives some supply which ascends from the abdomen. The venous drainage is mainly to the azygos system and thus ultimately to the inferior vena cava (the "caval" venous system). However, as in the case of the arterial supply, the venous drainage of the lowest (abdominal) portion of the esophagus is associated with that of other abdominal digestive organs. Therefore, at this site, there is a dual venous drainage, with a potential for anastomosis between them. The venous drainage of the digestive system is highly specialized, in that it does not pass directly to the heart. Instead, as will be studied shortly, it drains to the liver, where the products of digestion are further processed. This is known as the "portal" venous system (portage means to carry from one waterway to another). Because certain products of digestion if not properly processed in the liver can be toxic to the nervous system, "short-circuiting" of the liver through portal-caval venous anastomoses carries the danger of toxicity to the nervous system. Such anastomoses develop when the pressure in the portal system rises abnormally, for example in such liver diseases as that caused by chronic alcoholism. In addition, increased pressure in the veins of the portal system can lead to esophageal bleeding.

Examine the curvature of the domes of the diaphragm and study the sites of attachment of the diaphragm to the sternum, ribs, costal cartilages, vertebrae, and adjacent structures. Study the structures which pass through the diaphragm between the thorax and abdomen and verify the descriptions previously given: the descending aorta (page 120, point 4), the esophagus (page 121, point 5), the inferior vena cava (page 121, point 6), the thoracic duct (page 106), and the splanchnic nerves (page 106).

Complete your study of the mediastinum by reviewing its subdivisions and the contents of each of these.

Clinical Session 13

■ **TOPICS**

Abdominal Regions
Abdominal Wall

In this session you will study bony and muscular features of the body wall in the abdominal region. Recall that the abdomen is one of the five major regions into which the human body may be divided. For this session your model should be stripped to the waist, wearing bathing trunks, bikini, or underwear. Work in a well warmed room.

1. The **abdominal region,** as defined previously, includes the **pelvis** (containing the organs of reproduction and excretion), and **perineum** (containing the external genital organs, and the external openings of the above mentioned two systems). Thus the full extent of the abdominal region is from the boundary with the thorax, superiorly, to the most inferior extent of the perineum. Review the brief summary given on page 4.

You will recall that the abdomen is limited from the thorax superiorly by the diaphragm, which attaches (page 120) anteriorly to the posterior surface of the xiphoid process, and posteriorly to the upper lumbar vertebrae. Palpate the xiphoid process, and, posteriorly, the first two lumbar vertebrae. You should identify the latter by counting the vertebral spines from above, in the **median spinal furrow** (fig. 64, page 128). Mark the spine of L1 with a wax skin marker. On the anterior skin of the abdomen, mark the transpyloric plane. Its position should be identified, as previously (page 120, point 3), by finding the midpoint between the symphysis pubis and the suprasternal notch. Confirm, by comparison with your marking of the spine of L1, that the transpyloric plane passes just inferior to the latter (i.e., at the level of the intervertebral disc between L1 and L2).

2. Palpate the costal margin (page 86). Mark the most inferior point of the margin on left and right sides, and then draw a horizontal line through these points. This line marks the **subcostal plane,** which passes through the body of L3. Which costal cartilages did you palpate in finding the lower edge of the costal margin? (see page 86).

3. Beginning in the midaxillary line (page 85), palpate the abdominal musculature inferior to the subcostal margin, and proceed downwards until your palpating fingers encounter the hip bone. The superior edge of this bone is known as the **crest (edge)** of the **ilium (iliac crest).** The ilium of each side is the second of the three paired bones which fuse to form the bony pelvis (page 4). The pubic bones (page 120) have previously been encountered.

4. Find, by palpation, the highest (i.e., the most superior) point of the iliac crest on each side. Draw a horizontal plane running through these two highest points. This plane is known as the **intercristal plane,** which passes through the body of L4.

5. Slightly anterior to the highest point on each iliac crest, a thickening of the bone of the crest can be felt. This is known as the **tubercle** of the **iliac crest.** Draw in the horizontal plane which passes between the tubercles of the right and left iliac crests **(the intertubercular plane).** This plane corresponds to the level of L5. Confirm this by palpation of the vertebral spines, in the posterior midline.

Precise identification of the fourth and fifth lumbar vertebrae has a practical significance in the performance of **lumbar puncture.** This procedure is used clinically, to obtain a sample of **cerebrospinal fluid (CSF),** which is made in the **ventricles** of the brain. Recall (page 5) that the central nervous system (CNS) is tubular. In the brain, the lumen of the tube is subdivided into regions called ventricles. CSF leaves the ventricles to enter a space which surrounds the entire central nervous system. Because it is closed in by the **arachnoid** membrane, this space is called the **subarachnoid space.** Because the spinal cord of an adult extends down no further than the transpyloric plane, but the subarachnoid space extends down to the third sacral level (S3), a specimen of CSF can be obtained by lumbar puncture by inserting a needle into the subarachnoid space between L4 and L5 (or between L3 and L4).

6. You have previously used the **pubic symphysis** (page 120), the joint between the **pubic bones,** as a landmark to find the transpyloric plane. Review palpation of the pubic bones and their symphysis by palpating in the anterior abdominal wall, from the **umbilicus** (navel or belly button) progressively inferiorly toward the perineum. The palpating hand encounters a bony prominence with a palpable midline groove between the right and left (pubic) bones. Consult atlas illustrations to assist you in visualizing

the features described here and in the next two points, as you perform the prescribed palpation.

7. Superiorly, the pubic bone presents, at its anterior edge, a palpable ridge known as the **pubic crest,** which can be felt extending laterally for about 2 cm from the symphysis.

8. At the lateral extremity of the pubic crest, a small, rounded elevation, about 2 mm in diameter, can also be palpated. This is known as the **pubic tubercle.**

9. Palpate again the iliac crest, and follow it as far anteriorly as you can, where you will palpate its rounded anterior extremity, the **anterior superior iliac spine (ASIS).** The horizontal plane through the right and left ASIS is the **interspinous plane,** which passes through the level of S2.

10. Between the ASIS and the pubic tubercle, a groove is seen in the skin, demarcating the inferior border of the abdomen from the upper border of the thigh. This groove is known as the **inguinal sulcus.** Deep to the inguinal sulcus, the **inguinal ligament** extends from the ASIS to the pubic tubercle. The inguinal ligament is mainly formed by the lowermost fibers of the **aponeurosis** (flat tendon) of the **external oblique muscle of the abdomen.** In muscular subjects, the external oblique muscle is seen as a bulge, above the iliac crest, on either side (fig. 61, page 128). Superiorly it attaches to the lower eight ribs, by digitations which **interdigitate** with serratus anterior (pages 85 and 87; fig. 34, page 82 and fig. 63, page 128).

It is of morphological interest that the muscles of the abdominal body wall are, like those of the thoracic body wall, arranged in three layers (page 81). Review the names of the muscles of the three layers in the thorax. In the abdomen the three layers are known as the **external oblique,** the **internal oblique,** and the **transversus muscles** of the abdomen. These three layers furthermore correspond to the three **scalene muscles** of the neck, of which you have encountered one, **scalenus anterior,** in relation to the subclavian artery (page 122).

11. Now palpate the iliac crest as far back posteriorly as you can, until, at its posterior extremity, you reach the **posterior superior iliac spine (PSIS),** 3 to 4 cm from the midline. Overlying the PSIS, a more or less obvious dimple is very often seen (fig. 64, page 128). Thus the position of the PSIS can, in such subjects, be identified without palpation, and merely confirmed by palpating with a single finger in the dimple. A horizontal plane passing through the PSIS of each side passes through the second sacral spine (S2). Thus ASIS and PSIS are on the same horizontal plane (see point 9). Confirm this by palpation of the sacral spines, using the spine of L5 as your starting point, identified by means of the intertubercular plane. The five **sacral vertebrae** are fused together to form the **sacrum** (page 4).

12. Observe in the anterior abdominal wall a vertical sulcus (groove) in the anterior midline, extending from the tip of the xiphoid process to the pubic symphysis. The sulcus is due to a deep-lying fibrous band, the **linea alba,** which connects these two skeletal structures.

13. About 4 cm lateral to the sulcus of linea alba, a groove, curved with a slight convexity laterally, is seen on either side (especially in lean subjects). This groove is called **linea semilunaris** (fig. 33, page 82). It extends from the ninth costal cartilage to the pubic tubercle, and marks the lateral edge of a vertical muscle, the **rectus abdominis** (page 87), which extends from the costal margin and xiphoid process to the pubic crest. You have previously, in your dissection of the anterior thoracic wall, observed the medial branch of the internal thoracic artery, the **superior epigastric artery;** (fig. 43, page 90) pass into the fibrous sheath of the rectus abdominis muscle, posterior to the muscle, and noted that, in this plane, it anastomoses with the **inferior epigastric artery** (page 92 and fig. 62, page 128).

14. Draw in on your model the vertical planes passing tangentially through linea semilunaris on each side. These planes, known as the **left** and **right lateral planes,** also pass through the midpoint between the symphysis pubis and the ASIS. This midpoint is known as the **femoral point,** because it marks the surface projection of the **femoral artery.** Palpate the pulse of the femoral artery over this point.

15. Extended superiorly, the left and right lateral planes pass through the tips of the ninth costal cartilages on each side, in the transpyloric plane.

16. For descriptive purposes, the abdominal region is divided into 9 subsidiary regions, as viewed from the anterior aspect. The subcostal plane (L3) and the intertubercular plane (L5) create upper, middle, and lower regions, and the lateral planes create left, central, and right regions. The upper three regions are called the **right hypochondriac,** the **epigastric,** and the **left hypochondriac** regions. The regions of the middle zone are called the **right lateral** region (also known as the **right lumbar** region), the **umbilical region,** and the **left lateral region (left lumbar**

region). The regions of the inferior zone are the **right inguinal (right iliac), pubic** (also known as **hypogastric),** and **left inguinal (left iliac) regions.**

17. The fundamental pattern of segmentation of the human body (pages 81 and 86) is most fully retained in the thoracic region, and is considerably modified in other parts of the body. For this reason, a thorough understanding of the concept is obtained by studying its expression in the segmentally arranged components of the thoracic region. In the anterior abdominal wall, a trace of segmentation can be observed in the rectus abdominis muscle. With your model lying on his or her back, and raising the lower limbs (held straight at the knees) a few centimeters, palpate the **rectus abdominis muscle** at the level of the umbilicus. You will be able to feel a horizontal depression in the muscle, caused by a **tendinous intersection** (fig. 33, page 82) which interrupts the muscle and attaches it to the anterior wall of the muscle's sheath. The rectus abdominis muscle of each side is enclosed in a tough **fibrous sheath.** The anterior and posterior layers of the sheath of right and left muscles meet in the midline to form the **linea alba** studied in point 12 above. At the lateral edge of each muscle, the anterior and posterior layers of the sheath again fuse with each other, and also with the anterior continuations of the three muscle layers constituting the rest of the abdominal muscular wall. The superior epigastric artery lies within the rectus abdominis sheath, posterior to the muscle and anterior to the posterior layer of the sheath (see point 13). In addition to the tendinous intersection palpable at the level of the umbilicus, two other such tendinous intersections can also be observed in the muscle, situated between the umbilicus and the muscle's superior attachment to costal cartilages (fig. 33, page 82). Attempt to palpate all three tendinous intersections. In lean individuals these intersections may, in addition to being palpable, also be visible. In as much as their existence is a remnant of the segmental pattern of development, it is evident that the four portions of the rectus abdominis muscle between the three intersections develop from different **body segments:** the uppermost from T6, the second from T7 and T8, the third from T9 and T10, and the lowest from T11 and T12. This implies that thoracic nerves sweep around in the abdominal wall, supplying the latter. The nerve supplying the umbilical region is that of T10.

18. You have previously noted (page 94, point 5) a difference between males and females with respect to one observable feature of the abdominal wall, the functional trait of abdominal respiration which is more emphasized in men, whereas thoracic respiration is more emphasized in women. Reconfirm this point on your model. Other sexual differences exist in the anterior abdominal wall. Note that the abdomen is relatively **broader** below the level of the umbilicus **in females,** in conformity with shape and size requirements of the **pelvis** for the **function of childbearing.** Another obvious sexual difference is in the pattern of distribution of **pubic hair.** In both sexes, the upper limit of this hair is in an approximately horizontal line a few centimeters above the level of the pubic symphysis. However, in males, this line tends to be convex upwards, and the hair tends to extend in a peak toward the umbilicus. In females, the upper **hairline** tends not to extend in such a peak, and may even be concave upwards.

19. A clinically very important sexual difference exists in the inguinal region. Reidentify the pubic crest and tubercle (points 7 and 8 above) and the inguinal sulcus and ligament (point 10). Immediately superior to the medial one third of the inguinal ligament, a deficiency exists in the anterior abdominal muscles, through which the testis evaginates in fetal life. (E means out; vagina means sheath). **Evagination** is the converse of invagination, and implies acquiring a sheath-like covering by pushing **out of** a sac-like structure. By comparison, **invagination** implies acquiring a sheath by pushing **into** a sac-like structure. Prior to the testis traversing the body wall, a covering for it, comprising attenuated portions of the anterior abdominal muscles and skin is evaginated. This covering pouch, into which the testis subsequently descends anteriorly, is called the **scrotum** (see fig. 89, page 176). Its evagination is accompanied by and possibly caused by the growth into it of a fibromuscular band, the **gubernaculum,** which extends from the testis, in the abdomen, to the scrotum. As the testis descends the gubernaculum shortens, evidently playing a role in the descent. The deficiency of the abdominal wall which results from the above-mentioned evagination is known as the **inguinal canal.** This canal passes through the wall obliquely. It commences at an internal opening, the **internal (deep) inguinal ring,** situated about 1 cm superior to the femoral point (page 126). From there the canal slopes downwards, medially, and forwards to the **external (superficial) inguinal ring,** situated at a point medial to the pubic tubercle, just above the pubic crest. In adult life, the communications between the testis and internal, pelvic structures pass through this deficiency in the anterior abdominal wall. These communications, collectively called the **spermatic cord,** include the duct which carries the sperm cells from the testis into the pelvis, and the arteries, veins, lymphatic vessels, and nerves which supply the contents of the scrotum.

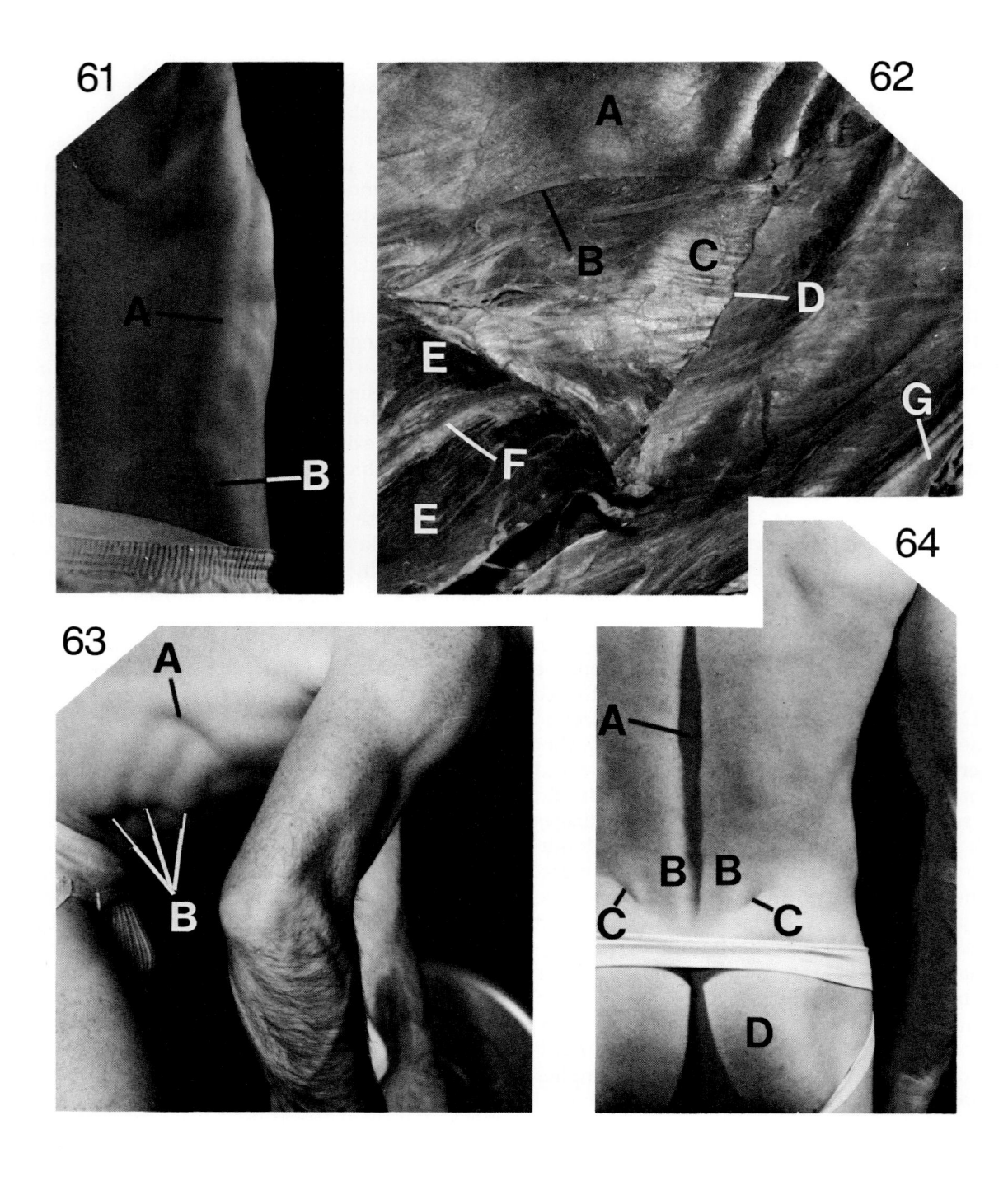

128 YOUR PATIENT'S ANATOMY: A CLINICAL VIEW OF HUMAN MORPHOLOGY

Figure 61. Right posterolateral aspect of trunk.
A. Lateral border of latissimus dorsi muscle
B. Posterior edge of abdominal external oblique muscle

Figure 62. Dissection of left side of anterior abdominal wall (viewed from front).
A. Posterior layer of rectus sheath
B. Arcuate line
C. Transversalis fascia
D. Cut edge of anterior layer of rectus sheath
E. Left rectus abdominus muscle (reflected downwards)
F. Left inferior epigastric artery
G. Left inguinal ligament

Figure 63. Right aspect of trunk, in "bent over rowing" motion.
A. Lateral border of latissimus dorsi muscle
B. Digitations of abdominal external oblique muscle

Figure 64. Posterior aspect of trunk.
A. Median spinal furrow
B. Sacrospinalis muscles
C. Dimples over posterior superior iliac spines
D. Gluteus maximus muscle

In females, the outer (large) lip of the external genitalia, the **labium majus** (see fig. 90, page 176) is the **homologue** (morphological equivalent, in the sense of common developmental and therefore genetic origin) of the scrotum. Labium means lip; majus means large. In the female a defect also exists in the anterior abdominal wall, and a structure called the **round ligament of the uterus** passes through it from the abdomen to the labium majus. However, unlike the spermatic cord of the male, this structure is not a functional organ, but is merely a **vestigial** portion of the **gubernaculum** (page 127). (Vestige means remnant of a structure present in embryonic life but not further developed in the adult.) Because its content is vestigial, the inguinal canal is much smaller in females.

Now palpate the opening which, as will be evident from the previous descriptions, is easier to feel in males than in females. In males, using a palpating index finger, invaginate the scrotum gently and push the finger gently upwards to the pubic tubercle. Medial to the tubercle, the pubic crest and the spermatic cord itself are easily palpable, the latter being freely mobile. Proceed to palpate laterally, superior to the medial third of the inguinal ligament, identifying the ligament itself. With the finger just superior to the ligament, at the site of the **internal inguinal ring** (page 127), ask your model to cough. You will feel an impulse on your finger at the site of deficiency in the abdominal musculature. In lean females, the internal inguinal ring can be palpated by gently invaginating the skin superolateral to the labium majus. In other females, the cough impulse may be felt by palpation without invagination of the skin. In both sexes the impulse may be visible as well as palpable.

As implied by the fact that an impulse is felt on coughing, any increase in the pressure within the body cavity will place great stress on the weak sites of the body wall, such as the inguinal canal. In such states of increased pressure, structures within the abdominal cavity may be forced out through the inguinal canal and other such sites. Such a protrusion of an internal organ (a **viscus** [plural **viscera**]) through a defect in the walls surrounding its natural position is known as a **hernia.** As implied by the above description, inguinal hernias are more frequent in males than in females.

20. A frequently used clinical test of the integrity of the nervous system involves **reflexes** (page 26) of muscles of the anterior abdominal wall. Your model should be in the **supine** position (i.e., lying on his or her back). To test for the so-called **"abdominal reflexes,"** you should use a fairly sharp but not abrasive instrument, such as the blunt end of a (clean) pair of forceps, the blunt end of a pen or pencil, etc. Stroke the skin of the abdomen horizontally from the lateral plane toward the midline, in the epigastric, umbilical, and hypogastric regions. In each case, contraction of the deep-lying muscles will cause them to deviate the linea alba toward the side stimulated by the stroking action. In the three regions mentioned, you would be testing the integrity of the nervous pathway that extends from the **motor area of the brain** (i.e., the area controlling voluntary motion or movement). Nerve fibers originating from this area pass through the **pyramidal tract** (pathway) from the brain, through the **spinal cord** and, through synaptic connections, to the **peripheral nerves** which begin as **intercostal nerves** and end in the anterior abdominal wall. The specific segmental nerve pathways being tested in this procedure are: T6, T7, and T8 in the epigastric region, T9 and T10 in the umbilical region, and T11 and T12 in the hypogastric region (point 17, page 127). In lesions of the pyramidal tract, these reflexes are absent.

Complete this session by reviewing the emphasized key words.

Laboratory Session 13

In the ensuing dissection of the anterior abdominal wall, the left side should be skinned according to the technique described on page xv, and illustrated in fig. 5, page 16. Superficial nerves and vessels should be preserved, and their segmental pattern demonstrated. On the right side, skin and subcutaneous tissue should be removed together, in order to allow rapid access to the deeper layers, (i.e., the abdominal musculature). Begin, then, by making a vertical skin incision from the xiphisternum to the symphysis pubis. On each side, skirt around the external genitalia, passing onto the medial aspect of the thigh and then onto the anterior aspect, 2 cm inferior to the inguinal sulcus (page 126, point 10) as you extend the incision out to the midaxillary line. From the incision previously made when skinning the thoracic wall (page 86), start at the junction of the costal margin and the midaxillary line. Make a vertical incision downwards, to meet the one already completed. Reflect skin on the two sides, as indicated.

On the left side, dissect and identify, with the help of atlas illustrations, the ventral rami (intercostal nerves) of T7 to T12, which, on reaching the anterior end of their respective intercostal spaces, continue into the anterior abdominal wall (page 127, point 17; page 130, point 20). In addition, ventral ramus L1 lies in sequence in the lower part of the abdominal wall where it divides into two branches: the **iliohypogastric** and **ilioinguinal** nerves. Recall (page 86) that the **neurovascular plane** of the body wall lies between the innermost and middle layers of musculature. As in the case of the intercostal nerves in the thoracic wall (page 86) the abdominal ventral rami become cutaneous as they approach the midline, and also give off lateral cutaneous branches. Superficial arteries and veins will also be encountered in this dissection.

On the right side, display the **external oblique muscle** of the abdomen (page 126; figs. 61 and 63, page 128; and fig. 34, page 82). Follow the muscular fibers downwards and inferiorly, and observe their transition into the external oblique aponeurosis. Carefully dissect superficial fatty and connective tissue from the aponeurosis, down to the **inguinal sulcus** at which level the aponeurosis, by folding inwards, like a gutter, forms the **inguinal ligament** (page 126). Follow the ligament down to its attachment to the pubic tubercle, and carefully display, in the male, the spermatic cord (page 127, point 19) as it emerges through the **external inguinal ring** (page 127) covered by a thin layer of tissue derived from the external oblique aponeurosis. The external inguinal ring is, in fact, the site in the external oblique aponeurosis where the spermatic cord protruded in fetal life; the thin layer derived from the aponeurosis is thus attached to the edges of the ring. This layer is usually called the **external spermatic fascia**. However, since a homologous structure exists in the female, it should be renamed the **external inguinal fascia**.

Superior to the external inguinal ring, trace the external oblique aponeurosis medially. In the lower part of the abdomen, a tough fibrous layer lies anterior to the **rectus abdominis muscle** (page 126, point 13; page 127, point 17), and this anterior layer of the **rectus sheath** is formed by the fusion of the medial ends of aponeuroses of all three muscle layers of the abdominal wall: the **external oblique, internal oblique,** and **transversus** muscles of the abdomen (page 126). At this site (i.e., **lower** abdomen) there is no posterior fibrous layer associated with rectus abdominis. However, above the level of the **umbilicus**, the aponeurosis of the middle muscle layer of the abdominal wall, the internal oblique, splits into an anterior and posterior layer, which respectively pass anterior and posterior to the rectus abdominis. Thus at **mid**-abdomen and above, the rectus sheath has anterior and posterior layers. At this level, the aponeurosis of external oblique reinforces the anterior layer of the rectus abdominis sheath, and the aponeurosis of the transversus abdominis reinforces the posterior layer (fig. 62, page 128). In the midline, the anterior and posterior sheaths meet and fuse, and meet with those of the opposite side in the linea alba (page 126, point 12, and page 127, point 17). Thus, in summary, the rectus abdominis muscle, in its upper portion, is surrounded by a sheath anteriorly and posteriorly, but in its lower extent, the posterior layer of the sheath is absent.

Identify, by inspection and palpation, linea semilunaris (page 126, point 13), the lateral edge of the rectus abdominis muscle. Returning to the external inguinal ring (pages 127 and 131), detach the external spermatic fascia from the ring, and separate the superior edge of the ring (i.e., external oblique aponeurosis) from the deep-lying internal oblique, by scissors dissection (page xv). Make a vertical scissor cut upwards from the superior edge of the ring, extending it superiorly by alternately separating external oblique aponeurosis from the deeper layer, and cutting it, until you have a vertical incision through the aponeurosis, parallel to linea semilunaris. Reflect the medial flap of the external oblique medially, demonstrating its fusion into the anterior layer of the rectus sheath. Reflect the lateral flap of the external oblique aponeurosis laterally, exposing the internal oblique muscle and aponeurosis. Note the direction of muscle fibers of internal oblique. Study the contribution of the internal oblique to the coverings of the spermatic cord (page 127, point 19). In addition to contributing fibers to the **cremaster muscle,** which extends over the cord, other fibers of the internal oblique, taking origin from the lateral two-thirds of the inguinal ligament, arch over the spermatic cord and then pass downwards to insert into the pubic crest (page 126, point 7). Some fibers insert lateral to the pubic crest and tubercle, on the **pectineal line** of the **superior pubic ramus,** the bar of bone which extends superiorly, posteriorly, and laterally, from the **body** of the pubic bone (page 120). Now identify, on a skeleton, the pectineal line, which extends from the pubic tubercle (page 126, point 8), posterolaterally on the upper surface of the superior pubic ramus and trace these fibers of the **internal oblique** in your cadaver. Posteriorly, the fibers that arch over the cord are reinforced by fibers from the deepest of the three muscle layers of the anterior abdominal wall, the **transversus abdominis.** The combined insertion of fibers from transversus abdominis and internal oblique is known as the **conjoined tendon.** Note that the fibers of internal oblique and transversus abdominis, which form the conjoined tendon, have their origin from the lateral part of the inguinal ligament, anterior to the spermatic cord, but after arching over the spermatic cord, medially, end up by lying posterior to the cord.

For completeness, note that deep to the transversus abdominis muscle, a layer of fascia, called the **fascia tranversalis** (fig. 62, page 128), forms an internal lining of the abdominal muscular wall. This fascia also is prolonged over the spermatic cord, as a result of the evagination of the testis and cord in embryological life (page 127, point 19). This covering of the spermatic cord is called the **internal spermatic fascia (internal inguinal fascia;** page 131).

After identifying the above mentioned components of the coverings of the spermatic cord, and visualizing the inguinal canal and its passage through the abdominal wall, and the components of its anterior, posterior, superior, and inferior boundaries, make a vertical incision into the internal oblique muscle deep to your previous vertical incision in the external oblique (i.e., extending up to the level of the umbilicus). Reflect the internal oblique medially toward the rectus sheath. Then incise the transversus abdominis at the same site, and reflect its contribution to the rectus sheath medially, thus confirming the triple nature of the anterior layer of this sheath. Mobilize the rectus abdominis muscle, and study the posterior layer of the sheath.

Recall that transversus abdominis aponeurosis passes into the anterior layer of the sheath inferiorly, but contributes to a posterior layer of the sheath at about the region of the umbilicus. As a result, an inferior free edge of the posterior sheath is created where transversus fibers pass posterior to the rectus abdominis. This edge is called the **arcuate line.** The **inferior epigastric artery** ascends on the posterior aspect of the rectus abdominis muscle, and passes anterior to the arcuate line, to enter the plane between the posterior layer of the sheath and the muscle itself (fig. 62, page 128). Superiorly, the superior epigastric artery, a branch of the internal thoracic (page 92), descends in this same plane. Thus the superior and inferior epigastric arteries are so placed that they are able to anastomose behind the rectus abdominis muscle.

The inferior epigastric artery is a branch of the external iliac artery, given off just superior to the inguinal ligament and medial to the deep (internal) inguinal ring (page 127, point 19). (The external iliac artery changes its name below the inguinal ligament, to the femoral artery [page 126, point 14].) The inferior epigastric artery passes upwards and medially, and the spermatic cord (page 127) hooks around the artery, in passing anteriorly from the pelvis to turn laterally and enter the inguinal canal. In females, the round ligament of the uterus, a ligamentous structure which passes from the uterus on the lateral wall of the pelvis to the inguinal canal, follows the same route as the spermatic cord of the male (see fig. 88, page 164). The round ligament is, as previously mentioned (page 130, point 19), a vestigial structure. Whereas in the male the spermatic cord passes to the scrotum, the round ligament of the uterus, although not the homologue of the spermatic cord, passes to the female homologue of the scrotum, the labium majus (page 130, point 19). Further details of these structures, and their respective true developmental homologies, will be dealt with subsequently, in study of the reproductive organs. In anticipation of this, attempt to resolve the apparent paradox described here.

Trace fibers of the external oblique muscle upwards, laterally and posteriorly, to appreciate the extent of the muscle, and its bony attachments, which should be studied with the assistance of atlas illustrations. Similarly, by dissecting more deeply, at selected sites, visualize the extent and attachments of the internal oblique and transversus abdominis muscles.

To complete your study of the anterior abdominal wall it will be incised, for the purpose of studying its posterior aspect, as well as for obtaining access into the abdominal cavity, for subsequent study of the internal organs. Make a transverse incision from the right to the left midaxillary lines, at the level of the umbilicus. Your incision should pass through the umbilicus, in order to avoid damage to internal structures which are related to the umbilicus superiorly and inferiorly, on the internal surface of the anterior abdominal wall. Several of these structures have in common that they are remnants of communications between the fetus and its maternal environment during prenatal life. Since communication between the mother and the fetus is mediated through the umbilical cord, which attaches to the umbilicus, it will be understandable that the structures which enter and leave the fetus at this site, on the external surface of the abdominal wall, will be found converging at this site, on the internal surface.

From the ends of the transverse incision thus made, make vertical incisions on both sides, upwards and downwards, in the midaxillary lines. This will create superior and inferior flaps of the anterior abdominal wall.

Reflect the inferior flap downwards. Observe the smooth, glistening inner lining of the wall, the **parietal peritoneum** (compare with parietal pleura, page 87). Identify, on the posterior surface of the anterior abdominal wall, a midline, cord-like structure, which ascends to the umbilical cord. This is the **urachus,** a remnant of the **allantois,** which in fetal life constitutes an extension from the bladder. You will, in a subsequent dissection, be able to trace the urachus downwards on the anterior abdominal wall, to the anterior wall of the pelvis, where it curves posteriorly to meet the bladder. On either side of the urachus are two further cord-like remnants, the **obliterated umbilical arteries,** converging on the umbilicus from below. These will be subsequently dissected. In fetal life, deoxygenated blood from the fetus was returned in these arteries for reoxygenation in the placental circulation. Lateral to the obliterated umbilical arteries, and somewhat further inferiorly, the inferior epigastric arteries, which you have just studied (page 132), should be reidentified now, converging from a point just medial to the internal inguinal ring, upwards and medially toward the arcuate line of the posterior layer of the rectus sheath. Having identified these folds of the lower part of the anterior abdominal wall, you may obtain an easier access to study the lower part of the abdomen by making a vertical incision, to the left of the urachus, from the umbilicus to the pubic crest.

Turn the superior flap of the anterior abdominal wall upwards. As with the lower half of the wall, the innermost layer that now presents for examination is the smooth and glistening parietal layer of peritoneum. In the midline, identify the **falciform ligament** (fig. 67, page 140), attached anteriorly to the anterior abdominal wall (falx means a sickle; falciform means shaped like a sickle; ligament means a fold in a serous, coelomic membrane [page 96]). The falciform ligament has an inferoposterior free edge, which is thickened, in that it contains a cord-like structure. This cord-like structure represents the **obliterated left umbilical vein** of fetal life, called the **ligamentum teres** (teres means round), and should be traced now from the umbilicus to the lower edge of the **liver.** The umbilical veins of the fetus convey oxygenated blood from the maternal circulation, via the liver, to the heart. On entering the right atrium through the inferior vena cava, the valve of the inferior vena cava diverts the blood through the foramen ovale (page 115). Thus oxygenated blood gets short-circuited from the right to the left side of the heart, without going through the lungs, which, of course, are nonfunctional in the fetus. In this way, oxygenated blood gets circulated to the fetus via the left side of the heart and the aorta.

Observe the visible apparent site of attachment of the ligamentum teres, and thus of the posteroinferior free edge of the falciform ligament, to the inferior edge of the anterior surface of the liver. Investigate the further extent of the falciform ligament, by following, superiorly, its attachment on the anterior surface of the liver, as this attachment converges, superiorly, with the anterior attachment to the anterior abdominal wall, in the midline. Thus the sickle shape of the falciform ligament is defined by an inferoposterior free edge, containing the ligamentum teres; a posterior edge attached to the anterior surface of the liver; and an anterior edge attached, in the midline, to the posterior surface of the anterior abdominal wall. After visualizing the falciform ligament, make vertical, paramedian incisions in the anterior abdominal wall, about 2 to 3 cm to each side of the falciform ligament, thus allowing you to fold the anterior abdominal wall upwards and laterally to either side to gain further access to the upper reaches of the abdominal cavity, as you previously did inferiorly. To further facilitate this access, extend the vertical incisions previously made in the midaxillary lines of either side, upwards, into the thoracic region. Using an electric saw, under guidance, cut through the rib cage in the midaxillary line on each side, and then horizontally in the xiphisternal plane, to meet your previous skin incision (page 86).

Finally, view the component structures of the abdominal portion of the digestive tract, as exposed in your cadaver, and as seen in other anatomical specimens provided. You should briefly observe the **stomach** and the first portion of the **duodenum,** both of which should be visible immediately on exposure by reflection of the flaps of the anterior abdominal wall. Gently lift the large fold of fatty peritoneum, called the **greater omentum;** which appears to hang down like an apron in front of other abdominal structures. You will then be able to visualize those portions of the **small intestine** which are in sequence with the duodenum, i.e., the **jejunum** and **ileum.** Note, in particular, the **mesentery,** the tissue paper-thin double membrane that extends from the posterior edge of the jejunum and ileum to the posterior abdominal wall. The mesentery arises by invagination of the **peritoneal sac** (of which you have already seen the parietal layer [page 133]) in the same way as the **mesocardium** arises by invagination of the serous pericardial sac (page 111). Near the lower right corner of the abdominal cavity, the small intestine continues into the **cecum,** the first portion of the **large intestine.** From the cecum, the **colon** (large intestine) commences, as the **ascending colon** on the right side, the **transverse colon** looping downwards and toward the left, and the **descending colon** on the left side. Finally, the descending colon connects, by way of the **sigmoid colon** (sigmoid means S-shaped), with the **rectum** (straight portion), which is situated in the pelvis, just anterior to the sacrum, and, finally, the **anal canal,** opening to the exterior at the **anus** (page 4).

Clinical Session 14

■ **TOPICS**

Upper Abdominal Structures
Upper Abdominal Digestive Tract
Esophagus, Stomach, and Duodenum
Spleen
Pancreas
Liver
Gallbladder

In this session you will study some upper abdominal organs which are accessible to examination through the anterior abdominal wall. Special attention will be paid to the upper part of the abdominal portion of the digestive tract. The projection of the tract will be drawn on the anterior abdominal skin surface. For examination of the abdomen, the model should be in a recumbent (supine) position. In order to obtain maximal relaxation of abdominal musculature, the lower limbs should be slightly bent at both the hips and the knees. The most comfortable way of doing this is to place a cushion under the knees of your subject.

1. Begin by marking the following planes: xiphisternal plane (page 84, point 7); subcostal plane (page 125, point 2); intertubercular plane (page 125, point 5); and right and left lateral planes (page 126, point 14). Which nine regions have you now demarcated (page 126, point 16)? Also mark the intercristal plane (page 125, point 4); transpyloric plane (page 125, point 1); and, finally, the **interspinous plane,** through the anterior superior iliac spines (ASIS). This plane is at the level of S2. Recall (page 126, point 11) that the posterior superior iliac spine (PSIS) also lies at this level. Thus one way of orientating the pelvis in the anatomical position is to place ASIS and PSIS in the same horizontal level.

2. None of the organs of the upper abdominal regions are visible to inspection, nor are they palpable, under normal conditions, although they may become so when diseased. Percussion of these organs is possible (see the following paragraph); auscultation is not.

You will recall (page 94, point 7) that lung tissue is typically **resonant** to percussion, because of its content of air, whereas more solid structures are **dull** to percussion. The air-containing, tubular components in the digestive system give, for similar reasons, an even more resonant note, known as a **tympanitic sound** (tympanum means drum) to percussion. Confirm this by percussing the **stomach** on the left, below the level of the sixth costal cartilage. Superior to these levels, a **dull sound** will usually be elicited, on the right because of the **liver,** which is situated in the hypochondrium above the costal margin, and on the left because of the **spleen.** The dullness due to the latter organ will most easily be detected lateral to the midclavicular line. However, the extent to which **spleen dullness** to percussion will be influenced by the overlapping lung tissue, and abdominal viscera, is variable. The liver extends from the right hypochondrium across the epigastrium, toward the left, and for this reason the dullness to percussion may also be elicited in the superior portion of the epigastrium.

3. Draw in the surface markings of the abdominal part of the **esophagus** and of the **stomach**. The latter organ varies greatly in size, shape, and position, as functions of age, body build, tone of abdominal muscles, and numerous other factors. The following description of the surface markings of this organ are therefore a stylized, theoretical average.

The esophagus passes through the **esophageal aperture** of the diaphragm opposite the tip of the left 8th costal cartilage, behind the costal margin (page 121, point 5, and page 124). Starting at this site, mark the abdominal portion of the esophagus, 1 cm in diameter and 1 cm in length, curving sharply to the left. The left side of the stomach, known as the **greater curvature,** may be drawn as a line which curves upwards and to the left from the left side of the end of the esophagus reaching the 6th costal cartilage a centimeter to the left of the lateral plane. The greater curvature then curves downwards to run with a lateral convexity in approximately the midclavicular line, to reach the intercristal plane. Here the line indicating the greater curvature should continue to the right, cross the midline and ascend to the transpyloric plane, about 2 cm to the right of the midline. This is the site of the **pylorus** (fig. 70, page 144), the junction of the stomach with the first part of the **small intestines,** the **duodenum.** The **lesser curvature** of the stomach should be indicated as commencing on the right side of the end of the esophagus. From there a line with a very slight convexity to the left should be drawn extending downwards to a point between the subcostal and the intercristal planes, 2 to 3 cm to the left of the midline. From here this margin of the stomach takes a sharp bend upwards and to the right, creating the **angular notch** of the lesser curvature. This border of the stomach then ascends toward the pylorus, at which point the stomach and its junction with the duodenum has a diameter of about 3 cm. Note that the last portion of the stomach, called the **pyloric part,** passes posteriorly as it ascends toward the pylorus: the stomach is an anterior abdominal structure and the duodenum is posterior. The portion of the stomach above the level of the esophageal aperture is called the **fundus;** the major part of the stomach, between the fundus and the pyloric part, is called the **body.**

4. The first portion of the small intestines, the **duodenum,** should now be drawn in. Closely adherent to the posterior abdominal wall, its course describes an incomplete circle which commences in the transpyloric plane (at the pylorus, the junction between the stomach and the duodenum), about 2 to 3 cm to the right of midline. From the pylorus, the **first part of the duodenum** passes to the right for about 2.5 cm. The duodenum then curves downwards for about 7.5 cm **(second part of the duodenum)** to about the subcostal plane (L3) where, just medial to the lateral plane, it turns to the left. From here it passes almost horizontally from right to left across the vertebral column **(third part of the duodenum)** until, just to the left of the abdominal aorta, it turns upwards and ascends **(fourth part of the duodenum)** to the **duodenojejunal flexure,** the junction with the **jejunum,** 2 to 3 cm to the left of the midline and 1 to 2 cm below the transpyloric plane.

The next part of this session will be devoted to ascertaining the situation, and surface markings, of the large glands of the digestive tract (i.e., the liver and gallbladder and the pancreas), and a nondigestive organ, the spleen.

The spleen is not a gland of the digestive system, but is a **hemopoietic organ** (hem means blood; poiesis means production; thus hemopoietic means blood producing) of the **hematic** system (page 6), as well as an organ of the **immunolymphatic** system (page 6). The spleen is thus not functionally, but only anatomically, related to the glands of the digestive tract.

5. Draw the surface markings of the **spleen** as follows. Mark, on the left side, the upper border of the ninth rib and the lower border of the eleventh rib, in each case from the scapular line (page 94, point 7, and fig. 41, page 90) to the midaxillary line (page 85, point 9, and fig. 35, page 82). The spleen occupies a posterior position, in close relationship to the area outlined above. It thus lies just anterior to the posterior portion of the diaphragm, and it lies posterior to the stomach. It is somewhat elongated in shape, its long axis being parallel to the tenth rib, and its outer (posterior) surface is convex. Viewed from the front, the **superior pole of the spleen** lies at about the level of T12, the **inferior pole** lies just below the level of the transpyloric plane, and the long axis of the organ, running from superior to inferior pole, slopes downwards and laterally, lateral to the midclavicular line. Posteriorly the spleen has a **diaphragmatic surface** (i.e., one that faces the diaphragm), and anteriorly a **visceral surface** (i.e., one that faces other viscera).

6. The **pancreas** has a head, body, and tail. The surface markings should now be drawn. The **head of the pancreas** occupies the concavity of the C-shaped curve made by the duodenum. The **body of the pancreas** is about 4 cm in superoinferior height, and extends from about the middle of the head toward the left, and slightly upwards. Its upper margin should be drawn from the lower border of the pylorus, and its lower margin from the central point of the head of the pancreas. Thus it overlaps a small prolongation of the head, the **uncinate process,** which lies just above and to the right of the duodenojejunal junction. The body continues into the **tail of the pancreas,** which tapers down to a point, ending on the visceral surface of the spleen, a little above the transpyloric plane and just lateral to the midclavicular line (fig. 71, page 144).

7. The outline of the **liver,** as seen projected onto the anterior abdominal wall, should now be drawn as follows. Starting in the xiphisternal plane in the midline, draw a horizontal line extending 2 to 3 cm to the right, then continuing into a curve, convex upwards and to the right, which follows the right dome of the diaphragm. This convexity will pass through the upper edge of the fifth rib in the lateral plane and will then curve down to descend with a slight convexity to the right, as far as the tip of the eleventh costal cartilage. From this point the border will follow the right costal margin to the tip of the ninth costal cartilage and will then leave it to traverse the position of the pylorus (i.e., in the transpyloric plane, 2 to 3 cm to the right of the midline). From here, the inferior border will curve to the left and upwards (with a slight convexity to the left and downwards) to cross the tip of the eighth left costal cartilage. The inferior border meets the superior border at an acute angle at this point. The superior border extends from the xiphisternal plane just to the left of the midline, curves upwards (with

the left dome of the diaphragm) to the lower edge of the fifth rib in the left lateral plane, and then down to the eighth left costal cartilage.

8. The **gallbladder** extends just beyond the costal margin, at the tip of the ninth right costal cartilage, just inferior to the inferior edge of the liver and lateral to the lateral edge of the rectus abdominis muscle. The protruding portion should be drawn in as a semicircle of 1 cm in diameter, centered at the tip of the ninth costal cartilage.

9. Review and repeat the percussion described in point 2, page 135, this time over the surface markings, thus confirming the situation and extent of the organs studied.

Review the emphasized key concepts of this session.

Laboratory Session 14

In this laboratory session your objective is to study the topographical anatomy of upper abdominal organs and their relationships with the peritoneum.

You should begin your examination by identifying, first, the large fold of peritoneum, usually laden with a considerable amount of **adipose tissue** (fat), and which appears to hang down like an apron, covering much of the abdominal content. This apron-like fold of peritoneum is called the **greater omentum**. Gently lift the greater omentum from posterior structures and follow it to its superior end, where its anterior surface will be found to be in contact with the **greater curvature** of the **stomach** (page 135). On its posterior surface, the greater omentum will also be found to be adherent to the anterior surface of both the **transverse colon** (a part of the **large intestine** which loops across from right to left hypochondrium), and its **mesocolon** (the "**transverse mesocolon**" (see mesentery, page 134). Using the description of the position of the stomach previously given (page 135, point 3), and the greater omentum as your marker for the greater curvature, identify the other named parts of the stomach, and verify the position and lie of the stomach. Bear in mind, as stated in the description previously given, that the position of the stomach is highly variable in the living state. In the cadaver, it will tend to lie more superiorly, may be somewhat rotated in position, and will vary considerably in size and shape, from cadaver to cadaver.

By gently mobilizing the stomach, and following its distal portion (the pyloric part [page 136, point 3]), identify the first part of the **duodenum.** While exercising great care to avoid tearing any membranous folds of the peritoneum, or damaging any organs, confirm, by gently tugging, that the stomach is basically a mobile organ (though the effects of embalming might well render the tissues tough, and considerably reduce mobility as compared to that of the living state). However, by comparison, only the first portion of the first part of the duodenum (page 136, point 4) is mobile; the remainder of the duodenum is immobile, due not merely to post-mortem changes of fixation, but to the fact that it is firmly bound down to the posterior abdominal wall by a tough layer of peritoneum.

Confirm, by identifying its second, third, and fourth parts (page 136), that the rest of the duodenum is firmly plastered down to the posterior abdominal wall. Note that the **transverse mesocolon,** mentioned above as being fused onto the greater omentum, appears to be attached posteriorly along a transverse line which crosses the second and fourth parts of the duodenum. Thus examination of the duodenum necessitates that you alternately palpate above and below the transverse mesocolon. In so doing, identify the **pancreas.**

Using the description previously given (page 136) palpate the **head of the pancreas** within the concavity of the duodenum, and follow the **neck, body,** and **tail** out to the left, as far as possible, without damaging peritoneal membranes at this stage. Note that the line of attachment of the transverse mesocolon also crosses the pancreas. The origin of these peritoneal relationships will become evident in due course.

In order to understand the relationships of the abdominal organs to each other and to the peritoneum, you should get a clear grasp of the concept that in embryologic life, much of the abdominal cavity can be pictured as being occupied by the peritoneal sac, which may be likened to a large balloon. Anteriorly and laterally, the balloon may be pictured as being in contact with the abdominal wall: for this reason, on cutting through the abdominal wall, and reflecting it, the wall is found to be lined on its inner surface by the smooth, glistening peritoneal membrane, a fact which you have previously observed (page 133) and should reconfirm now. Whereas the sac is in **direct** contact with the abdominal wall anteriorly and laterally, in the remaining relationships (superiorly, posteriorly, and inferiorly), various abdominal organs **intervene.** In the course of embryologic development, some of these organs invaginate into the peritoneal sac, whereas others do not. Those remaining entirely uninvaginated, for example the kidneys, situated on the posterior abdominal wall, are referred to as **primarily retroperitoneal organs** (retro means behind; primarily retroperitoneal indicates an organ that is retroperitoneal by virtue of never invaginating the peritoneum.

Other organs invaginate the peritoneal sac, partially or completely. When a solid object, such as a discrete organ, invaginates a coelomic sac such as the peritoneum, it acquires for itself a visceral layer, intimately applied to the surface of the organ itself. This situation resembles that previously described for the invagination of the pleural sac by the lung (page 93). Alternatively, a tubular organ may invaginate the sac, for example from posteriorly, and, in addition to obtaining a visceral covering for the organ itself, may drag a double sheet of the membrane behind it. This process creates, for example, the **mesocardium** of the heart (page 111), the **mesentery** of the jejunum and ileum (page 134), or the transverse mesocolon (page 138).

A further embryological mechanism may be mentioned here, to facilitate understanding of some of the relationships you will be examining later in this session: after invaginating the sac and acquiring for itself both a visceral layer of peritoneum, and a mesentery, an organ may "swing" back to the posterior parietal peritoneum, with the result that both the organ and its mesentery flatten themselves onto that layer of peritoneum. In this event, an actual fusion takes place between the layers of peritoneum that come in intimate contact with each other. Thus the posterior layer of the visceral peritoneum, and the posterior layer of mesentery, fuse with the parietal layer of peritoneum of the posterior abdominal wall. As a result of this fusion, the organ **appears** to be retroperitoneal: the parietal peritoneum lateral to the organ, the visceral peritoneum covering the organ, and the mesentery medial to the organ all appear to be fused into one layer. Organs which come to lie deep to such a fused layer of peritoneum are called **secondarily retroperitoneal organs** (compare with primarily retroperitoneal organs, described above). An example, to be discussed in greater detail below, is the duodenum, much of which, as you have already seen, is "plastered" to the posterior abdominal wall.

The relative mobility of the stomach and immobility of the duodenum exemplify two of the possible contrasting results of invagination of loops of the digestive tract described previously. Consider the abdominal portion of the digestive tract as an approximately midline tubular structure, extending from the lower end of the esophagus, as it passes through the diaphragm, to the lower end of the sigmoid colon, where the latter continues into the rectum. This tube grows considerably in length, and will tend to bulge forward, to accommodate the increase in size. It is convenient to consider the growth and protrusion forward of one section of the gut tube at a time. To start with, one may think of a first section, consisting of the stomach, and a second, consisting of the duodenum. The duodenum should be thought of as bulging forward in the sagittal plane, as a loop, prestaging the C-shaped loop which it will form in the adult. As it bulges forward, in its initial, embryological state, it carries a visceral covering of peritoneum, as well as a mesenteric double-fold of peritoneum, with it. Thereafter, the C-shaped loop of duodenum "swings" to the right, rotating about a vertical axis, thus changing the position of the loop from one which bulges forward, into the peritoneal cavity, to one which loops out to the right side. Thus the duodenum is brought into its characteristic adult position, to create the well known C-shape which you previously studied (page 136, point 4). Fusion of peritoneal membranes brings most of the duodenum to lie secondarily retroperitoneal. The **head of the pancreas,** which develops in the mesoduodenum, is similarly brought into this position.

Consider now the fate of the embryological stomach. The future greater curvature is developed from the **dorsal** surface of the embryological stomach (to which the mesentery is attached). The stomach also undergoes a "swing," or rotation about a vertical axis, which brings the greater curvature from a dorsal position to a left-sided position. However, the stomach does not fuse onto the posterior abdominal wall. Instead, the lower portion of the mesentery is subjected to an extensive outpouching, in an inferior direction, such that it comes to hang downwards, like an apron. This is, of course, the future greater omentum, and this origin of the greater omentum accounts for the fact that, in the adult, the greater omentum is found to attach to the greater curvature of the stomach.

In the ensuing dissection, you will be examining the anatomy of the liver, the pancreas, and the spleen. The former two are, by virtue of their function and embryological origin, regarded as being major glands of the digestive system. The spleen, although not part of this system, is anatomically and developmentally related to it, and is appropriately considered at this time.

Review the position, shape, and outline of the **liver,** as previously studied in projection on the living body (page 136, point 7), identifying the features described, in your cadaver specimen. In particular, first study the inferior border of the liver, and its relationship to related structures, including the rib cage. Identify the gallbladder (page 137, point 8), and likewise confirm the relationships described for it on your specimen.

By palpating the surface of the liver, note its smooth, shiny surface, due to the closely applied visceral layer of peritoneum. From this exercise, you can draw the conclusion that the liver is, like other organs of the digestive tract, invaginated into the peritoneal sac. In fact, the liver invaginates the sac from above and behind, making the invagination incomplete so that a true equivalent of a mesentery is not formed.

Place a palpating hand on the **anterior surface** of the liver, to the right of the falciform ligament (page 133 and fig. 67 page 140), with the palm of the hand following the curve of the liver, and the dorsal surface of the hand therefore in contact with the inferior surface of the diaphragm. Starting at the right aspect of the falciform ligament, follow, with the tips of your fingers, the anterior layer of the peritoneal fold, called the **coronary ligament,** which curves to the right from the falciform ligament

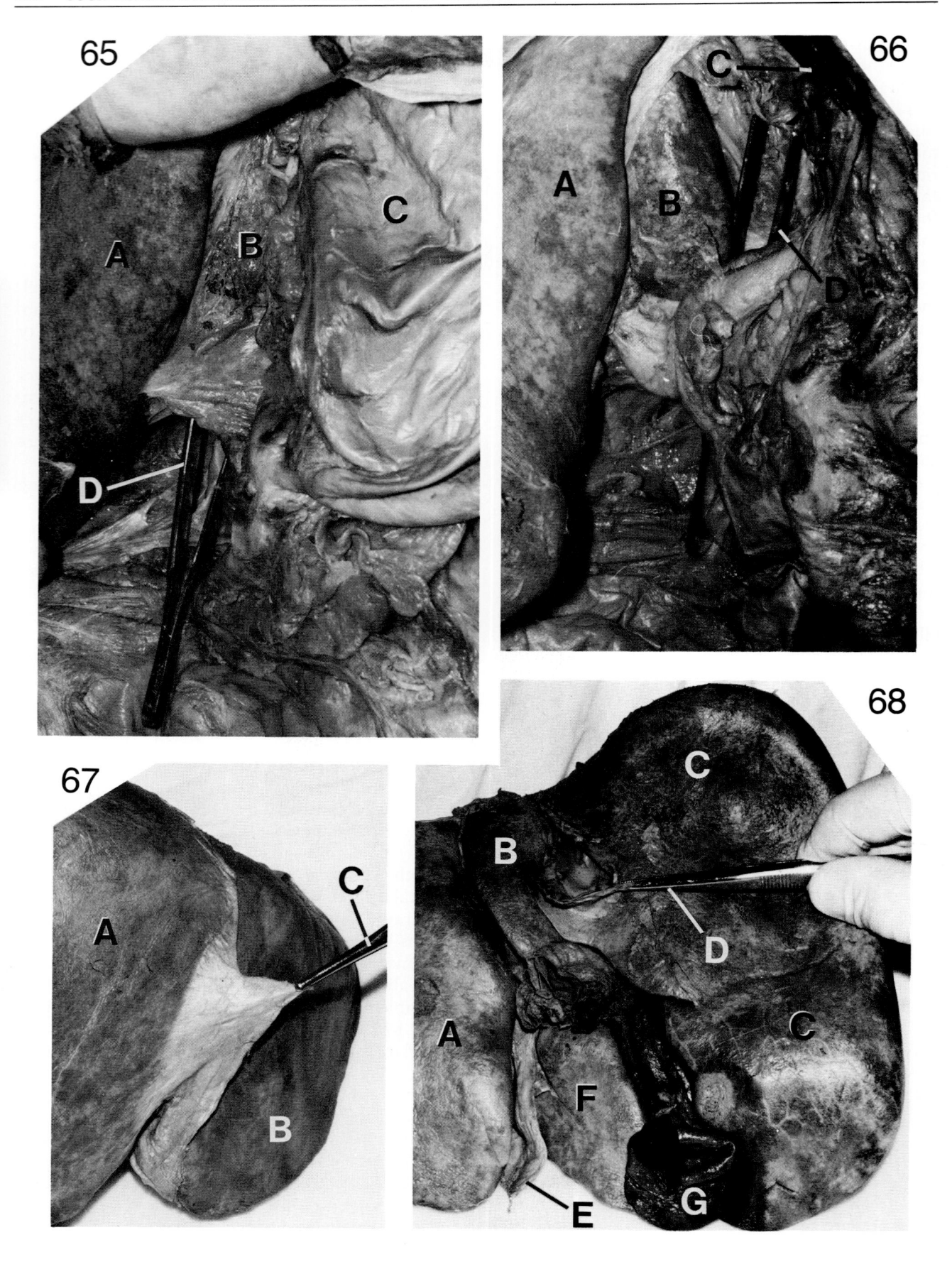

140 YOUR PATIENT'S ANATOMY: A CLINICAL VIEW OF HUMAN MORPHOLOGY

Figure 65. Lesser omentum, anterior view.
A. Right lobe of liver
B. Lesser omentum
C. Stomach
D. Forceps passing through epiploic foramen, to enter lesser sac

Figure 66. Similar view of lesser omentum, lesser sac exposed.
A. Right lobe of liver
B. Caudate lobe of liver
C. Forceps gripping cut edge of lesser omentum
D. Second pair of forceps, passed through epiploic foramen into lesser sac, and exposed by cutting lesser omentum

Figure 67. Liver, right anterolateral view.
A. Right lobe of liver
B. Left lobe of liver
C. Forceps gripping falciform ligament

Figure 68. Visceral (posterior) surface of the liver.
A. Left lobe of liver
B. Caudate lobe of liver
C. Right lobe of liver
D. Forceps holding wall of inferior vena cava
E. Ligamentum teres
F. Quadrate lobe of liver
G. Gallbladder (opened)

(fig. 69, page 144). By palpating posterosuperiorly, it can be ascertained that the visceral peritoneum of the liver recurves anteriorly onto the inferior surface of the diaphragm, to become parietal peritoneum. Thus you are in this way able to establish the incompleteness of the invagination of the liver into the peritoneal cavity. Repeat this exercise on the left side of the falciform ligament, where a similar reflection of peritoneum, also curving away from the falciform ligament, this time to the left, can be palpated. This is the anterior layer of the **left triangular ligament** (fig. 69), and it, similarly, reflects onto the inferior surface of the diaphragm.

Elevate the inferior edge of the liver. While maintaining traction on the liver, to pull it upwards and forwards, identify the **lesser omentum** (fig. 65, page 140), a peritoneal membrane that extends from the lesser curvature of the stomach, and the first part of the first portion of the duodenum to pass upwards and attach on the posterior surface of the liver. The part from stomach to liver is called the **gastrohepatic ligament**, and that from duodenum to liver is the **duodenohepatic ligament**. Identify the right free edge of the duodenohepatic ligament. In this edge, palpate the contained **bile duct** (a duct which carries bile from the liver and gallbladder to the second part of the duodenum); the **portal vein** (a vein which carries blood from most of the digestive tract, to the liver (page 123)); and the **hepatic artery** (which carries oxygenated blood to the liver).

Gently insert a finger behind the right free edge of the duodenohepatic part of the lesser omentum, passing your finger from right to left. Your finger has entered the **epiploic foramen** (fig. 65, page 140). Anterior to your finger lies the portal vein, and the other two vessels mentioned previously; posterior to your finger lies the **inferior vena cava;** inferior to your finger is the first portion of the duodenum; and superior to your finger is a part of the liver.

The epiploic foramen opens into the **lesser sac** (also called the **omental bursa**), which your finger will enter if you push it further to the left. Here your finger will be posterior to the lesser omentum, which thus forms an anterior limit of the lesser sac. Make an incision in the lesser omentum for further exploration of the lesser sac (fig. 66, page 140).

At this stage, examine an anatomical specimen of the liver, to identify structures not yet visible in your own dissection. Note first, on the **superior surface** (fig. 69, page 144), the lines of attachment of the ligaments you palpated previously: the **left triangular ligament** and the **coronary ligament** (and the extension to the right of the latter, the **right triangular ligament**). On the **posterior surface** of the liver (fig. 68, page 140), identify the attachment of the lesser omentum, noting the right-angle bend between the attachment of the gastrohepatic and duodenohepatic portions, corresponding to the right-angle bend between the lesser curvature of the stomach, and the superior surface of the first part of the duodenum. Also observe that the anterior layer of the lesser omentum is continuous with the posterior layer of the left triangular ligament.

Returning to your cadaver, turn to the left side of the thoracic cavity. With the lung removed, make a peripheral, semicircular circumferential incision in the **left dome** of the diaphragm, close to the diaphragmatic attachments to ribs, leaving intact the connection of the dome to the **central tendon** of the diaphragm (page 120). Through the opening thus created, you can now palpate the immediate inferior relations of the left dome of the diaphragm, prior to reflecting the dome medially (which will sever the left triangular ligament). Using a bimanual approach, with one hand through the diaphragm from the thorax and the other hand in the abdominal cavity, palpate these infradiaphragmatic structures, referring to page 136 (spleen) and page 135 (stomach). Palpate the lesser omentum bimanually, confirming its continuity with the left triangular ligament described previously. Pass your "intrathoracic" hand far posteriorly, behind the **diaphragmatic surface** of the spleen (page 136), and your "intra-abdominal" hand through the opening you made in the lesser omentum (and thus into the lesser sac). On the **visceral surface** of the spleen (page 136), you can now palpate bimanually an apparent attachment of a peritoneal membrane which limits the lesser sac laterally (to the left), and which thus separates your right and left hands. As will be explained in the following paragraphs, this membrane is part of the mesentery of the stomach, also called the **dorsal mesogastrium**. From the site of attachment on the spleen (the **hilus** of the spleen) the membrane extends anterolaterally to the greater curvature of the stomach; posteromedially, the membrane can be traced from the hilus of the spleen to the posterior abdominal wall, where it can be felt to attach in front of the left kidney. In this latter part of the membrane, the tail of the pancreas can be palpated, as it reaches the spleen (page 144 and fig. 71, page 63).

The spleen develops between the two layers of the dorsal mesogastrium, at a position midway between the greater curvature of the stomach and the dorsal attachment of the mesentery to the posterior abdominal wall. It thus divides the dorsal mesogastrium into two components: a ventral half, which passes from the spleen to the greater curvature of the stomach, known as the **gastrosplenic** or **gastrolienal ligament** (gaster means stomach; lien means spleen; ligament means peritoneal fold [page 96]);

and a posterior portion, which contains the body and the tail of the pancreas. Recall (page 139) that because of its situation, the **head** of the pancreas is adherent to the posterior abdominal wall, as a secondarily retroperitoneal organ. Tracing the pancreas to the left, the **body** of the pancreas crosses the midline, and in so doing, enters the dorsal mesogastrium. This posterior part of the dorsal mesogastrium becomes, in part, secondarily fused with the posterior parietal peritoneum, in front of the **left kidney,** a **primarily retroperitoneal structure** (page 138). As a result, the adjacent part of the dorsal mesogastrium appears to extend from the left kidney to the spleen, and is acccordingly known as the **lienorenal ligament** (renal meaning of the kidney).

It should be evident from this and previous descriptions, and should be confirmed now by palpation, that the lienorenal ligament, the gastrolienal ligament, and the greater omentum are parts of the same fold of peritoneum, and are in continuity with each other.

Palpate the anterior layer of the left triangular ligament of the liver, as previously, on the superior portion of the anterior aspect of the left lobe of the liver (page 142). With a bimanual approach, utilizing, once again, the access through the left dome of the diaphragm, also examine the posterior layer of the triangular ligament and define its apex, to the left. Now cut through the anterior and posterior layers of the ligament, by careful scalpel dissection, carrying your incisions through the anterior layer across to the falciform ligament, and your incisions through the posterior layer medially to the upper end of the lesser omentum. Note again that the posterior layer of the triangular ligament is continuous, medially, with the anterior layer of the lesser omentum. Returning to the incision previously made in the lesser omentum, continue this incision upwards to meet the cut through the posterior layer of the left triangular ligament, which you just made. Inferiorly, carry your incision in the lesser omentum laterally, to cut through the structures of the free edge of the lesser omentum (i.e., the portal vein, bile duct, and hepatic artery).

Make an incision in the right dome of the diaphragm, comparable to that previously made in the left. Using a combined superior and inferior approach, identify the coronary ligament (page 139), and its lateral extremity, the **right triangular ligament,** and cut through the anterior and posterior layers of these ligaments, as you previously did in the case of the left triangular ligament. By carrying your incision in the posterior layer of the coronary ligament medially, you will be skirting around the inferior vena cava, posterior to the liver, and will ultimately reach the posterior layer of the lesser omentum. The liver will thus be freed of its ligamentous attachments, and can now be removed from the abdomen, by cutting the **hepatic veins** as they enter the **inferior vena cava.**

LIVER

With the aid of atlas illustrations, and anatomical specimens, study the surface features of the liver, identifying, in the first instance, the cut edges of peritoneal folds and reflections, which you made in order to remove the liver. You should commence by identifying and reviewing the coronary ligament and right triangular ligament, the falciform ligament, the left triangular ligament, and the cut edges of the lesser omentum and the structures of the **porta hepatis** (i.e., the bile duct, portal vein, and hepatic artery). Identify the **right** and **left lobes** of the liver, and the two intermediate lobes: the **caudate** and **quadrate** lobes (fig. 68, page 140). Find the ligamentum teres (page 133), the **gallbladder,** and the impressions, on the **visceral (posterior) surface** of the liver, which are created by the impinging, related organs: the gastric impression, esophageal impression, duodenal impression, colic impressions, and renal impression. Identify the **fissure for the ligamentum venosum,** into which the two layers of the gastrohepatic portion of the lesser omentum pass. The ligamentum venosum is a deep lying, vestigeal, cord-like remnant of a vein which connects the portal venous system with the hepatic, in the fetus (see also page 133). In the adult liver, it can be traced from the left branch of the portal vein to the left hepatic vein.

Study the region in the abdominal cavity from which you removed the liver, identifying the structures most intimately related to the liver, and making a special note of the remaining portions of the lesser omentum and other peritoneal folds. Then replace the liver **in situ,** noting the relationships of impinging viscera, and the impressions they make on the liver. The exercise of replacing and removing the liver should be repeated several times, until a clear, three-dimensional image is obtained of these important anatomical relationships.

The large glands of the digestive tract, the liver and the **pancreas,** develop from the digestive tube as tubular outgrowths in much the same way as the respiratory system develops from the upper portion of the digestive tract (page 69).

The future liver develops in part from a bud that grows from the second part of the duodenum, in the concavity of the C-shape. Before the loop of the duodenum rotates to the right, the liver bud grows cranially (to the right of the first part of the

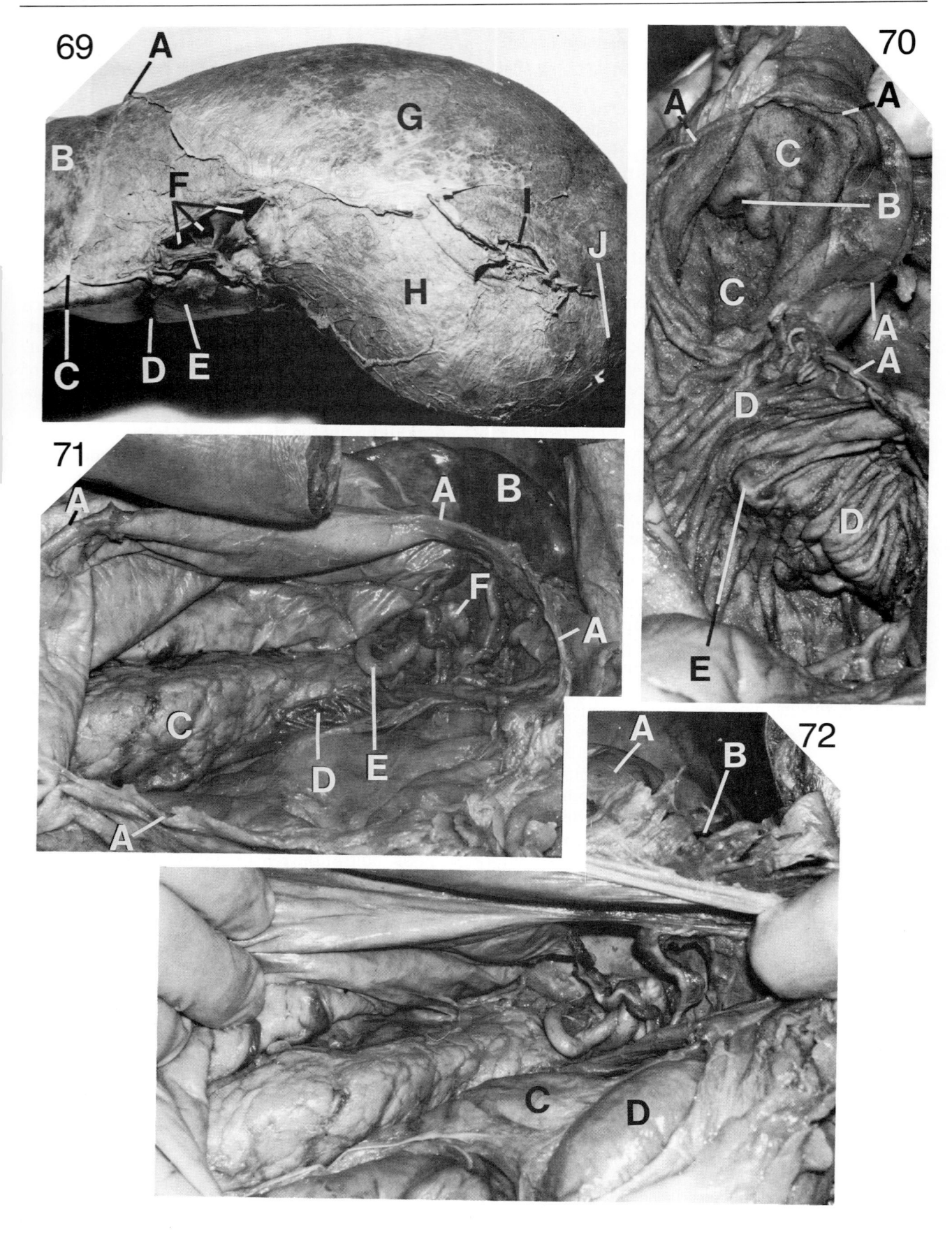

144 YOUR PATIENT'S ANATOMY: A CLINICAL VIEW OF HUMAN MORPHOLOGY

Figure 69 Superior aspect of liver.
A. Falciform ligament
B. Left lobe
C. Left triangular ligament
D. Fissure for ligamentum venosum
E. Caudate lobe
F. Hepatic veins (near site of entry into inferior vena cava)
G. Right lobe
H. Bare area
I. Anterior layer of coronary ligament
J. Right triangular ligament

Figure 70. First and second parts of duodenum (cut open on anterior aspect).
A. Cut edge of duodenal wall
B. Pyloric opening (from stomach into first part of duodenum)
C. Mucosa of first part of duodenum
D. Plicae circulares in second part of duodenum. In mucosa and plicae circulares, villi can be seen as small dots
E. Duodenal papilla

Figure 71. Anterior view of lesser sac, seen through opening in greater omentum.
A. Cut edge of greater omentum
B. Spleen
C. Pancreas
D. Splenic vein
E. Splenic artery
F. Tip of tail of pancreas in contact with spleen

Figure 72. Related view of lesser sac, showing additional details.
A. Left dome of diaphragm
B. Incision in diaphragm, through which spleen is approached from thorax (see text)
C. Fused posterior layer of greater omentum and transverse mesocolon
D. Loop of transverse colon

duodenum) and enters a mass of mesoderm, known as the septum transversum, situated ventral to the stomach. In visualizing the invagination process of the duodenal loop, and of the stomach, into the peritoneal sac, you should picture the developing liver, and the duct from which it arose, as also being involved in the peritoneal invagination. The liver and the stomach invaginate from a cranial and dorsal position, and carry before them peritoneal folds as do other organs. The duct from which the liver arises (known, in the adult, as the **bile** duct), carries behind it a double fold of mesentery, as with any other tubular structure invaginating the peritoneum. This "bile duct mesentery" is the **lesser omentum.** When the duodenum rotates to the right, and the dorsal aspect of the stomach rotates to the left (page 139), the mesentery of the bile duct, unlike that of the duodenum, remains unfused to the posterior abdominal wall. As a result, it is represented in the adult as a mobile fold, with a right free edge, within which the bile duct lies.

It should also be apparent to you from the descriptions that have preceded this examination of the omental bursa, that the omental bursa itself will extend into the fold of mesentery formed behind the stomach, and which pouches out to become the greater omentum. Thus, the lumen of the pouch enclosed by the greater omentum is an extension of the omental bursa. However, the extent of this pouch is limited by fusion of the double membranes originating from the mesentery of the stomach. In addition after this fusion of the two layers, the greater omentum also fuses onto the transverse mesocolon (page 138).

For completeness, it should also be mentioned that the lesser omentum, because it constitutes a mesentery which lies ventral to the stomach, is often referred to as the ventral mesentery of the stomach (although it really is a mesentery of the bile duct rather than of the stomach). The true, dorsal mesentery of the stomach is therefore specifically designated the dorsal mesentery. The terms ventral mesogastrium and dorsal mesogastrium are also used (gaster means stomach).

GALLBLADDER

Identify the **fundus** (rounded, blind end), **body,** and **neck** of the gallbladder. Incise the organ to study its mucous membrane. Note that the fundus extends beyond the inferior border of the liver and the costal margin (page 137), thus coming in contact with the peritoneum covering the anterior attachments of the diaphragm (pages 120 and 124). In cholecystitis (inflammation of the gallbladder), the patient often experiences **referred pain** in the right shoulder, the nerve supply to which is through the same cervical segments as those that supply the diaphragm (page 106).

The gallbladder and its related duct system are components of the excretory and exocrine apparatus of the liver. An important function of the liver is detoxification, both of digestive products, reaching it via the portal system (page 123), as well as substances brought in the systemic circulation. In addition, among numerous other functions, the liver contributes to the digestive process in the intestines. Within the liver, secretory and excretory products leave the liver cells, to be collected in a network of minute **bile canaliculi.** These unite into **ductules** of increasing size, finally forming **right** and **left hepatic ducts** which leave the liver and unite to form the **common hepatic duct** at the porta hepatis. The **cystic duct** connects the gallbladder to the common hepatic duct, and the resulting common pathway is called the **bile duct,** which you have identified in the free edge of the lesser omentum (page 142). The hepatic secretion (bile) collects in the gallbladder and then passes back, via the cystic duct, to the bile duct. As explained in the embryological account above, the bile duct empties into the duodenum (in common with the pancreatic duct, as described in the following section).

PANCREAS

Reidentify the **pancreas** (page 136, point 6). Further study will be facilitated by making an opening into the omental bursa through the greater omentum, inferior to the greater curvature of the stomach but above the transverse colon (figs. 71 and 72, page 144). The **head** of the pancreas occupies the concavity of the C-shape of the duodenum; the **body** extends across the midline to the left, and the **tail** reaches as far across to the left as the spleen.

Between the head and the body, a region, usually called the **neck,** lies anterior to the origin of the **superior mesenteric artery.** Another important relation is the **coeliac artery,** which takes its origin from the anterior surface of the aorta just above the neck of the pancreas. The uncinate process of the pancreas is an extension from the lower part of the body, just above the third part of the duodenum, which projects upwards, posterior to the neck of the pancreas. The superior mesenteric artery (and the accompanying superior mesenteric vein, to the right of the artery), lie anterior to the uncinate process. After crossing the anterior surface of the aorta, the pancreas extends to the left, lying anterior to the left kidney. The tip of the tail of the pancreas reaches the spleen (fig. 71, page 144), by passing in the lienorenal ligament (page 143). In order to understand the latter fact, bear in mind that the head and body of the pancreas, lying in the concavity of the C-shape of the duodenum, and therefore in the region of the mesoduodenum, follow the duodenum when it rotates to the right, adhere to the posterior parietal peritoneum and become secondarily retroperitoneal. However, that portion of the pancreas which extends to the left passes into the mesogastrium, just above the site where the omental bursa develops. By extending upwards and to the left, the tail of the pancreas, therefore, comes to lie in that portion of the mesogastrium which becomes the gastrolienal ligament.

The liver and pancreas have in common that they function as glands of the digestive tract: they secrete their respective products, through ducts, into the lumen of the gut, and their secretion products play a role in digestion. This process fulfills the criteria of exocrine secretion (i.e., that kind of secretion which involves deposition of the gland's product on a free, epithelial-lined surface, as described on (page 9). For completeness, it should be mentioned that, in addition to their exocrine function, being considered here, both the liver and the pancreas also function as endocrine glands (page 7). For example, embedded within the tissue of the pancreas are small **islets,** which secrete the hormone insulin.

As is frequently the case when organs share a common role in their function (in this case, as major exocrine glands of the digestive tract), the liver and pancreas also have developmental and structural features in common. Embryologically, part of the pancreas develops from a bud which grows out of the liver diverticulum (bud [pages 143 to 146]). Accordingly, in adult life, the duct of the liver (the bile duct) and the pancreatic duct have a common opening into the duodenum, at the site of the original outgrowth of the liver bud (fig. 70, page 144). A second pancreatic diverticulum develops, independently of the first, from the duodenum. The two buds fuse, in the dorsal mesoduodenum, and their ducts connect, such that the major secretion of the pancreas, in the adult, is through the common hepatopancreatic duct.

Reidentify the bile duct, in the free edge of the lesser omentum, accompanied by the portal vein and the hepatic artery. In a previous dissection, you cut through this triad of structures, in freeing the liver of its peritoneal attachments. Using scissor dissection, the inferior stump of the bile duct should now be followed inferiorly. Remove pancreatic tissue as necessary, to trace the duct posterior to the pancreas, and to its destination, as it curves laterally to enter the second part of the duodenum. To aid this dissection, mobilize the duodenum from the posterior abdominal wall, from a right-sided approach. In removing pancreatic tissue, care should be taken to avoid damaging the pancreatic duct, which you should identify at its junction with the bile duct, close to the duodenal wall. Trace the hepatopancreatic duct to the site where it actually enters the duodenum (in the concavity of the C-shape, somewhat on the posterior aspect of the second part of the duodenum).

DUODENUM

Having traced the duct to the duodenal wall, incise the anterior wall of the second part of the duodenum, creating a flap which can be folded to the right, as a "window." Use this window to examine the internal surface of the duodenal mucosa. If necessary, clean the lumen and lining mucous membrane by pouring wetting fluid onto your specimen. Note the texture and consistency of the mucous membrane, identifying any folds that may be present, and establishing which directions such folds may have. The site of opening of the hepatopancreatic duct should be observed, on the tip of the **duodenal papilla** (fig. 70, page 144) (compare with parotid papilla [page 26]; sublingual papilla [page 29]). Note also the **plicae circulares** (circular folds) of the duodenal mucosa (fig. 70). Observe the proximal communication of the duodenum with the stomach at the **pylorus** (page 135 and fig. 70), extending your incision if necessary.

SPLEEN

Review the descriptions previously given of the situation and anatomical relationships of the spleen (pages 136 and 142 and 143), correlating these descriptions, and your previous findings on the living body, with the features now observable in your cadaver specimen. Utilize, for this purpose, the dual approaches of the diaphragmatic incision you made (page 142), and the anterior access that is possible, through the opening into the omental bursa which you previously made (page 142), and through which you were able to study the body and tail of the pancreas (fig. 71, page 144). Reidentify the tip of the tail of the pancreas on the visceral aspect (page 136) of the spleen. Identify the posterior relations of the spleen, including its relationships to ribs, as well as the visceral relationships of the neighboring organs. Observe the splenic artery and vein (fig. 83, page 160), and related vessels. Reidentify the lienorenal ligament (page 143), and review its embryological origin. When all of these important relationships have been reviewed, with the spleen **in situ,** cut the spleen free of its remaining tethers, and remove it for further study. With the aid of atlas illustrations and anatomical specimens, identify the surfaces and specific regions of the spleen, replacing it within the abdominal cavity as required, for reestablishing relationships.

STOMACH

Observe the stomach, and study its relationships, including the posterior relationships visible by inspection and palpation within the omental bursa. Make a "window" into the anterior wall of the stomach, and examine the internal surface of the lining mucous membrane, noting the direction and nature of folds of the mucous membrane.

ESOPHAGUS

Review the abdominal portion of the esophagus (page 135) including its relations, replacing the liver and revisualizing peritoneal relationships to achieve the latter. The venous drainage of this part of the esophagus is to the portal system (page 142). Recall the clinical importance of anastomosis, at this site, between portal and systemic veins (page 123).

Clinical Session 15

■ **TOPICS**

Abdominal structures
Abdominal Descending Aorta
Cecum
Jejunum
Ileum
Colon: Ascending; Transverse; Descending; Sigmoid

In this session your objective is to study lower abdominal structures, in particular the abdominal digestive tract distal to the level of the organs previously examined (page 148). For optimal accessibility to abdominal examination, your model should be positioned as previously (page 135).

1. Start by marking the relevant vertical and horizontal planes employed previously (page 135, point 1).

2. Individual abdominal organs are not usually visible to inspection as such, with the occasional exception of visible pulsation of the **abdominal descending aorta.** Examine the anterior abdominal wall in the midline, and note whether a pulsation can be seen. You are more likely to observe this feature in a lean subject.

3. The abdominal descending aorta is more often detectable on palpation. Attempt by deep, but gentle, palpation to feel the pulsation of this large vessel in the midline. The abdominal aorta is the continuation of the thoracic descending aorta, and becomes abdominal as it pierces the diaphragm. It descends almost vertically, in the midline, on the anterior surfaces of the bodies of lumbar vertebrae L1 to L4. At L4 it bifurcates, slightly to the left of the midline, into two large branches, the **right** and **left common iliac arteries** (see fig. 82, page 160).

4. The only portion of the digestive tract that is palpable in the normal state is the **descending colon.** The **colon,** a portion of the **large bowel,** has **ascending** (right), **transverse,** and **descending** (left) parts. The surface markings of the descending colon extend from the left hypochondriac region of the anterior abdominal wall to the left inguinal (also called **left iliac**) region. The lower portion of the descending colon lies just deep to the anterior abdominal wall, a little above the intertubercular plane. Place the fingers of the palpating hand on the anterior abdominal wall in this position, and the other hand posteriorly, just above the iliac crest. The descending colon can be rolled between the fingers when the abdominal muscles are relaxed.

5. Draw in the surface marking of the entire **large intestine** (large bowel). The large intestine begins as the **cecum,** which lies in the right inguinal region **(right iliac region).** Therefore its surface marking lies within the triangle formed by the intertubercular plane, the right lateral plane, and the right inguinal sulcus. It should be drawn in with its right border adjacent to the right lateral plane. Its inferior border is about 6 to 7 cm below the intertubercular plane, and the breadth of the cecum is also about 6 to 7 cm, so that its lateral border would be that distance from the right lateral plane.

Mark the **appendix,** a small, finger-like process extending from the medial border of the cecum, in the interspinous plane, and passing in an inferior direction. Because of the frequent occurrence of appendicitis it is of clinical importance that the position and size of the appendix are enormously variable.

Superiorly the cecum is continuous, in the intertubercular plane, with the **ascending colon.** The latter and subsequent portions of the colon should be drawn in with a slightly smaller diameter than the cecum, and with **haustrations** (sacculations), separated by puckers (slight constrictions) in the wall of the organ, at intervals of 3 or more centimeters (fig. 74, page 152). The ascending colon extends to a point 1 to 2 cm inferior to the transpyloric plane, where it bends to the left at the **hepatic flexure** (so named because of its relationship to the liver). Here the large intestine continues to the left as the **transverse colon,** which has a diameter somewhat less than that of the ascending colon. The transverse colon, in the anatomical position, hangs down in a loop which reaches the intertubercular plane in the midline. It then loops upwards to the left, and reaches the **splenic flexure** (compare with hepatic flexure) in the left hypochondrium, at the level of the 8th costal cartilage. The **descending colon** can be drawn in from this point and downwards, lateral to the lateral plane and passing to the interspinous plane on the left. Here it loops medially toward the lateral plane, where it ends by continuing into the **sigmoid colon,** a content of the pelvis.

6. As mentioned, the descending colon continues

distally into the sigmoid colon. The latter, in turn, continues into the rectum and the tract terminates in the anal canal. These last portions of the **gut** will be studied in the pelvis and perineum. **Proximally,** the large intestine is continuous with the small intestine. You should now mark the junction between the latter and the cecum, in the right lateral plane, about 2 cm inferior to the intertubercular plane. At this latter point, the 3rd of the three components of the **small intestine (small bowel),** the **ileum,** enters the cecum from its left aspect. The ileum and the second portion of the small bowel, the **jejunum,** are connected to the posterior abdominal wall by a fan-shaped **mesentery** the root of which attaches to the posterior abdominal wall along a diagonal line sloping downwards and to the right, as follows. The upper extremity of the posterior attachment of this mesentery is to the left of the second lumbar vertebra. This point is thus about 2 to 3 cm to the left of the midline, and about 1 to 2 cm inferior to the transpyloric plane. At this point, the first part of the small bowel, the **duodenum,** continues into the jejunum. From this **duodenojejunal flexure,** (page 136) the attachment of the mesentery extends to the **ileocecal junction,** previously marked in, to the left of the cecum. Draw in the attachment of the mesentery as two straight, parallel lines, a millimeter from each other.

In embryological life, the abdominal portion of the digestive tract invaginates the **peritoneal sac** from behind (i.e., from the dorsal aspect) in a comparable manner to the way the lungs and heart invaginate the pleural and serous pericardial sacs, respectively (pages 93 and 111 to 114). A **mesentery** is the double layer of **peritoneum** which a loop of the digestive tract carries in behind it, on invaginating the peritoneal sac, comparable to the **mesocardium** of the cardiac loop (page 111). Unlike the mesocardium, the mesenteries of the various loops of digestive tract do not degenerate but are by and large retained in adult life. Gut invagination takes place, as mentioned, from behind. The parietal (parietes mean wall; see page 93) peritoneum subsequently adheres posteriorly to the posterior body wall. As a result, the mesenteries attach to the posterior body wall, inasmuch as each of the two layers continues, on its respective side, into the parietal peritoneum. Thus if the mesentery is cut away, a double cut "line of attachment" would be visible, analagous to the cut edge of the sheath of pericardium which can be observed surrounding the oblique sinus, after the heart is cut away (page 119).

The mesentery is "fan-shaped" in the sense that this posterior attachment is relatively very small (i.e., corresponding to the handle of the fan) in comparison to the extent of the jejunum and ileum, which is very extensive (i.e., corresponding to the outer edge of the fan). The length of the entire small intestine is about 6.5 m, of which the duodenum constitutes about 25 cm.

7. Because of the extent of the jejunum and ileum, they do not lie merely in front of the attachment of their mesentery, but loops of these parts of the bowel lie also to either side of it, mainly inferior to the transverse colon. Thus the jejunum and ileum are both situated close to the abdominal wall, the jejunum largely to the left and the ileum largely to the right. On the left, the small intestines reach to approximately the midaxillary line, and on the right to the right lateral line.

8. Examine the nine regions of the abdomen (page 126, point 16) by percussion. Review point 2, page 135, and note that most of the bowel will, under normal conditions, give a tympanitic tone. However, to a variable extent, the **ascending, transverse,** and **descending** parts of the **colon** may give a degree of **dullness** to percussion.

9. The digestive tract can be auscultated, and in normal individuals **bowel sounds** are heard. Bowel sounds consist of a low, gurgling sound, and are due to the normal, **peristaltic contractions** of the muscles within the wall of the intestines. These muscular contractions are responsible for moving the contents of the bowel through the digestive tract, during the process of digestion.

The presence of normal bowel sounds is an important clinical sign of normality of the state of the abdominal contents. For example, in rupture of an abdominal viscus, acute infection within the abdominal cavity, or following trauma, with extravasation of blood or other fluids, the peristaltic movements of the digestive tract become paralyzed. Thus absence of bowel sounds would be one indication, among others, of a need for active surgical intervention. Conversely, in cases of **obstruction** of the **lumen** of the digestive canal, bowel sounds may become abnormal in volume and in quality.

Using a stethoscope, auscultate the abdomen of your model at each of the nine regions. Patience is needed for this exercise, as there are frequently long intervals between audible peristaltic bowel sounds.

Complete this session by reviewing the emphasized key concepts.

Laboratory Session 15

In following the description given in Laboratory Session 14, you pictured the abdominal portion of the developing digestive tube as an approximately midline structure extending from the last part of the esophagus superiorly, to the beginning of the rectum inferiorly. Invagination of the peritoneal sac by the stomach, the duodenal loop, and related structures has been considered. Now leaving these proximal structures, with the duodenum in place, adherent to the posterior abdominal wall, move your attention to the sigmoid and descending parts of the colon (page 149). In your cadaver specimen, identify the descending colon, on the left side of the abdominal cavity, and trace it downwards into the sigmoid colon, then trace the sigmoid colon down to the rectum. The sigmoid colon is relatively mobile, since it has a mesentery. The descending colon, however, has undergone the same development fusion of its mesentery as described previously for the duodenum. Thus the descending colon swings to the left, and its mesentery fuses with the posterior abdominal parietal peritoneum. Trace the descending colon upwards with your fingers, to the splenic flexure (page 149).

Returning now to the last part of the duodenum, visualize the loop of gut which remains unattached (at the time in development when the duodenum superiorly, and the descending colon inferiorly, have each become adherent to the posterior abdominal wall, and therefore secondarily retroperitoneal). The remaining free loop of bowel, between the fourth part of the duodenum, slightly to the right of the midline, and the splenic flexure, on the left, now rotates in an anti-clockwise direction, about a posteroanterior axis, and then "swings" to the right. The loop can be pictured as falling in such a way, that the portion immediately after the fourth part of the duodenum comes to run downwards and to the right, toward the right iliac fossa (page 150, point 6). The next portion of the tube lies on the right side of the abdominal cavity (and becomes the ascending colon). The subsequent portion crosses from the right to the left hypochondrium, and becomes the transverse colon. The portion between the fourth part of the duodenum and the right iliac fossa develops into the jejunum and ileum, and retains a free mesentery. The portion which becomes the ascending colon loses its mesentery by posterior fusion; and the transverse colon retains its mesentery, and hangs down in the abdomen. Confirm these features of the peritoneal development, by careful examination in your cadaver. A clinically important consequence of the midline origin of the gut is that, although rotation brings the appendix to a definitive position in the right inguinal region (page 149), the pain of appendicitis is initially in the midline.

It was mentioned previously (page 146) that the extension of the omental bursa into the greater omentum is limited by a partial fusion of the layers of the omentum with each other, a well as fusion of the omentum to the transverse mesocolon. You are now in a position to visualize, that, as a result of the anti-clockwise rotation, and "swing" of the bowel, which leads to the formation of the transverse colon and its mesentery (the transverse mesocolon), the latter mesentery is placed in a situation just posterior to the position of the greater omentum. Fusion takes place between the posterior layer of the greater omentum, and the anterior surface of the transverse mesocolon. For this reason, on lifting the greater omentum, you will find that the transverse colon and its mesocolon are attached, and will also be lifted (page 138).

Embryologically, the digestive tract may be divided into the foregut, midgut, and hindgut, each of which receives its individual arterial blood supply. The artery of the foregut is the **coeliac artery**, that of the midgut the **superior mesenteric artery** (fig. 73, page 152), and that of the hindgut the **inferior mesenteric artery** (fig. 76, page 152). Each of these is an independent, anterior branch of the descending abdominal aorta. It should be noted that the aorta gives off two kinds of branches: **somatic** and **visceral**. This is in keeping with the classification of bodily structures into somatic and visceral (page 5). The somatic branches of the aorta are segmentally arranged arteries which follow the body wall. In the thorax, these constitute the intercostal arteries (pages 86 and 123). In the abdomen, segmentally arranged **lumbar arteries** are given off from the aorta, and these follow a course comparable to that of the intercostal arteries in the neurovascular plane (pages 86 and 131) of the abdominal body wall. Visceral branches of the aorta include anterior, midline branches which supply the digestive tract. These are the three arteries of the gut, mentioned previously. In addition, anterolateral visceral arteries are given off, on either side, to viscera which are not gut derivatives, but have their developmental origin from the intermediate cell mass of mesoderm of the embryo (the urogenital ridge). These include the kidneys, the suprarenal glands, and the gonads (figs. 82, page 160 and 164).

With the aid of atlas illustrations, identify and dissect the celiac, superior mesenteric and inferior

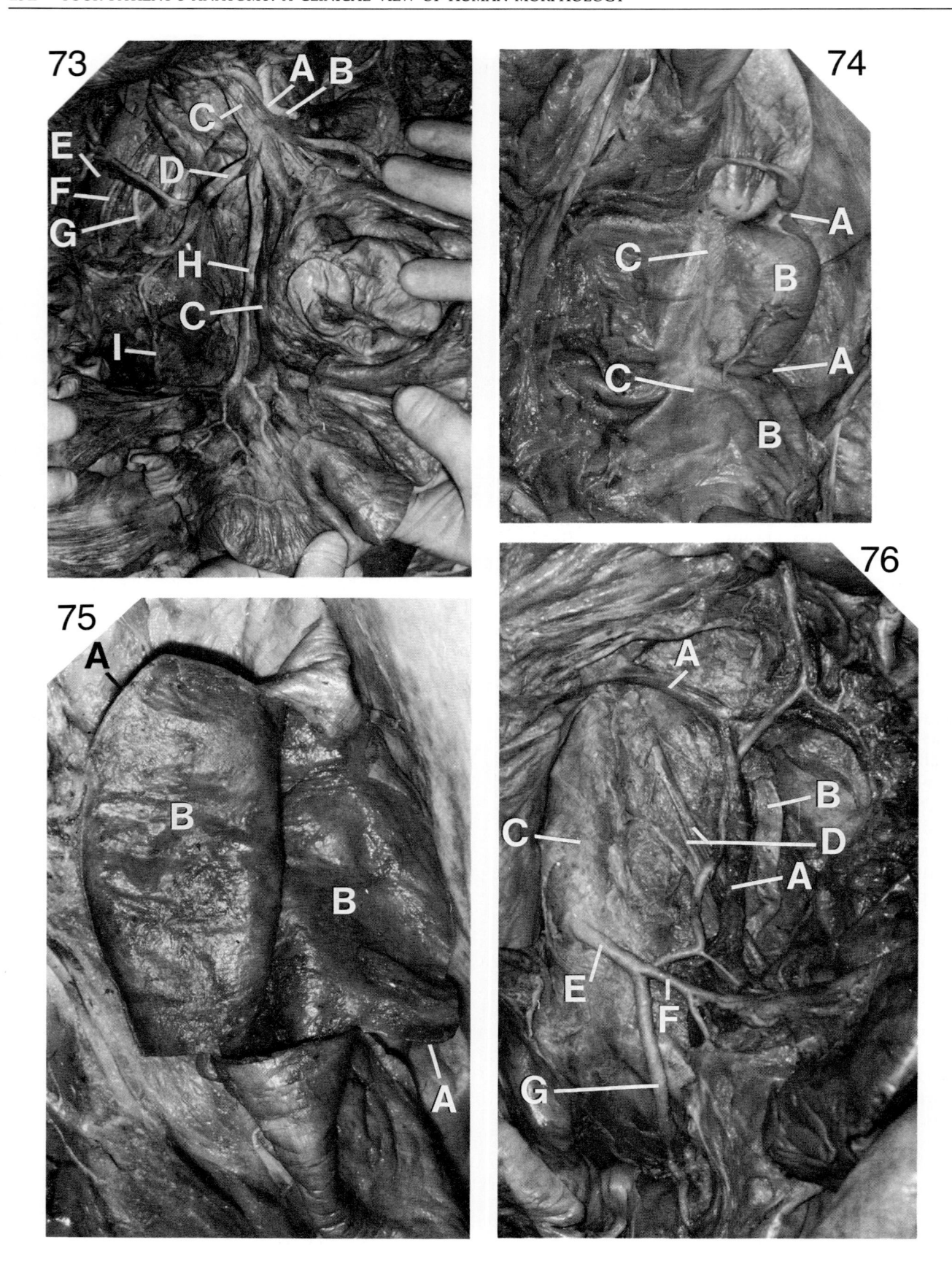

152 YOUR PATIENT'S ANATOMY: A CLINICAL VIEW OF HUMAN MORPHOLOGY

Figure 73. Superior mesenteric vessels and relations.
A. Middle colic artery, coming off, unusually, from jejunal branch of superior mesenteric artery
B. Jejunal branch of superior mesenteric artery
C. Superior mesenteric vein
D. Right colic artery
E. Right psoas muscle
F. Right gonadal vessels
G. Right ureter
H. Superior mesenteric artery
I. Ileocolic artery, coming off, unusually, from right colic artery

Figure 74. External appearance of colon.
A. Puckers (wrinkles) between haustrations
B. Haustrations (sacculations)
C. Taenia coli

Figure 75. Internal appearance of colon.
A. Cut edge of wall of colon
B. Mucosal surface of colon

Figure 76. Inferior mesenteric vessels and relations.
A. Inferior mesenteric vein
B. Left ureter
C. Abdominal descending aorta
D. Left gonadal vessels
E. Inferior mesenteric artery
F. Left colic artery
G. Superior rectal artery

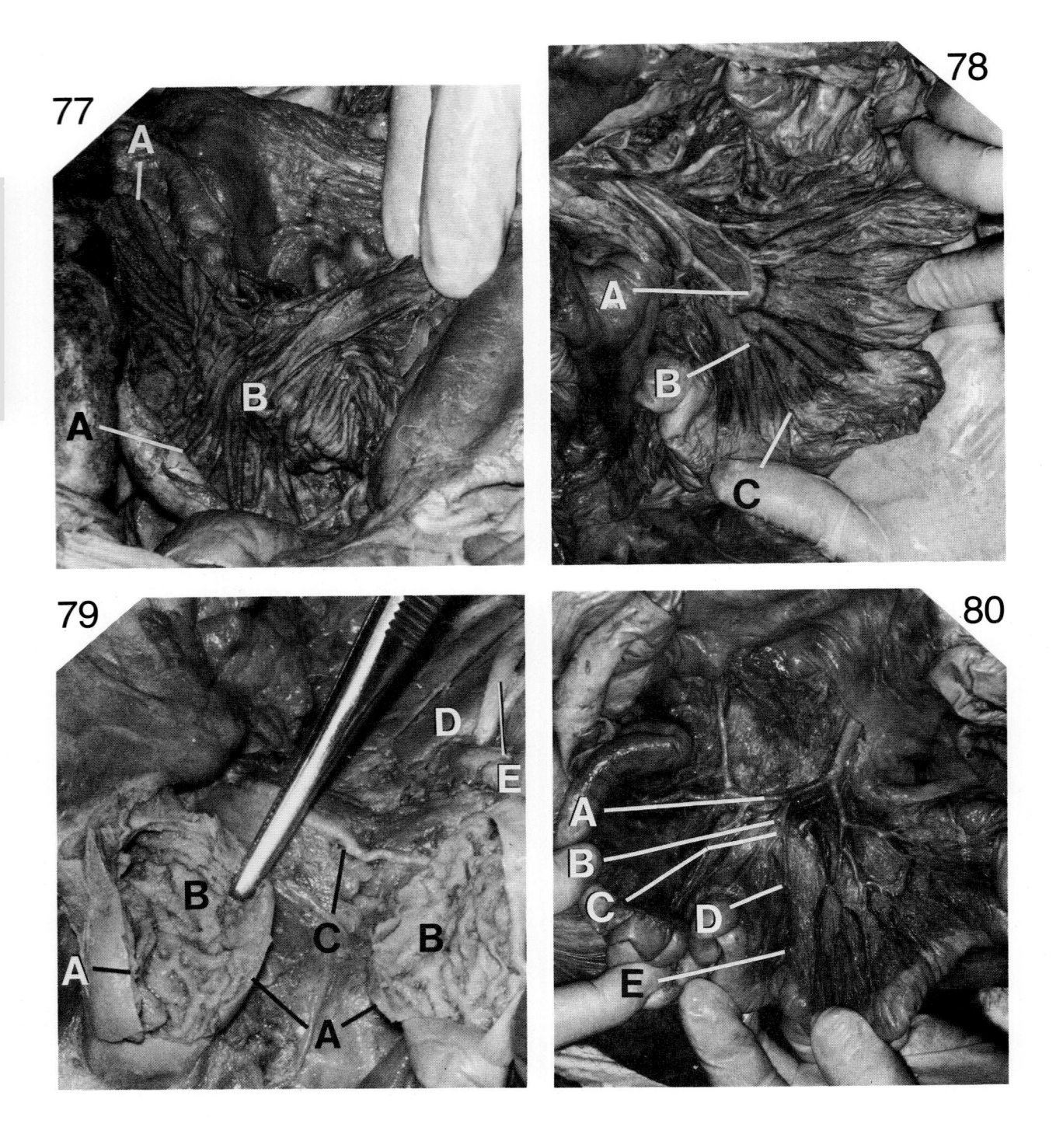

154 YOUR PATIENT'S ANATOMY: A CLINICAL VIEW OF HUMAN MORPHOLOGY

Figure 77. Mucosal surface of jejunum.
A. Cut edge of jejunal wall
B. Plicae circulares. Villi are visible as small dots

Figure 79. Mucosal surface of terminal portion of ileum.
A. Cut edge of ileum
B. Ileal mucosa
C. Ileocolic artery
D. Right psoas muscle
E. Right ureter

Figure 78. Arterial arcades of jejunum.
A. First "generation" arcade
B. Second "generation" arcade
C. Third "generation" arcade

Figure 80. Arterial arcades of ileum.
A. First "generation" arcade
B. Second "generation" arcade
C. Third "generation" arcade
D. Fourth "generation" arcade
E. Fifth "generation" arcade

mesenteric arteries, and their branches. To aid in your understanding of the courses and relationships of these structures, you should bear in mind that these arteries originate from a dorsally situated structure, the descending abdominal aorta, and pass forwards to reach parts of the gut, invaginated into the peritoneal cavity. As a loop of bowel, or any other organ, invaginates the peritoneal sac, it carries its vascular supply (and its nerve supply) with it. As a result, vessels and nerves will be found between the two layers of a mesentery, or, in the event of the mesentery becoming secondarily retroperitoneal, these structures will be found embedded in the posterior peritoneal layer.

The individual courses which arteries and veins take, in reaching their destinations, are best understood in terms of the particular regions of the mesentery they run in, and the folds or fusions which these mesenteries make. The celiac artery has three main branches: the **splenic** (fig. 83, page 160), **left gastric,** and **common hepatic arteries.** The former two run in the dorsal mesogastrium, the splenic entering the lienorenal ligament to reach the spleen, the left gastric attaining the abdominal esophagus (page 123) and the stomach itself, in a more cranial portion of the dorsal mesogastrium. From the splenic artery, the **short gastric arteries** run to the fundus (page 136), and the **left gastroepiploic artery** runs to the greater curvature, both these sets of arteries lying in the gastrosplenic ligament. The common hepatic artery, on the other hand, passes to the right, and follows the plane of the mesentery of the duodenal loop. It accordingly gets plastered onto the posterior abdominal wall, becoming secondarily retroperitoneal. From that site, the **hepatic artery proper,** a branch of the common hepatic, attains the free edge of the lesser omentum, and ascends to the liver (page 142). The common hepatic artery also gives off the **right gastric** artery, which passes to the lesser curvature of the stomach. The **gastroduodenal artery** is the remaining branch of the common hepatic, and it passes behind the first part of the duodenum, branching into the **right gastroepiploic** which passes around to the greater curvature of the stomach, and the **superior pancreaticoduodenal artery** which runs in the concavity of the "duodenal C," to supply both duodenum and pancreas.

The **superior mesenteric artery** is the artery of the midgut, and it supplies that loop of the gut which undertakes the anti-clockwise rotation of the gut and subsequent "swing" to the right, described previously (page 151). Indeed, the superior mesenteric artery may be regarded as the axis about which this loop of gut and its mesentery rotate. As a result of the mesentery of the ascending colon adhering posteriorly, the superior mesenteric artery comes to cross the third part of the duodenum anteriorly. It gives off branches (fig. 73, page 152) to the right, running in that part of the mesentery which becomes secondarily retroperitoneal, as well as to the left, in that part of the mesentery which remains free, and is related to the jejunum and ileum (page 150). An **inferior pancreaticoduodenal** artery runs upwards in the concavity of the duodenum, between the latter and the pancreas, to anastomose with the superior artery of the same name. This anastomosis marks the junction of mid- and foregut arteries, and therefore the junction of these two portions of the digestive tract. Numerous **jejunal** and **ileal branches** pass in the free portion of the mesentery, to the respective parts of the small bowel (fig. 73, page 152); figs. 78 and 80, page 154). The **ileocolic artery** (fig. 73) passes downwards and to the right, supplying the ileum, cecum, and first portion of the ascending colon. A **right colic artery** passes to the right and supplies the ascending colon, and a **middle colic artery** supplies the right-sided portion of the transverse colon. The transverse mesocolon remains partially free: posteriorly it adheres on the anterior surface of the pancreas, leaving the remainder of this part of the mesocolon free to hang downwards. Therefore, the transverse mesocolon appears to be attached posteriorly to the pancreas; it is, as mentioned previously (page 151), adherent anteriorly to the greater omentum.

The **inferior mesenteric artery** (fig. 76, page 152) comes off the aorta at the level of the third lumbar vertebra and passes downwards and to the left to supply the descending colon. The **left colic branch** passes upwards and to the left and supplies the left portion of the transverse colon and the upper portion of the descending colon. Below the origin of the left colic artery, **sigmoid branches** are given off from the inferior mesenteric artery. The inferior mesenteric artery terminates by entering the pelvis, and supplying the rectum, where it changes its name to the **superior rectal artery** (fig. 76, page 152).

The veins of the abdominal portion of the digestive tract, by and large, follow the arteries, and are similarly named. However, their destination is a final, common channel, the **portal vein,** which enters the liver (pages 123, 142 and 143). Thus the products of digestion absorbed through the walls of the digestive tract and transported into the veins supplying the canal, are carried directly to the liver for further metabolism. The portal vein is formed by the union of the **superior mesenteric vein** and the **splenic vein.** The **inferior mesenteric vein** (fig. 76, page 152) drains into the splenic vein (fig. 83, page 160). The portal system anastomoses with systemic veins in the lower esophagus (page 123) and in the rectum.

Study now the internal structure of the digestive tract as follows.

JEJUNUM

Identify a coil of jejunum, fairly close to the duodeno-jejunal junction, and similarly make a window into the jejunum as you previously have done in both the stomach and the duodenum. As previously, examine the mucous membrane, in particular, with respect to its folds (fig. 77, page 154). Using a hand lens and a pocket torch (flashlight), study the surface of the mucous membrane in detail, noting the minute projections, called **villi** (singular: **villus**). The villi subserve the purpose of increasing the surface area of the mucous membrane, an important feature in the function of absorption of the products of digestion.

ILEUM

Isolate a small loop of ileum, by two string ties, about 10 cm apart, in order to restrict outflow of ileal contents, on opening the ileum. When your string ties are securely in place, make a window into the ileum, wash the mucous membrane using wetting fluid, and study the mucous membrane of the ileum, comparing it with that of the jejunum, and noting what differences are observed (fig. 79, page 154).

Study the externally observable differences between the jejunum and the ileum, with respect to the nature of the arterial blood supply. The arteries branch more, creating more "generations" of arcades, and therefore shorter terminal branches, in the ileum than in the jejunum (figs. 78 and 80, page 154).

COLON

Identify the cecum, correlating the features of its position with the description given on page 149, point 5. Follow the cecum superiorly, as it gives way to the ascending colon (page 149). On the surface of the ascending colon, identify three externally visible muscular bands: the **taenia coli** (figs. 74, page 152, and 81, page 160). These constitute localized bands of the external, longitudinal layer of musculature of the large bowel. Recall (page 61) that the longitudinal muscles of the digestive tract are, in general, arranged externally, whereas the circular or sphincter muscle layer is, in general, internal to the longitudinal. The pharynx constitutes an exception to this rule. By following the three taenia coli inferiorly, onto the surface of the cecum, it is often possible to identify the **vermiform appendix,** the small projection from the cecum (page 149), in that the three taeni coli tend to converge on the appendix.

Returning to the ascending colon, note, in addition to the taenia coli, other characteristic features of the external appearance of the large bowel, as opposed to that of the small bowel. The large bowel is puckered, in **haustrations** (sacculations) (page 152 and fig. 74, page 152). Note, too, the **appendices epiploicae** (see fig. 81, page 160). These are small accumulations of fat, contained within the visceral peritoneum of the large bowel.

Follow the large bowel through ascending, transverse, and descending parts of the colon, and into the sigmoid colon (page 149). Select a suitable loop of large bowel, from the descending colon, and, isolating it with string ties, as previously done for the ileum, make a window into the large bowel, and compare the mucous membrane of it (fig. 75, page 152) with that previously studied in the duodenum, jejunum, and ileum.

BLOOD SUPPLY OF THE DIGESTIVE TRACT

Review the blood supply of the abdominal portion of the digestive tract, identifying any of the major vessels which you previously were unable to study because of peritoneal coverings. Use atlas illustrations to assist you in reviewing the names of the vessels. Review also the differences between the arterial supply of the jejunum and ileum, with respect to the number of "generations" of branches and resulting anastomotic arcades, and the lengths of the terminal branches (figs. 78 and 80, page 154).

Clinical Session 16

■ **TOPICS**

Posterior Abdominal Structures
Kidneys
Suprarenal Glands
Abdominal Aorta

In this session the posterior abdominal wall, and some abdominal structures which are closely related to it, will be studied. Begin by examining the **posterior abdominal wall** from the posterior aspect.

1. Identify, by inspection, the **median spinal furrow** (page 125), a median groove within which the spinous processes of the vertebrae (fig. 85, page 164) are palpable. Palpate the spinal processes, and, by counting from above downwards, identify the **lumbar spines.** Mark the 1st to 5th lumbar spines.

2. Palpate the crest of the ilium, and draw the intercristal plane on the posterior surface, confirming that it corresponds to the level of L4. Identify, by palpation, the **twelfth rib**, and mark its entire extent. Identify the posterior superior iliac spine (PSIS), by inspection (in its dimple) (fig. 64, page 128) and by palpation.

By inspection, and then by palpation, identify the powerful, vertical column of muscles, the **sacrospinalis** (fig. 84, page 160), which lies just lateral to the median spinal furrow on either side. The name derives from the fact that the muscle originates on the **sacrum** (page 4), and extends upward along and parallel to the **spinal column.** Medially, the sacrospinalis forms a lateral boundary for the furrow, and laterally, the lateral edge of sacrospinalis forms a groove which extends from the iliac crest, just lateral to the PSIS, upwards to the 12th rib, above which level it becomes less distinct due to other, overlying muscles. In lean subjects, this lateral edge is both visible and palpable; in less lean subjects, it may only be palpable, with difficulty.

Inferiorly, the median spinal furrow expands into an inverted triangle, the apex of which points downwards into the upper end of the **gluteal cleft** (the cleft between the **buttocks**). The inverted triangle described lies over the upper part of the posterior surface of the **sacrum.** The sacrospinalis muscle is so named because it takes its origin in part from this surface of the sacrum, and extends upwards on either side of the **spinal column,** all the way up to the **skull.** Palpate the inferior attachments of the sacrospinalis muscle to the iliac crest and to the sacrum. In order to bring the muscle into play, and to study one of its functions, ask your model to stand on one foot at a time, changing repeatedly from right to left and back to right and thus shifting the weight of the body from one to the other lower limb. During this exercise, palpate the muscle and its inferior attachments. The sacrospinalis column of muscle is also known as **erector spinae** (which means straightener of the spinal column).

3. Just lateral to the lateral border of the sacrospinalis muscle, between the iliac crest and the 12th ribs, the posterior wall of the abdomen is formed by another muscle, the **quadratus lumborum** (figs. 84, page 160 and fig. 85, page 164). (Quadrate means foursided; lumborum means of the **lumbar region** or small of the back.) This rectangular muscle gives attachment at its lateral edge, to the deepest (internal) of the three layers of the triple-layered abdominal musculature which you previously have studied in the anterior abdominal wall. The lateral edge of the quadratus lumborum may be palpable. Draw in the lateral border of quadratus lumborum, from the iliac crest vertically upwards to a point in the middle of the lower border of the 12th rib. Medially, the quadratus lumborum extends to the lumbar **transverse processes** and thus lies anterior to sacrospinalis, and reinforces it in the **posterior abdominal wall.**

4. The **kidneys** are posteriorly placed abdominal organs (fig. 82, page 160). Viewed from the posterior aspect, the kidney lies deep to the sacrospinalis muscle, overlapping its lateral margin, and separated from it by the quadratus lumborum muscle inferiorly, and the 12th rib, diaphragm, and pleura (pages 97 and 124) superiorly. Mark in the 11th and 12th thoracic spinous processes, in sequence above the first lumbar spinous process. The **superior pole** of the **left kidney** is at the level of T11 spinous process, about 3 cm from the midline, and the **inferior pole** is at the level of L3 spinous process, about 6 to 7 cm from the midline. The **hilus** is at L1 (i.e., slightly above the transpyloric plane) and about 3 to 4 cm from the midline. The outline of the left kidney should be drawn in, with the long axis tilted slightly obliquely to correspond with the fact that the upper pole is slightly more medial than the lower. The right kidney can similarly be drawn in: it lies, however, at a slightly lower level. The upper pole is between the levels of T11 and T12 spinous processes; the lower pole is between L3 and L4; and the hilum

(hilus) is at L2. Thus, posteriorly, the right kidney extends down to a distance of 1 to 2 cm from the iliac crest, whereas the left kidney is some 3 cm above the crest.

5. Mark in, on the **anterior abdominal wall,** the projection of the lumbar portion of the vertebral column, as follows. Recalling that the transpyloric plane passes through the intervertebral disk between L1 and L2, draw the upper margin of L1 in appropriate distance above the transpyloric plane, with the line extending about 1.5 cm to either side of the midline. Below the inferior border of L5 should be drawn a transverse line slightly broader than that of L1 (because the lumbar vertebrae increase in size in descending order) a little below the intertubercular plane (since this plane passes through the **body** of L5). Join the ends of the markings for the upper border of L1 and the lower border of L5, to indicate the lumbar vertebral column. Now mark in the bodies of the lumbar vertebrae 1 to 5, and their disks, remembering that each vertebra is larger than the one above, and that the disks comprise approximately one third of the length of the lumbar column.

6. Superior to the body of L1, draw in the body of T12 (appropriately, slightly smaller in size than L1). From the right and left sides of T12, draw in, at an angle of 45° to the horizontal, the 12th rib, about 10 cm in length.

Draw in the lateral edge of **quadratus lumborum** muscle, extending from the middle of the lower border of the 12th rib almost vertically (slightly inferolaterally) to the level of the intercristal plane, where quadratus lumborum attaches to the iliac crest (fig. 87, page 164). Superiorly, it attaches to the medial half of the lower border of the last rib, and medially it attaches to the tips of the transverse processes of all the lumbar vertebrae. As mentioned above (point 4), quadratus lumborum is a posterior relation of the kidney. Bear this fact in mind while performing palpation of the kidney, and marking the kidney on the anterior abdominal surface in the following two points (7 and 8).

7. With your subject in the supine position, palpate the kidney of first the right and then the left side. To perform this exercise, the abdominal muscles must be relaxed. Remember (page 135) that this is best achieved by placing a cushion under your model's knees. The lower end of the kidney may be palpable between two hands, one being placed in the small of the back and the other being placed over the anterior abdominal wall, at the level of the inferior pole of the kidney. The model is asked to breathe deeply and slowly, and in deep inspiration, as the posterior palpating hand lifts quadratus lumborum muscle anteriorly, the anterior palpating hand feels deeply for the inferior pole of the kidney. Because the right kidney lies at a slightly lower level than the left (fig. 82, page 160), it is usually easier to palpate the kidney on the right side. (Why should this palpation be performed in deep inspiration?)

8. You should now draw in the kidney as projected onto the **anterior abdominal wall.** To do this, mark in the transpyloric plane, the subcostal plane and the xiphisternal plane. The hilus of the right kidney is a little below the transpyloric plane, about 4 to 5 cm from the midline; the inferior pole is in the subcostal plane and the superior pole is approximately midway between the xiphisternal and transpyloric planes. The hilus of the left kidney is slightly above the transpyloric plane, about the same distance from the midline as the right kidney's, and the superior and inferior poles of the left kidney are similarly slightly higher than on the right. Bearing in mind the oblique tilt of the long axes of the kidneys, draw in both right and left kidneys projected onto the anterior body surface.

At the superior pole of both kidneys, the **suprarenal gland (adrenal gland)** (fig. 82, page 160) can be drawn in. On the right side, the gland caps the upper pole of the kidney, perched superiorly; on the left, the gland should be drawn a little further down on the medial aspect of the kidney.

9. Reexamine, by inspection and palpation, the **abdominal descending aorta** (page 149, points 2 and 3), and review its anatomical situation.

Now draw in the abdominal aorta. It commences anterior to the upper edge of L1, is about 2 cm in breadth, and extends down to the lower part of the body of L4, slightly to the left of the midline, in the intercristal plane, where it bifurcates into the **common iliac arteries** (fig. 82, page 160).

10. Referring back to page 136, point 4, draw in the duodenum. The incomplete circle which it forms commences in the transpyloric plane, at the pylorus, just to the right of the midline. The first part of the duodenum passes to the right and mounts the medial border of the kidney anteriorly. The second, or descending, part of the duodenum lies anterior to the hilus and medial border of the kidney. The third portion turns medially to leave the kidney and cross the midline, passing anterior to the aorta. The fourth part ascends to the left of the aorta.

11. Now refer to page 136, point 6, and draw in the pancreas. The head of the pancreas occupies

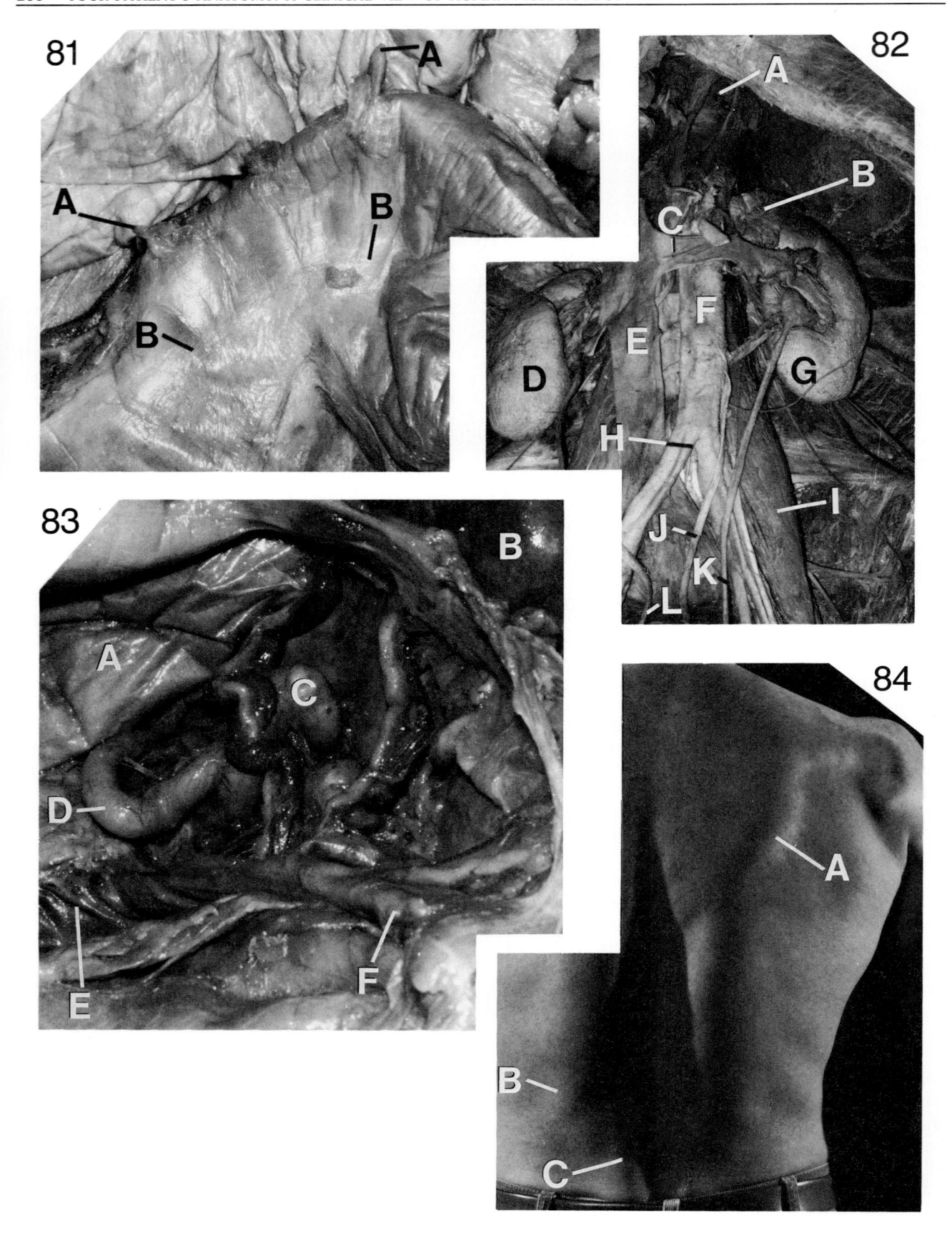

160 YOUR PATIENT'S ANATOMY: A CLINICAL VIEW OF HUMAN MORPHOLOGY

Figure 81. Loop of colon.
A. Appendices epiploicae
B. Taenia coli

Figure 82. Posterior abdominal structures.
A. Esophageal opening (in right crus of diaphragm)
B. Left suprarenal gland
C. Left renal vein
D. Right kidney
E. Inferior vena cava
F. Abdominal descending aorta
G. Lower pole of left kidney
H. Aortic bifurcation
I. Left psoas muscle
J. Inferior mesenteric artery
K. Left ureter
L. Right ureter

Figure 83. Details of branches and tributaries of splenic artery and vein.
A. Stomach
B. Spleen
C. Tip of tail of pancreas
D. Splenic artery
E. Splenic vein
F. Tributary of splenic vein from greater omentum

Figure 84. Muscles of the back.
A. Inferolateral border of right trapezius muscle
B. Left quadratus lumborum
C. Sacrospinalis

the concavity of the incomplete circle (or C-shape) of the duodenum, lying anterior to the aorta. The body of the pancreas passes upwards and to the left, in front of and then lateral to the aorta. The tail of the pancreas continues out to the left, crossing the left kidney anteriorly at its hilus and then occupying a strip on the anterior surface of the kidney, finally extending onto the visceral surface of the spleen (fig. 71, page 144).

12. The purpose of reviewing the positions of the **duodenum** and the **pancreas** at this time is to study and to understand the relationships of these structures to the **kidneys,** the **abdominal aorta,** and the **vertebral column,** and, in so doing, to understand the **posterior situation of these structures.**

To complete the session, review the emphasized key words.

Laboratory Session 16

You have, in the course of your studies of the peritoneal cavity, encountered several structures which become secondarily retroperitoneal, by virtue of fusion of one surface of their visceral peritoneum, and of their mesentery, to the posterior abdominal parietal peritoneum. However, it was also previously mentioned that some structures are primarily retroperitoneal, in that they never do invaginate the peritoneal sac. In the following, some of the primarily retroperitoneal organs, and other posterior abdominal structures, will be dissected.

Identify the right and left **kidneys,** their respective **suprarenal glands,** and the related blood vessels (fig. 82, page 160), removing, where necessary, the posterior abdominal parietal peritoneum, and surrounding connective tissue and fat. The **fascia transversalis,** which surrounds the entire parietal peritoneum, lying between it and the inner layer of abdominal musculature (page 132 and fig. 62, page 128), splits posteriorly into two layers which enclose the kidneys. Within these two layers, perirenal fat surrounds the kidneys.

Study the positional relationships of the kidneys, using the descriptions previously given (pages 158 to 159, points 4 to 8) as your guideline. Identify the landmarks described, by palpation: the eleventh and twelfth ribs; the subcostal margin (to identify the subcostal plane, and therefore L3); the highest point of the iliac crests (for the intercristal plane, L4); the iliac tuberosity (for the intertubercular plane, L5); the pylorus (for the transpyloric plane, L1–L2); and the vertebral bodies, in the midline. Identify, by palpation, the lateral border of the **quadratus lumborum** muscle (page 158, point 3, and page 159, point 6). Note the relationships of the kidneys to the quadratus lumborum muscle, confirming the descriptions given in the above mentioned points. Also identify the **psoas major** muscle which lies medial to the medial edge of the kidney on each side, and is responsible for tilting this edge and the hilus of each kidney, slightly anteriorly. The psoas major takes its origin from the sides of the lumbar vertebrae, and transverse processes, and passes inferiorly, as a thick band on either side of the vertebral column, into the pelvis and toward the lower limbs (fig. 87, page 164).

Study the anterior relationships of the right and left kidneys, by observing, and replacing **in situ** where necessary, the various organs related to this surface. On the right, the suprarenal gland is perched at the upper part of kidney. The convexity of the C-shape of the duodenum crosses the hilus of the kidney, and the pancreas is accordingly also related to the hilus. Above and to the left, the liver is the main anterior relation of the right kidney. Inferiorly, the **hepatic flexure** of the colon is related to the anterior surface of the kidney. (Hepatic means of the liver; flexure means bend, in this case the site where the ascending colon bends to the left to become the transverse colon. This flexure is called the hepatic flexure because it is situated just inferior to the liver.) At the lower pole of the right kidney, coils of small intestine are usually also an anterior relation.

On the left side, the suprarenal gland is a relation of the superior pole, though often placed somewhat on the medial aspect of the kidney. Above and to the left, the spleen is a relation of the kidney (fig. 82, page 160). Crossing the middle of the kidney, the pancreas extends across the hilus to the left, where (page 136) the tail of the pancreas reaches the spleen. Superior to the pancreas, the stomach is related to the kidney, and inferiorly, coils of small intestine encroach from the medial aspect. Laterally, the descending colon is in contact with the lower part of the lateral border of the left kidney.

Identify the **ureters** (fig. 82, page 160), the ducts which lead from the kidneys to the bladder, in the pelvis. The ureters lie on either side of the vertebral column, and when followed superiorly, toward the kidney, are seen to originate in an expanded **renal pelvis,** which emerges from the hilus of the kidney (fig. 86, page 164). Inferiorly, each ureter can be traced down to where it crosses the **brim of the pelvis,** the upper boundary of the true pelvis, which should be identified now on a skeleton. The further course of the ureter, in the pelvis, and the pelvic portions of the urinary system, will be studied subsequently.

Identify the **renal arteries and veins.** The veins lie anterior to the arteries and drain into the inferior vena cava. The left renal vein usually receives the **gonadal (testicular or ovarian) vein** as a tributary. Identify, on both sides, the gonadal vessels (figs. 73 and 76, page 152) and those of the suprarenal glands.

Remove the kidney on the right side, cutting the ureter where it joins the urinary pelvis, the renal artery and vein close to the hilus, and separating the suprarenal gland from the superior part of the kidney (but leaving the gland intact). Using a large dissecting knife, under guidance, cut a frontal section through the kidney, in order to study its internal structure. On the cut surface, identify the subdivisions of the urinary pelvis, called the **major calyces** (singular: calyx). Into each major calyx, several **mi-**

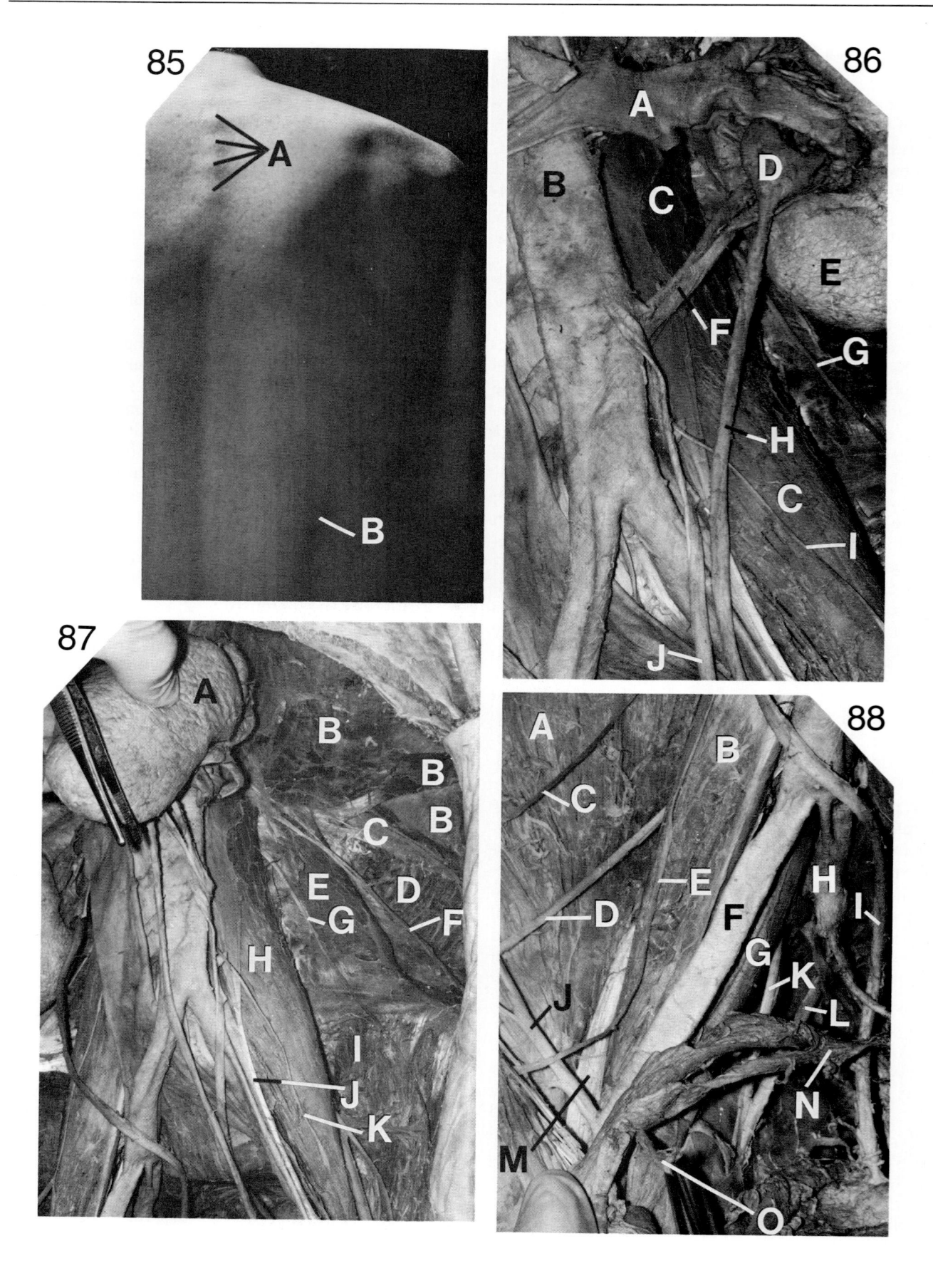

164 YOUR PATIENT'S ANATOMY: A CLINICAL VIEW OF HUMAN MORPHOLOGY

Figure 85. The back.
A. Thoracic vertebral spinous processes
B. Right quadratus lumborum

Figure 86. Relationships of hilar structures of left kidney.
A. Left renal vein
B. Abdominal descending aorta
C. Left psoas muscle
D. Left renal pelvis
E. Inferior pole of left kidney
F. Accessory left renal artery
G. Iliohypogastric nerve
H. Left ureter
I. Left genitofemoral nerve
J. Inferior mesenteric artery

Figure 87. Posterior relations of left kidney.
A. Left kidney reflected toward right
B. Left dome of diaphragm
C. Transversus abdominis muscle
D. Abdominal internal oblique muscle
E. Quadratus lumborum muscle
F. Iliohypogastric nerve
G. Ilioinguinal nerve
H. Psoas major muscle
I. Iliacus muscle
J. Psoas minor muscle
K. Genitofemoral nerve

Figure 88. Right side of pelvis.
A. Right iliacus muscle
B. Right psoas major muscle
C. Iliohypogastric nerve
D. Ilioinguinal nerve
E. Genitofemoral nerve
F. Right external iliac artery
G. Right external iliac vein
H. Right internal iliac artery
I. Right ureter
J. Circumflex iliac artery
K. Obturator nerve
L. Obturator artery
M. Femoral nerve
N. Round ligament of uterus
O. Inferior epigastric artery (cut and grasped in forceps) nor calyces open, and into the

nor calyces open, and into the minor calyces, the apex of a **pyramid** impinges. The pyramids of the kidney collectively comprise the **medulla,** whereas the rim of tissue surrounding the pyramids, and the **columns** separating them, constitute the **cortex.** Each pyramid and its related parts of cortex constitute a lobe of the kidney: in fetal life, the surface of the kidney is lobulated, but in the adult, the pyramids represent the only visible trace of a lobular structure in the kidneys. Within each lobe, numerous microscopic tubules filter the blood, and the filtrate is collected in successively larger ducts, which empty at the **papilla** of each pyramid (the "apex" of the pyramid, which opens into the minor calyces).

Before removing a suprarenal gland, for further study of its internal structure, identify the nerve fibers that enter the glands from the **celiac plexus,** situated anterior to the aorta, close to the origin of the celiac artery (pages 146 and 151). This celiac plexus, and the nerve fibers entering the suprarenal gland, are mainly of sympathetic origin: the **splanchnic nerves** (page 106, and fig. 47, page 102) contribute fibers to this plexus. Many of the connector neurons conveyed in the splanchnic nerves are destined to synapse in ganglia situated in the celiac plexus. A few fibers continue through the celiac plexus, to the suprarenal gland, without synapsing, and then finally terminate by synapsing within the medulla of the gland.

Remove the right suprarenal gland now, retaining, if possible, some of the nerve fibers entering it. Make a knife incision through the gland, to observe the internal structure, comprised of cortex and medulla. If possible, trace nerve fibers into the medulla.

The celiac plexus, is one of several small plexuses, situated in relation to the abdominal aorta. Recall (page 151) that the branches of the aorta may be classified into the **somatic (segmental) branches,** which pass out laterally into the body wall; the **anterior (median) visceral branches,** to the digestive tract (i.e., the celiac, superior mesenteric, and inferior mesenteric arteries); and **anterolateral visceral branches,** destined for nondigestive viscera, including the kidney, suprarenal glands, and so forth. The digestive and nondigestive visceral branches of the aorta carry extensions of the sympathetic plexuses with them, to their respective organs of supply. In this way, sympathetic nerve fibers reach the various abdominal viscera.

In addition to identifying the celiac plexus, and tracing the splanchnic nerves to it (from the thorax via an opening in the posterior attachment of the diaphragm), you should now trace the sympathetic trunk, similarly passing from the thorax, through the diaphragm, to the abdomen. Identify the trunk, medial to the psoas major muscle, lying anterior to the transverse processes of the lumbar vertebra. Trace the trunk downwards, to where it crosses the pelvic brim to enter the pelvis. Note that the abdominal portion of the sympathetic trunk, just like the thoracic portion, gives off small branches which follow the segmental spinal nerves, to supply such structures as arteries and sweat glands. In addition, medial branches pass, as they do in the thorax, to supply visceral structures. Branches from the lumbar sympathetic trunk contribute to the plexuses, described above, which follow the blood vessels to supply the abdominal viscera.

Examine the descending abdominal aorta (page 149, points 2 and 3; page 159, point 9 and fig. 82, page 160). Define it in its entire length, and identify its branches (page 151). Reexamine the branches to the digestive tract: the celiac, and the superior mesenteric and inferior mesenteric arteries. Review the branches and distributions of these arteries. Thereafter, study the anterolateral visceral branches, supplying the suprarenal glands, the kidney, and the gonads. Note asymmetrical distributions. Finally, identify the segmental branches to the body wall, represented in the abdomen by the lumbar arteries (which morphologically correspond to the intercostal arteries of the thorax).

Study the veins that correspond to the branches of the aorta. With the important exception of the veins of the digestive tract, the veins which accompany branches of the aorta, by and large, drain into the **inferior vena cava.** Study the position and extent of this important large vessel (fig. 82, page 160). Note the exceptions to the preceding rule, and the asymmetrical pattern of venous drainage which these create. Review the venous drainage of the digestive tract: study the inferior mesenteric vein, the splenic vein, the superior mesenteric vein, and the portal vein (page 156), as well as the terminal course of the latter. What is the significance of the fact that these veins do not drain into the inferior vena cava?

The lumbar veins, corresponding to the lumbar arteries, drain into **ascending lumbar veins** on either side. On the right, the ascending lumbar vein continues into the thorax, as the azygos vein. The left ascending lumbar vein becomes the inferior hemiazygos vein in the thorax.

On the posterior abdominal muscle wall, the ventral rami of spinal nerves of the lumbar region can be seen, emerging in relation to the psoas muscle, to participate in the **lumbar plexus.** This plexus participates in nerve supply of the abdominal body wall and the lower limbs.

Both upper and lower limbs are developmentally derived from **limb buds** with grow out from the **somatic body wall** (page 5) **ventral** to the **coronal morphological plane** (page 86). Their innervation is therefore derived from **ventral rami** (page 86) and the limb plexuses such as the lumbar plexus, cervical plexus (page 55), and others to be studied in later sessions, are entirely derived from ventral rami.

Two nerves of the lumbar plexus, the iliohypogastric and the ilioinguinal, have previously been seen, after they have curved around in the abdominal wall, to emerge in the inguinal region (page 131). Identify the nerves of the lumbar plexus at the lateral and medial edges of the psoas muscle. On the medial aspect find the **obturator nerve** and laterally, the **femoral nerve** (fig. 88, page 164), and the above-mentioned branches as they emerge, to lie on the quadratus lumborum muscle (fig. 87, page 164). Review this muscle (page 163 and page 159, point 6). At the lateral edge of quadratus lumborum, the fascia covering its anterior surface fuses with a similar layer from its posterior surface (these two layers constituting the anterior and middle layers of the **thoracolumbar fascia.** A posterior layer of this fascia covers the posterior surface of the sacrospinalis group of muscles [page 158]). From the fused anterior and middle layers of the thoracolumbar fascia, observe the origin of the transversus abdominis muscle (pages 126, 132, and 158, point 3), the innermost of the three layers of abdominal musculature (fig. 87, page 164).

Complete your dissection of the posterior abdominal wall by studying the posterior origins of the diaphragm. On either side of the upper lumbar vertebrae, a muscular crus (plural : crura) originates, curving upwards and forwards into the main muscle sheet. Lateral to each crus, a medial arcuate ligament curves over the psoas muscle, and lateral to this again, a lateral arcuate ligament curves over the corresponding quadratus lumborum muscle. Structures which pass between thorax and abdomen posteriorly, do so by passing between the crura, or through a crus, or deep to a medial or a lateral arcuate ligament. Review the passage, between the thorax and abdomen, of the aorta, the azygos vein, the thoracic duct, the splanchnic nerves, the sympathetic trunk, as well as, more anteriorly, the esophagus (fig. 82, page 160) and the inferior vena cava (pages 108, and 121).

Review the course of the thoracic duct in the thorax (pages 106, 107, and 123). Recall (page 124) that the duct enters the thorax by passing through the diaphragm. The duct begins in the abdomen, originating from an enlarged lymphatic channel, the **cisterna chyli,** which lies on the upper lumbar vertebrae to the right of the aorta. The cysterna chyli receives lymph vessels from the digestive tract, and the lower abdomen and lower limbs. Thus the thoracic duct, as stated previously (page 106), is the final common pathway for much of the lymph of the entire body.

Clinical Session 17

■ TOPICS
Pelvis

In this session, certain bony features of the pelvis will be palpated, and the surface projection of pelvic organs will be studied. Although not essential to perform at this time, a description will be given of the methods for examination of some pelvic organs by per vaginam and per rectum examination.

The abdominal region includes the pelvis and the perineum inferiorly. The total extent of the abdominal region is, therefore, from the boundary between the thorax and the abdomen superiorly, to the most inferior limit of the perineum.

1. Begin by palpating four bony landmarks which demarcate the perineum inferiorly. Anteriorly, you have previously located the **symphysis pubis** (page 120 and page 125, point 6). and this should now be repalpated. When palpated from the anterior aspect, the symphysis and the right and left pubic bones on either side of the symphysis slope slightly inferoposteriorly and have an extent of about 2 cm. Posteriorly, the tip of the **coccyx** (page 4), the most inferior componet of the vertebral column, should be palpated. You have previously noted (page 158, point 2), that the posterior surface of the sacrum can be palpated in an area with the shape of an inverted triangle, the apex of which triangle points into the upper extremity of the gluteal cleft. By palpating inferiorly in the upper part of the gluteal cleft, the lower end of the sacrum can be felt articulating with the coccyx, which is usually about 3 cm in vertical extent. Palpate down to the inferior tip of the coccyx. Note that the inferior tip of the coccyx, posteriorly, and the superior end of the symphysis pubis, anteriorly, are approximately in the same horizontal plane.

Now palpate the **ischial tuberosity** on both sides. These are the large bony prominences which the body weight rests upon in the sitting position. Each tuberosity is part of the ischium, the third of the three bilateral, fused bones, which form the bony pelvis (page 4). The other paired bones of the pelvis are the pubis (pages 120, 125, and 132) and the ilium (page 125). Identify the **gluteal fold,** the rounded, inferior border of the buttock which continues medially up into the gluteal cleft. The ischial tuberosity may be located by palpating deep to the gluteal fold. The pubic symphysis, the tip of the coccyx, and right and left ischial tuberosities constitute the diamond-shaped bony boundaries of the perineum. Note, however, that whereas the tip of the coccyx and the superior end of the symphysis pubis are on the same horizontal level, the ischial tuberosities are several centimeters inferior to that horizontal level. Thus one may conveniently think of the four points as constituting two triangles set at about right angles to each other: an anterior **urogenital triangle** (the three points of which are the symphysis pubis and the two ischial tuberosities) at about 45 degrees to the horizontal plane; and a posterior **anal triangle** (the three corners of which are at the tip of the coccyx and the two ischial tuberosities), also at about 45 degrees to the horizontal plane.

2. Referring to pages 125 to 126, points 3 to 11, reidentify the following bony features previously palpated: the iliac crest, including the posterior superior iliac spine (PSIs), the tuberosity of the ilium, and the anterior superior iliac spine (ASIS); the pubic tubercle (and the inguinal ligament extending from the ASIS to the pubic tubercle); and the pubic crest. Now palpate the **inferior pubic ramus.** Start at the inferior end of the pubic bone just lateral to the pubic symphysis. This ramus extends posterolaterally toward the ischial tuberosity on either side, becoming continuous with the **ischial ramus:** this connection between the **inferior ramus of the pubic bone** and the **ischial ramus** is also sometimes known as the **ischiopubic ramus.** The right and left inferior pubic rami meet at approximately a right angle (slightly less in males, slightly more in females) below the symphysis pubis, to form the **pubic arch.** The arch and rami can be palpated on either side of the scrotum in males, or lateral to the **labia majora** (larger, outer lips of the external genitalia) in females. (The **superior pubic ramus** [page 132] can not be palpated.)

Note, by visualizing the bony pelvis, that the bony structures palpated in points 1 and 2 in this session constitute, with associated ligaments, the boundaries of the obstetric **pelvic outlet** through which, in the female, a baby must pass at birth.

3. The **pelvic part of the abdominal region** may be considered as extending from the lumbosacral disk, the iliac crest, the inguinal ligament, and the pubic crest and symphysis, superiorly, to the **pelvic diaphragm** inferiorly. The pelvic diaphragm is a

muscular sheet, convex inferiorly, which forms a floor for the pelvis, extending from behind the symphysis pubis anteriorly to in front of the coccyx posteriorly. It is evident from this description that the anterior wall of the pelvis is much shallower than the posterior wall, because the anterior wall extends from the upper end of the symphysis pubis superiorly to the lower end of the symphysis pubis inferiorly, whereas the posterior wall extends from the upper limit of the bony pelvis (i.e., the upper edge of the sacrum) superiorly, to the coccyx inferiorly. Palpate these limits of the anterior and posterior walls of the pelvis to appreciate this point. The region inferior to the pelvic diaphragm constitutes the perineum.

4. From the above description of the pelvis and its floor, the pelvic diaphragm, it will be evident that the pelvis resembles, in shape, a shallow basin (truncated obliquely, so that it is more shallow anteriorly than posteriorly). Indeed, the word pelvis means basin. The **pelvic organs** may therefore be thought of as being supported, in the anatomical position, on the diaphragmatic floor of this basin.

5. Anteriorly within the pelvis, the **bladder** is a prominent organ. The superior surface of the bladder is usually more or less horizontal, slightly above the level of the superior end of the symphysis pubis. The bladder can be identified by percussion, which you should now do: it is relatively dull to percussion, in comparison to the loops of bowel which normally lie superior to it, and which are resonant. Identify the level of the superior border of the bladder, by percussion, and draw in a horizontal line on the anterior abdominal surface to mark this position.

6. Draw in the inferior end of the abdominal aorta. You should refer to page 159, point 9. The aorta bifurcates at the level of L4, slightly to the left of the midline. Identify this level by using the intercristal plane, and mark the point of bifurcation of the artery. Reidentify the femoral point on either side (page 126, point 14). Palpate the pulse of the femoral artery, as you did previously. Now connect the point of bifurcation of the abdominal aorta to the femoral point on either side, by double lines to indicate the two large branches that the aorta divides into, about a centimeter in diameter each. For the first one third of the distance between the point of bifurcation and the femoral point, these lines represent the **common iliac artery** (fig. 82, page 160). The common iliac artery then branches into an external and an internal branch. The lower two thirds of the lines proceeding to the femoral point represent the **external iliac artery** on each side (fig. 88, page 164). (At the femoral points, the external iliac arteries change their name and become the **femoral arteries**.) The **internal iliac arteries** (fig. 88, page 164) pass inferomedially from the site of branching to cross the **pelvic brim** and enter the deepest portion of the pelvis.

On a skeletal specimen note that the pelvic brim divides the bony pelvis into an upper, expanded portion (the **greater pelvis**) and a lower, narrower portion (the **lesser or true pelvis**). The brim commences posteriorly at the **promontory** of the sacrum (the upper, anterior edge of the first sacral vertebra) and ends anteriorly at the pubic crest. Laterally, between these two components, it is represented by a ridge on the lateral wall of the bony pelvis, part of which is formed by the **pectineal lines** (page 132). At the posterolateral end of the pectineal line on each side, identify the small, rounded **iliopubic (iliopectineal) eminence,** which marks a developmental site of fusion of the ilium (page 125) and the pubis (page 168). Right and left ridges, and the promontory which connects them, together circumscribe an approximate heart-shape. In obstetrics, the brim of the female pelvis is referred to as the **pelvic inlet,** though which the baby's head passes some time before delivery (compare with **pelvic outlet,** page 168).

7. Draw in the kidneys on the right and left sides, referring to page 159, point 8 and fig. 82, page 160. The middle of the kidney hilus is approximately at the transpyloric plane (a little below on the right and a little above on the left). From the hilus of each kidney, about 3 cm from the midline, the **ureters** should be drawn in, passing inferiorly to the point of bifurcation of the common iliac artery (fig. 88, page 164), at which point the ureter turns medially and inferiorly to cross the brim of the pelvis, and proceeds to the posterior surface of the bladder.

8. Per vaginam (PV) examination is an examination performed by passing, under the hygienic conditions described on page x, a lubricated finger (or two fingers) into the **vagina**. In such an examination, the lower portion of the **cervix** of the **uterus** can be palpated projecting into the upper part of the vagina. On bimanual examination (i.e., placing an external hand on the anterior abdominal wall while the other hand is palpating within the vagina), it is possible to palpate the **fundus** (superior portion) of the uterus. The external hand should be placed several centimeters above the symphysis and pressure directed inferiorly. The major part of the uterus, between the fundus superiorly and the cervix inferiorly, is called the **body** of the uterus.

9. On per rectum (PR) examination (an examination performed by passing, under hygienic conditions, a lubricated finger through the **anus** [ring or external opening of the digestive tract], into the **anal canal** and **rectum**), the tone (spontaneous tension) of the **external anal sphincter muscle** can be felt as the muscle contracts around the examining finger. This tone is an important clinical indication of the integrity of the nerve supply to the muscle (i.e., the **perineal branch of S4** and the **inferior rectal branches** of the **pudendal nerve** [S2 and S3]). In the very frequent condition of degeneration of an intervertebral disk ("slipped disk"), these nerve roots may be affected by pressure.

In the female, the cervix of the uterus can be felt through the anterior wall of the rectum. Posteriorly, the sacral promontory may be felt far superiorly, and the ischial tuberosity may be felt laterally, and, superior to the ischial tuberosity, the **ischial spine** is also palpable. (In some subjects the latter may also be discernable externally by deep palpation medial and superior to the is chial tuberosity.)

In the male, the sacral promontory, ischial tuberosity, and ischial spine may similarly be palpated. In addition, anteriorly, the **prostate**, which is situated in the midline, on the posterior surface of the bladder, can be palpated. Lateral to the prostate on either side, the **seminal vesicles** are rarely palpable.

The **rectum** extends from the middle of the sacrum (S3), following the concavity of the sacrum, to the level of the coccyx where, as it passes through the pelvic diaphragm, making a sharp posterior bend, it becomes continuous with the **anal canal.** The latter is thus not a true pelvic, but a perineal organ. The rectum is superiorly in continuity with the last part of the colon, the sigmoid colon, which establishes continuity between the descending colon and the rectum, as it passes from the left iliac region of the abdomen toward the pelvis (page 149).

To complete this session, review the emphasized key words.

Laboratory Session 17

In the following dissection of the pelvis you should have an articulated skeleton available. Begin by identifying on the cadaver (making reference to the skeleton where necessary) the bony landmarks you previously palpated on the living body. Access to some of these structures is considerably facilitated by study of a median sagittal section through the pelvis. For this reason your cadaver will be prosected in this manner, after the bowel has been appropriately tied off and cut. In addition, a transverse section will be made on one side, providing you with a half pelvis which remains attached to the trunk and one lower limb, and a severed second half pelvis which is attached to the other lower limb. You should study both halves, to obtain an integral impression of the region.

Palpate, both externally, through the skin, and internally, from the abdominal cavity, the **symphysis pubis,** the **coccyx,** and the **ischial tuberosities** on each side (page 168). Similarly palpate the **body of the pubic bone** and the **ischiopubic ramus,** by combined internal and external approach in the cadaver (page 168, point 2).

Palpating from within the abdominal cavity, identify the inferior limits of the pelvis (i.e., the **pelvic floor**), formed by the muscular sheet, concave upwards, called the **pelvic diaphragm.** The actual muscular pelvic diaphragm will not be visible at this time, as it is covered by peritoneum, and related connective tissue, and also obscured by the pelvic organs situated superior to it. Nevertheless, by observation and palpation, it is possible to obtain an impression of the basin shape of the pelvis, and of its shallow, curved floor. By inspection of the sagittal section, and the description of anterior and posterior attachments of the pelvic diaphragm, given on page 168, point 3, identify this thin sheet of muscle as it curves from the symphysis pubis to the coccyx.

Identify the major pelvic organs, obtaining, at the same time, an impression of their peritoneal relationships. In order to understand these, it is simply necessary to bear in mind that these organs, as is the case for other abdominal organs, begin their existence outside of the peritoneal sac. Some remain extraperitoneal, and others invaginate the peritoneum, partly or completely, to thus acquire the equivalent of a ''mesentery'' (page 134), or ''mesocardium'' (page 111). In the abdomen, organs that remain outside of the peritoneal sac are mainly situated dorsal to the sac, and are accordingly called retroperitoneal (retro means behind). In the pelvis, organs will invaginate the peritoneum in part from behind, but also in part, and in some cases mainly, from the caudal (inferior) aspect. Organs which only partially invaginate, or do not invaginate at all, may thus be better described as infraperitoneal, rather than retroperitoneal.

Referring to the description given on page 169, point 5 as a guideline, identify the **bladder,** noting that it partly invaginates the peritoneal cavity from below. Using both halves of the sagittally sectioned pelvis, follow the lumen of the bladder anteroinferiorly, into the **urethra.** The urethral exit from the bladder is situated at the inferior angle of a triangular region of the posterior wall of the bladder, the **trigone,** the mucous membrane of which is smooth, in contrast to the folded appearance of the lining of the rest of the bladder. At the superolateral angles of the trigone, identify the openings of the right and left **ureters** (page 163).

As the urethra passes the plane of the pelvic diaphragm, it enters the region known as the perineum (page 168, point 1). The further course of the urethra in the perineum will be studied later. Posterior to the bladder, identify, in the female, the **uterus** (page 169, point 8). In the sagittal section, identify the **vagina,** posterior to the urethra. By placing the index and middle fingers of the right hand into the vagina, and the left hand onto the anterior abdominal wall, simulate a PV examination (page 169, point 8), thus palpating the **cervix** of the uterus within the vagina, and the **fundus** externally. (Note that surgical removal of the uterus [hysterectomy] is not an infrequent operation, and should be considered as a possibility, in the event that you have difficulty in identifying the uterus in your female cadaver.)

Posterior to the uterus, identify the rectum, establishing continuity of it with the sigmoid colon (pages 149, 134, and 157). Perform a PR examination (page 170, point 9), and identify the structures indicated in the description given in the above mentioned guideline. In the male, observe and palpate, PR, the **prostate gland,** situated at the junction of the bladder with urethra and the seminal vesicles.

In the female, examine the **uterine tubes,** extending laterally from the junction of the **body** and **fundus** of the uterus (page 169, point 8). On either side, the uterine tubes extend out laterally, toward the lateral wall of the pelvis. The uterus and tubes may be thought of as having invaginated the peritoneal cavity from below. The uterus is in continuity, inferiorly, with the vagina, which is virtually not invaginated into the peritoneal sac at all. Thus the uterus acquires for itself a visceral covering, which reflects onto the parietal peritoneum posteriorly, just in front of the rectum, and anteriorly, onto the superior surface of the bladder. Laterally, the uterine tubes, being attached near the superior end of the uterus, carry with them, on invaginating the peritoneal sac, a double layer of peritoneum which forms a "mesentery" for the uterine tubes. The term **mesosalpinx** is applied to the portion of the double fold adjacent to the uterine tubes (salpinx means tube: compare with salpingopharyngeus, page 60, point 10). The remainder of the broad fold of peritoneum, raised, in a more or less vertical (coronal) plane, is called the **broad ligament.** Study the mesosalpinx and the broad ligament. The **ovary** is situated, in development, between the two layers of the broad ligament, far out laterally, adjacent to the lateral wall of the pelvis. In the adult, the ovary is seen to have evaginated the posterior layer of the broad ligament, carrying with it a double layer, called the **mesovarium.** Because of the close proximity of the ovary to the lateral pelvic wall, the mesovarium reflects onto the parietal peritoneum of this wall, and therefore serves to convey to the ovary its blood vessels (the ovarian artery and vein), passing between the pelvis and the abdomen proper (page 163).

By palpating the broad ligament between the index finger and the thumb, attempt to identify a cord-like structure, the **round ligament of the ovary,** running from the ovary to the lateral surface of the uterus, somewhat inferior to the uterine tube. Reexamine the inguinal canal, on the exterior of the abdominal wall. Reflecting the anterior abdominal wall back, identify, on the interior, the **round ligament of the uterus** (page 130 and fig. 88, page 164), the vestigial structure which runs in the inguinal canal in the female, ending in the labium majus. On the pelvic wall, trace by palpation, the round ligament of the uterus posteromedially to the uterus. This ligament meets the uterus close to where the round ligament of the ovary meets it. Ebryologically, the round ligament of the ovary and the round ligament of the uterus represent a continuous cord, running from the gonad to the **labioscrotal fold** (the embryological fold which in the male develops into the scrotum, and in the female develops into the labium majus). The male homologue of this cord, called the **gubernaculum,** plays a role in guiding the testis into the scrotum, in the descent of the testis through the inguinal canal, late in fetal life (pages 127 to 130). In the female, the cord is also important in positioning of the ovary. The uterus secondarily attaches to the cord, dividing it into the round ligament of the ovary and the round ligament of the uterus. This embryological explanation accounts for the round ligament of the uterus constituting the contents of the inguinal canal, and its termination in the labium majus, in the female.

In the male, identify the spermatic cord in the inguinal canal, on the external surface of the abdominal wall. Reflect the abdominal wall back, to follow the spermatic cord on the internal surface of the pelvis. Trace it inferomedially, by palpation through the peritoneum, to the back of the bladder, where it passes over the ureter, to reach the posterior surface of the bladder. At this site, palpate the seminal vesicle (page 170, point 9), just lateral to the terminal portion of the vas deferens.

Turn now to the disarticulated half of your pelvis specimen, and reflect the peritoneum to expose the following structures. In the female, study the uterus, uterine tube, ovary, and ovarian vessels. In addition, identify the round ligament of the ovary and the round ligament of the uterus, running toward the internal inguinal ring. In the male, study the **vas deferens** in its course from the internal inguinal ring onto the lateral pelvic wall. Confirm that after looping over the ureter, it converges with the duct of the opposite side, close to the midline, on the posterior surface of the bladder. Lateral to the vas, identify the **seminal vesicle.** The small duct of the latter gland combines with the terminal portion of the vas, forming the **ejaculatory duct.** The latter enters the posterior surface of the prostate, and ultimately opens into the **prostatic urethra** (the first portion of the urethra, which is surrounded by the prostate, as it leaves the neck of the bladder).

Review (page 171) the concept of partial invagination by pelvic organs of the peritoneal sac **from the caudal aspect.** As elsewhere, this results in double folds of visceral peritoneum ("mesenteries") and spaces or pouches adjacent to these folds. Such pouches have a special clinical significance in the pelvis, however, because the upright stance determines that these pelvic pouches are gravitational targets, with resulting susceptibility to infection.

In the female identify, anterior to the broad ligament, the **paravesical fossa** on each side of the bladder, and the **vesicouterine pouch** between the bladder and the uterus. Posterior to the uterus find the **recto-uterine pouch (of Douglas)** between uterus and rectum, and the **pararectal fossa** on either side of the rectum. In the male find paravesical and pararectal fossae, and the **rectovesical pouch.**

Reidentify the descending abdominal aorta (pages 149, points 2 and 3, and page 159), and the inferior vena cava (page 166). Identify the bifurcation of these vessels (page 149, point 3), and follow the common iliac arteries, and their branches, the external and internal iliac arteries. Trace the external iliac artery to the point where it passes deep to the inguinal ligament, where it changes its name to the femoral artery (page 169, point 6). Reidentify the inferior epigastric branch of the external iliac artery (page 132), given off just above the inguinal ligament, and around which the spermatic cord or round ligament of the uterus turns, at the internal inguinal ring (page 132 and fig. 88, page 164).

The internal iliac artery crosses the **pelvic brim** (page 169, point 6), to enter the true pelvis (fig. 88). It gives off important branches to the pelvic viscera. Identify the **obliterated umbilical artery** (page 133), and the **vesical** (of the bladder), **uterine,** and **middle rectal** branches. In the female one of the vesical branches is replaced by a **vaginal** branch. (Recall that the superior rectal artery is the termination of the inferior mesenteric artery [page 156 and fig. 76, page 152].) The **inferior rectal artery** is given off in the perineum, from a further branch of the internal iliac artery: the **internal pudendal artery.** This branch leaves the pelvis above the **ischial spine** (page 170: also review the spine on a skeleton now). The artery then recurves, inferior to the ischial spine, to pass forwards into the perineum. Other branches given off by the internal iliac artery include the **obturator artery** (fig. 88, page 164), which leaves through the **obturator foramen** of the bony pelvis, to enter the lower limb, and **superior** and **inferior gluteal arteries,** which leave through the space above the ischial spine, to pass to the region of the buttock (gluteal means of the buttock; compare with gluteal cleft, page 158, and gluteal fold, page 168). The iliolumbar and lateral sacral arteries comprise the remaining named branches of the internal iliac artery. These structures will be studied further in connection with the lower limb.

The veins of the pelvis follow, in general, the pattern of the arteries. The **superior rectal vein** drains, through the inferior mesenteric vein, into the portal vein (page 156). The **middle** and **inferior rectal veins,** on the other hand, drain into the **internal iliac vein.** Because of the portosystemic anastomosis thus formed in the region of overlap between superior and middle rectal veins, increased portal vein pressure can cause swelling of rectal veins, called hemorrhoids (pages 123 and 156).

On the posterior pelvic wall, corresponding to the concave anterior surface of the sacrum (which should be observed on the skeleton), identify the sacral spinal nerves, which combine to form the **sacral plexus.** With nerves from the lumbar plexus (pages 166–167), this constitutes the lower limb plexus, derived from ventral rami of spinal nerves. Recall that this origin of the nerve supply to the limbs is due to the fact that the limb buds develop ventral to the coronal morphological plane (pages 166 and 86).

Some nerves of the sacral plexus accompany various branches of the internal iliac artery, which you have observed. Thus, for example, gluteal nerves accompany the gluteal arteries to the buttock. Of particular importance is the **pudendal nerve,** which accompanies the **internal pudendal artery** around the ischial spine, to reenter the perineal portion of the pelvis (page 170). Although situated in the pelvis, the sacral plexus is mainly concerned with nerve supply of the lower limbs, and is therefore appropriately further studied with that region.

Of relevance to the nerve supply of the viscera of the pelvis are the **superior** and **inferior hypogastric sympathetic plexuses,** and the **parasympathetic nerves of S2 and S3** (page 106) that form the **pelvic splanchnic nerve.** The inferior hypogastric plexus is situated in the plane above the pelvic floor (pelvic diaphragm), and is joined by the pelvic splanchnic (parasympathetic) nerve. The plexus extends up onto the sides of the viscera, including the rectum, seminal vesicals and prostate, urinary bladder, uterus, and upper part of the vagina. The pelvic splanchnic component supplies the visceromotor fibers to the rectum, the bladder, and the sigmoid, descending and the terminal part of the transverse colon, and vasodilator fibers to the erectile tissue of the penis and clitoris. Male ejaculation is mediated by sympathetic fibers supplying visceral musculature in the vas deferens, seminal vesicals, and prostate. Somatic nerve supply to certain perineal muscles (see later) is also involved in the reflex.

Reidentify the pelvic diaphragm, as seen in sagittal section, on the disarticulated half of your cadaver specimen. Trace the muscle fibers laterally, to their origin. Study a skeleton, as well as your cadaver, to establish the line of origin of this complex muscle: the line extends from the posterior surface of the body of the pubis, to the ischial spine. From this line of origin, fibers insert into a median raphe, which attaches posteriorly to the tip of the coccyx. The diaphragm is interrupted in the midline by several structures which pass through it from pelvis to perineum. As a result, the insertion of left- and right-sided fibers into the median raphe is interrupted by these structures. In some cases, fibers insert into the side walls of the organs passing through the diaphragm, or, as in the case of the rectum, fibers loop around the posterior surface of the organ. Because of the complexity introduced by the passage of organs through the muscle diaphragm, as well as phylogenetic heterogeneity (i.e., differing evolutionary origin), different parts of the pelvic diaphragm have a degree of independence. Thus the following parts of the diaphragm are recognized: in the male, a **levator prostatae;** and in the female, a **sphincter vaginae; puborectalis** (which loops around the rectum as it bends posteriorly to become the anal canal), and **iliococcygeus** (which takes its origin from the border between the ischium and ilium, on the lateral pelvic wall, and inserts into the lower portion of the coccyx and the raphe between the tip of the coccyx and the anorectal flexure). The pelvic diaphragm is sometimes collectively called the **levator ani** muscle, because of a function it carries out in suspending and elevating the anal canal.

Note that the line of origin of the pelvic diaphragm crosses the region of the obturator foramen, which in the living organism is closed by a tough membrane, on the internal surface of which the **obturator internis** muscle is attached. The tendon of this muscle leaves the pelvis posteriorly, to act on the lower limb, and the muscle should therefore be studied with the lower limb rather than the pelvis. However, it should be identified at this time, by palpation, as forming part of the lateral wall of the pelvis.

Clinical Session 18

■ **TOPICS**

Perineum
Urogenital Triangle
Anal Triangle

In this session, the anatomical structures of the perineum will be studied in both sexes.

The perineum (see figs. 89, 90, 91, and 92, page 176) is the most inferior portion of the abdomen. It is comprised of an anterior, traingular region, called the **urogenital triangle,** and a posterior, triangular region called the **anal triangle.** The urogenital triangle faces anteroinferiorly, at an angle of 45° with the horizontal, and the anal triangle faces posteroinferiorly, also at an angle of 45° to the horizontal (see page 168, point 1). The two triangular regions share a common base, the imaginary line connecting the **ischial tuberosities.** The apex of the urogenital triangle is at the **symphysis pubis,** and the apex of the anal triangle is at the tip of the **coccyx.** The perineum is most easily examined with the model in the supine position.

1. Commence by reidentifying the bony landmarks which delimit the perineum and form the **pelvic outlet** (page 168, points 1 and 2).

2. In the male, the urogenital triangle contains the **scrotum** and the **penis.** A **median raphe** (seam) passes from the front of the anus (in the anal triangle), to the inferior surface of the scrotum and onto the inferior aspect of the penis (fig. 89, page 176).

3. The scrotum is the pouch of skin, subcutaneous tissue, and other coverings which are evaginated from the anterior abdominal wall, preceding the descent of the testis in fetal life (page 127, point 19). A midline **septum** separates right and left halves of the scrotum.

In the evagination of the anterior abdominal wall some fibers of the **internal oblique** and **transversus** muscles are carried into the scrotum as the **cremaster muscle** (page 132). With the model in the supine position, test the **cremasteric reflex** by stroking the medial aspect of the thigh with a sharp, but nonabrasive instrument (such as the blunt end of a pen) and observe contraction of the cremaster muscle, as evidenced by movement of the scrotum. This reflex tests the integrity of the first lumbar nerve.

4. Within each half of the scrotum, the testis should be palpated with care. The long axis of the testis is approximately vertical. On the posterolateral aspect of the testis, the **epididymis** (the organ into which sperm cells pass on leaving the testis) should be gently palpated. An expanded, superior pole of the epididymis, the **head,** can be distinguished from the central portion or **body** and the inferior segment or **tail.** From the epididymis, the sperm pass to the **ductus deferens (vas deferens).** At its inferior pole, the epididymis continues into the vas deferens, where the tail of the former organ bends back on itself to connect with the vas. The vas passes onto the medial aspect of the epididymis (posteromedial aspect of the testis) and can be gently palpated there. Superiorly, the vas continues above the testis, within the scrotum, toward the inguinal canal, where it has been previously palpated (as a component of the spermatic cord: page 127, point 19). Refer to the latter description again, and reexamine the **spermatic cord** and the **inguinal canal** as described on pages 127–130. After entering the abdominal cavity through the inguinal canal, the vas deferens passes to the posterior aspect of the bladder. Here the duct of the seminal vesicle unites with the vas to form the ejaculatory duct (page 172). The right and left ejaculatory ducts enter the **urethra,** a duct which commences at the base of the bladder, surrounded by the **prostate gland** (page 170, point 9), and passes from there into the **penis.**

5. The **body** or **shaft** of the **penis** is situated anterior to the scrotum. The latter partly conceals the **base** of the penis. However, the **bulb** of the penis (part of the base) can be palpated through the scrotum, in the midline, just anterior to the posterior border of the urogenital triangle (i.e., just anterior to an imaginary line connecting the two ischial tuberosities). The bulb of the penis continues forwards into the **corpus spongiosum,** which contains the penile part of the **urethra** (point 4 in this session), and which can be palpated in the midline on the inferior (posterior) surface of the penis. The remaining parts of the base of the penis are the right and left **crura.** Each **crus** is firmly applied to the ischiopubic ramus (page 168), and each continues anteromedially to become continuous with its corresponding **corpus cavernosum.** On the superior (anterior) surface of the penis, the two **corpora cavernosa** can be felt, one on either side of the midline. The two anteriorly

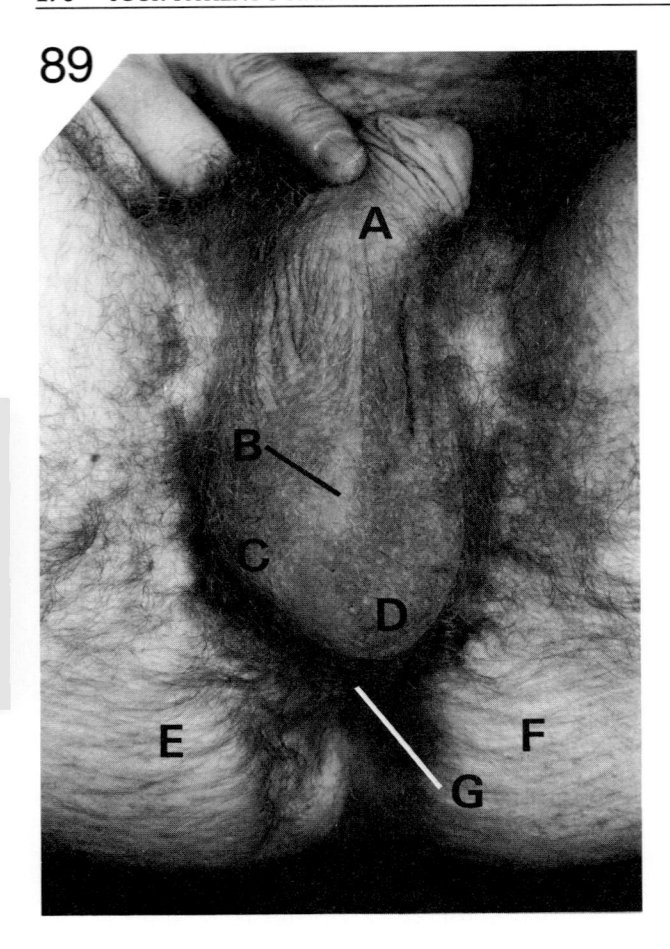

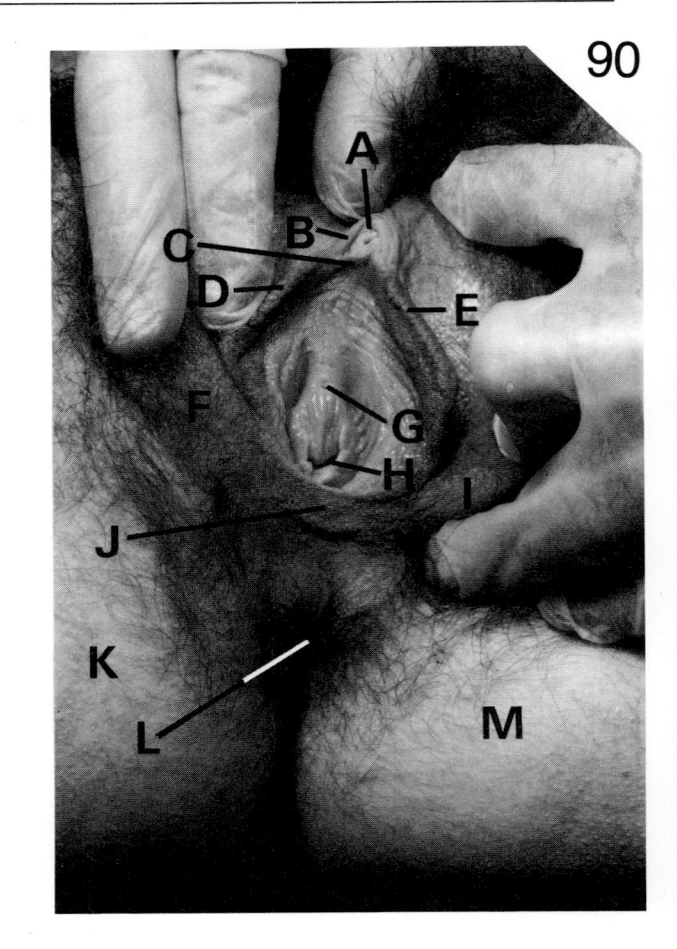

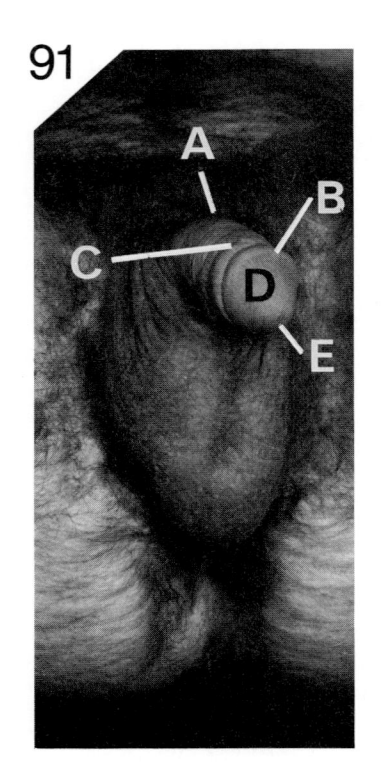

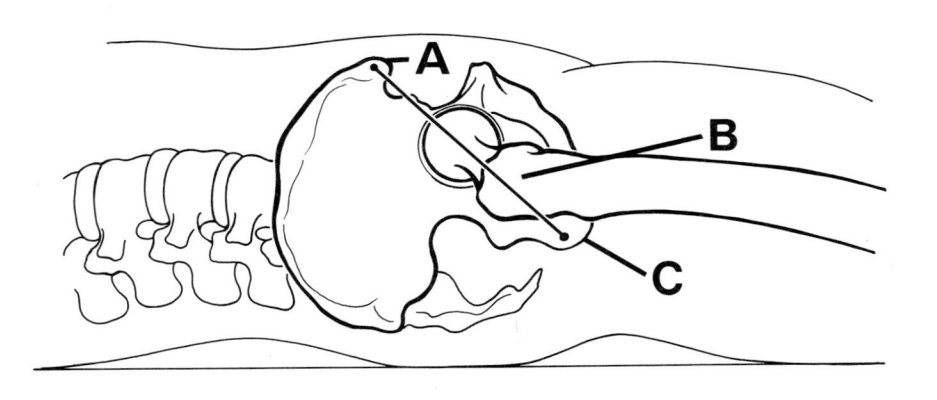

176 YOUR PATIENT'S ANATOMY: A CLINICAL VIEW OF HUMAN MORPHOLOGY

Figure 89. Male perineum, showing posterior (inferior) aspect of penis.
A. Body (shaft) of penis
B. Median raphe of penis
C. Bulge of right testis within scrotum
D. Bulge of left testis within scrotum
E. Position of right ischial tuberosity
F. Position of left ischial tuberosity
G. Anus

Figure 90. Female perineum with labia majora separated.
A. Clitoris
B. Prepuce of clitoris
C. Frenulum of clitoris
D. Right labium minus
E. Left labium minus
F. Right labium majus
G. Urethral opening
H. Vaginal opening
I. Left labium majus
J. Posterior commissure
K. Postion of right ischial tuberosity
L. Anus
M. Position of left ischial tuberosity

Figure 91. Male perineum showing anterior (superior) aspect of penis.
A. Position of symphysis pubis
B. Corona
C. Anterior (superior) surface of penis
D. Glans
E. Urethral opening

Figure 92. Nelaton's line in supine subject (see page 282).
A. Anterior superior iliac spine
B. Greater trochanter of femur
C. Ischial tuberosity

placed corpora cavernosa and the single, midline, posterior corpus spongiosum together constitute the **tripartite** (three-part) erectile **body** of the penis. The penis ends in the **glans,** which has a flared rim, called the the **corona.** The glans is continuous with the corpus spongiosum, and the urethra opens to the exterior on the tip of the glans (figs. 89 and 91, page 176).

6. In the female, the **labia majora** are **homologous** to the scrotum of the male, but instead of being fused in the midline, they enclose an elliptical opening, the **pudendal cleft.** Anteriorly, the labia majora unite in a median eminence, the **mons pubis.** Posteriorly, the labia join at the **posterior commissure.** Medial to the labia majora, the **labia minora** are seen as elongated folds of skin. The male homologues of the labia minora contribute to the **urethra** in the penis. Anteriorly, each **labium minus** divides into a pair of folds which enclose the glans of the **clitoris.** The space between the two labia minora is the **vestibule.** Posterosuperiorly, the vestibule contains the **vaginal** opening (fig. 90, page 176).

7. The **clitoris** is the homologue of the penis, but does not contain the **urethra.** The latter opens separately in the female, between the clitoris and the vaginal opening.

As in the penis, the clitoris is a tripartite, erectile structure. The **body** contains right and left **corpora cavernosa,** each continuous with a **crus** which, as in the male, is attached to the corresponding ischiopubic ramus. The homologue of the **bulb** of the penis is the **bulb of the vestibule,** which, to accommodate the midline vaginal opening, is a bilateral structure situated deep to the labia minora. Anteriorly, each half tapers toward the **glans** of the clitoris, the erectile homologue of the glans of the corpus cavernosum of the penis.

8. In both sexes, the anal triangle contains the **anus,** in the midline between the ischial tuberosities, anterior to the **tip of the coccyx.** The anus (ring) is the caudal opening of the gut, the last portion of which is the **anal canal.**

To complete this session, review the emphasized key concepts.

Laboratory Session 18

Study the perineum in your cadaver, making frequent reference to a skeleton. Review the subdivision of the perineum into the urogenital triangle and the anal triangle (pages 175 and 168). Reidentify the bony landmarks which delimit the perineum. Skin the anal and urogenital triangles (page 175) and study their contents on the detached half pelvis and lower limb.

The anal triangle contains the anus, the external opening of the anal canal, which is the last portion of the digestive tract. By definition, the anal canal is the portion of the tract which extends from the pelvic diaphragm to the anus. The rectum, which follows the lower part of the curvature of the sacrum (which you should observe now on a skeleton), extends downwards and anteriorly. At the anorectal flexure, the canal turns sharply backwards, so that the anal canal is directed inferoposteriorly, at about a 90 degree angle to the rectum. The flexure is partly brought about by the **puborectalis** muscle, the fibers of which form a "sling," posterior to the canal, about which the flexure is formed (page 173).

Surrounding the anal canal is a fat-filled space, the **ischioanal fossa** (commonly but erroneously called the ischiorectal fossa). The fossa is the space between the walls of the anal canal, medially, and the ischium, laterally. By carefully dissecting away the fat, study the contents of the fossa. The pudendal nerve and internal pudendal vessels (page 173), which left the pelvis above the level of the pelvic diaphragm by passing above the ischial spine, return into the perineum by running in the lateral wall of the ischioanal fossa. In the fossa, branches are given out medially, to the anal canal. The fossa has an extension upwards, deep to the inferior edge of the gluteus maximus muscle. (The inferior edge of the muscle is responsible for the gluteal fold [page 168].) Surrounding the lower portion of the anal canal, an external anal sphincter muscle can be identified. Review the nerve supply of this functionally important muscle (page 170).

In the **urogenital triangle,** identify the external genitalia in both male and female cadavers, referring to the descriptions given on pages 175 to 178. In the male, incise the scrotum anteriorly and study its layers, which correspond to those of the abdominal wall, as evaginated in connection with descent of the testis (page 127, point 19). In addition, a tubular prolongation of the parietal peritoneum precedes descent of the testis into the scrotum. Subsequently the testis invaginates this prolongation, which consequently is called the **tunica vaginalis testis.** The connection of the tunica with the peritoneal sac normally becomes obliterated, but if this fails to occur, herniation into the scrotum may occur by this route.

Examine the position of the testis and epididymis within the scrotum (page 175, point 4). At the inferior pole of the testis, a remnant of the **gubernaculum** can be observed. In fetal life, the testis develops in the abdomen, and the gubernaculum extends from the abdominal testis, through the inguinal canal, to the scrotum. In the descent of the testis through this course, the gubernaculum appears to play a role in guiding the testis to its destination (pages 127 to 130). Incomplete descent of the testis is a frequent and important developmental problem in boys. As mentioned previously (pages 130 and 172), in the female, the labium majus is homologous to the scrotum, and the gubernaculum in the female extends morphologically from the ovary to the labium majus. The uterus becomes secondarily attached to the gubernaculum, dividing it into a round ligament of the ovary and a round ligament of the uterus. The latter is the vestigial structure which passes through the female inguinal canal (fig. 88, page 164).

Make a longitudinal section through the testis, displaying its subdivision in lobes, and the minute seminiferous tubules, within which the sperm cells develop. Identify the epididymis, posterolateral to the testis, to which the sperm cells pass on leaving the testis. At its inferior pole, the epididymis continues into the vas deferens, which ascends to the external inguinal ring, enters the inguinal canal, and follows the further course, within the pelvis, which you previously studied (page 172).

Study the **penis** by dissecting its three erectile components; the **corpus spongiosum** on the inferior (posterior) surface, and the two **corpora cavernosa** on the superior **(anterior)** surface. Separate these structures from the fibrous sheath, called the **tunica albuginea,** which surrounds them. To appreciate the relationships, study both halves of the bisected perineum. The corpus spongiosum contains the **urethra** (pages 175 to 178, point 5), and is expanded at its anterior tip, into the **glans.** In the urogenital triangle, the other end of the corpus spongiosum is expanded into the **bulb of the penis** (page 175, point 5), which is attached to the inferior surface of the **perineal membrane.** This membrane fills the triangular area of the urogenital triangle, being attached laterally to the ischiopubic rami, and having a posterior free edge extending between the ischial tuberosities. Anteriorly, the membrane is incomplete, allowing passage of nerves and vessels at the apex of the triangle. Each corpus cavernosum continues, proximally, into a **crus** of the penis, which, like the bulb of the penis, is attached at the urogenital triangle. Each crus is attached to the corresponding ischiopubic ramus: in the flaccid state of the penis, therefore, there is an angle of approximately 90 degrees between the crus and the distal part of the corpus cavernosum.

Deep blood vessels pass within the corpora cavernosa, and superficial vessels run outside of the surrounding sheath of deep fascia. In the urogenital triangle (inferior to the perineal membrane), each crus is covered by a small muscle, the **ischiocavernosus,** and the bulb of the penis is similarly covered by the small **bulbospongiosus.** These three muscles run forward to insert on the sides of the penis, and, through somatic innervation derived from the **pudendal nerve,** are involved in the ejaculation reflex.

Deep to the perineal membrane, and visible in the sagittally cut surface, is a thin layer of muscle which fills out the triangle between the ischiopubic rami. The muscle layer and its investing superior and inferior fascial layers are together known as the **urogenital diaphragm.** The perineal membrane forms the inferior fascia for this layer of muscle and thus the inferior layer of the diaphragm. The portion of the diaphragm which surrounds the urethra is known as the **sphincter of the urethra.** This portion of the urethra is the **membranous urethra** (the other parts being the **prostatic urethra,** above the level of the urogenital diaphragm [i.e., where the urethra is within the prostate]; and **penile urethra,** below the level of the urogenital diaphragm and perineal membrane [i.e., that portion of the urethra within the penis]). In addition to the urethral sphincter, a small posterior portion of the urogenital diaphragm is made up of the **deep transverse perineal muscles,** which extend medially from the ischial tuberosity to the midline. Superiorly, the muscles are covered by a poorly defined superior fascia. **Superficial transverse perineal muscles** lie inferior to the perineal membrane, in the same plane as the ischiocavernosus and bulbospongiosus muscles.

The **pudendal nerve** and **vessels,** after entering the perineum on the medial aspect of the ischium, pass anteriorly, giving off superficial branches which run forward in the **superficial perineal space,** the region superficial to the perineal membrane. The nerve and artery then enter the **deep perineal space** (the space, superior to the perineal membrane, within which the urogenital diaphragm is situated). Here, branches are given off that supply the muscles of the diaphragm, as well as the penis, and other terminal branches emerge anteriorly through the gap at the apex of the triangular perineal membrane to enter the dorsal surface of the penis.

In the female, the homologue of the penis is the **clitoris.** Although much reduced in size, a right and left crus of the clitoris are similarly situated on the ischiopubic rami, and unit anteriorly, in the body of the clitoris (page 178). The homologue of the bulb of the penis is the **bulb of the vestibule.** To accommodate the **vagina,** which lies in the midline, the bulb of the vestibule is "split" into a bilateral structure. Thus one bulb is found on either side of the vagina, covered by the thin **bulbospongiosus** muscle. The female homologues of the penis, including the bulb of the vestibule, are, as their male counterparts, composed of erectile tissue, the function of which is under control of the parasympathetic nervous system, via the pelvic splanchnic nerve (S2 and S3) (page 173).

Contents of Section 3

Cranial Structures

19. Neurocranium; Nervous System; Cranial Nerves
   Clinical Session 19 183
   Laboratory Session 19 195

20. Eye; Eyebrows; Eyelids; Conjunctiva; Lacrimal Apparatus; Associated Soft Tissue Structures
   Clinical Session 20 201

Laboratory Session 20A 207
Laboratory Session 20B 212

21. External Ear; Middle Ear; Inner Ear
   Clinical Session 21 214
   Laboratory Session 21 218

Clinical Session 19

■ **TOPICS**

Neurocranium
Nervous System
Cranial Nerves

In this clinical session you will identify several external features of the neurocranium (page 14) on the living subject, and attempt to correlate these landmarks with internal features, to be identified on the isolated skull. You will, in broad outline, visualize the brain as it occupies the cranial cavity (the space within the neurocranium, in which it is situated). In addition, you will visualize the cranial nerves and their sites of attachment to the brain, as well as the cranial foramina through which they pass. Finally, you will study the manner in which the cranial nerves are examined clinically, and you should correlate this knowledge with your visualization of the anatomical situation and relationships of these nerves. (Note that gloves **must** be worn when handling brain specimens.)

Recall that the brain is the expanded, cranial portion of the central nervous system (CNS) and the spinal cord is the extended, caudal, narrow portion. The basic pattern of the nervous system is that of the **spinal segment** (page 86 and figs. 51 and 52, page 104). In the brain, a phylogenetically determined increased organizational complexity is superimposed on this pattern. The system thus exhibits a morphological gradient of **craniocaudal specialization.** This may be thought of as being related, in an evolutionary sense, to the development of the sense organs at the head end of the primitive organism, as mechanisms to assist in localization and intake of food, also at the head end. The cranial ends of the nervous and the digestive systems accordingly develop in intimate relationship to each other (page 14). Craniocaudal specialization is a fundamental morphological trait, and other organ systems also exhibit it in both sequence and complexity of development.

NEUROCRANIUM

With your subject's head placed in the Frankfurt plane (page 3), identify by inspection and palpation the following features of the neurocranium (referring to figs. 93, 94, 95, and 96, page 184; figs. 97, 98, 99, and 100, page 188; and figs. 105 and 107, page 202). In five regions, one on either side, two anteriorly and one posteriorly, the skull bulges out, as seen from above, in what are known as: the right and left **parietal,** right and left **frontal,** and the **occipital eminences.** Prominence of these eminences may vary in individuals, creating a skull shape predominantly long (in the anteroposterior dimension) or broad (in the lateral dimension). Both individual and racial variation play a role. At the maximal point of prominence of the occipital eminence (as seen from above), identify the **maximum occipital point.** Anteriorly, identify the **glabella,** a slight midline projection in the frontal bone, just above its articulation with the nasal bones. The maximum occipital point is the point on the occipital bone farthest from the glabella.

On either side, palpate the two **superciliary arches,** which radiate from the glabella as ridges above their respective orbits. At the lateral extremity of each arch, reidentify the zygomatic process of the frontal bone and the anterior commencement of the **temporal line** from this process (page 42), and palpate the superior temporal line on your subject as far superoposteriorly as possible. Palpate in the depression between the glabella and the midline ridge of the nose (the angle made by the articulation of the two nasal bones), and identify on your subject, and on the isolated skull, **nasion,** the intersection of the internasal and frontonasal sutures (fig. 105, page 202). Note the highest point of the skull in the anatomical position, the **vertex.** Identify on the skull the intersection between the **coronal** and **sagittal sutures,** called **bregma,** and compare the positions of bregma and the vertex. Identify on the lateral aspect of the skull the region called **pterion,** in which the frontal, parietal, squamous portion of the temporal, and greater wing of the sphenoid bones meet (fig. 95, page 184). Pterion is usually H-shaped, but may occasionally be X- or I-shaped. Pterion is an important landmark for the internal site of a clinically significant artery, the **middle meningeal** branch of the maxillary artery (page 48), which may be damaged and cause intracranial bleeding following trauma to this region. Posteriorly, inferoanterior to the maximum occipital point (see above), identify the **external occipital protuberance,** the most prominent projection on the inferior aspect of the occipital eminence. The external occipital protuberance is a rough area on the skull, situated on or just above the intersection between a horizontal and an anteroposterior ridge. The horizontal ridge is the **superior nuchal line** and the anteroposterior ridge is the **external occipital crest,** both of which should be identified

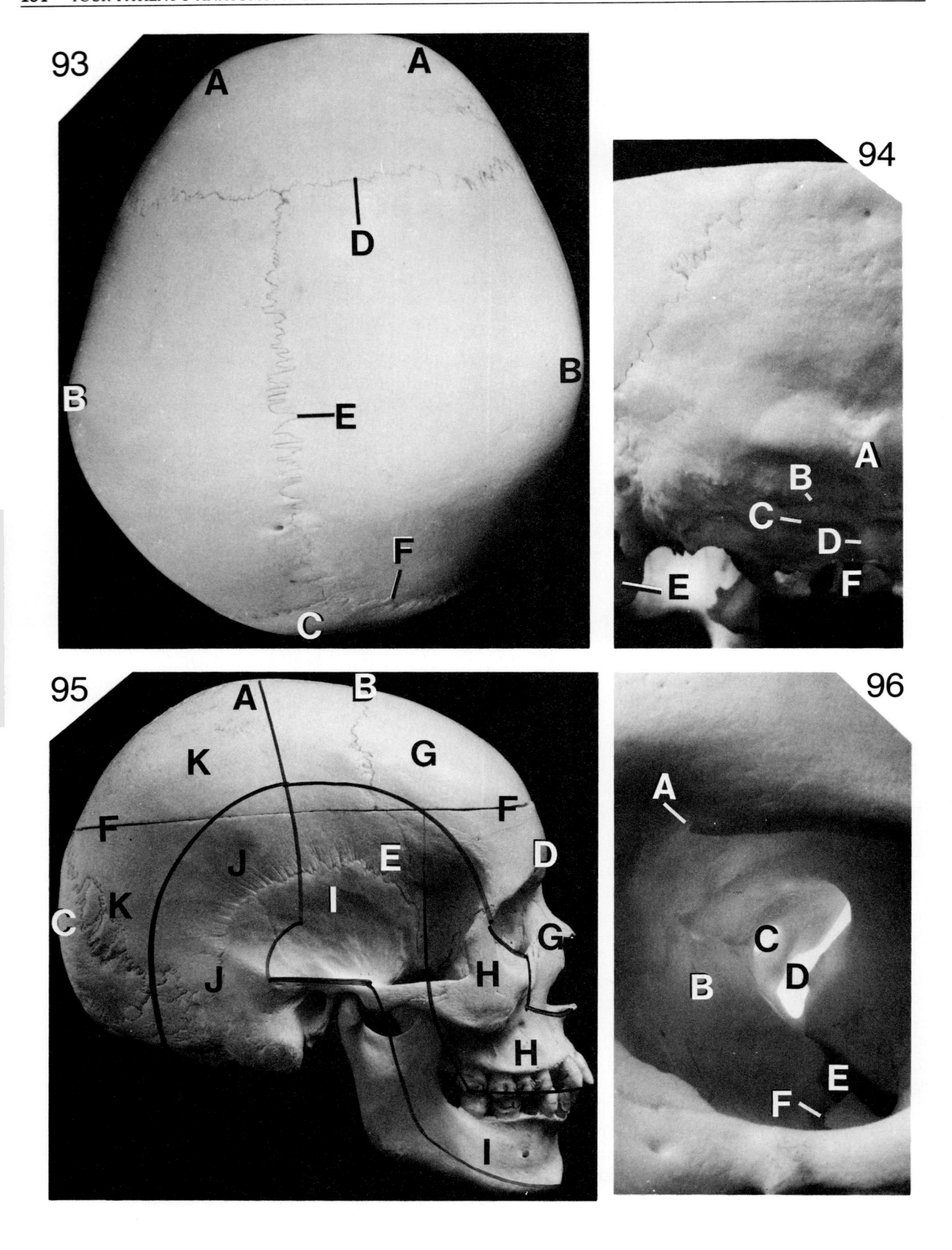

184 YOUR PATIENT'S ANATOMY: A CLINICAL VIEW OF HUMAN MORPHOLOGY

Figure 93. Immature skull viewed from above.
A. Frontal eminences
B. Parietal eminences
C. Occipital eminence
D. Coronal suture
E. Sagittal suture
F. Lambdoidal suture

Figure 94. Posterior aspect of left half of skull (calotte removed).
A. Inion
B. Superior nuchal line
C. Inferior nuchal line
D. External occipital crest
E. Left mastoid process
F. Foramen magnum

Figure 95. Right lateral aspect of skull.
A. Vertex
B. Bregma
C. Maximal occipital point
D. Glabella
E. Pterion (H-shaped)
F. Horizontal saw-cut separating calotte ("skull-cap," above) from calvaria ("brain-box," below)
G. Region of supply of ophthalmic division of trigeminal nerve
H. Region of supply of maxillary division of trigeminal nerve
I. Region of supply of mandibular division of trigeminal nerve
J. Region of supply of cervical ventral rami
K. Region of supply of cervical dorsal rami

Figure 96. Left orbit, from front.
A. Supraorbital notch
B. Orbital plate of ethmoid bone
C. Optic foramen
D. Superior orbital fissure
E. Inferior orbital fissure
F. Infraorbital groove

94, page 184). The tip of the external protuberance is known as the **inion.**

Making reference to anatomical specimens, and aided by atlas illustrations, identify, on the surface of the human brain, the folded surface of the **cerebral hemispheres,** characterized by ridges or **gyri** (singular: gyrus) separated by grooves or **sulci** (singular: sulcus). Identify (fig. 101, page 196) the lobes of the cerebral hemisphere, known as the **frontal, temporal, occipital,** and **parietal lobes.** Also identify the **cerebellar hemispheres,** situated inferior to the occipital lobes of the cerebral hemispheres, and characterized by much finer ridges and grooves. Examine the interior of the cranial cavity (fig. 99, page 188) and visualize the situation of the frontal lobes in the **anterior cranial fossa;** the temporal lobes in the **middle cranial fossa;** and the cerebellar hemispheres (overlayed by the occipital lobes) in the **posterior cranial fossa.** Note that on the inferior (ventral) surface of the brain, cranial nerves are attached, and a glance at the corresponding surface of the cranial cavity indicates numerous holes (foramina) through which the cranial nerves pass. In the following, clinical examination of the function of the cranial nerves will be described. Before proceeding you should, however, recall the reason why the inferior surface of the brain is morphologically ventral, by reviewing the morphological note on page 14, which explains this important point, and fig. 38, page 88, which illustrates it.

CRANIAL NERVES

I. The first cranial nerve is called the **olfactory nerve** (olfaction means sense of smell). The neural pathway for the sense organ of olfaction is unique, in that the sensory **olfactory cells,** situated in the **olfactory epithelium** of the **olfactory region** of the nasal cavity (page 60) also constitute the afferent neuron which conveys the impulse centripetally to the CNS (pages 5 to 6). Thus one and the same cell serves as sensory cell and afferent neuron. The nerve cell fibers of these neurons pass, in small bundles, through the **cribriform plate** (cribrum means sieve) of the ethmoid bone, in the roof of the nasal cavity and floor of the **anterior cranial fossa.** Identify the cribriform plate on an isolated skull, and ascertain the situation of the **olfactory bulbs** on the inferior surface of the frontal lobes of the cerebral hemispheres, to confirm that these bulbs overlie the cribriform plate. The olfactory nerves terminate and synapse in the olfactory bulbs.

The function of the olfactory nerve is tested by ascertaining whether the patient can identify the aroma of different characteristic and easily identifiable odors. Test each nostril (and therefore right and left olfactory nerves) separately, using, for example, oil of cloves, oil of peppermint, and so forth. One should bear in mind that an abnormality of the sense of smell can also be caused by local inflammation in the mucous membrane of the olfactory region of the nasal cavity. Absence of sense of smell is known as **anosmia. Parosmia** refers to incorrect experience of odor, for example, where offensive smells appear to be pleasant and vice versa. Note that hallucinations of olfaction may constitute a component of the prelude to an epileptic seizure.

II. The second cranial nerve, or **optic nerve,** mediates the sense of sight. The sensory receptor cells are situated in the **retina,** which constitutes the innermost of the three layers of the wall of the approximately spherical eyeball. Impulses are conveyed from the retina through fibers of the optic nerve, which leave the orbit through the **optic foramen,** the round hole observable in the posteromedial aspect of the orbit (fig. 96, page 184). Identify this foramen on the disarticulated skull, and gently pass a wire probe through it, into the optic canal, to identify the opening of the latter into the anterior portion of the **middle cranial fossa** (fig. 100, page 188). Identify the transverse **optic groove,** connecting the two optic canals, and delimiting the central part of the middle cranial fossa anteriorly. On an anatomical brain specimen, identify the **optic nerves** and their site of junction in the **optic chiasma,** visualizing that the optic chiasma is situated, in the living subject, in the optic groove of the middle cranial fossa (fig. 111, page 208).

Testing the optic nerve involves investigation of **visual acuity** (sharpness of vision) and **fields of vision,** as well as the ability to **discern colors.** In cases of suspected poor vision, the patient is asked to cover one eye with a hand, and is then asked to count the fingers of the examiner's hand as two, three, or four fingers are held up for observation. For more precise estimation of **visual acuity,** charts of letters of different sizes are placed for the patient to view. The largest letters are readable by normal individuals at a distance of 60 m, and the smallest at a distance of 6 m. The patient is placed at a distance of 6 m, and if he or she can read the smallest letters (i.e., those normally readable at 6 m) the subject's visual acuity is 6/6, or normal. If, however, the subject at 6 m can only read the size of letters normally readable at 60 m, his or her vision is said to be 6/60.

Test the **field of vision** of your subject as follows. Seat yourself facing the patient and at a distance of about half a meter from him or her. Test the right eye first, asking the patient to place his or her hand over the left eye and to look with the right eye at your left eye, which you should keep focused on the patient. Close your own right eye, and hold your left hand in a frontal plane midway between the subject's face and your own and at full arm's length out to the left, and your right hand in the same plane but half as far out to the right. Moving the fingers of your left hand, bring it slowly in toward the midline, until you can (out of the "corner of your eye") observe the movement of your fingers. Ask the patient to indicate as soon as he or she is able to see the movement of the fingers, keeping his or her gaze on your own left eye. Keep moving your fingers in toward the center until the patient indicates that he or she can see the movement. Repeat the process with your left hand, moving it also inwards toward the center. Similarly, bring the left hand up from below and the right hand down from above to test the extent of the field of vision in these directions as well. Finally, repeat the procedure for the patient's left eye, keeping your own right eye open and left eye closed. When the eye is fixed on an object, as your patient's eye has been fixed on yours, the object fixed on is observed by a central area of the retina, a couple of millimeters in diameter, called the **macula.** This region of the retina is specialized for detailed vision. The eye is, in addition to **macular vision,** capable of **peripheral vision,** which means that other areas of the retina, outside of the macula, are capable of seeing, albeit less sharply, objects which the eye is not actually "focused" upon. The test that you have just performed delimits the field of peripheral vision.

Color vision can be tested by several techniques. One is to ask the subject to match various objects of defined colors against each other, placing similar colors together. Other more sophisticated tests of color vision, such as the Ishihara color charts, and the Pseudoisochromatic charts, can be used when necessary. Total color blindness is rare, but red–green blindness (the tendency to be unable to distinguish between reds and greens) is a common, genetically determined (sex-linked) trait.

III, IV, and **VI.** The third cranial nerve (the **oculomotor nerve**), the fourth **(trochlear nerve),** and the sixth **(abducens nerve)** are usually examined together, since they are all concerned with movements of the eyes. Identify, in the posterior aspect of the orbit of the isolated skull, lateral to the optic foramen, the oblique slit known as the **superior orbital fissure** (fig. 96, page 184), through which cranial nerves III, IV, and VI pass between the orbit and the middle cranial fossa. Pass a probe gently through the fissure, to identify the site of opening in the cranial cavity (fig. 100, page 188). Identify the third and sixth cranial nerves attaching on the inferior (ventral) surface of a brain specimen, and the fourth cranial nerves as they appear on the inferior surface, having originated on the superior (dorsal) surface.

The movements of each eyeball are controlled by six muscles, four of which have a more or less "straight," anteroposterior course from the back of the orbit to their attachment on the eyeball, and are accordingly called the rectus muscles (rectus meaning straight). These muscles are placed one above, one below, one medially, and one laterally in relationship to each eyeball. They are known respectively as the **superior rectus,** the **inferior rectus,** the **medial rectus,** and the **lateral rectus muscles.** It will be evident that the superior rectus muscle is mainly capable of elevating the eye to look upwards, the inferior rectus will move the eye to look downwards (these movements being rotations about a horizontal axis), the medial rectus will turn the eye inwards toward the opposite side, and the lateral rectus outwards (rotations about a vertical axis). In addition, there are two muscles which run an oblique course. The **inferior oblique muscle** arises from the floor of the orbit at its anteromedial corner and passes posterolaterally to attach to the inferior surface of the eyeball, far back in the posterolateral quadrant of that surface.

The lateral rectus muscle is supplied by cranial nerve VI. Thus to test the integrity of this nerve, ask your subject to follow your finger with his or her eyes, as you move the finger out laterally, first to the left (testing the left cranial nerve VI), and then to the right.

The remaining muscles mentioned thus far are supplied by cranial nerve III. Move your finger upwards to test the function of the superior rectus muscle; downwards to test the function of the inferior rectus; medially to test the medial rectus; and superolaterally to test the inferior oblique muscle. Test each eye separately.

The **superior oblique muscle,** supplied by cranial nerve IV, originates at the back of the orbit, passes forward to the upper anteromedial corner of the orbit, and then turns sharply backwards and laterally, in a pulley formed by connective tissue, to insert in the posterolateral quadrant of the upper surface of the eyeball. Its action is therefore mainly to pull the eyeball in a downwards and lateral direction. Test this function, and therefore the integrity of the fourth cranial nerve, by asking your model to follow the movement of your finger as you move it downwards and laterally in relation to first the one and then the other eye. (Note that trochlea means pulley; hence the name of the fourth cranial nerve.)

In the above description, the functions of the various extrinsic muscles of the eyes are described in broad outline. A more detailed description of these

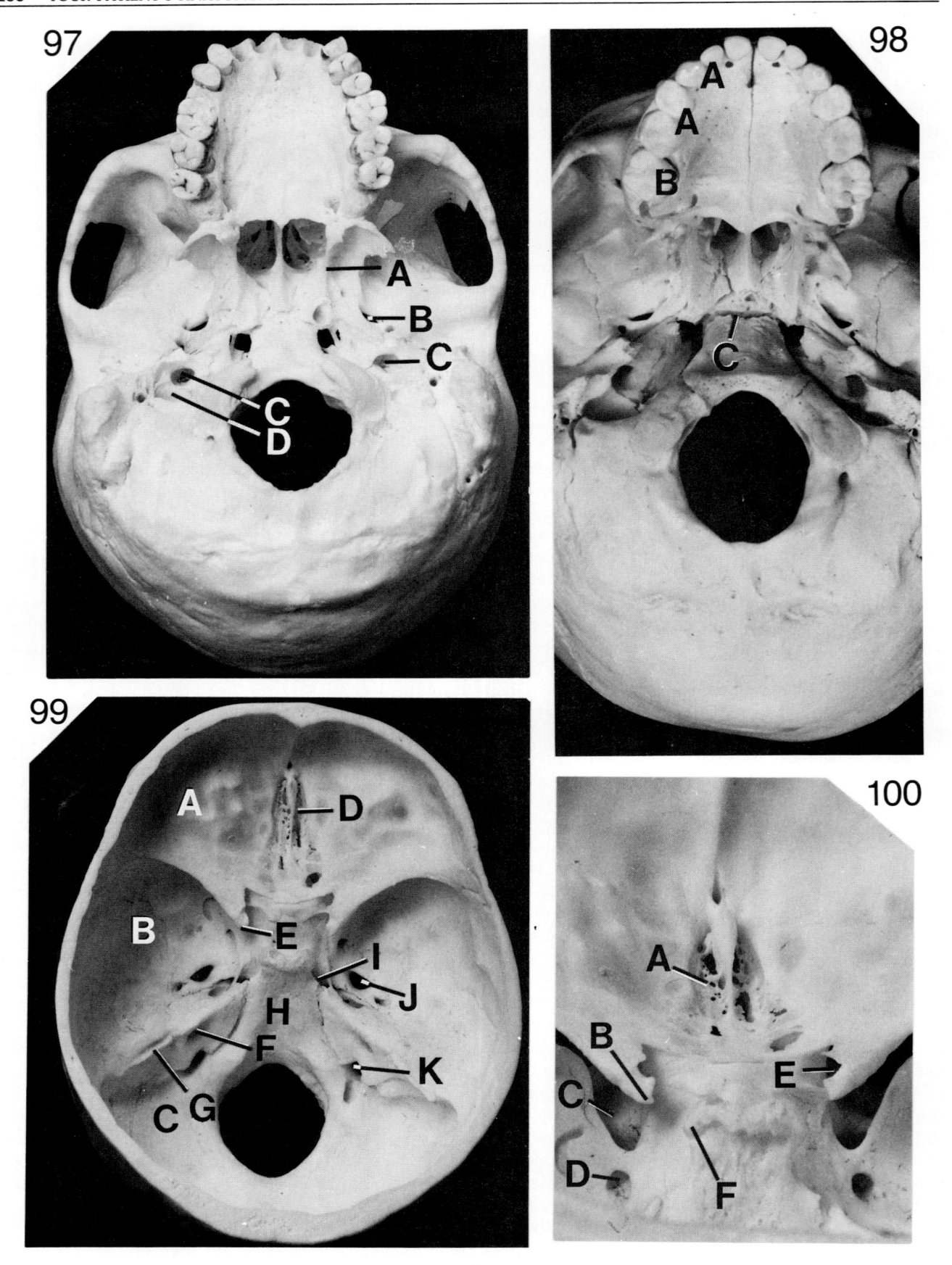

188 YOUR PATIENT'S ANATOMY: A CLINICAL VIEW OF HUMAN MORPHOLOGY

Figure 97. Skull viewed from below.
A. Medial pterygoid plate
B. Foramen ovale
C. Carotid canal
D. Jugular foramen

Figure 98. Immature skull viewed from below.
A. Deciduous ("milk") dentition
B. First permanent molar
C. Spheno-occipital synchondrosis

Figure 99. Internal aspect of base of cranium, from above.
A. Anterior cranial fossa
B. Middle cranial fossa
C. Posterior cranial fossa
D. Cribriform plate
E. Anterior clinoid process
F. Internal auditory meatus
G. Superior border of petrous portion of temporal bone
H. Clivus
I. Foramen lacerum
J. Foramen ovale
K. Jugular foramen

Figure 100. Detail of internal aspect of base of cranium, from above and behind.
A. Cribriform plate
B. Left anterior clinoid process
C. Left superior orbital fissure
D. Left foramen rotundum
E. Right optic canal
F. Left posterior clinoid process

functions will be given in a subsequent session, when the eye is studied in greater detail.

The following additional test of the function of cranial nerve III should be performed. Shine the light from a pocket torch (flashlight) into the left eye, and observe the pupil of the left eye. Repeat the procedure of shining light into the left eye, and observe the reaction of the right pupil. Then repeat both exercises by shining light into the right eye and observing first the right and then the left pupil. The pupil is surrounded by a minute **sphincter** muscle, which is innervated by parasympathetic nerve fibers which are carried in the third cranial nerve, and that synapse in the **ciliary ganglion** (page 106), which is situated in the orbit, just behind the eyeball. **Short ciliary nerves** carry postganglionic fibers to the eye. Communications between the right and left pathways, in the brain, account for the **crossed reflex** observed when the right pupil contracts on stimulation of the left retina, and vice versa. (A second minute muscle, the **dilator** of the pupil, opposes this action through sympathetic innervation; see also pages 101–106.)

V. The fifth cranial nerve is called the **trigeminal nerve** (tri means 3; gemini means twins; trigemini means triplets) because, almost immediately upon emerging from the brain, it divides into three major **divisions.** All three divisions supply sensory nerve fibers to the face (page 20). You previously studied branches of these nerves and approximately delineated the distribution of the three divisions as follows (page 25). The first or **ophthalmic division of the trigeminal nerve** supplies the upper face (i.e., forehead, eye region, and nose); the second or **maxillary division** supplies the upper jaw and adjacent region; and the third or **mandibular division** supplies the lower jaw, extending up into the temporal region. These three regions of distribution should now be more precisely demarcated with a skin marker. Refer to Figure 95, page 184, as you follow the ensuing instructions. (The irregularity of the areas of supply in the head and neck region, as contrasted with the very regular, segmental pattern seen in the thorax, is due to the complexity of morphogenetic movements and tissue migrations during development, at the head end of the organism. This, in turn, reflects the phylogenetic predominance of the head end of the organism as the intake end of the digestive tube and, as a result, the site of specialization of the food-sensing system [i.e., the special senses and nervous system].)

An imaginary line passing upwards from the right ear, over the top of the head in a coronal plane, to the left ear, marks the dividing line between the region of distribution of the trigeminal nerve, anteriorly, and distribution of cervical spinal nerves, posteriorly. This imaginary line should be continued downwards, halfway around the posterior margin of the **outer ear,** or **auricle,** to include the upper half of the skin surface of the auricle. Thereafter, a line should be drawn extending anteriorly in a horizontal plane through the position of the posterior root of the zygomatic process of the temporal bone (page 39). At the anterior root of the process (page 40) the line turns downwards to pass parallel to and midway between the anterior and posterior borders of the ramus of the mandible, then continues inferoanteriorly, at first parallel to the base of the mandible, but finally sloping anteriorly to the tip of the chin.

To demarcate the three divisions of cranial nerve V from each other, start at the midpoint of your original imaginary line extending from the upper edge of the auricle to the top of the head. From this midpoint draw a line which passes anteroinferiorly, parallel to the curve of the top of the neurocranium and the forehead, as seen from the side, extending down to the lateral "corner of the eye" (the **lateral canthus**). The line should then follow the upper border of the lower eyelid medially to the **medial canthus,** then pass inferiorly to include most of the nose, and then veer anteroinferiorly, excluding only the lateral tip of the ala (or wing, page 59) of the nose, to end at the upper border of the upper lip, which the line should follow medially. The region anterosuperior to this line is that supplied by the **ophthalmic division.**

Draw a line which extends from the angle of the mouth (page 11) curving with a slight posterior convexity to pass upwards and meet the previous line a little posterior to the forehead. The region enclosed anterior to this line is that of the **maxillary division.** The remaining portion of the trigeminal area is the region supplied by sensory nerves of the **mandibular division.**

The course of the ophthalmic division of the trigeminal nerve should now be visualized. After leaving the brain (figs. 101 and 102, page 196) the trigeminal nerve passes anterolaterally, from the posterior cranial fossa into the middle cranial fossa (fig. 99, page 188). The ophthalmic division passes forwards through the superior orbital fissure, to enter the orbit. Within the orbit it divides into **lacrimal, frontal,** and **nasociliary** branches. All three of these end by giving smaller branches to the surface of the skin. Branches of the lacrimal nerve emerge from the orbit near the lateral canthus; the frontal nerve divides into **supraorbital** and **supratrochlear** branches, which pass upwards into the forehead; and **infratrochlear** and **external nasal** terminal branches of the nasociliary nerve supply the side of the nose. (Trochlea means pulley; the term refers here to the pulley about which the superior oblique muscle turns [page 187]. This accounts, as mentioned, for the name of the fourth cranial nerve which supplies this muscle, and also for the names of the supratrochlear branch of the frontal nerve, and the infratrochlear branch of the nasociliary nerve.) **Long ciliary nerves,** carrying sensory fibers from the cornea, leave the back of the eyeball to reach the nasociliary nerve.

Integrity of sensory nerves is tested by asking the patient to close his or her eyes, and gently touching the skin with cotton wool (to test the modality of **light touch**), or with the point of a pin or needle (to test the modality of **pain**). Perform these tests (gently!), examining the different branches of the three divisions of the trigeminal nerve separately.

The ophthalmic division of the trigeminal nerve contains only somatic sensory nerve fibers. However, at certain sites along the courses of some of its branches, certain autonomic nerve fibers "hitchhike." These fibers will be studied more closely in connection with the orbit.

The maxillary division of the trigeminal nerve passes forwards through the middle cranial fossa, to exit through the **foramen rotundum,** which can be seen far forwards in the middle cranial fossa (fig. 100, page 188) just inferior to the superior orbital fissure. Gently insert a wire probe through the foramen rotundum of a skull, to establish its continuity with the pterygopalatine fossa (pages 46 and 63). The maxillary division passes through the foramen rotundum and through the pterygopalatine fossa, to gain access to the infraorbital groove (fig. 96, page 184), which continues into the infraorbital canal and opens, through the infraorbital foramen, onto the face. You previously palpated this foramen (page 37), and dissected the terminal branches of the maxillary division of the trigeminal nerve, here called the infraorbital nerve, as they emerge through the foramen (page 25). The infraorbital nerve supplies the region of the face below the eye, extending down to the upper lip. In addition, the maxillary nerve gives off, in its course through the floor of the orbit (as the infraorbital nerve), a **zygomatic branch** which passes through a foramen on the orbital surface of the zygomatic bone. You should now identify this on a skull. Within the bone, the zygomatic nerve divides into a **zygomaticotemporal** and a **zygomaticofacial** branch, and each emerge through corresponding foramina on the outer surface of the bone, and pass to the regions indicated by their respective names. Using the technique described above, test for light touch and pain sensation in the distribution of these branches of the maxillary division of the trigeminal nerve.

An additional sensory (afferent) modality is that of **position sense.** In cases of abnormal **occlusion** (closure) of the jaw due, for example, to a filling that is too high, a vicious cycle, involving overactivity of masticatory muscles, may lead to symptoms in these muscles or the temporomandibular joint.

As in the case of the ophthalmic division, the maxillary division is also entirely sensory, but certain autonomic fibers travel with it at certain sites. You previously exposed the pterygopalatine fossa (page 63) and observed the sphenopalatine ganglion. Although the ganglion is suspended from the maxillary nerve, and some of its postganglionic fibers accompany branches of that nerve, the preganglionic fibers originate from the facial nerve (cranial nerve VII). Postganglionic branches accompany the zygomatic branch of the maxillary nerve, before transferring to the lacrimal branch of the ophthalmic division (page 190), to supply the **lacrimal gland** (the tear gland, situated in the orbit above the lateral canthus of the eye) with secretomotor fibers. Other postganglionic fibers pass in nasal and nasopalatine branches to the mucous membrane of the nasal cavity, and in palatine branches to supply palatine glands. Thus abnormality of tear secretion, or of glandular secretion in the nose or palate, may indicate abnormality in these nerve pathways.

The mandibular division of the trigeminal nerve contains, in addition to sensory fibers, motor fibers that mainly supply the muscles of mastication. The function of these fibers can therefore be tested by asking the subject to clench his or her teeth, while the examiner palpates the masseter, temporalis, and medial pterygoid muscles (pages 38–39) through the skin. Palpation of the mylohyoid muscle, during swallowing (page 4) also tests this function. In the latter palpation, it may be impossible to distinguish between the mylohyoid and anterior bellies of the digastric muscles, but since all are supplied by the mandibular division of the trigeminal, this point is academic in this test. In addition to this, the lateral pterygoid muscle can be palpated internally (page 38), but in practice the external examination is usually regarded as satisfactory. It should also be noted that the tensor veli palatini muscle (page 61, page 66) is also supplied by the mandibular division. Therefore, in a lesion of this nerve, the function of the soft palate in swallowing will also be affected. Finally, for completeness, note that a small muscle in the ear, the tensor tympani, is also supplied by this nerve. The mandibular division exits the skull through foramen ovale (fig. 99, page 188).

VII. The seventh cranial nerve, or **facial nerve,** is, first and foremost, the motor nerve of supply to the muscles of facial expression. Therefore in order to test the integrity of this cranial nerve, the patient is asked to perform the various actions which these muscles execute (page 19). In practice, it is common to concentrate efforts on asking the patient to raise eyebrows, close eyes tightly, and smile widely, showing the teeth. Review (page 19) the different muscles being tested by these exercises. In addition, and importantly, the function of buccinator should be tested (page 27).

For completeness, it should be mentioned that in addition to the muscles of facial expression, the facial nerve also supplies the posterior belly of the digastric muscle, the stylohyoid muscle (page 52); and the stapedius, a minute muscle of the middle ear.

The facial nerve probably carries no somatic sensory fibers, although some authors have claimed that it supplies a small area of the lateral aspect of the auricle, and an even smaller area of the cheek. However, the facial nerve does have parasympathetic fibers that pass to the pterygopalatine ganglion, the branches of which are associated with those of the maxillary nerve, and which are therefore usually tested with the latter (see above).

Identify the seventh cranial nerve on the base of the brain, and visualize its path through the **internal auditory meatus,** in the lateral wall of the **posterior cranial fossa** (fig. 99, page 188 and figs. 117 and 118, page 220) into the **inner ear.** Within the ear the nerve passes further laterally and then makes a right angle bend, the **genu** ("knee"), to turn posteriorly. Thereafter it makes a second right angle bend inferiorly (see fig. 116, page 216), to pass downwards and exit from the skull through the **stylomastoid foramen,** which you should now identify on the inferior surface of the skull, between the styloid and mastoid processes. You previously (page 48) studied the deep portion of the parotid gland and its relationship to these structures, including the facial nerve, which, as it leaves the foramen, enters the substance of the gland. It curves forward within the gland, dividing into its several branches, which radiate out from the anterior edge of the gland, into the face. You also previously (page 25) dissected these branches by removing portions of the gland. Branches of the facial nerve could be traced to individual facial muscles (page 25).

Parasympathetic fibers of the facial nerve leave it at the **genu** (see above), and pass anteromedially as the **greater petrosal nerve,** through a minute foramen on the anterior surface of the petrous temporal bone. The greater petrosal nerve then passes anteromedially in the middle cranial fossa, to attain the **foramen lacerum** (fig. 99, page 188). The fibers enter the **pterygoid canal,** which can be seen in an "exploded" skull to lead from the foramen lacerum to the pterygopalatine fossa where, as previously described (page 191), they synapse in the spheno-palatine ganglion. Other parasympathetic fibers of the facial nerve leave the main trunk of the nerve in the last part of its course through the ear, just as it enters the stylomastoid foramen. These fibers form the **chorda tympani** nerve which passes forwards on the inner surface of the tympanic membrane (page 60). After passing through the middle ear in this situation, it emerges through the minute **petrotympanic fissure** (see fig. 114, page 216). Visualize this course on the isolated skull, and the subsequent course as the chorda tympani nerve passes on the medial aspect of the **spine of the sphenoid bone,** to join onto the lingual nerve, as the latter branches off the mandibular division of the trigeminal (figs. 19 and 20, page 44). These fibers accompany the lingual nerve until they leave it to enter the submandibular ganglion (page 33). The fibers synapse in this ganglion, and the postganglionic branches pass to the submandibular, sublingual, and small salivary glands of the region.

Finally, the facial nerve mediates the sense of **taste,** through special sensory (afferent) fibers from taste buds in fungiform and foliate papillae (pages 14–15) on the anterior two thirds of the tongue. Four modalities of taste exist; salt, sour, bitter, and sweet. Using the four solutions provided, test your subject's ability to distinguish between these, by placing a drop of each first on the tip of the tongue, then on the sides, then the middle, and finally the back of the tongue. Taste fibers from the anterior two thirds of the tongue enter the lingual nerve and pass centripetally via the chorda tympani to the seventh cranial nerve.

The vallate papillae (page 15) contain taste buds which are innervated by cranial nerve IX, but it is convenient to examine taste function in the latter nerve simultaneously with that of the facial nerve.

Attempt to distinguish whether the four different tastes are appreciated specifically in different regions of the tongue. Note that the subjective appreciation of the sense of taste comprises combinations only of the four modalities mentioned. Components of taste which cannot be accounted for by these four are derived from the sense of smell. To confirm this perhaps surprising fact, you should attempt to "taste" a gourmet delicacy while consciously closing the communication between oropharynx and nasopharynx (page 62).

VIII. The eighth cranial nerve is known as the **statoacoustic nerve** (stato means concerned with position; acoustic means concerned with sound). As the name implies, the nerve mediates both the sense of position and balance, as well as the sense of hearing. The nerve is also known as the **vestibulocochlear nerve,** because the sense of position and balance is mediated through a minute organ within the **vesti-**

bule of the inner ear, and the sense of hearing is mediated through a sense organ situated in the **cochlear** (shell) of the inner ear. Identify the eighth nerve on the base of the brain, and visualize its passage, through the internal auditory meatus (accompanying cranial nerve VII) to the inner ear.

The sense of hearing should be tested in each ear separately (i.e., the ear not being tested should be covered by the patient's own hand). The patient should be asked to close his or her eyes, as his or her ability to detect a standardized sound is ascertained. For example, a ticking sound can be simulated by the examiner's fingernails. Alternative sounds such as rapid rubbing of fingers and so forth, can also be employed. Starting about a meter away from the patient's ear, the source of sound should be gradually moved toward the patient, who is asked to indicate when he or she hears a sound, and to identify the sound being heard. The second ear should be tested using the same sound. Acuity can thus be compared from right to left sides, and between successive examinations, by stating the distance at which the sound is first heard.

Abnormalities of vestibular nerve function may cause loss of balance, or cause vertigo (a sense of rotation of one's surroundings about oneself, or of rotation of one's body in relation to the surroundings). To test the nerve, ask the subject to stand (in normal anatomical position) with his or her eyes closed, and to turn the head sharply to the left, to the right, downwards (flexion), and backwards (extension). These exercises stimulate the vestibular sensory organs, which detect movement in all three dimensions. In impairment of vestibular nerve function the subject will experience disturbance of balance during this test. A second method to assess this function is for the subject to attempt to walk a few steps in a straight line, with his or her eyes closed.

It should be noted that further tests, which are beyond the scope of this session, would be needed to distinguish between impairment of vestibular nerve function and disturbances in function of other neural pathways in the brain to which the vestibular pathways connect. Such disturbances of function can be caused by drugs, including alcohol.

IX. The ninth cranial, or **glossopharyngeal nerve,** supplies, as the name implies, the tongue (posterior one third) and the pharynx. Nerve fibers to the pharynx are responsible for sensory supply to the mucous membrane, and are therefore the fibers which mediate choke or gagging reflexes (page 15). In addition to supplying sensory fibers to the pharynx (including the tonsils) and the posterior third of the tongue, taste fibers to the vallate papillae also run with this cranial nerve (page 72). Sensory fibers of cranial nerve IX can thus be tested by eliciting the gagging reflex (by placing a spatula on the posterior third of the tongue), while taste function is usually tested simultaneously with the seventh nerve (see above).

The glossopharyngeal nerve also contains motor fibers which supply the stylopharyngeus muscle and, possibly, the pharyngeal constrictors. Integrity of this function of the nerve is thus tested by asking the patient to swallow, and observing by inspection and palpation the movements which accompany deglutition (page 70) including elevation of the larynx (into which stylopharyngeus also inserts).

The glossopharyngeal nerve contains parasympathetic fibers which are secretomotor to the parotid gland. These and other secretomotor fibers to salivary glands are tested in eliciting the salivary reflex which is obtained by the lemon juice test (page 26). The glossopharyngeal nerve also mediates cardiac and respiratory reflexes through its carotid branch, which carries afferent fibers from the carotid sinus and carotid body (sensory organs situated in the wall of the common carotid artery, at the site of its bifurcation). Impulses originating in pressor receptors (in the carotid sinus) and chemoreceptors (in the carotid body) pass in this branch to the trunk of the glossopharyngeal nerve and hence into the brain.

Identify the origin of the glossopharyngeal nerve from the base of the brain, and trace its pathway through the jugular foramen (fig. 99, page 188) to the inferior surface of the base of the skull. The nerve passes first posterior to, and then lateral to, the stylopharyngeus muscle (page 72), and thus gains the gap between the superior and middle pharyngeal constrictor muscles. Shortly after its emergence on the inferior surface of the skull, the glossopharyngeal nerve gives off the **tympanic branch,** which curves upwards through a minute foramen situated between the external openings of the jugular foramen and the carotid canal (fig. 97, page 188). From here the tympanic branch passes into the middle ear or **tympanic cavity,** and forms the **tympanic plexus,** which supplies sensory fibers to the middle ear and the auditory tube. In addition, this plexus gives off the **lesser petrosal nerve.** The lesser petrosal nerve leaves the middle ear by passing forwards, through a minute foramen on the anterior surface of the petrous temporal bone, into the middle temporal fossa, to pass anteromedially in the latter, just lateral to the **greater petrosal nerve** (page 192). The lesser petrosal nerve runs to the **foramen ovale,** through which it passes to reach, and synapse in, the **otic ganglion** (page 48). Secretomotor postganglionic fibers originating in the ganglion pass to the **auriculotemporal nerve** and are carried to the parotid gland (page 48).

X and **XI.** It is convenient to consider the **vagus nerve** (cranial nerve X) and the **accessory nerve** (cranial nerve XI) together. The vagus (wanderer) is so named because of its very extensive, meandering course through the body. The accessory nerve is so named because part of it (the **cranial accessory nerve**) is, in essence, an accessory part of the vagus nerve. The other part of this nerve (the **spinal accessory nerve**) arises from the cervical spinal cord and ascends through the foramen magnum into the skull (fig. 104, page 196) to join the cranial accessory nerve, with which it leaves the posterior cranial fossa through the jugular foramen. Immediately after exit, however the cranial part of the accessory nerve joins the vagus nerve and is distributed with it.

The vagus nerve supplies motor fibers, mainly originating from the cranial accessory nerve, to the muscles of the pharynx and the muscles of the soft palate. These fibers are clinically examined by asking the patient to open his or her mouth and say "ah," while the examiner observes movement of the soft palate (page 30). [Note, however, that the tensor veli palatini muscle receives its innervation from cranial nerve V (page 191].) In addition, the patient is asked to swallow, and the function of the muscles of deglutition are observed (page 70). Branches to the palate and pharynx leave the vagus in the **pharyngeal branch,** which passes between the internal and external carotid arteries to reach the surface of the pharynx, where it participates in the pharyngeal plexus together with branches of the glossopharyngeal nerve and the sympathetic trunk. Most authors believe that the vagus nerve supplies most of the motor fibers to the pharyngeal constrictors, whereas the glossopharyngeal nerve supplies stylopharyngeus (and sensory nerves to the pharynx). However, some authors claim that IX also contributes fibers to the constrictor muscles.

Through its recurrent laryngeal branch (page 76) the vagus nerve innervates all the intrinsic muscles of the larynx, except for the cricothyroid muscle, which is supplied by the external laryngeal branch of the superior laryngeal nerve.

The vagus nerve also carries sensory fibers. It has a small meningeal branch, and an auricular branch which enters a minute foramen in the jugular fossa on the base of the skull and passes through the temporal bone to emerge behind the external auditory meatus and supply sensory fibers to the medial surface of the auricle, as well as to the external auditory meatus and tympanic membrane. The internal laryngeal branch of the superior laryngeal nerve enters the larynx through the thyrohyoid membrane (page 76) and supplies sensory fibers to the part of the larynx superior to the vocal cords. The recurrent laryngeal nerve (page 76), in addition to supplying intrinsic muscles of the larynx, also supplies sensory fibers to the vocal cords and the larynx inferior to these.

Finally, the vagus is an important carrier of parasympathetic fibers (page 101–106). Vagal fibers of this modality pass to the heart (page 100) in cardiac branches; to the lungs in pulmonary branches; and to the esophagus, stomach, and digestive tract in other appropriately named branches. As in the case of other parasympathetic innervation, the ganglia are situated close to, or within, the organ of supply (page 101).

As mentioned, the fibers of the cranial portion of the accessory nerve are distributed with the vagus. The spinal part of the accessory nerve, on leaving the skull through the jugular foramen, runs posterolaterally to enter the deep surface of the sternocleidomastoid muscle, to subsequently pass through the muscle and emerge on its lateral surface or close to its posterior border. It supplies the sternocleidomastoid muscle and the trapezius muscle of the back.

Function of the vagus nerve may be tested by examining movements of the palate and of the muscles of deglutition (see above). In addition, it mediates sinus arrhythmia (page 100); and its integrity can also be tested by auscultating normal bowel sounds (page 150). The spinal accessory nerve can be tested by examining the function of the sternocleidomastoid muscle (page 49), and by testing the superior portion of the trapezius muscles, which pass from the occipital region of the skull to the scapulae. The latter is tested by asking the subject to raise his or her shoulders, while the examiner applies resistance by holding the shoulders down.

XII. The twelfth cranial nerve, or **hypoglossal nerve,** makes its exit from the skull through the **anterior condylar foramen** (on the occipital condyl) and passes inferiorly and anteriorly in a course which you have previously studied (pages 56 and 24). It supplies the intrinsic and extrinsic muscles of the tongue (page 15). The integrity of this cranial nerve is therefore tested by eliciting contraction of these muscles (page 18).

Laboratory Session 19

Your objective for this dissection is to study the cranial cavity. To do so, use the detached half of the head, and commence study on the sagitally sectioned surface. (Depending on how accurately your specimen has been sagittally sectioned, some of the structures to be mentioned in the following may be better identified on the attached half. You should therefore examine both halves, while proceeding with the dissection on the detached half.) In addition, you should have a skull available, for reference, while you perform this dissection, and refer to Figures 85 to 88, 97, 98, 99, and 100, page 188 and Figures 89 to 92, 101, 102, 103, and 104, page 196.

To get your bearings, first identify the frontal and occipital lobes of the cerebrum and the cerebellum, and, in the skull, the corresponding regions in which these parts of the brain are situated. Thus the frontal lobe lies in the anterior cranial fossa (figs. 99, page 188 and 101, page 196), and the occipital lobe and cerebellum are situated in the posterior cranial fossa. Note, on the isolated skull, the horizontal, curved groove in the posterior fossa which delimits a superior region for the occipital lobes from an inferior region for the right and left cerebellar hemispheres. This groove, and other similar grooves on the internal surface of the cranial cavity, contain large venous channels known as the dural sinuses. Identify, on the cadaver specimen, the **transverse sinus** (fig. 103, page 196) in the horizontal groove. Note that the sinus is completely surrounded by a tough layer of fibrous tissue, the **dura mater**, or outermost of the three membranes which surround the brain. At the sagittally sectioned transverse sinus it will be possible for you to observe an important general principal of the way these sinuses are situated, (i.e., that one layer of dura is tightly adherent to the internal surface of the skull bone, and a second layer surrounds that aspect of the sinus which faces the brain). Superiorly and inferiorly the two layers of dura fuse, to continue around the outer surface of occipital lobe above and cerebellum below. Anteriorly, the inner layer of dura may be observed to fold forwards, as the **tentorium cerebelli**, or tent of the cerebellum, so called because it is tent-shaped, with the peak of the tent being in the midline where you are observing the cut edge of the tentorium. Using the handle of a pair of forceps, gently separate, by blunt dissection, the cerebellum from the inferior surface of the tentorium cerebelli, and observe the slope of the tentorium downwards and laterally toward the lateral part of the groove for the transverse sinus. The tentorium has an **attached edge** at this groove and extending to the superior border of the petrous temporal bone. Anteriorly, gently follow the tentorium forwards to its **free edge.**

Superiorly, observe a similar fold of dura, this time in the sagittal plane. This is the **falx cerebri** (falx means sickle, a descriptive term which indicates the shape of this fold of dura) (fig. 103, page 196). Note that the falx cerebri is attached in the sagittal plane anteriorly and superiorly, and similarly encloses a venous channel, the **superior sagittal sinus.** Identify the groove for the superior sagittal sinus in the skull, anteriorly and on the inner surface of the calotte (the upper portion of the skull which may be separated from the rest of the skull by a horizontal saw cut). Trace the superior sagittal sinus as far posteriorly as possible. Gently remove blood clot from the interior of the superior sagittal sinus, and identify openings of **superior cerebral veins** and of **venous lacunae** into the sinus, and small projections, the **arachnoid granulations,** through which **cerebrospinal fluid (CSF)** (page 125) drains from the subarachnoid space, into the sagittal sinus. The **arachnoid mater** is the middle of the three membranes surrounding the brain, and the **subarachnoid space** is the space between it and the innermost of the membranes, the **pia mater,** which is closely applied to the surface of the brain. The CSF circulates in the subarachnoid space, and is drained from it largely through the arachnoid granulations.

After noting the shape and position of the falx cerebri, its attachments to the skull, and its situation in relation to the cerebral hemispheres and the corpus callosum, separate the falx cerebri from the cerebrum by blunt finger dissection. Then use scalpel dissection to separate the falx from its bony attachments in the floor of the anterior cranial fossa (the **crista galli** of the perpendicular plate of the **ethmoid bone**), and from the sagittally placed **frontal crest,** in the anterior wall of the anterior cranial fossa. The superior attachment of the falx cerebri need not be separated at this time: the falx can be reflected upwards, to allow of access to the superior surface of the cere brum. Using the handle of a pair of forceps, and blunt, finger dissection, gently separate the superior surface of the cerebral hemisphere from the inner surface of the cranial dura mater. Your fingers and forceps should be entering the plane of cleavage between the arachnoid mater and the dura mater. The arachnoid mater is easily identifiable by its characteristic, "spider web" appearance (arachnoid means of a spider). You should gradually work your fingers laterally and posteriorly, freeing the parietal and occipital lobes first. Work your fingers forwards, over the frontal lobes. As you approach

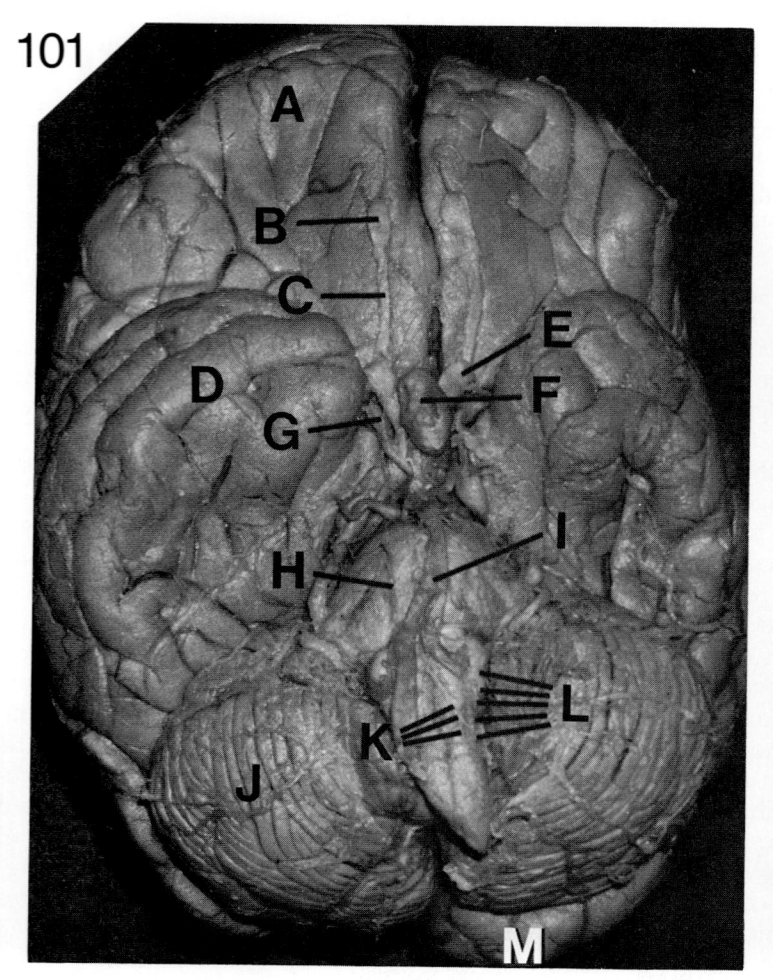

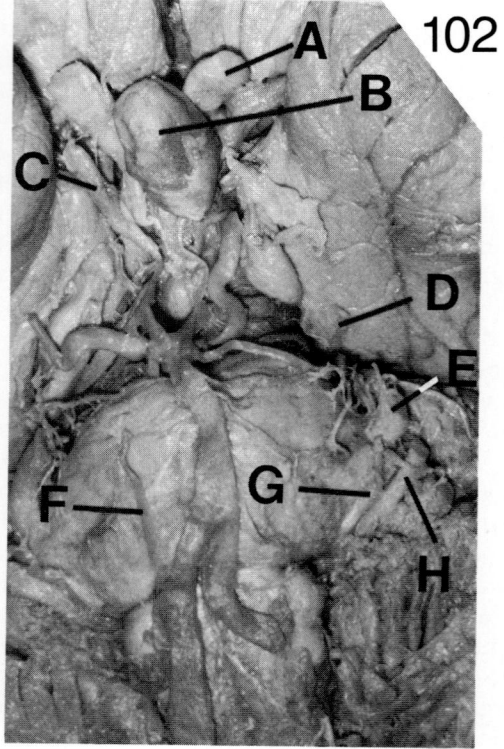

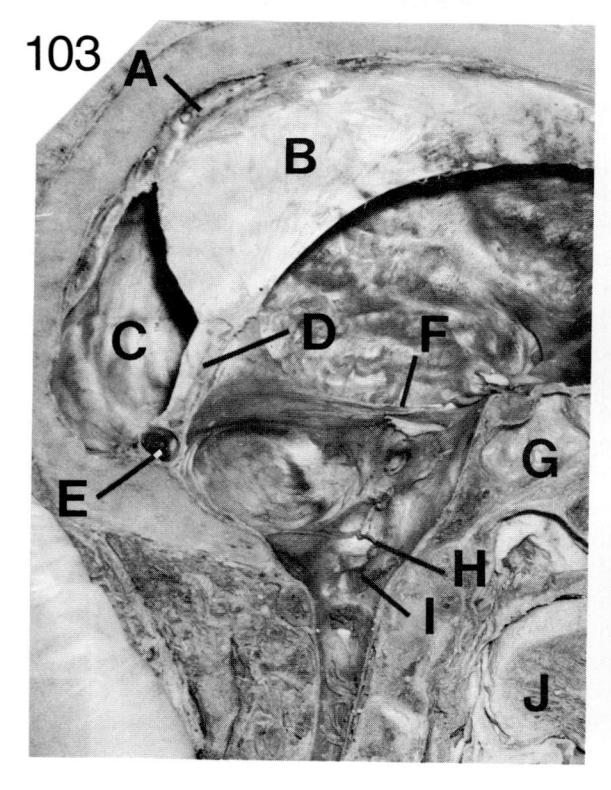

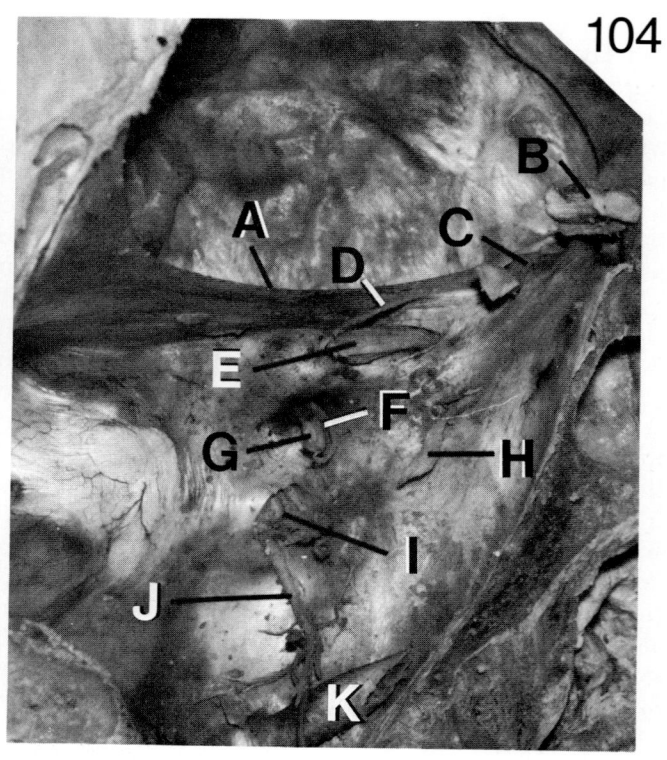

196 YOUR PATIENT'S ANATOMY: A CLINICAL VIEW OF HUMAN MORPHOLOGY

Figure 101. Inferior aspect of the brain.
A. Frontal lobe
B. Olfactory bulb
C. Olfactory tract
D. Temporal lobe
E. Left optic nerve
F. Pituitary gland
G. Right oculomotor nerve
H. Right abducent nerve
I. Basilar artery, on ventral surface of pons
J. Right cerebellar hemisphere
K. Site of origin of rootlets of left hypoglossal nerve
L. Site of origin of rootlets of left glossopharyngeal, vagus, and cranial accessory nerves
M. Occipital lobe

Figure 102. Detail of inferior aspect of the brain.
A. Left optic nerve
B. Pituitary gland
C. Right oculomotor nerve
D. Left trochlear nerve
E. Left trigeminal nerve
F. Right abducent nerve
G. Left facial nerve
H. Left vestibulocochlear nerve

Figure 103. Median view of left side of cranial cavity.
A. Superior sagittal sinus
B. Falx cerebri
C. Window cut in falx cerebri
D. Straight sinus
E. Transverse sinus
F. Free edge of tentorium cerebelli
G. Sphenoid sinus
H. Hypoglossal nerve
I. Vertebral artery, entering through foramen magnum
J. Tongue

Figure 104. Detail of median view of left side of cranial cavity.
A. Free edge of tentorium cerebelli
B. Optic nerve
C. Oculomotor nerve
D. Trochlear nerve
E. Trigeminal nerve
F. Facial nerve
G. Vestibulocochlear nerve
H. Abducent nerve
I. Glossopharyngeal, vagus, and cranial accessory nerves
J. Spinal accessory nerve
K. Vertebral artery

the frontal pole of the brain, you are about to begin the next phase of this dissection: the elevation of the inferior surface of the brain from the internal surface of the base of the skull. This procedure should be performed with care, observing the structural relationships in detail step by step. The cranial nerves and their passage through the foramina are best observed as the elevation proceeds, and each nerve has to be cut, after having been observed, as you proceed to elevate the brain and expose the next cranial nerves.

Starting anteriorly, gently lift the frontal lobe of the brain from the anterior cranial fossa, and identify the **olfactory bulb** (fig. 101, page 196), and the minute **olfactory nerves** passing through the cribriform plate (figs. 87 to 88, page 188). Cut the **olfactory tract** about 1 cm posterior to the olfactory bulb. This will allow further elevation of the frontal lobe to reveal the **optic nerve,** emerging through the optic foramen. The optic nerve should be traced posteriorly, to the site where it joins the **optic chiasma** (page 186 and figs. 102, page 196, and 111, page 208). After identifying the situation of the chiasma on a skull, gently lift the nerve and chiasma to reveal the **internal carotid artery.** At this site the artery, after having passed in the carotid groove anterior to foramen lacerum (fig. 99, page 188) recurves upwards and then posteriorly, inferior to the optic nerve (fig. 111, page 208); It gives off a branch, the **ophthalmic artery,** which passes inferior to the optic nerve, into the optic canal. The internal carotid artery terminates by dividing into the anterior and middle cerebral arteries, near this site.

Cut the optic nerve close to the optic chiasma, leaving a short stump of the nerve in contact with the chiasma. Then gently proceed with further elevation of the inferior surface of the brain from the cranial surface. Observe, on either half of your sagittally sectioned specimen, the **hypophysis (pituitary gland)** (figs. 101 and 102, page 196), in the **hypophyseal fossa,** between the carotid grooves in the central portion of the middle cranial fossa (shown, but not labeled, in fig. 99, page 188). Identify in your cadaver specimens, and in a dried skull, the anterior and posterior **clinoid processes.** Using scissors, cut through the internal carotid artery as far forward (anteriorly) as you can. Now identify the prominent **oculomotor nerve** (figs. 101, 102, 103, and 104, page 196), about 2 mm in diameter, passing toward the superior orbital fissure (fig. 100, page 188), visible in the skull but not in your cadaver dissection at this stage. To reach this fissure, the nerve pierces a fold of dura mater, made by the **free edge of the tentorium cerebelli** (page 195), at a site posteromedial to the palpable anterior clinoid process (figs. 99 and 100, page 188 and 103 and 104, page 196).

Cut the oculomotor nerve with scissors, leaving a stump attached to the brain, and a stump visible as it passes through the dural fold. Finger and blunt forceps dissection should now be gently used to elevate the temporal lobe of the brain from the lateral part of the middle cranial fossa.

Return again close to the sagittal plane, where the brainstem (specifically, the pons) reclines on the **clivus** (fig. 99, page 188), the surface of the sphenoid and basioccipital bones which slopes from the posterior clinoid processes down to the anterior margin of foramen magnum. Gently elevate the pons and medulla from the clivus. This will reveal first the small **abducens nerve** (cranial nerve VI) (figs. 102 and 103, page 196), close to the midline, passing from the junction of pons and medulla, forwards and laterally, to pierce the dura about 1.5 cm inferior to the posterior clinoid process. Cut the nerve with scissors, and proceed to elevate the pons further laterally. The large trunk of the **trigeminal nerve** will come into view. Superior to it, the very thin **trochlear nerve** will be seen to pierce the free edge of the tentorium cerebelli, just posterior to the oculomotor nerve (fig. 104). Cut the trochlear nerve, retaining, as usual, an identifiable proximal and distal portion of it, for subsequent study. Similarly cut cranial nerve V.

Identify, on the skull, the **internal auditory meatus** (fig. 99, page 188), on the posterior surface of the petrous part of the temporal bone. Now proceed laterally with elevation of the pons, to reveal cranial nerves VII and VIII, the **facial** and the **vestibulocochlear** (fig. 104, page 196), passing laterally through the internal auditory meatus. Cut these nerves and elevate the brain gently to now expose cranial nerves IX, X and XI, the **glossopharyngeal, vagus,** and **accessory** (fig. 104, page 196). Identify the large **jugular foramen** in the skull (figs. 97 and 99, page 188), and observe the pathway of these nerves toward that foramen. Take special care to identify nerve fibers which pass upwards from the spinal cord, through the foramen magnum, coming to lie anterior to the cerebellum, to join the most posterior of the bundle of cranial nerve fibers just identified (fig. 104, page 196). This is the **spinal accessory nerve** joining the **cranial accessory nerve** to leave the cranium, through the jugular foramen, in association with the latter. Recall, however, (page 194) that shortly after emerging from the cranial cavity, the cranial part of the accessory nerve joins with, and is distributed with, the vagus nerve, whereas the spinal accessory proceeds on an independent course, to supply somatic motor fibers to muscles of the neck and back. Preserving the spinal accessory nerve, cut through the bundle of IX, X, and XI fibers. Finally, posteroinferior and medial to the site of emergence of these three cranial nerves, find the **hypoglossal nerve** (XII), passing from the medulla toward the **hypoglossal canal (anterior condylar canal)** through which this nerve emerges (page 194). The canal should be identified, in your dissection specimen and in an isolated skull, just above the lateral edge of the foramen magnum. Note that the twelfth cranial nerve usually arises by several rootlets, which may pierce the dura independently.

Before leaving the region of the last four cranial nerves, identify the **vertebral artery** (figs. 103 and 104, page 196) as it enters the foramen magnum. Study its relationship to the spinal accessory nerve, and identify the branches of the vertebral artery and their relationships. Note the remnants of the denticulate ligament, and its position in relationship to the spinal accessory nerve, and to the upper dorsal and ventral cervical spinal nerve roots. Gently lift the posterior part of the cerebellum from the posterior cranial fossa, to further identify the tentorium cerebelli (page 195). Now complete removal of the half brain.

Returning to the tentorium cerebelli, now completely exposed, note again that the sagittally cut edge which you initially identified (page 195), is continuous, when followed forwards, with the intact, curving **free edge,** which sweeps out laterally and then anteriorly, toward the anterior clinoid process. Reidentify the distal stumps of the oculomotor and trochlear nerves which you left intact, in relation to the free edge, when you cut these nerves in freeing the brain. Preserve these stumps with great care, as they will play an essential role in assisting your understanding of anatomical relationships in subsequent dissections. By visualizing the cut edge of the tentorium in relationship with the corresponding cut edge of the other half of your dissection specimen, as well as by studying prosections, appreciate the shape and the double-layered nature of the tentorium cerebelli. Review the **attached edge** of the tentorium. It is attached posteriorly along the grooves of the transverse sinus (page 195), the two layers corresponding to those which enclose the sinus. Anterolaterally, the tentorium is attached to the superior border of the petrous temporal bone. Observe carefully the relationship of the free and attached edges at the site where cranial nerves III and IV pierce the free edge.

Note the opening through the tentorium between the free edge and the clivus, through which rostral (front: see fig. 38, page 88) and caudal parts of the brainstem are continuous. With the aid of an atlas, study the dural folds, and the dural venous sinuses, identifying, in addition to those you have already observed (the transverse and superior sagittal sinuses), also the inferior sagittal, the straight and the sigmoid sinuses. Your study of dural folds should include identification of the **diaphragma sellae,** a fold that roofs over the **hypophyseal fossa,** and that has a central opening through which the stalk of the pituitary gland connects the hypophysis with the base of the brain.

For further study, the half calotte of your dissection specimen should now be removed, with the cautious aid of an electric hand saw. Cut through the bones of the neurocranium with a semicircular cut passing just superior to the transverse sinus, posteriorly, and the crista galli anteriorly.

Your objective in the ensuing dissection is to study the **cavernous sinus** which, like the other dural venous sinuses, lies between two layers of dura mater. It is situated on the sides of the body of the sphenoid bone. Anteriorly, it receives, as tributaries, **ophthalmic veins** from the orbit, reaching through the superior orbital fissure, and also the sphenoparietal sinus and certain veins from the brain. Posteriorly, the cavernous sinus is continuous with the inferior petrosal sinus, which drains ultimately, through the jugular foramen, into the internal jugular vein. Several cranial nerves, and part of the internal carotid artery, are situated "within" the cavernous sinus (i.e., are situated between the two layers of dura that enclose the sinus, and partially invaginate the endothelial wall of the venous channel, although remaining morphologically outside of the lumen of the sinus).

Exercising the utmost care, trace the minute trochlear nerve anteriorly, by incising, with a scalpel, the fold of the free edge of the tentorium cerebelli which the nerve pierces just behind the posterior clinoid process. Gently dissect surrounding dural tissue away, to expose the nerve as it passes forwards. The free edge of the tentorium cerebelli should be traced forwards to the anterior clinoid process, and carefully detached from the latter, reflecting the dura downwards and laterally to expose the lateral aspect of the cavernous sinus. As you dissect carefully into the sinus, clotted blood will reveal that you are entering the sinus space. Reidentify the trigeminal nerve, and insert the tip of a pair of forceps alongside the nerve as it passes anterolaterally through the dura. Your forceps are entering the **trigeminal cave,** a little pocket of dura raised by the trigeminal nerve and its afferent ganglion (homologous with the dorsal root ganglion of spinal nerves), which you should identify as a swelling, as you cut into the roof of the trigeminal cave. The dura should be stripped, to expose the trigeminal nerve and its three divisions. This dissection may be difficult, because the dura may be extremely adherent to both the nerve and the underlying cranial bone. Careful scalpel and scissors dissection are required.

As the clotted blood within the cavernous sinus is carefully removed, the **internal carotid artery** will be revealed. In addition, the sixth cranial nerve will be seen to be lying lateral to the internal carotid artery. This part of the course of the abducens nerve should be exposed, and its continuity with the stump of the sixth nerve that you previously identified passing through the dura on the clivus, should be established, by stripping dura away. To facilitate this dissection you should use a pair of fine pointed bone scissors, to cut off the anterior clinoid process with care, and also the adjacent portion of the posterior border of the lesser wing of the sphenoid bone to better expose the superior orbital fissure and the pathway of the cranial nerves III, IV, V (1) (the ophthalmic division) and VI through this fissure into the orbit. Trace also cranial nerve V (2), the maxillary division, to its site of exit, the **foramen rotundum** (fig. 100, page 188), and V (3), the mandibular division of the trigeminal nerve to the **foramen ovale.**

To complete your study of the internal surface of the cranial base, identify the **greater** and **lesser petrosal nerves** (see pages 192 and 193), emerging through hiatuses on the anterior surface of the petrous part of the temporal bone. The greater petrosal nerve, which is a parasympathetic branch of the facial nerve (VII), passes toward the foramen lacerum (fig. 99, page 188), at which it enters the pterygoid canal, together with the **deep petrosal nerve** (derived from the sympathetic plexus of the internal carotid artery). The **nerve of the pterygoid canal,** thus formed, passes to the **pterygopalatine ganglion** (pages 63 and 191). The **lesser petrosal nerve,** comprising parasympathetic fibers derived from the **glossopharyngeal nerve** (IX), emerges just lateral to the greater petrosal nerve, and passes to the foramen ovale, through which it emerges to reach, and synapse in, the otic ganglion (pages 48, 101, and 193).

Clinical Session 20

■ **TOPICS**

Eye
Eyebrows
Eyelids
Conjunctiva
Lacrimal Apparatus
Associated Soft Tissue Structures

The eye, or organ of sight, is situated in the front part of the orbit and is approximately spherical in shape. More precisely, the anterior segment, constituting about one sixth of the eye, has the curvature of a smaller sphere, and the larger, posterior segment has the curvature of a larger sphere. The central point of the anterior curvature is called the **anterior pole** and the central point of the posterior curvature is called the **posterior pole** of the eye. Thus defined, one can imagine a pole to pole **primary axis** of the eye, which is useful in describing the situation of various other structures.

Commence your examination of the eye by observing the eyebrows, which are situated superficial to the superciliary arches (page 183), and are richly endowed with short thick hairs. The frontalis muscle (page 19) and fibers of the orbicularis oculi muscle (page 19) insert into the skin of the eyebrows.

The eyelids (**palpebrae**) are folds, separated by the **palpebral fissure,** which protect the eye, and can be moved up and down by muscular action. Medially and laterally, upper and lower eyelids meet at the medial and lateral angle, or **canthus,** respectively.

Examine the free edges of the eyelids of your model, noting that **eyelashes** are attached to these edges. Observe how many rows of eyelashes are present. Identify, a few millimeters from the medial canthus, on each eyelid, a small opening, the **punctum lacrimale** (fig. 106, page 202), situated on the apex of a small elevation, the **lacrimal papilla.** Each punctum lacrimale is a minute opening of a canal, the **lacrimal canaliculus,** each of which leads to the small **lacrimal sac.** Examine a disarticulated skull, and identify the **lacrimal groove,** in the inferior anteromedial corner of the orbit, in which the lacrimal sac is situated. The groove (fig. 107, page 202), is limited anteriorly by the anterior lacrimal crest, on the frontal process of the maxilla and posteriorly by the posterior lacrimal crest, on the small **lacrimal bone,** the limits of which should be identified on the skull. Using a thin piece of insulated wire, gently probe the inferior extremity of this groove, and identify its communication, through a bony canal, with an opening into the inferior meatus of the nose (pages 59 and 63). The lacrimal sac is the upper, expanded end of the **nasolacrimal duct,** which passes through this canal, and empties tears into the nasal cavity.

In the medial angle of the eye, a small, triangular red structure called the **caruncula lacrimalis,** and the adjacent **plica semilunaris,** should be identified (fig. 106, page 202).

Examine the inner surface of the lower eyelid by simply pulling the lid gently downwards. This surface is lined by a moist, glistening membrane, the **conjunctiva,** through which vertical ridges, formed by the **tarsal glands,** can be seen (fig. 106). These glands are so named because they lie, deep to the conjunctival membrane, in grooves in the **tarsus** (tarsal plate), a sheet of firm connective tissue within the eyelid, which serves to retain the form of the lid. The tarsus of the upper eyelid is larger, firmer, and more easily examined. Gently, but firmly, hold a cotton wool swab at the uppermost part of the upper eyelid with one hand, and then grasp the eyelashes of that lid in the other hand and evert the lid by rotating it around the cotton wool swab. In this way, because of the firmness of the tarsus, the lid can be made to remain in a stable position for several minutes, with its inner surface turned outwards. Observe again, more prominently in the upper lid than in the lower lid, the vertical markings of the tarsal glands. Minute openings of the tarsal glands can be seen, posterior to the lower free edge of the eyelid (fig. 108, page 202). The upper lid is, like the lower lid, lined with conjunctival membrane. When the eyes are closed, the conjunctiva constitutes a sac, the anterior wall of which is slit in the horizontal plane, corresponding to the palpebral fissure. The posterior wall of the sac abuts onto the anterior surface of the eyeball, although it is replaced, in the center portion, by the **cornea,** the transparent structure which lies in front of the colored part of the eye, or **iris.** Accordingly, when a patient gets a foreign body "in his or her eye," the foreign body is, in the usual course of events, within the **conjunctival sac.** The sac has an upper and a lower **fornix,** where the surface of the conjunctiva which lines the eyelid reflects back onto the surface of the eyeball.

Tears are secreted into the conjunctival sac at the upper anterolateral corner of the orbit, where the major part of the **lacrimal gland** is situated in a fossa in the wall of the orbit at this site. The lacrimal

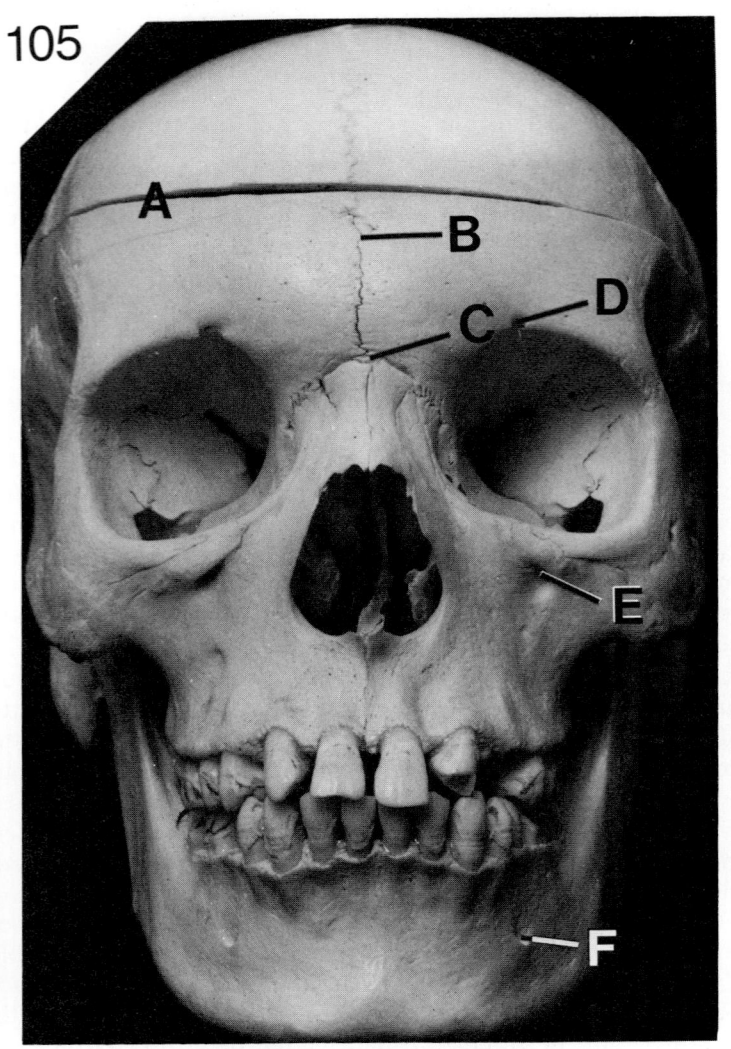

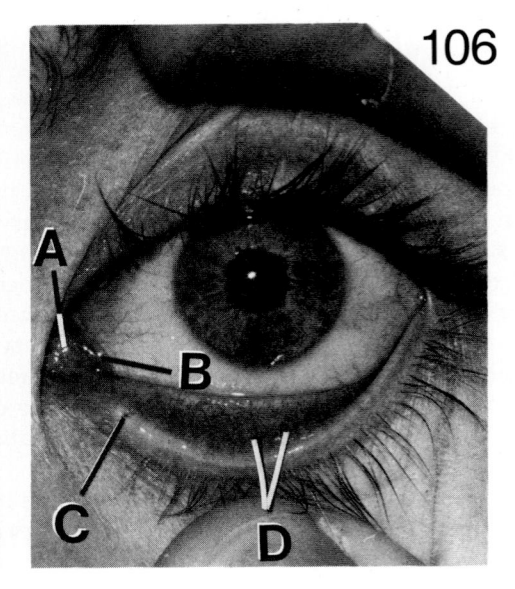

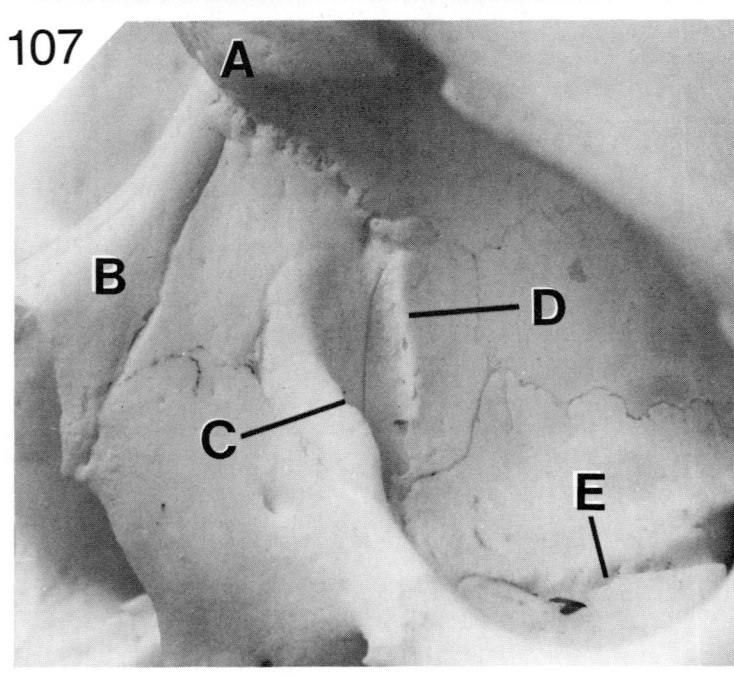

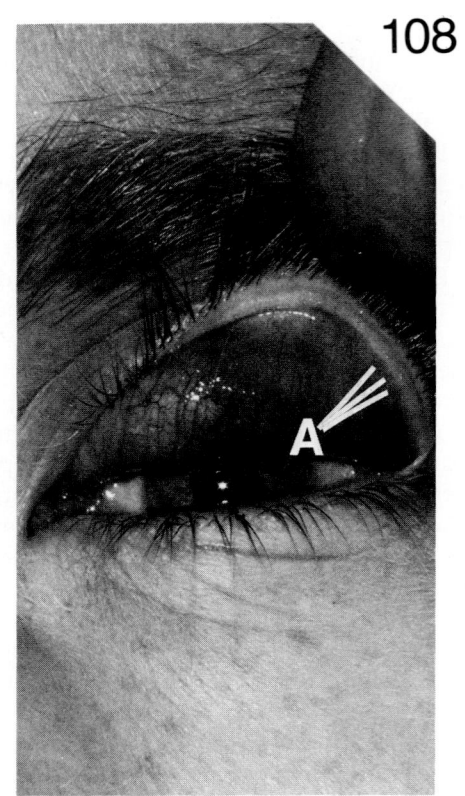

202 YOUR PATIENT'S ANATOMY: A CLINICAL VIEW OF HUMAN MORPHOLOGY

Figure 105. Anterior aspect of skull.
A. Saw-cut separating calotte (above) from calvaria
B. Metopic (frontal) suture (usually closed by 8 years of age)
C. Nasion
D. Supraorbital notch
E. Infraorbital foramen
F. Mental foramen

Figure 106. Left eye.
A. Caruncula lacrimalis
B. Plica semilunaris
C. Punctum lacrimale
D. Tarsal glands

Figure 107. Anterolateral view of left orbit.
A. Glabella
B. Left nasal bone
C. Anterior lacrimal crest of frontal process of maxilla
D. Posterior lacrimal crest of lacrimal bone
E. Infraorbital groove

Figure 108. Left eye, upper eyelid everted.
A. Openings of tarsal glands

gland is superolateral and posterior to the conjunctival sac, and tears enter the sac through minute ducts which open through the conjunctival membrane, into the sac. In fact, the gland has an **orbital part,** lodged in the **lacrimal fossa,** which you should now observe on the medial side of the zygomatic process of the frontal bone, and a **palpebral part,** which extends into the lateral part of the upper eyelid. Thus tears sweep across the eye from the superolateral corner to the puncta lacrimalia, where they are drained.

EYEBALL

Begin your examination of the eyeball by inspection, noting that the visible part of the eyeball, within the palpebral fissure, comprises a central, dark, circular region known as the **pupil;** a colored rim around the pupil, called the **iris;** and the surrounding white region, referred to as the **sclera,** which is usually visible lateral and medial to the iris, and sometimes visible normally (and always observable by retracting the eyelids), above and below the iris. Careful scrutiny will indicate that the pupil is a hole through the middle of the iris, and the iris lies somewhat deep to the surface of the eye, covered by the transparent, disk-shaped **cornea,** which is continuous with the sclera. The posterior wall of the conjunctival sac is fused onto the sclera, but is lacking over the cornea. Thus the conjunctiva ends at, and is fused to, the margin of the cornea. The cornea and sclera are indeed part of the same structure: the **outermost** of the three coats or layers of the wall of the eyeball. The sclera comprises dense bundles of fibrous tissue, whereas the cornea, derived from the same material, it made up of highly orientated molecules that allow for transparency. The curvature of the cornea is mainly responsible for refraction of the rays of light which pass through the pupil to enter the eye and create an image on the light sensitive, **innermost** of the three layers of the eye, the **retina.** The iris itself comprises part of the **middle** of the three layers of the eye, known as the **vascular layer.**

Inspection of the eye can reveal information relating to the health of the patient, as well as local health of the eye. For example, the degree of prominence of the eye may frequently provide information concerning certain illnesses. To mention one, exophthalmos (displacement forwards) is a sign frequently observed in metabolic disturbance caused by overfunction of the thyroid gland. The state of the conjunctiva, and its color, may also be an indicator of health or illness. For example, in jaundice, the conjunctiva is discolored yellow.

On palpation, tension of the eyeball can be estimated, and where necessary, can be accurately measured by appropriate instrumentation. In a condition called glaucoma, intraocular tension is increased, whereas in certain conditions, such as dehydration, it may be decreased.

Review, (pages 190 and 186–187) the methods of examination of the pupils, and of acuity and fields of vision. If the subject wears glasses, it might be of interest to know whether the patient is myopic (nearsighted) or hypermetropic (farsighted). Holding the patient's glasses in front of your own eyes, observe an object through the glass and move the spectacles from side to side. If the object moves in the opposite direction to the glass, the user is hypermetropic, and if it moves with the glass, myopic.

The eye is, as mentioned, a sphere, the wall of which comprises three layers. You have observed two parts of the outer layer: the cornea and the sclera: You can also see the iris and its central hole, the pupil; parts of the middle layer; and mention has been made of the innermost layer, the light sensitive retina. It has also been mentioned that the cornea is largely responsible for refracting light rays, which subsequently pass through the pupil, to reach the retinas. It should be mentioned, in addition, that posterior to the iris there is a **lens** which provides additional refraction of light rays. The lens is capable of slight changes in shape, which can increase or decrease its convexity and therefore play a role in shifting the focus of the refracted rays, a process known as **accommodation.** In addition, fluids within the eye that fill the spaces surrounding the lens, and posterior to it, also contribute to refraction.

Under guidance, using an ophthalmoscope, examine the interior of the eye, with special reference to the retina, to identify the following features of it. Examination is best performed when the patient's pupil is dilated, in a dark room or with the use of a mydriatic substance, which dilates the pupil by inhibition of the **sphincter** muscle (page 190) or activation of **dilator** muscles of the pupil. The subject should be asked to keep his eye still and look straight ahead, focused on a particular object, even if the observer's head gets into his line of vision.

The retina typically has a reddish color, and at several sites, arteries and veins are visible. In the retina, the arteries are thinner and the veins thicker, and they converge, posteriorly, on an approximately centrally placed, pale pink structure, the **optic disk.** In fact, the optic disk is slightly medial to the **posterior pole** of the eye (page 201), and slightly superior to it. The optic disk is the site at which optic nerve fibers converge, as they leave the retina, to pass through the posterior wall of the eye and form the **optic nerve per se.** The nerve fibers convey to the brain impulses created by light rays impinging on the retina. The optic nerve leaves the orbit through the optic foramen (page 186). Identify the optic disk, and observe its size, shape, color, and outer edge, which is sharp when the disk is healthy. As mentioned, the disk should be sought superomedial to the posterior pole of the eyeball, and a useful way of finding it is to follow the blood vessels, which converge toward it.

Slightly inferolateral to the posterior pole of the eye, about two disk diameters away from the disk itself, a darkish yellow region of the retina, the **macula lutea,** may be observed (macula means spot; lutea means yellow). It is much smaller than the optic disk, its diameter being less than a quarter that of the disk, and it has a central, lighter area known as the **fovea centralis.** The fovea centralis is the region of the retina in which visual acuity is greatest. In focusing sharply on an object, the eye is moved in such a way that the image becomes focused on the fovea centralis; regions of the retina peripheral to this give a less sharp image, hence the term "peripheral vision" (page 187).

Examination of the optic disk is of vital importance as part of a general medical examination, not only because illnesses of the disk or optic nerve manifest themselves in abnormal appearances, but also because certain features of general health similarly do so. Of major significance among these is increase in the intracranial pressure. Because the optic nerve carries a sheath of dura mater with it, increases in intracranial pressure get transmitted along the course of the optic nerve, and can result in swelling of the optic disk, known as papilloedema. Examination of the retina is also of importance because it is unique in being a site at which blood vessels can be directly observed. In various diseases, such as hypertension (increased arterial blood pressure), certain changes in the walls of blood vessels occur, and these give a characteristic appearance to the retinal blood vessels, thus aiding in diagnosis and prognosis of the condition.

The mydriatics used to dilate the pupil do so by stimulating the sympathetic innervation of the dilator of the pupil or by inhibiting the parasympathetic innervation of its sphincter (page 190). Note the side effects that this medication has on your model, including constriction of the scleral blood vessels, and retraction of the upper eyelid (due to sympathetic innervation of the **levator palpebrae superioris** muscle).

ORBIT

Consult Figures 96 (page 184) and 107 (page 202), and atlas illustrations, and study the component bones in the walls of the orbit.

The shape of the orbit is that of a four-sided pyramid, the apex being situated posteriorly and the base anteriorly, at the orbital opening. Note the angle which the **axis** of the pyramid, from the apex to the center of the base, makes with the anteroposterior (parasagittal) plane. This point is of importance in understanding the functions of the extrinsic eye muscles, to be discussed below.

Examine the orbital opening, reidentifying the supraorbital margin, formed entirely by the frontal bone. At the junction of the lateral two thirds and the medial one third, identify the supraorbital notch (or foramen) through which the supraorbital nerve and vessels pass (page 190). The lateral margin is formed by the frontal process of the zygomatic bone, and the zygomatic process of the frontal bone. Palpate the lateral margin on your subject to identify the suture, which is usually easily located. Review palpation of the infraorbital margin and, just inferior to it, the infraorbital foramen (page 37). Examine the medial margin, formed above by the frontal bone and below by the lacrimal crest of the frontal process of the maxilla (which also demarcates the anterior limit of the fossa for the lacrimal sac [fig. 107, page 202]).

The superior wall or roof of the orbit is formed by a thin sheet of bone which separates the orbit from the anterior cranial fossa. Anteriorly, the roof separates into two layers, containing the frontal sinus, and medially it may similarly contain part of the ethmoid sinus. The roof is part of the frontal bone, and anterolaterally it presents the lacrimal fossa for the orbital part of the lacrimal gland (page 204). Posteriorly, identify the suture with which the frontal bone articulates with the greater wing of the sphenoid, and examine continuity of this surface of the greater wing of the sphenoid with other parts of that bone which you have previously studied (pages 27, 39, 40, and 183). Posteromedially, at the apex of the pyramid, identify the small portion of the lesser wing of the sphenoid, through which the optic canal passes (fig. 96, page 184).

The medial wall is formed partly by the lesser wing. Anteriorly, the major part of the medial wall is formed by the orbital plate of the ethmoid bone. The latter articulates anteriorly with the lacrimal bone, on the surface of which the posterior lacrimal crest should be observed. This delimits, posteriorly, the fossa for the lacrimal sac, this fossa being bounded anteriorly by the anterior lacrimal crest, on the frontal process of the maxilla (fig. 107, page 202). At the superior border of the orbital plate of the ethmoid bone, observe the anterior and posterior ethmoidal foramina, through which important nerves and vessels pass from the orbit.

The lateral wall of the orbit is mainly formed by the greater wing of the sphenoid, which separates the orbit from the middle cranial fossa. The lateral wall and roof are separated by the superior orbital fissure posteriorly (see fig. 96, page 184). This important fissure transmits nerves and blood vessels between the orbit and the middle cranial fossa. The anterior part of the lateral wall is formed by the zygomatic bone, which contains minute foramina through which the zygomaticotemporal and zygomaticofacial nerves (page 191) pass, to reach the face.

The floor of the orbit is mainly formed by the orbital surface of the maxilla and, anterolaterally, by the zygomatic bone. Posteriorly, a small area of the floor is formed by the orbital process of the palatine bone. The inferior orbital fissure (fig. 96, page 184) is situated between the floor and the lateral wall. It transmits the maxillary nerve and infraorbital vessels. Observe the infraorbital groove and canal, the latter opening anteriorly at the infraorbital foramen.

Review (page 187) the clinical examination of the extraorbital muscles and the cranial nerves which supply them, the oculomotor (III), trochlear (IV), and abducent (VI). The main functions of the various muscles were previously described, and it was stated at that time that several of the muscles have additional actions. In the following, a more detailed description of the functions of the extraocular muscles is given.

The major actions of the extraocular muscles are apparent from consideration of the origins and insertions of these muscles. Thus, for example, since the four recti muscles have their origins in a tendinous ring surrounding the optic foramen, posteriorly in the orbit, and their insertions are distributed anteriorly at sites corresponding to the twelve o'clock, three o'clock, six o'clock, and nine o'clock positions of the anterior aspect of the eyeball, it is clear that the superior rectus muscle will pull the gaze upwards, and the other muscles will similarly pull in directions appropriate to their origins and insertions. However, the discrepancies between this simple plan, and the actual function of these muscles, arises from the fact that this description presupposes the primary axis of the eyeball (page 201) to be parallel to that of the orbit. In fact, the primary axis of the eyeball is in the anteroposterior plane when the eye is gazing toward the horizon, whereas the axis of the orbit is not in this plane, but deviates from it by about 23°. Thus, for the extraocular muscles to function in a simple manner, according to the predictions that could be made from their origins and insertions, the eye should be gazing anterolaterally, such that its primary axis is parallel to that of the corresponding orbit.

It is important to understand that although the actions of each of the extraocular muscles are usually described independently, no muscle acts in isolation, and coordinated function of all muscles of an eye, and, indeed, in binocular vision, of both eyes, takes place in all eye movements.

Follow the ensuing description with the aid of a skull and atlas illustrations, to visualize the position of the eye and its muscles, while testing movements on your subject. The movements should be pictured as taking place about three axes through each eyeball, corresponding to the three planes in space: a vertical axis, a horizontal axis, and an anteroposterior axis, each passing through the midpoint of the eyeball.

The function of the medial and lateral rectus muscles is not much affected by the difference in anteroposterior axis of the eyeball and the orbit, and therefore their function corresponds closely to the theoretical expectation. Thus the medial rectus muscle rotates the eye medially, and the lateral rectus rotates it laterally. However, with the eye gazing at the horizon, the direction of pull of the superior rectus at its point of insertion is not merely upwards and backwards, but is also medially. Therefore, contraction of the superior rectus will not only cause the eye to rotate upwards about the transverse axis, but will also cause some medial rotation about the vertical axis. In addition, some rotation about the anteroposterior axis is also produced; this movement is called **torsion**. Consider a point at the twelve o'clock position on the cornea of the right eye, as viewed from the front. This point will be caused to rotate about the anteroposterior axis in a direction toward three o'clock, a movement known as **intorsion**. Consideration of the inferior rectus muscle as described above will indicate that its functions, in addition to drawing the gaze downwards, include medial rotation and **extorsion**.

The superior and inferior oblique muscles have comparable directions of pull, since the superior oblique must be regarded as pulling from the site of the trochlear, and the origin of the inferior oblique is approximately vertically below that, on the floor of the orbit. Both muscles insert into the posterior, lateral quadrant of their respective surface of the eye, and both pass posterior to the vertical axis, to reach their insertions. Therefore, the pull of the superior oblique is upwards, forwards, and medially, and produces a downward movement about the transverse axis of the eye, a lateral rotation about the vertical axis, and a degree of intorsion. The inferior oblique correspondingly produces upward gaze, lateral rotation, and extorsion.

Laboratory Session 20A

Because freshly fixed human eyes are rarely available for dissection, and the eye of a routinely embalmed cadaver is usually not suitable for detailed study, it is customary to use a pig or cow eye, both of which are sufficiently similar to the human to make this exercise a worthwhile learning experience. The following description is given for dissection of a pig eye. Important differences between the pig and human eye will be mentioned as they are encountered.

Examine the eye externally. Begin by identifying the anterior and posterior poles (page 201), confirming that the eyeball appears to be comprised of an anterior portion with the curvature of a smaller sphere, and a posterior, larger portion with the curvature of a larger sphere. Identify the stump of the optic nerve, posteriorly, and the cornea, through which you can see the iris and pupil, anteriorly. You should attempt to recognize the insertions of any of the extraocular muscles which you may be able to find, though a detailed identification of the individual muscles may not be possible. If eyelids are present, these should be examined, and the nature and position of the tarsal plates (page 201) should be studied. Attempt to dissect a small portion of the conjunctiva from the inner surface of the eyelid, and from the anterior surface of the sclera, to appreciate the nature of the conjunctival sac. Identify any other of the accessory structures of the eye which may be present in your specimen (pages 201–204). Perform an initial external examination to identify the sclera (page 204), the cornea (page 204), a stump of the optic nerve (page 204), short and long ciliary nerves (pages 190 and 191), and the insertions of any of the extraocular muscles you may be able to find. Presence of blood vessels on the external surface should also be noted. Finally, examine the accessory structures of the eye, as follows.

The eyeball is surrounded by thick **bulbar fascia** which you will need to reflect and remove from the surface of the eyeball, in order to study underlying structures. Starting posteriorly, identify the optic nerve. Using a sharp razor blade, cut the nerve close to the eyeball. Now examine the cut surface of the stump of the optic nerve in the dissecting microscope. Attempt to identify the minute **central artery of the retina,** which enters the optic nerve obliquely, close to the point of attachment to the eye, and supplies the nerve in the retina. Note that in the human eye, the central artery enters the optic nerve further away from the eyeball. Observe also, under the dissecting miscroscope, bundles of nerve fibers within the optic nerve. Now proceed to cut slices from the stump of the optic nerve, until you actually slice into the outer coat of the eyeball. After taking a few thin slices of the sclera, you will then slice into the next layer of the eye, the choroid, which you will recognize by its pigmented nature. Study the appearance of the optic nerve as it passes through the choroid.

Reflect bulbar fascia from the posterior half of the surface of the eyeball, to the plane of the equator. (In keeping with the analogy of calling the anterior and posterior maximal points of curvature of the eyeball the **anterior** and **posterior poles** [page 201], anatomical terminology also identifies an **equator** of the eyeball, between the anterior and posterior poles. In addition, a line passing over the surface of the eyeball, the shortest distance from one pole to another, and intersecting the equator at right angles, is known as a **meridian.)** Using a **slicing** action, cut through the eyeball in the plane of the equator, to separate a posterior hemisphere from an anterior. Note, as you cut through the coats of the eye, that a somewhat gelatinous substance, the **vitreous humor,** will escape. The vitreous humor is a refractile medium (page 204), which fills the **vitreous chamber** of the eye (i.e., that part of the internal space of the eyeball which lies posterior to the lens). Now study the posterior hemisphere of the eyeball.

Identify the **optic disk** by the criteria you used in studying this structure in a living subject, during clinical examination (pages 204 and 205). Note the convergence of blood vessels toward the disk. The edge of the disk may be less clear in the pig eye than that you observed in the human. Confirm, by studying alternately the internal and external surface of the posterior hemisphere, the relationship of the stump of the optic nerve, which you sliced, with the optic disk as seen from the inside of the eye. Search also for the **macula,** which may also be inconspicuous. If you are able to find it, that will enable you to orientate the eye in respect of lateral and medial (pages 204 and 205).

Study the cut edge of the hemisphere of the eyeball in the dissecting microscope. Beginning on the internal surface, you will see a gray, somewhat folded layer, which is easily separated from an underlying black sheet. Both the easily removable gray and the underlying adherent black layers are parts of the **retina,** the black portion representing that which is known as the **pigmented layer of the retina,** and the gray comprising the **sensory retina,** including the light-sensitive layer of **rods** and **cones.** Carefully remove part of the gray layer of the retina with a pair of fine forceps. Then elevate the black, pigmented layer of the retina from the outer, less

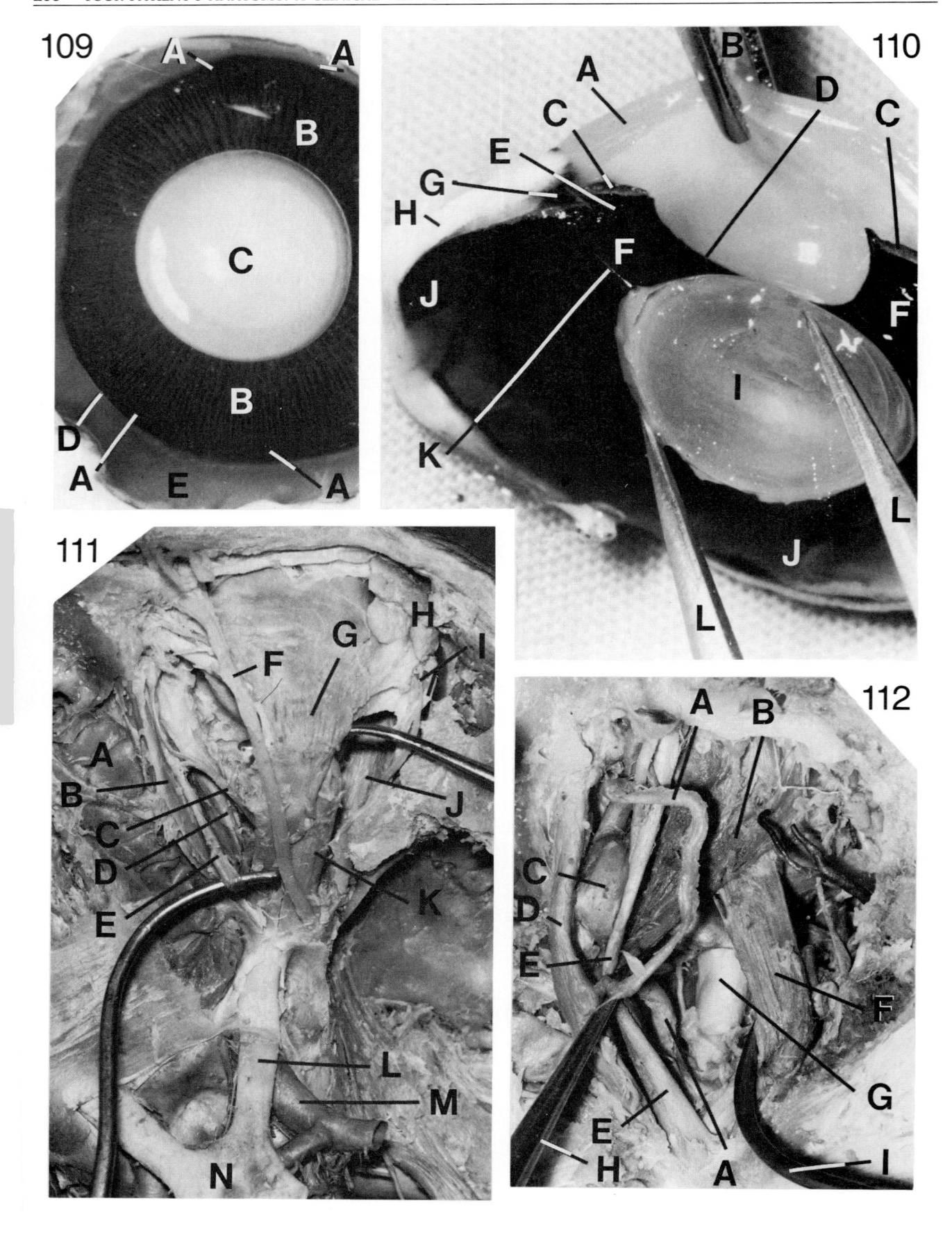

208 YOUR PATIENT'S ANATOMY: A CLINICAL VIEW OF HUMAN MORPHOLOGY

Figure 109. Interior aspect of anterior half of eyeball.
A. Ciliary ring
B. Ciliary folds
C. Lens
D. Ora "serrata"
E. Retina

Figure 110. Anterior quadrant of eye, viewed from cut surface, with lens being gently retracted.
A. Cut edge of cornea
B. Forceps gripping cornea
C. Cut edge of iris
D. Rim of pupil
E. Cut edge of corona ciliaris
F. Posterior aspect of corona ciliaris
G. Ciliary muscle
H. Cut edge of sclera
I. Lens
J. Retina
K. Zonular fibers
L. Forceps grasping lens

Figure 111. Right orbit, superior aspect.
A. Ethmoid sinus
B. Superior oblique muscle
C. Nasociliary nerve
D. Medial rectus muscle
E. Ophthalmic artery (elevated on probe)
F. Frontal nerve
G. Levator palpebrae superioris muscle
H. Lacrimal gland
I. Lacrimal nerve and artery
J. Lateral rectus muscle
K. Superior rectus muscle
L. Right optic nerve
M. Right internal carotid artery
N. Optic chiasma

Figure 112. Right orbit: deep view to show anterior extent of optic nerve, just posterior to eyeball.
A. Ophthalmic artery
B. Levator palpebrae superioris, displaced medially
C. Superior ophthalmic vein, displaced medially
D. Superior oblique muscle
E. Frontal nerve
F. Superior rectus muscle, displaced laterally
G. Optic nerve
H. Probe, displacing A, B, D, and E medially
I. Probe displacing superior rectus muscle laterally

darkly pigmented layer, the **choroid**. At the cut edge of the hemisphere, the pigmented choroid layer can be seen to give way to the white, thick **sclera**. Gently attempt to separate the choroid from the sclera. The adherence between these two layers is tighter than the previous, and the separation is accordingly more difficult. It is in this plane, however, that you will observe small branches of the **long ciliary nerves** (pages 191 and 207) and **arteries**, and the dissection is therefore worth attempting.

Place the anterior hemisphere of the eyeball under the dissecting microscope, with the internal surface facing upwards (fig. 109, page 208). Observe the extent of the grayish sensory retina, and note that it appears to end at a sharp border, where it is in contact with the black **ciliary ring** or **orbiculus ciliaris** (orbiculus means ring). The ciliary ring and the adjacent black folds (the **ciliary folds**, also known as **pars plicata**, or **corona ciliaris**), which project centrally toward the lens of the eye, are part of the **ciliary body**, a component of the middle layer of the eyeball. The apparent sharp border between the retina and the ciliary ring is known as the **ora serrata** in the human eye, because it is scalloped (serrated) in humans, although smooth in certain other species. The border is, in addition, apparent, and not real, as a very thin layer of the sensory retina continues on the posterior aspect of the ciliary body. With a fine pair of forceps, gently attempt to displace the lens, and note that it is fairly firmly attached to the adjacent ciliary body. The minute **zonular fibers** responsible for this attachment will be studied in the following dissection.

Invert the anterior hemisphere of the eyeball, so that the cornea faces upwards. Using a small pair of forceps, a sharp pair of scissors, and a sharp blade, cut a central disk-shaped portion of the cornea out, to expose the **anterior chamber** of the eye. Explore the space posterior to the cornea to establish that the anterior chamber, limited anteriorly by the cornea itself, is bounded posteriorly by the iris, the pigmented structure that gives the eye its color, and which is perforated in the center by the pupil (page 204). Study the iris and the pupil, bearing in mind that the human pupil is usually circular, whereas that of other species may be somewhat oval. Now replace the anterior hemisphere in a petri dish, with the internal surface of the eye facing upwards. Take a sharp blade, and place it over the lens, so that by using downward force with the blade, you bisect the lens in a meridional plane and, completing your incision, divide the entire hemisphere into two quadrants.

Turn the cut surface of one quadrant upwards to examine it in the dissecting microscope. Note the layered structure of the lens (fig. 110, page 208).

Using a pair of fine forceps, gently grasp the lens and apply traction in a posterior direction, and observe, at the periphery of the lens, the band of minute **zonular fibers** that attach the periphery of the lens to the posterior surface of the ciliary body (fig. 110). Gently separate the ciliary folds (the pars plicata) from the posterior aspect of the iris. The space between these two structures is known as the **posterior chamber** of the eye. The anterior and posterior chambers of the eye are in free communication through the pupil, and both contain **aqueous humor** which, like vitreous humor (page 207), is a refractile medium (page 204). However, aqueous humor and vitreous humor are not in free communication with each other: a hyaline membrane limits the vitreous humor around its entire surface, including that abutting posteriorly onto the lens and ciliary body. The aqueous humor is produced by a process of diffusion across capillary walls, in the ciliary folds. It is drained through minute channels which exist in the angle between the cornea and the iris **(the iridiocorneal angle)**. The channels unite to form the **canal of Schlemm**, which drains into veins of the eyeball. The aqueous humor is a nutrient to both the lens and the cornea, neither of which has blood vessels. The fluid is also essential in maintaining the normal tension of the eyeball. In certain abnormal states, drainage may be hindered and a state of raised intraocular pressure may arise, causing the condition of glaucoma. If not treated, this can seriously impair and ultimately ablate eyesight, through pressure on the sensitive retina.

The continuity of the sensory retina over the posterior surface of the ciliary body can be established by carefully elevating the cut edge of the retina, in a quadrant of the anterior hemisphere, and lifting it with a fine pair of forceps close to the ora serrata. By gentle manipulation, the sensory retina can be separated from the ciliary ring and ciliary processes, quite far forward. Although it is unlikely that you can establish it in this dissection, the retina actually continues forwards onto the **anterior** surface of the ciliary body and then folds back onto the iris. In the iris, the small **dilator muscles of the pupil**, which run in a radial direction from the edge of the pupil out toward the periphery of the iris, are developmentally derived from this fold of the retina.

Examine further the **iridiocorneal angle** of the meridionally cut section of the anterior hemisphere. Identify the small projection from the cornea, to which the iris is attached (the **corneal spur**), and attempt to identify, in the angle, the minute channels into which the aqueous humor drains. Consult atlas illustrations to aid you in your search.

Also by examination of the same cut meridional

surface, study further the components of the ciliary body: the **ciliary ring,** the **ciliary processes** and, finally, the third component of the ciliary body, the **ciliary muscle** (fig. 110, page 208). The muscle is situated in the crevice between the ciliary ring and the sclera (i.e., at the periphery of the ring). It has three components, which cannot be identified at the magnification of a stereoscopic dissection microscope. Nevertheless, you should attempt to appreciate that it has fibers which are placed **radially** (in relation to the ciliary ring and ciliary processes); **circular** fibers (in the same circular plane as the ciliary ring, which therefore act as a sphincter muscle); and **meridional** fibers (which pull in a pole to pole direction and therefore alter the position of the ciliary body in an anteroposterior direction). All of these fibers are capable of altering the position and shape of the lens, through the pull transmitted through the zonular fibers that suspend the lens from the ciliary body (page 210). Note that the lens has an inherent elasticity which tends to maintain its spherical shape. Pull on the periphery of the lens, through the zonular fibers, tends to flatten the lens. Contraction of the circular, sphincteric fibers of the ciliary muscle will tend to relax this pull and therefore to increase the spherical shape of the lens. It is through this mechanism that the process of accommodation takes place.

Laboratory Session 20B

Previously, you dissected the cavernous sinus and the contents thereof, and traced cranial nerves III, IV, and VI, and the first and second divisions of cranial nerve V, into the sinus. In order to dissect the orbit, you should now proceed with following these cranial nerves forward, through the **superior orbital fissure,** and into the orbit. This will be done by chipping the superior bony border of the superior orbital fissure away with bone forceps, and continuing this piecemeal removal of the roof of the orbit (floor of the anterior cranial fossa) anteriorly, to expose the entire orbit from above. Start by extending laterally the dissection of the posterior border of the lesser wing of the sphenoid bone from the site where you removed the anterior clinoid process previously (page 200). Starting laterally in the roof of the orbit, removal of the roof will expose the periosteum. Further medially, you may encounter, intervening between the periosteum and the layer of bone you are chipping, extensions of both the **frontal** and **ethmoid sinuses.** These should be studied and their three-dimensional relationships appreciated, before they are removed. Thereafter, gently incise the periosteum, and reflect it, to expose the orbital contents, comprising fatty tissue in which embedded muscles, nerves, and vessels are found (figs. 111–112, page 208). Follow first the **trochlear nerve** forwards into the orbit from the site at which you previously dissected it in the cavernous sinus (page 200). By gently removing fat, using careful scissors and scalpel dissection, the nerve can be followed to its destination as it sinks into the superior surface of the **superior oblique muscle** (page 187). To reach this muscle, the trochlear nerve veers medially, crossing superior to the oculomotor nerve. Lateral to the trochlear nerve, at the superior orbital fissure, the ophthalmic division of the trigeminal nerve can be seen to divide into branches, of which the most prominent and superiorly placed is the **frontal nerve,** which should now be dissected forwards until it is observed to divide into its terminal branches, the **supraorbital** and **supratrochlear** nerves. Recall (page 190) the subsequent distribution of these branches. By tracing the frontal nerve posteriorly, the origin of the **lacrimal branch** of the **ophthalmic division** of the trigeminal nerve may be found, running anterolaterally toward the lacrimal gland at the superior, anterolateral corner of the orbit. Further exposure of this region may be obtained using bone cutting scissors, as necessary. Trace the lacrimal nerve to the lacrimal gland (fig. 111, page 208). The lacrimal nerve conveys secretomotor fibers to the gland (page 191), as well as sensory fibers, the distribution of which you have previously considered (page 190).

Cut the frontal nerve with a pair of scissors, reflecting proximal and distal ends upwards. This will expose the **levator palpebrae superioris muscle,** which lies just inferior to the frontal nerve. Mobilize the muscle carefully from surrounding fat and other structures and trace it forwards as far as possible toward its destination in the superior eyelid. When well mobilized, carefully cut it across its belly and reflect proximal and distal halves posteriorly and anteriorly, respectively. By reflecting the proximal half very carefully, its nerve supply, entering it on its inferior surface, can be preserved, and should be traced inferiorly. Scissor dissection will aid in tracing the distal half of the levator palpebrae superioris muscle into its tendinous expansion,. Inferior to the latter muscle, identify the **superior rectus,** which similarly should be mobilized carefully from its surroundings. Cut the belly of this muscle with scissors, again reflecting the two halves. Once again, the proximal half should be carefully freed, and the little nerve supplying it and the previous muscle should be traced to its source, the superior branch of the **oculomotor nerve.** Careful scissor dissection will allow you to visualize the continuity of this branch with the portion of the oculomotor nerve that you previously dissected in the cavernous sinus (page 200). Further scissor dissection in this region will also reveal the **ophthalmic artery** and the **nasociliary branch** of the ophthalmic division of cranial nerve V, curving over the superior surface of the **optic nerve,** from lateral to medial (fig. 111, page 208). In due course it will be possible to establish continuity of the ophthalmic artery with the internal carotid artery, and continuity of the nasociliary nerve with the ophthalmic division. At this time, carefully dissect the ophthalmic artery and the nasociliary nerve anteromedially. To facilitate this, reidentify the superior oblique muscle (see above), and reflect it medially. The nerve and artery each give off small **posterior ethmoidal** branches, which pass through the posterior ethmoidal foramen in the medial wall of the orbit. Reidentify this foramen, and the anterior ethmoidal foramen, on a skull (page 205). The nerve and artery continue forwards, giving **anterior ethmoidal branches** through the latter foramen. The anterior ethmoidal branch is one of the terminal branches of the nasociliary nerve, the other being the **infratrochlear nerve,** the ultimate course of which you have previously studied (page 190). Furthermore, the anterior ethmoidal nerve, after a complicated pathway through the anterior cranial fossa and the nasal cavity, emerges below the inferior edge of the nasal bone, to become superficial as the **external nasal nerve** (page 190).

Returning to the superior rectus muscle, reflect the anterior half of the cut muscle belly upwards, identifying the large superior orbital vein, embedded in orbital fat. Trace the superior rectus muscle forwards to its insertion into the tough white fibrous outer coat (the sclera) of the eyeball. In so doing, be on the lookout for the insertion of the tendon of the **superior oblique muscle,** posterolateral to the insertion of the superior rectus. The fibers of the superior oblique run, as the name implies, obliquely (i.e., posterolaterally, from the superior anteromedial corner of the orbit, where the tendon of this muscle makes an acute bend through its pulley or trochlea). Mobilize the superior oblique tendon. Then return to the **belly** of the superior oblique muscle where you previously identified it and establish continuity between the muscle belly and its tendon, by gently tugging with a pair of forceps, first on the tendon, observing movement in the muscle belly, and then vice versa. Inferomedial to the belly of the superior oblique muscle, identify the **medial rectus,** and follow its tendon anteriorly, to its insertion into the sclera.

With careful scissor dissection in the fat of the orbit, superior to the eyeball, carefully trace the optic nerve anterolaterally to its site of emergence from the posterior aspect of the eye. Observe the numerous small branches of the ophthalmic artery, including those to muscles, to the eyeball, and accompanying the various nerves of the orbit. Carefully remove fat piecemeal on the lateral aspect of the optic nerve, exposing the region between the latter nerve and the medial surface of the lateral rectus muscle. If necessary, cut and reflect the superior orbital vein. By reconstruction (i.e., by folding back bisected structures into their original positions, and then reflecting them out again), reidentify the nerve to the superior rectus and levator palpebrae superioris muscles. By tracing this nerve proximally, you will now be able to establish continuity with the superior division of the oculomotor nerve, and similarly identify the inferior division of this nerve, and its branches. These should be traced forwards, to their destinations in the **inferior rectus muscle,** to be identified inferior to the eyeball, and the **inferior oblique muscle,** running a course on the inferior surface of the eyeball which corresponds to that of the superior oblique muscle along the superior surface. Also by this reconstruction process, reidentify the nasociliary nerve, and complete the process of establishing its posterior continuity. The **abducens nerve** (page 200) should be reidentified in the cavernous sinus, and traced forwards to the medial surface of the **lateral rectus muscle** (page 187). Also, in the space between the optic nerve and the lateral rectus muscle, at the site where the ophthalmic artery arches over the optic nerve, identify the **ciliary ganglion** (pages 151 and 190). It is about a millimeter in diameter, and it is identifiable by the numerous short ciliary nerves (pages 190 and 207) which pass from its anterior surface toward the back of the eyeball. Posteriorly, it has connections with both the oculomotor nerve and the nasociliary nerve. It is the relay station for parasympathetic nerve fibers to the **sphincter muscle of the pupil** (page 190) and to the **ciliary muscle** which adjusts the shape and therefore focusing power of the lens of the eye (page 211). However, other fibers than parasympathetic fibers (i.e., sensory fibers as well as sympathetic fibers), pass through the ganglion without synapsing, as do some similar fibers in several of the other parasympathetic ganglia.

Using bone cutters, expose the anterolateral aspect of the orbit by cutting through the frontal process of the zygomatic bone, and by removing the adjacent part of the lateral wall of the orbit. Reidentify the lateral rectus muscle and trace it to its insertion; then similarly identify, on the inferior surface of the eyeball, the belly and insertion of the inferior oblique muscle (pages 187 and above). This lateral approach will also enable you to make a more detailed study of the lacrimal gland and the eyelids. Dissect the gland, identifying the deep and superficial portions of the gland, and the relationships of the gland to the conjunctival sac (pages 201–204) into which the gland secretes. Trace the levator palpebrae superioris forwards, and follow its fibers into the superior **orbital septum.** Identify the posterior aspect of the orbicularis oculi muscle, and, by dissecting in the angle between the orbital septum and the muscle, identify the superior **tarsal plate,** the firm, fibrocartilaginous sheet present in the eyelid. By cutting through the lateral canthus of the eye, reflect the upper eyelid superiorly, and study the posterior surface of the tarsal plate and the relationship of the conjunctiva to it, and attempt to identify the minute **tarsal glands.** In the lower eyelid, similarly study the tarsal plate and conjunctival relationships, by dissecting in from the lateral canthus. Finally identify and, where necessary, dissect to demonstrate, the lacrimal papilla and punctum lacrimale, lacrimal canaliculus, lacrimal sac, and nasolacrimal duct, as described previously (page 201).

Clinical Session 21

■ Topics

External Ear:
Auricle
External Auditory Meatus
Middle Ear
Inner Ear

Review (pages 192 and 193) the methods for clinical examination of the auditory component of the vestibulocochlear nerve (cranial nerve VIII).

THE EXTERNAL EAR

The external ear comprises the auricle (which corresponds to the structure known colloquially as the ear), as well as the external auditory meatus. The auricle is a mainly cartilaginous structure, which is associated with soft tissues and which is covered by skin. The cartilage of the auricle is continuous with a cartilaginous portion of the external auditory meatus. The latter adjoins the bony external auditory meatus (page 3) forming a skin-lined canal which extends to the boundary between the external and middle ears, (i.e., the tympanic membrane) (page 60).

Identify (fig. 113, page 216) the following features of the auricle: the helix, antihelix, auricular tubercle of Darwin (not pronounced in fig. 113); scaphoid fossa, triangular fossa, crus, cymba conchae, tragus, antitragus, intertragic notch, and lobule. One important reason for knowing the normal features of the auricle is that this organ is very frequently the site of congenital variation or abnormality, in conditions of chromosomal or genetic anomaly. Such an abnormality may be part of a syndrome which may also affect such structures as the teeth, facial features, internal organs including the heart, and fingers and toes.

The auricle has several small extrinsic muscles which pass from the cartilage to the skull and scalp. These belong to the facial group of muscles, and have their nerve supply in common with other facial muscles of expression. In most individuals these muscles are vestigial, but in some, voluntary movement of the ears is possible. In addition to the extrinsic muscles, small intrinsic muscles, situated entirely on the surface of the auricle, are also present.

The external auditory meatus, about 2.5 cm long, passes medially toward the tympanic membrane. Approximately the lateral third is cartilaginous, and the medial two thirds of the canal constitute the bony external auditory meatus. The last (most medial) portion of the canal passes, in addition to medially, also forwards and slightly downwards. Therefore, to facilitate clinical examination of the tympanic membrane, the cartilaginous part of the meatus may be aligned with the terminal portion of the bony meatus by applying gentle traction on the auricle in a posterosuperior direction. Using an otoscope, examine, under guidance, the tympanic membrane of your subject. In normal health, the membrane has a grayish color, and is placed at an acute angle with the floor of the meatus. Near the center of the membrane, a concavity, called the **umbo**, is observed, due to the membrane being pulled medially by the attachment of the minute ear bone or ossicle, called the **malleus** (hammer) (fig. 115, page 216). The "handle" of the hammer is attached to the umbo, extends obliquely upwards and forwards, and can be seen through the tympanic membrane as a line. Superiorly, a **lateral process** of the malleus projects laterally (i.e., toward the observer), and an anterior and posterior fold curve away from this process.

Posterosuperior to the handle of the malleus, and much less clearly visible, is a second, shorter line, parallel to that of the handle of the malleus. This is the **long process** of the **incus** (anvil), the second of the ossicles (fig. 115). Although this cannot be discerned by clinical examination, the incus articulates with the malleus laterally. Medially, it articulates with the third of the ossicles of the middle ear, the **stapes** (stirrup) (fig. 116, page 216).

In disease, the tympanic membrane, instead of its normal gray color, may become inflamed. Diseases of the external ear may be diagnosed by changes in the appearance of the skin lining the external auditory meatus, and that of the tympanic membrane. However, inflammations of the middle ear, or **tympanum** (drum), are also often diagnosable by this method, since they also cause inflammation of the tympanic membrane. (Note that in the analogy of the middle ear to a drum, the tympanic membrane corresponds to the velum of the drum.)

THE MIDDLE EAR

The tympanum, or middle ear, is a drum-shaped space with the velum facing laterally, toward the external auditory meatus. It is open anteromedially, into the bony auditory tube (fig. 115, page 216).

You previously explored this communication on a disarticulated skull, using a wire probe, and you should repeat this exercise now (page 60). In addition, the middle ear also communicates posteriorly with a canal that extends into air spaces situated within the mastoid process (fig. 115, page 216), and that are known as the **mastoid air cells.** The tympanum is lined by a mucous membrane, that is continuous, through the auditory tube, with that of the pharynx, and that also extends posteriorly to line the mastoid air cells. Inflammation of the middle ear can extend into the mastoid air cells, causing the condition of mastoiditis. Examine an anatomical specimen, to visualize the shape of the middle ear cavity.

On the medial wall of the middle ear, a minute, **oval window** (fig. 119, page 220), closed by a membrane, separates the middle ear from the inner ear. The three ossicles of the middle ear, which articulate with each other, form a chain of levers which extends from the tympanic membrane laterally to the oval window medially. Thus the malleus, firmly attached to the tympanic membrane, is set in motion when the membrane vibrates in response to sound waves. This movement is transferred from the malleus to the incus, and from the incus to the stapes, the foot plate of which fits snugly into the oval window (fig. 116, page 216). The three ossicles communicate with each other by means of minute synovial joints (page 39), and create, as mentioned, a system of levers for the transference of movement to the inner ear.

Using a pocket torch (flashlight), examine the medial wall of the middle ear in a disarticulated skull, exploiting the fact that the tympanic membrane is absent in the dried state, and that the middle ear can therefore be visualized. Attempt to observe both the oval window, also known as the **fenestra vestibuli,** as well as the smaller **fenestra cochleae,** situated inferior to the former (figs. 116, page 216, and 119, and 120, page 220).

THE INNER EAR

Visualize the inner ear in a disarticulated skull, by viewing its lateral wall (i.e., the medial wall of the middle ear), from the lateral aspect, through the external auditory meatus, as well as observing its medial wall at the lateral end of the internal auditory meatus. The latter can be studied in the disarticulated skull, with the aid of a flashlight and a mirror. In addition, examine specially prepared anatomical specimens to study this aspect of the inner ear, and observe the sites at which fibers of the vestibulocochlear nerve pass through the bony wall, to communicate with the inner ear. In suitable anatomical specimens, you will be able to see a small **transverse crest,** above which a foramen can be observed, through which the facial nerve passes to enter the middle ear (page 192 and figs. 117 and 118, page 220). Posterior to that is the **superior vestibular area,** and inferior to the transverse crest is the **inferior vestibular area.** The vestibular areas are perforated, for the passage of small fibers of the vestibular part of cranial nerve VIII into the **vestibule** of the inner ear. The vestibule, and the **semicircular canals,** situated posterolateral to the vestibule, house the organ of balance. Review the manner in which function of this sense organ is tested clinically (page 193).

Also inferior to the transverse crest, and placed somewhat anteriorly, is the **tractus spiralis foraminosus,** a spiral of minute perforations which correspond to the turns of the **cochlea** (shell) (figs. 118, 119, and 120, page 220), and through which pass fibers of the acoustic or auditory part of cranial nerve VIII. Study anatomical specimens to visualize the position of the cochlea, and the minute bony canal which winds its way around the axis **(modiolus)** of the shell. A membranous canal, the **duct of the cochlea,** spirals around this groove, and contains the microscopic sense organs of hearing. Vibrations from sound waves are transmitted to these sense organs, from the tympanic membrane, via the ossicles and the fenestra vestibuli, to the fluid-filled duct of the cochlea.

In addition, study, in anatomical specimens, the vestibule and **semicircular canals** of the inner ear (figs. 117, 118, 119, and 120, page 220), and visualize their positions in the intact skull. Review the courses of cranial nerves VII and VIII through the internal auditory meatus, and the subsequent pathway of the facial nerve through the ear (page 192) and of the vestibulocochlear nerve to supply the cochlear and vestibular portions of the ear respectively (see above).

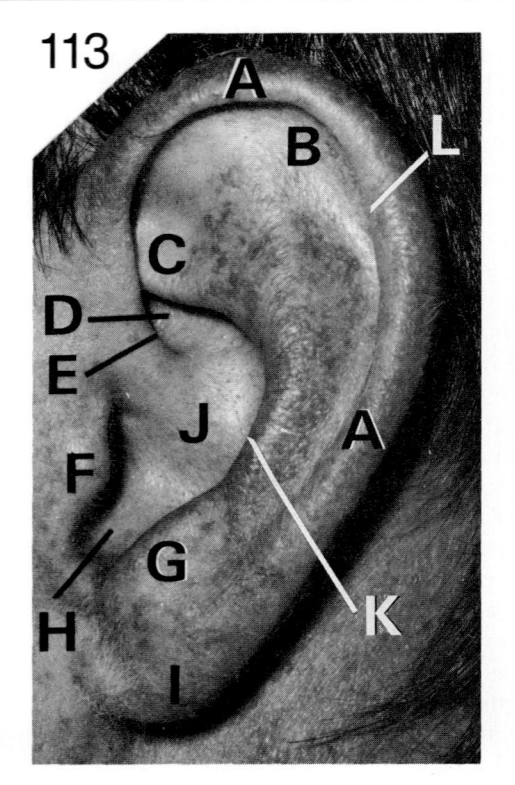

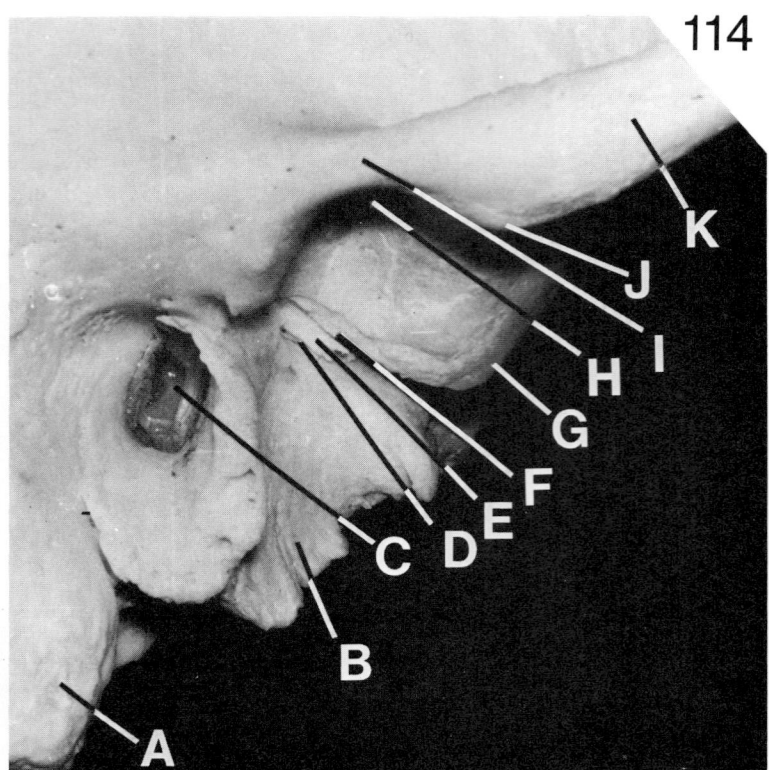

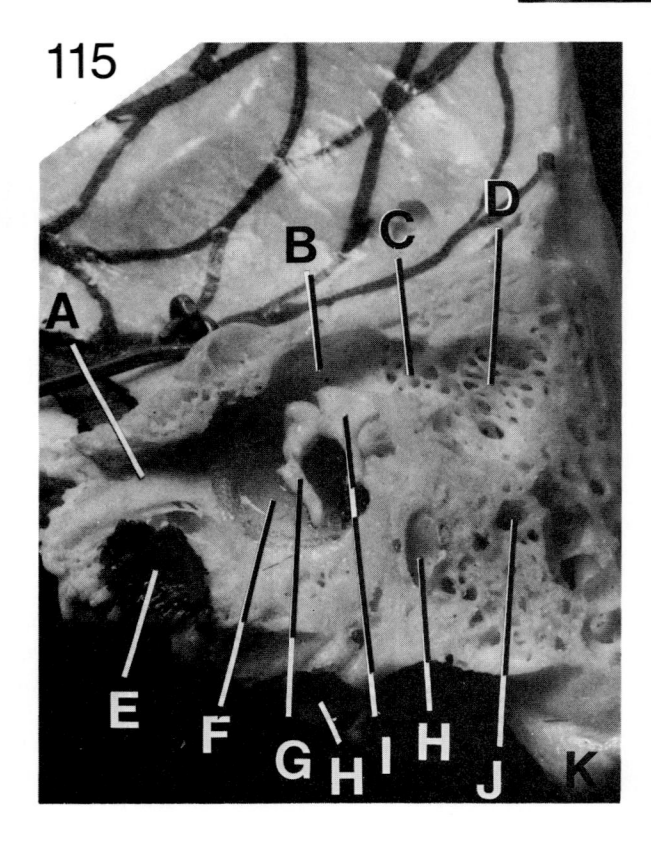

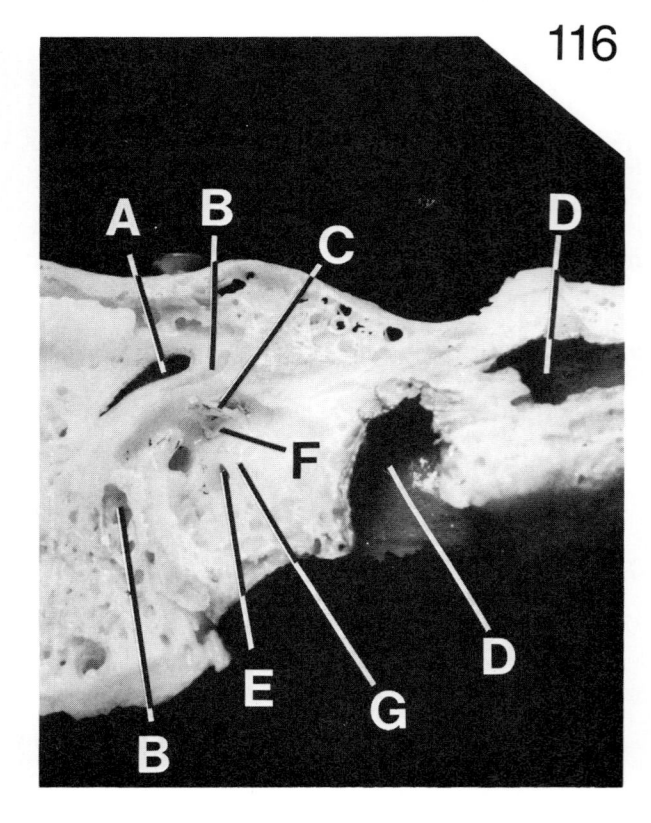

216 YOUR PATIENT'S ANATOMY: A CLINICAL VIEW OF HUMAN MORPHOLOGY

Figure 113. Lateral aspect of left auricle.
A. Helix
B. Scaphoid fossa
C. Triangular fossa
D. Cymba conchae
E. Crus
F. Tragus
G. Antitragus
H. Intertragic notch
I. Lobule
J. Concha
K. Antihelix
L. Darwin's tubercle

Figure 114. Lateral aspect of right bony external auditory meatus and relations.
A. Mastoid process
B. Tympanic plate of temporal bone
C. External auditory meatus
D. Petrotympanic fissure
E. Down-turned edge of petrous temporal bone
F. Squamotympanic fissure
G. Spine of sphenoid bone
H. Mandibular fossa
I. Posterior root of zygomatic process of temporal bone
J. Anterior root of zygomatic process of temporal bone (articular tubercle)
K. Zygomatic process of temporal bone

Figure 115. Medial aspect of lateral wall of right middle ear.
A. Auditory (pharyngotympanic) tube
B. Epitympanic recess
C. Aditus to antrum
D. Tympanic (mastoid) antrum
E. Carotid canal
F. Tympanic membrane
G. Malleus (hammer)
H. Canal for facial nerve
I. Incus (anvil)
J. Mastoid air cells
K. Medial aspect of mastoid process

Figure 116. Lateral aspect of medial wall of right middle ear.
A. Lateral semicircular canal
B. Canal for facial nerve
C. Anterior limb of stapes (stirrup)
D. Carotid canal
E. Fenestra cochleae (round window)
F. Base of stapes, in fenestra vestibuli (oval window)
G. Promontory

Laboratory Session 21

THE EXTERNAL EAR

Begin your laboratory study of the ear by identifying, on the detached half head specimen, the named features of the auricle (page 214 and fig. 113, page 216). Thereafter, make a vertical scalpel incision just behind the ear, extending from the saw-cut edge of the cranium, where you removed the **calotte**, downwards to a position 2 cm inferior to the tip of the mastoid process. Dissecting soft tissues away from bone, reflect a flap, including the auricle, forwards. In the process of separating soft tissues from bone, you will cut through the cartilaginous portion of the external auditory meatus. You should also clear soft tissues away from the surface of the mastoid process, reviewing anatomical relationships of these structures as you do so. Study the cut surface of the cartilaginous part of the external auditory meatus in the flap, after you have reflected it adequately for this purpose. Using a blunt probe, gently remove any **cerumen** (ear wax) which may be present from the proximal portion of the external auditory meatus, and identify the junction between cartilaginous and bony parts of this tubular structure. Using a flashlight, illuminate the meatus, and observe the **tympanic membrane,** noting its position, distance from the auricle, and angle of inclination (page 214). If necessary, take the entire specimen to a sink and flush out the meatus with running water under an open tap, and remove residual particles of cerumen, under visualization, using a torch (flashlight) and probe. Reidentify as many of the features of the lateral surface of the tympanic membrane, as you are able to, referring back to the clinical examination you performed on a living subject (page 214). Replace the flap that you have reflected, and attempt to identify the features of the tympanic membrane by examination from the exterior.

THE MIDDLE EAR

In preparation for further dissection of the ear, revert to the internal surface of the cranial cavity and strip the cranial layer of dura mater from the region related to the ear (i.e., in the posterior cranial fossa: the posterior surface of the petrous temporal bone and, especially, the sigmoid sinus, the contents of which should be flushed out and cleared completely away; and in the middle cranial fossa: the superior surface of the petrous temporal bone and adjacent part of the squamous part of the temporal bone). Review, as you remove dura from the superior surface of the petrous temporal bone, the situation of the greater and lesser petrosal nerves (page 200).

Review your specimen from above, and place a finger of one hand on the mastoid process and a finger of the other hand in the sigmoid sinus. Note that the sigmoid sinus is an intimate medial relation of the mastoid process. To further study this relationship, you should now use an electric saw to cut a vertical (coronal) section through the mastoid process (and therefore through the contained mastoid air cells [page 215]) and extend your incision medially into the sigmoid sinus. You should also extend your incision superiorly, to the previously cut edge. As usual, exercise caution in use of the electric saw: soft tissue should be cleared away prior to the use of the saw, as its presence increases the hazard of the saw slipping; and the specimen should be stably placed on an underlying firm surface, and no hands or fingers should be in the vicinity of the saw blade. If necessary, a second incision at right angles to the first, can also be made with the saw. This should be cut to extend posteriorly from the mastoid process to the back of the skull, allowing for removal of a portion of the skull and exposure of the mastoid air cells from behind. Examine the relationship of the mastoid air cells to the sigmoid sinus, and consider what clinical significance this intimate relationship may have.

Utilizing the posterior access you now have created through your bone incisions, dissect in the space immediately medial to the mastoid process. Identify the bulky transverse process of the atlas (first cervical vertebra), and mobilize this structure, if necessary cutting muscles and ligaments superior to it, to displace it slightly medially. Using a disarticulated skull to get your bearings, dissect toward the stylomastoid foramen, with a view to identifying the facial nerve as it emerges from that foramen to pass into the substance of the deep part of the parotid gland (pages 25 and 48). Be aware of the fact that your dissection brings you very close to the jugular foramen, and that you might well mistake the cranial nerves emerging through that foramen for the facial nerve. Your objective should be to stay close to the medial aspect of the mastoid process, and dissect anteriorly and slightly medially. Deep palpation of the styloid process will become possible after you have mobilized some of the soft tissues, and scissor dissection will reveal the facial nerve. After clearing soft tissues surrounding the facial nerve, dissect superiorly and clear the bony inferior cranial surface posterior to the site of emergence of the nerve. When you have performed this exercise, you should now make a careful partial incision (see below) with the electric saw, in a parasagittal plane, toward a point just lateral to the facial nerve. Your objective is to attempt to expose the middle ear. In so doing, you will be able to dissect the downward path of the seventh nerve, toward its site of exit at the stylomastoid foramen (page 192 and fig. 116, page 216). By extending your cut forward, you should also pass through the medial end of the external auditory meatus, thus allowing an approach into the middle ear through the tympanic membrane.

Accurate placement of the position of a saw cut is almost impossible, and furthermore, the saw is likely to damage some of the minute structures of the middle and internal ear. Therefore, although you should attempt to place your section as described you are more likely to get a satisfactory exposure of the region by making only a partial cut with the saw, and then completing the exposure with the help of a chisel and, if necessary, a hammer. Place the edge of a chisel into the partial saw cut, and lever the two bony surfaces apart. The plane in which the bone will cleave is, to a certain extent, a matter of chance. However, accurate placement of your saw cut increases the likelihood of the split being appropriate. Additional coronally placed saw cuts may be helpful.

Now, referring to Figures 115 and 116 (page 216) and Figures 117, 118, 119, and 120 (page 220), as well as atlas illustrations, study the middle ear as follows. Dissect the **tympanic membrane** carefully, to expose the "eardrum" of the **middle ear.** Three minute bones, the auditory ossicles (os means bone) will be encountered. The **malleus** (hammer) is adherent, by its "handle," to the tympanic membrane, and articulates, by its head, with the **incus** (anvil). One process of the anvil extends posteriorly, toward the mastoid air cells. Another process, the longer, extends downwards and posteriorly, behind and parallel to the handle of the malleus. The inferior end of this process curves inward (medially) to the third of the ossicles, the **stapes** (stirrup). The foot plate of the minute stirrup fits into the **fenestra vestibuli** or **oval window** (fig. 216) in the medial wall of the middle ear. Thus, vibration of the tympanic membrane, caused by soundwaves, is transmitted through the three ossicles to the oval window. From there, the vibration is transmitted into the cochlea of the inner ear.

Other features of the middle ear should be identified as follows. Inferior to the fenestra vestibuli, observe a round projection, the **promontory,** which is due to the first coil of the cochlea. Anteriorly, the ear drum opens into the auditory (also known as the Eustachian or pharyngotympanic) tube (fig. 115, page 216). In the dry skull, communication between the bony part of the auditory tube and the middle ear has previously been demonstrated (pages 60 and page 215). You should repeat this exercise on a skull now. Note that a thin ridge of bone divides the bony part of the auditory tube into two channels, the upper of which contains a minute muscle, the **tensor tympani,** that inserts into the handle of the malleus on the tympanic membrane and is able to tense the membrane and modify the intensity of the vibrations set up in the membrane. The nerve supply of the muscle is a branch of the mandibular division of the trigeminal nerve (page 191). Posteriorly, the tympanic cavity communicates with the **tympanic antrum,** a small space that, in turn, opens into the **mastoid air cells** (page 215 and fig. 115, page 216).

On the posterior wall of the tympanic cavity (ear drum), there is a small bony site for origin of yet another minute muscle, the **stapedius,** that inserts into the neck of the stapes. The nerve supply of the stapedius is the facial nerve (page 192).

THE INNER EAR

Place a second cut as follows, using the same procedure as previously, of initial sawing and subsequent cleavage with the aid of a chisel. Clear dura away from the superior surface of the petrous temporal bone, above the internal auditory meatus. Cut vertically through the bone from above, parallel to the meatus, so that when you complete the cleavage with the chisel your incision opens into the internal auditory meatus.

Dissect the course of the vestibulocochlear (VIII) and facial (VII) cranial nerves. The auditory fibers of the eighth cranial nerve pass into the cochlea, the shell-like structure of which may be evident as a result of your chisel cut (figs. 118, 119, and 120, page 220). The axis, or **modiolus,** of the cochlea points anterolaterally, with the base of the shell facing posteromedially. At this site the internal auditory meatus ends, and the auditory fibers of cranial nerve VIII enter the base of the cochlea through the **tractus spiralis foraminosis** (page 215). There are two-and-one-half coils in the shell, the largest turn being at the base; the smallest turn is therefore anterolateral. Identify the first, large coil, and correlate its position to that of the promontory, observed in the medial wall of the tympanic cavity. The vestibular fibers of cranial nerve VIII can be traced to a position posterior to that of the cochlear fibers, at the blind, lateral end of the internal auditory meatus. A minute, horizontal ridge, the **transverse crest,** divides this area into an upper and lower region. Small branches enter the superior and inferior vestibular areas (page 215)

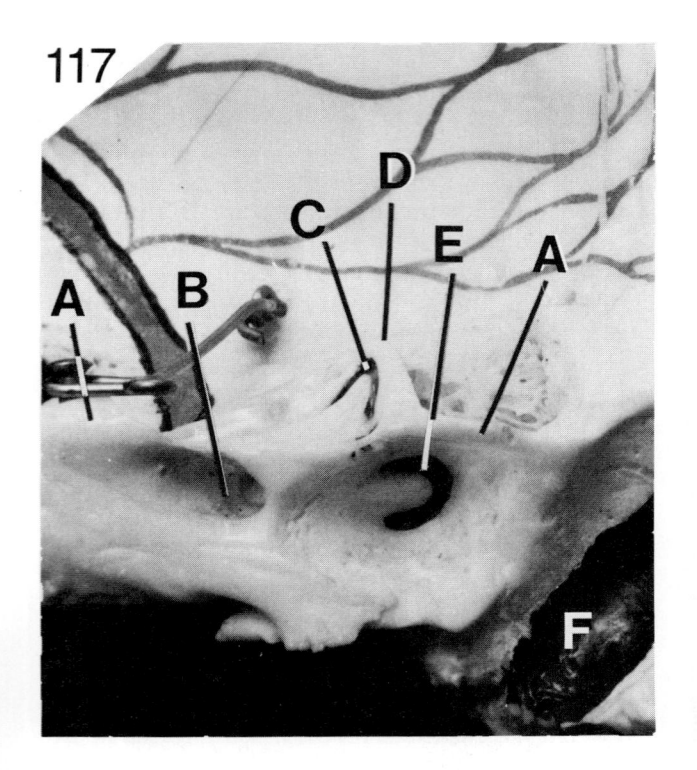

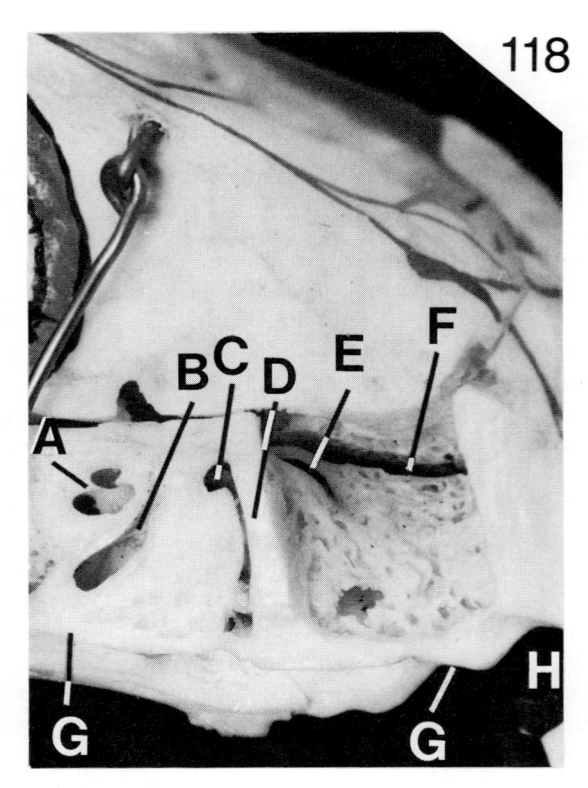

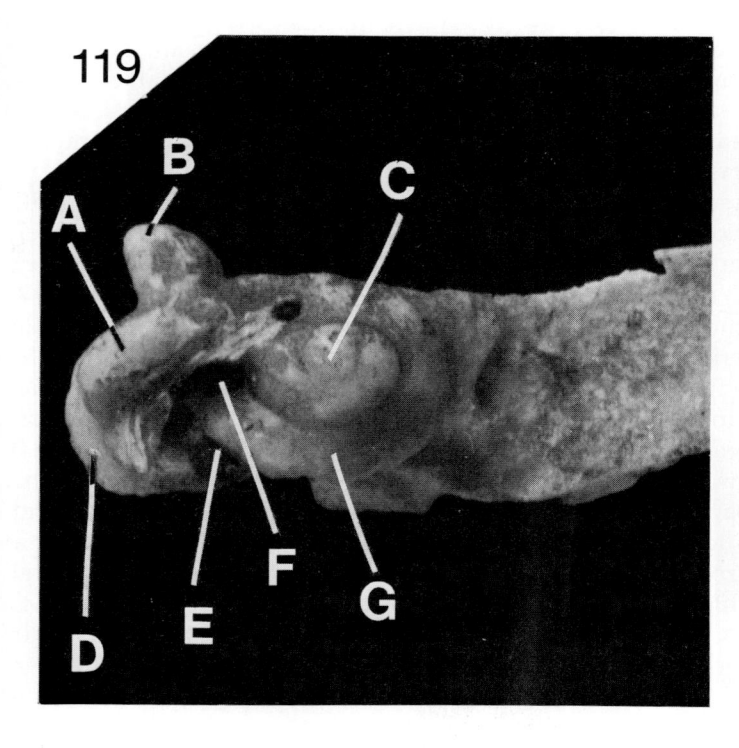

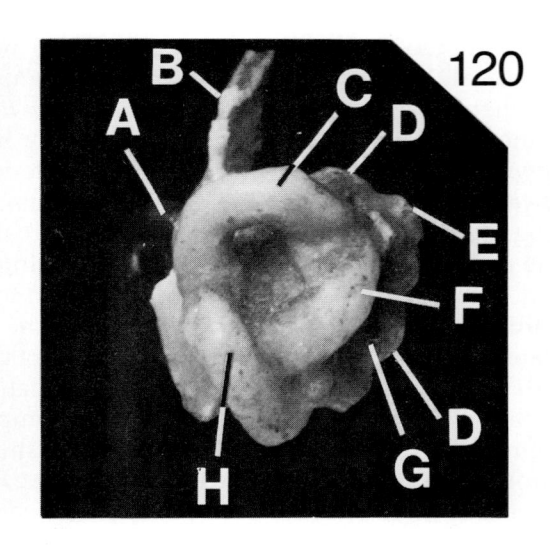

220 YOUR PATIENT'S ANATOMY: A CLINICAL VIEW OF HUMAN MORPHOLOGY

Figure 117. Posteromedial aspect of dissection of right bony inner ear.
A. Superior border of petrous temporal bone
B. Internal auditory meatus
C. Superior semicircular canal
D. Arcuate eminence
E. Posterior semicircular canal
F. Groove for sigmoid sinus

Figure 118. Superior aspect of dissection of right bony inner ear.
A. Cochlear
B. Canal for facial nerve
C. Superior semicircular canal
D. Arcuate eminence
E. Lateral semicircular canal
F. Saw cut in anteroposterior plane of tympanic cavity
G. Superior border of petrous temporal bone (see A in fig. 117)
H. Groove for sigmoid sinus

Figure 119. Anterolateral aspect of right bony labyrinth (inner ear), dissected free of surrounding bone.
A. Lateral semicircular canal
B. Superior semicircular canal
C. Cupola of cochlea
D. Posterior semicircular canal
E. Fenestra cochleae (round window)
F. Fenestra vestibuli (oval window)
G. First coil of cochlea

Figure 120. Posterolateral aspect of right bony labyrinth (inner ear).
A. Probe in internal auditory meatus
B. Posterior surface of petrous temporal bone
C. Superior semicircular canal
D. First coil of cochlea
E. Cupola of cochlea
F. Lateral semicircular canal
G. Round window
H. Posterior semicircular canal

above and below the crest, to pass into the vestibule of the inner ear, situated posterior to the cochlea of the inner ear.

By dissecting from both medial and lateral approaches, made possible by your bone incisions, study the course of the facial nerve. From the blind end of the internal auditory meatus it passes through a foramen above the transverse crest and anterior to the superior vestibular area. It continues laterally, into the region of the inner ear, then turns sharply posteriorly, this sharp bend being known as the **genu** (knee) (page 192 and fig. 116, page 216). At the bend, the **geniculate ganglion** is situated (homologous to the dorsal root ganglion of a spinal nerve, page 86). After the geniculate ganglion, the course of the nerve is posterior. At this site, the nerve lies just above the position of the "oval window" (fenestra vestibuli, fig. 116, page 216). At a position corresponding to the posterior end of the medial wall of the middle ear, it takes a sharp right angle bend downwards, toward its exit at the stylomastoid foramen. It is in this downward course that the nerve gives off its chorda tympani branch (page 192). At the geniculate ganglion, the facial nerve gives off the greater petrosal nerve (page 192). To demonstrate the course of the facial nerve through the inner and middle ear, use bone cutting scissors where necessary.

The vestibule (page 215) is a box-shaped space, the long axis of which lies parallel to the superior border of the petrous temporal bone (fig. 99, page 188). At the anterior end of the box-like vestibule, the cochlea makes communication with the vestibule. Thus the foot of the stapes fits into the oval window, in the vestibule, and through that communication, is in contact with the cochlea duct. At the posterior end of the vestibule, three **semicircular canals,** concerned with the sense of balance, are also in contact with the vestibule (figs. 117, 118, 119, and 120, page 220). Review and reconsider the clinical tests you performed (page 193) to assess this organ as you now study it in your specimen. There is a posterior, semicircular canal, which lies in a plane parallel to the posterior surface of the petrous temporal bone; a lateral semicircular canal, which lies in a plane parallel to the superior surface of the petrous temporal bone; and a superior semicircular canal, which lies in a plane at right angles to both of the previous two. The apex of this canal is responsible for the **arcuate eminence** on the superior surface of the petrous temporal bone (figs. 117 and 118, page 220). All three bony, semicircular canals contain membranous, fluid-filled canals that connect with two minute sac-like structures within the vestibule: the **sacculus** and the **utriculus.** Sense organs for balance are found within the sacculus, the utriculus, and at the ends of the semicircular canals. Branches of the vestibular part of the eighth cranial nerve reach these sense organs by passing through the vestibular areas at the blind end of the internal auditory meatus (pages 215 and 219), which is, as you should confirm on your specimen, adjacent to the sites of the semicircular canals and their communication into the vestibule. Attempt to see and to visualize as much of this system as you are able in your own specimen, and to correlate the tests you performed (page 193) with the specific parts of the organ of balance.

Before leaving your dissection of the ear, identify the internal carotid artery, and note its proximity to the middle and inner ear. Review the position of the auditory tube, and ensure that you are fully orientated with respect to all parts of the inner, middle, and outer ear.

Contents of Section 4

Somatic Structures of the Neck, Back, and Upper Limb

22. **Anterior Triangle of the Neck; Digastric Triangle; Carotid Triangle; Submental Triangle; Muscular Triangle; Posterior Triangle of Neck; Occipital Triangle; Subclavian Triangle; Supraclavicular Fossa**
    Clinical Session 22 225
    Laboratory Session 22A 232
    Laboratory Session 22B 236

23. **Posterior Aspect of the Neck; The Back; Bony Prominences; Muscles of Neck and Back; Movements of Neck and Back**
    Clinical Session 23 241
    Laboratory Session 23 248

24. **Shoulder, Arm, and Elbow; Bony Landmarks; Muscles of Shoulder and Arm; Vessels and Nerves; Functional Anatomy and Movements**
    Clinical Session 24 255
    Laboratory Session 24 262

25. **Forearm, Wrist, and Hand; Bony Landmarks; Muscles and Muscle Actions; Arteries and Nerves**
    Clinical Session 25 267
    Laboratory Session 25 275

Clinical Session 22

■ **TOPICS**

Anterior Triangle of the Neck
  Digastric Triangle
  Carotid Triangle
  Submental Triangle
  Muscular Triangle

Posterior Triangle of the Neck
  Occipital Triangle
  Subclavian Triangle
  Supraclavicular Fossa

The purpose of returning to the neck region to study its somatic structures at this stage is threefold. It reemphasizes the concept of somatic and visceral components of the body, important to the understanding of the somatic (voluntary) and visceral (autonomic) nervous systems, as well as of other features of morphology. By reviewing early sections of the course, other fundamental concepts introduced in the beginning will be reinforced. Finally it provides a link between the anatomy of the axial (trunk) structures and that of the appendicular (limb) components.

Your objective for this clinical session is to review the **anterior triangle** of the neck and its subdivisions (page 49 to 58); to further study some of the accessible contents of these subdivisions; and to examine the **posterior triangle** of the neck.

Commence by reidentifying the sternocleidomastoid muscle on the right and left sides of your model's neck. Activate the muscle by letting your subject turn his or her head against resistance to the opposite side. Both muscles can be activated simultaneously by your model flexing the head against resistance. **Flexion** is the act of moving a ventral surface toward another ventral surface. In the case of flexion at the neck, the ventral surface of the face is brought toward the ventral surface of the chest. Resistance to this movement is exerted by holding the subject's chin up as he or she attempts to bend the head down. The opposite of flexion is **extension,** in which one dorsal surface is moved toward another.

As described previously (page 49), the anterior border of the sternocleidomastoid muscle forms the posterolateral boundary of the anterior triangle of the neck. Review the other sides of this triangle. The posterior border of the sternocleidomastoid muscle forms the anterior boundary of the posterior triangle of the neck. Inferiorly the latter triangle is bounded by the clavicle, the sinuous (S-shaped) course of which you have previously palpated (page 84). Posteriorly the triangle is bounded by the superolateral border of the **trapezius** muscle (fig. 121, page 228). Palpate the trapezius muscle and review the exercise you previously performed (page 194) in testing the integrity of the spinal accessory nerve (cranial nerve XI).

ANTERIOR TRIANGLE OF THE NECK

For the purposes of reviewing the subdivisions of the anterior triangle of the neck, palpate the body and greater horn of the hyoid bone (page 49) and indicate, with a skin marker, the site of junction of the greater and lesser horns with the body. Also review at this time, the appearance of the hyoid bone on a skeleton. Now ask your model to depress the mandible against resistance and palpate the anterior belly of the **digastric** muscle and mark it in. Identify the mastoid process by palpation and identify the posterior belly of the digastric while your model swallows. Complete your marking of the digastric muscle by indicating the posterior belly and intermediate tendon.

Attempt to palpate the superior belly of the omohyoid muscle (page 53 and 55) as your model depresses the mandible against resistance. In this action, the anterior belly of the digastric muscle has the function of pulling the mandible downwards (i.e., its attachment on the mandible is the insertion, and the tethering of the intermediate tendon to the hyoid bone acts as the origin). In order that the hyoid bone can provide the **immobile** site of attachment for the muscle's **origin** (page 21), it is necessary that the hyoid bone, itself normally movable, needs to be stabilized. In the above action, the superior belly of the omohyoid muscle performs the function of stabilizing or "fixing" the position of the hyoid bone, and in this role the superior belly of the omohyoid muscle is acting as a **fixator**. The anterior belly of the digastric muscle is, when depressing the mandible, functioning as the **prime mover** or **agonist**. For completeness, it should be mentioned that muscles that are capable of opposing the action of the agonist, and that therefore are required to relax and lengthen while the agonist is contracting, are called **antagonists.** Which muscles are antagonists in the action being described here?

When you have identified the position of the superior belly of the omohyoid muscle, made palpable by its function as a fixator, mark its position to delimit the carotid triangle from the muscular triangle (pages 53 and 54).

Review the boundaries and contents of the digastric (submandibular) triangle. Review the palpation of the submandibular gland by bimanual examination (page 52), and review the three-dimensional structure of the gland and its relationships with surrounding structures, including the mylohyoid muscle and the facial artery. The latter artery should be palpated (page 38), as should the contraction of the mylohyoid muscle during swallowing (pages 4 and 52).

Review the carotid triangle (pages 53 to 54). Palpate the common carotid artery and identify by palpation of the relevant landmark (page 53) its site of bifurcation. Palpate also the greater horn of the hyoid bone and the space between the latter and the thyroid cartilage. What soft tissue structures (normally not palpable) lie in the vicinity of the space between these two skeletal landmarks?

Palpate in the submental triangle to identify the submental lymph nodes by bimanual examination (page 54). In some subjects the superficial **anterior jugular vein** may be visible close to and parallel to the midline (fig. 10, page 22).

Review the borders and contents of the muscular triangle (page 54). Palpate the body of the hyoid bone and, inferior to it, the thyroid cartilage (pages 53, 54, and 69. Inferior to the thyroid cartilage, palpate the cricoid cartilage and the tracheal rings (pages 69–70). Review the relationships of the larynx and trachea, including the palpable common carotid artery (page 70), and also review the palpation of structures in the floor of the mouth (page 4), the anterior surface of the neck, and the larynx during the three stages of deglutition (page 112).

Review the strap muscles of the muscular triangle (page 70) making use of anatomical specimens and palpation on your model. Palpate the thyroid gland (page 71).

With the aid of anatomical specimens and skeletal material examine the deep (posterior) relations of the structures of the anterior triangle of the neck. Attempt to appreciate that the visceral structures, which you have reviewed here and studied previously, lie just anterior to the cervical vertebral column. The respiratory tract (larynx and trachea) and its immediate posterior relation, the digestive tract (the pharynx and esophagus, pages 69, 72 to 73, and 77) lie anterior to the **vertebral bodies,** separated from these by a layer of **prevertebral muscles** covered by **prevertebral fascia.** The contents of the carotid sheath (pages 56 to 57), situated just lateral to the cervical viscera described above, lie immediately anterior to the cervical vertebral transverse processes. To attempt to appreciate this important point, examine the intact cervical vertebral column on a skeleton, and then gently perform deep palpation of the common carotid artery on your model, exerting gentle but firm pressure posteriorly, toward the cervical transverse processes. The latter can be palpated, and it is important to appreciate the proximity of the transverse processes to the carotid sheath. As an emergency measure in uncontrollable bleeding, the carotid arteries may be compressed against the cervical transverse processes in an attempt to stem the flow.

Review, on a skeleton and on anatomical specimens the **thoracic inlet** (page 122), the space through which structures pass between the thorax and the neck. The inlet is bounded by the first thoracic vertebra posteriorly, the first ribs curving round on either side, and the manubrium of the sternum anteriorly. Cervical viscera, such as the larynx and pharynx, are in continuity with their thoracic counterparts, such as the trachea and esophagus, through the central portion of this inlet. Other important structures, such as the thoracic duct, and various blood vessels and nerves, similarly traverse this central region. On either side the apex of the lung extends a short distance through the inlet, and **the cervical dome of pleura** accompanies the apex. Using a stethoscope, reaffirm that because of this fact normal breath sounds of the apex of the lung can be auscultated in the supraclavicular fossa (page 95). The cervical dome of pleura is, furthermore, covered on each side by a tough **suprapleural membrane,** which protects the pleura and lung and contributes to maintenance of the intrathoracic negative pressure.

Study anatomical specimens to visualize the relationships of the thoracic inlet (including the **root of the neck,** the region just superior to the thoracic inlet). Study also an isolated first rib and thoracic and cervical vertebrae.

POSTERIOR TRIANGLE OF THE NECK

Review the borders of the posterior triangle of the neck (page 225) making use of appropriate activation of the sternocleidomastoid and trapezius muscles of your model to define these boundaries. Follow the trapezius downwards and laterally to the **spine of the scapula** (shoulder blade). This spine lies approximately horizontally and can be traced laterally where it bends forward at a right angle to form the flattened **acromion process.** Palpate the superior surface and lateral edge of this process and its articulation, medially, with the clavicle.

The posterior triangle is subdivided into a superior **occipital triangle,** so named because its apex extends toward the occiput, where both trapezius and sternocleidomastoid have their superoposterior attachments; and an inferior **subclavian triangle,** so named because the subclavian artery is situated in it, just above the clavicle. (Sub means under; clavian means of the clavicle. In fact it is the continuation of the subclavian artery, called the **axillary artery,** and not the subclavian artery itself, which passes inferior to the clavicle, to reach the upper limb.)

The occipital and subclavian subdivisions of the posterior triangle of the neck are demarcated from one another by the **inferior belly of the omohyoid muscle.** Review the exercise you performed above (page 225) in order to activate and palpate the superior belly of omohyoid as it functions as a fixator in the action of depressing the mandible against resistance. The inferior belly of omohyoid is necessarily activated simultaneously in this function, and therefore you should repeat the exercise of asking your model to depress the mandible against resistance while you palpate in the inferior portion of the posterior triangle of the neck. At the appropriate site you will feel the cord-like inferior belly become activated. By asking the model to simultaneously move his or her shoulder back and forth while activating the muscle, you will be able to palpate movement in it. This is, of course, because the muscle is attached posteriorly to the scapula, or shoulder blade (omo means shoulder) (page 53). Identify the site of attachment of the inferior belly to the upper margin of the scapula, at the **suprascapular notch** (where the muscle is attached to the **suprascapular ligament** which bridges this notch, and the adjacent part of the superior border of the scapula). Now review the description of the superior belly of the omohyoid muscle previously given (pages 53 and 55) and visualize the entire extent of the muscle as now perceived through palpation of both its superior and inferior bellies when activated. Using a skin marker indicate the site of the inferior belly of the omohyoid muscle and thus of the occipital and subclavian triangles.

Note that the supraclavicular fossa (page 94 and fig. 33, page 82) mainly overlies the subclavian triangle but also extends somewhat into the occipital triangle. As you reconfirmed on page 226, structures passing through the thoracic inlet, such as the apex of the lung, are accessible to examination in this fossa.

Now examine the contents of the subclavian triangle. In the inferior angle of the triangle palpate deeply to identify the pulsation of the subclavian artery. At this site the artery crosses the first rib to enter the upper limb. Review, on a skeleton, the relationships of the clavicle and first rib to appreciate in three dimensions the situation of that part of the artery that you are able to palpate on your model. By pressing the artery against the rib attempt to stop your model's wrist pulse. This maneuver can be lifesaving in massive hemorrhage. Also, at this time, examine on the skeleton the relationship of the superior edge of the scapula to the clavicle and first rib. These three bony structures form the superior inlet to the **axilla.** You should review (page 85 and fig. 35, page 82) the fact that the anterior wall of the axilla is formed mainly by the pectoralis major muscle (reinforced by pectoralis minor) and the posterior wall of the axilla mainly by the latissimus dorsi muscle and teres major, whereas the medial wall of the axilla is formed by the ribs and its coverings. Given these boundaries of the axilla you will be able to understand that the inlet to the axilla is formed superiorly by the clavicle, to which pectoralis major attaches (fig. 34, page 82, and page 85); by the scapula, to which the latissimus dorsi muscle and the **teres major** muscle (figs. 123, page, 228, and 134, page 252) attach; and by the first rib, inferior to which the rest of the ribs continue in series to form the medial wall of the axilla. The lateral wall of the axilla is formed by the **humerus,** the bone of the arm. Superiorly, the humerus articulates at the shoulder joint where the clavicle and scapula converge.

The portion of the subclavian artery palpated here is developmentally derived from one of the serially arranged intersegmental branches of the descending aorta (most of which become intercostal and lumbar arteries). Morphologically the subclavian artery therefore lies in the neurovascular plane between the innermost and middle muscle layers of the body wall (page 86). The muscles in the neck that represent and are derived from the three layers of the body wall are the scalene muscles (page 126). The subclavian artery passes across the superior surface of the first rib between the **scalenus anterior muscle** that inserts into the first rib just anterior to the artery and that represents the innermost layer of muscle (page 122), and the **scalenus medius muscle,** that inserts into the rib just behind the artery and belongs to the middle layer. The **scalenus posterior muscle,** situated further posteriorly and inserting into the second rib, represents the outermost muscle layer. Review the developmental origin of the first part of the right subclavian artery (page 107).

The space between the scalene muscles is small, with the result that the subclavian artery is closely surrounded by the anterior and middle scalene muscles, in front and behind respectively. By activating the scalene muscles (flexing, rotating, or bending laterally the neck, or elevating the first rib in deep

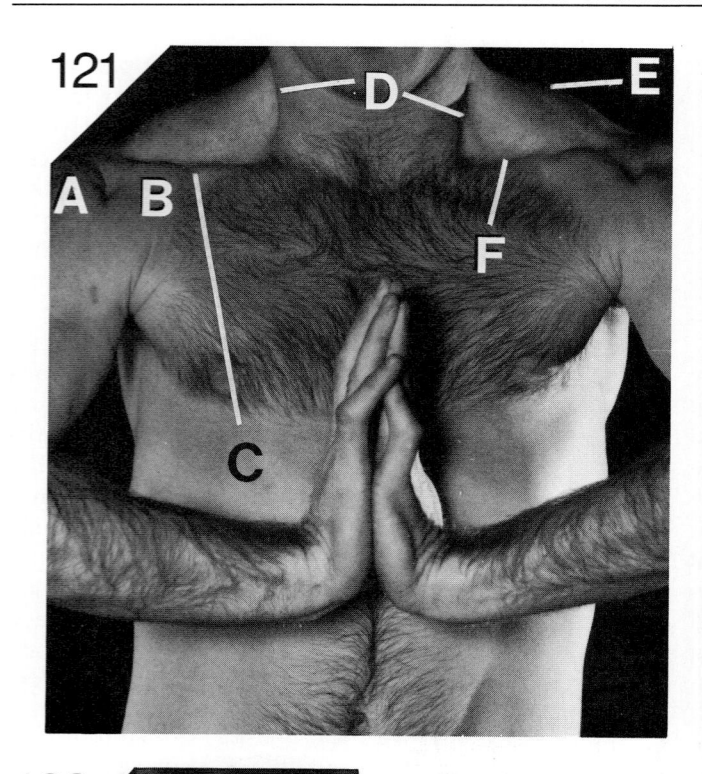

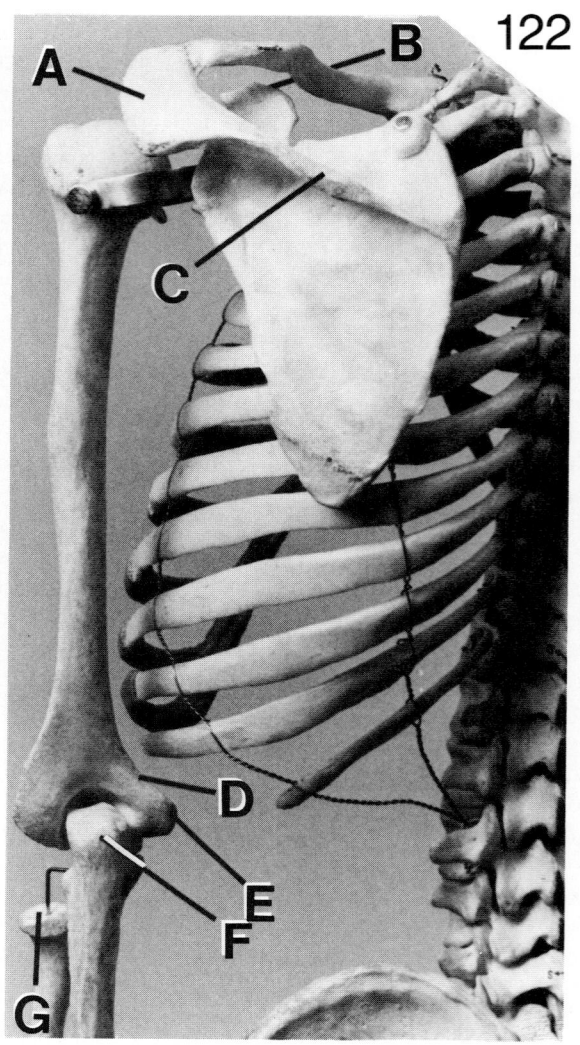

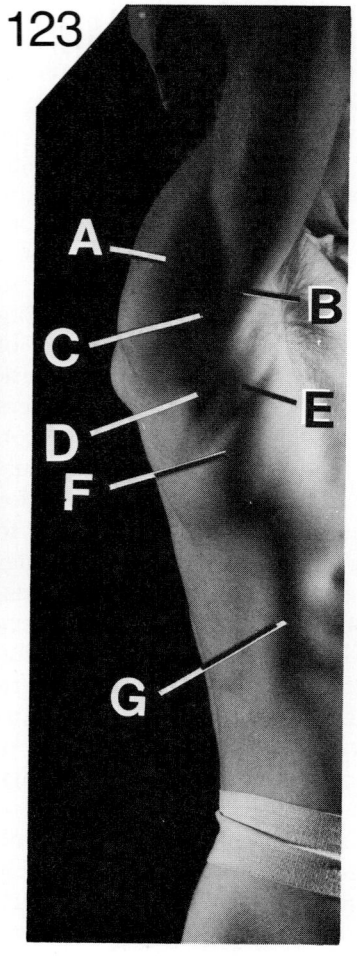

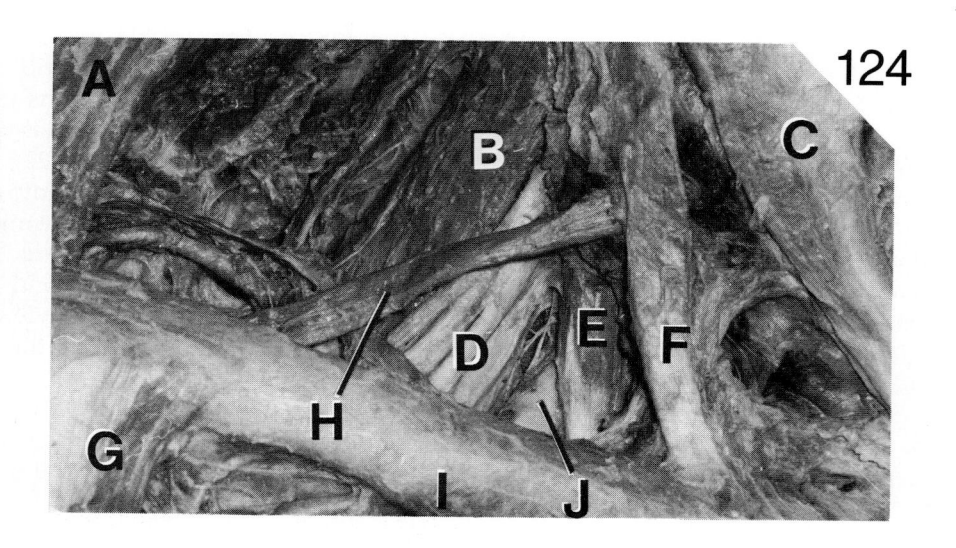

228 YOUR PATIENT'S ANATOMY: A CLINICAL VIEW OF HUMAN MORPHOLOGY

Figure 121. Boundaries and relations of posterior triangle of neck.
A. Acromial part of deltoid
B. Clavicular part of deltoid
C. Deltopectoral triangle
D. Posterolateral border of sternocleidomastoid
E. Superolateral border of trapezius
F. Clavicle

Figure 122. Posterior aspect of scapula, humerus, and elbow.
A. Acromion process
B. Coracoid process
C. Tubercle of spine of scapula
D. Medial supracondylar ridge
E. Medial epicondyle
F. Olecranon process
G. Head of radius (displaced inferiorly from its normal relationship to humeral capitulum)

Figure 123. Lateral aspect of shoulder and trunk.
A. Deltoid muscle
B. Long head of triceps
C. Infraspinatus muscle
D. Teres minor
E. Teres major
F. Upper border of latissimus dorsi
G. Inferolateral border of latissimus dorsi

Figure 124. Anterior view of lower part of posterior triangle of neck.
A. Clavicular part of trapezius
B. Scalenus medius
C. Sternal head of sternocleidomastoid muscle
D. Brachial plexus
E. Scalenus anterior
F. Clavicular head of sternocleidomastoid
G. Clavicular part of deltoid
H. Inferior belly of omohyoid
I. Clavicle
J. Subclavian artery

inspiration) the subclavian artery may be partially compressed. Palpate the artery while your model performs these movements to appreciate this clinically important point. Its significance is that in patients with an excessively narrow space between the scalene muscles compression may cause symptoms in the upper limb. One possible cause of decreased space in the **scalene triangle** (the triangle formed by the anterior and middle scalene muscles and the first rib) may be the presence of a cervical rib (i.e., a rib that attaches to the last cervical vertebra, thus lying above the first normal rib attaching to the first thoracic vertebra).

The narrowing of the space of the scalene triangle and the resulting compression that may arise because of the presence of a cervical rib may express itself clinically not only in **vascular** symptoms related to the **subclavian artery** but also in **neurological** symptoms related to the nerves of this neurovascular plane, (i.e., the **brachial plexus**) (brachium means arm; plexus means network). The nerves of this plexus, which are responsible for the entire nerve supply of the upper limb, are entirely derived from ventral rami of spinal nerves (page 86). This is because the embryological limb bud, destined to become the upper limb, grows out from the body wall ventral to the **coronal (frontal) morphological plane,** which demarcates the ventral from the dorsal part of the body. Because spinal nerves emerge through the intervertebral foramina (page 86), and then divide into ventral and dorsal rami, this plane passes through the intervertebral foramina. Ventral rami immediately pass ventrally and dorsal rami dorsally, to supply their respective portions of the body (page 86). Note that the lower limb similarly develops ventral to the coronal morphological plane, and its nerve supply, derived from the lumbar (page 166 to 167) and sacral (page 173) plexuses, therefore similarly is entirely made of ventral rami. Thus all the limb plexuses (brachial, lumbar, and sacral) and the adjacent cervical (page 55) and coccygeal plexuses, are made up of ventral rami. That the limb buds grow out from the body wall and, as it were, drag their nerve supply with them, also accounts for the fact that the regional areas of distribution of individual segmental nerves in the limbs appear not to follow the ordered and regular segmental pattern seen in the thorax. Compare this reason for apparent lack of segmental pattern with that given for the head and neck region on page 190.

The ventral rami of C5, C6, C7, C8, and T1 spinal nerves form the **roots** of the brachial plexus, and these can be palpated as they emerge between scalenus anterior and scalenus medius muscles and link up to form the **trunks** of the brachial plexus. With your model's upper limb in a relaxed position, palpate in the subclavian triangle (just above the clavicle) and feel the "stringy" or rope-like components of the brachial plexus passing, just posterior to the subclavian artery, across the first rib to enter the axilla. The roots derived from C5 and C6 combine to form the **upper trunk,** the root from C7 continues on its own as the **middle trunk** and the roots from C8 and T1 combine to form the **lower trunk** of the brachial plexus.

In the occipital subdivision of the posterior triangle the main superficial contents are branches of the **cervical plexus** of nerves (page 55), which emerge at the posterior edge of the sternocleidomastoid muscle and run superficially from there in various directions. These are not normally visible or palpable. The external jugular vein is usually visible (page 49 and fig. 33, page 82) crossing the sternocleidomastoid muscle just deep to the skin. Identify this vein on your model. If necessary, you can cause it to become more prominent by gently occluding its site of termination, in the region of the subclavian triangle. The vein runs an almost vertical course from an origin close to the angle of the mandible downwards and slightly posteriorly, to enter the subclavian triangle. Superiorly it is formed by the posterior auricular vein and the posterior division of the retromandibular vein. Recall that the anterior division of the retromandibular vein unites with the facial vein to form the common facial vein, which drains into the internal jugular vein (pages 55 and 56). The external jugular vein ends by passing deeply, in the subclavian triangle, to enter the subclavian vein. You should also be able to trace, in most subjects, the anterior jugular vein. It commences below the chin (fig. 10, page 22) and runs close to and parallel with the midline, down the anterior aspect of the neck, to the suprasternal space between the two sheets of the investing layer of deep cervical fascia (page 55). Here it curves laterally, to pass deep to the sternocleidomastoid muscle and terminate by draining into the external jugular vein, just before the latter ends in the subclavian vein. Two tributaries of the external jugular vein, the transverse cervical and the suprascapular veins, may be seen draining across the posterior triangle of the neck toward their destination.

Also usually not visible or palpable but an important content of the posterior triangle of the neck is the spinal accessory nerve. After emerging through the jugular foramen (page 194) this nerve curves downwards and posteriorly, deep to the sternocleidomastoid muscle, which it supplies. It emerges at the posterior border of the muscle, about half way down its length, to cross the triangle in its course toward the trapezius muscle that, as mentioned previously (page 194), it also innervates.

The deep surface of the posterior triangle of the neck is made up of several muscles. Inferiorly the scalenus anterior and medius muscles, which take origin from the anterior tubercles of transverse processes of cervical vertebrae, pass downwards and laterally to insert into the first rib (page 227). Superior to the scalenes, the **levator scapulae** muscle also arises from transverse processes of cervical vertebrae and passes downwards and posteriorly toward the superomedial angle of the scapula. By deep palpation in the occipital triangle, these muscle bellies can be felt and, in particular, their origins (transverse processes of cervical vertebrae) can be identified. The muscle bellies can be distinguished from each other by activating the respective muscles to perform their differing functions. Thus the scalene muscles will be palpably contracted when the model rotates the head, or bends it laterally or flexes it, whereas levator scapuli can be felt when the shoulder is shrugged against resistance.

Laboratory Session 22A

You have previously dissected superficial structures of the anterior triangle of the neck (pages 55–58). Some neck structures continue into the mediastinum (page 98) or are continued from origins in the mediastinum (pages 100 and 123). Your objectives for this dissection session are to review the structures previously studied in the anterior triangle of the neck; to study the continuity of neck structures and mediastinal structures, through the junction between these two regions; and to study deep structures of the anterior triangle of the neck, which you have not yet dissected.

Replace the sternocleidomastoid muscle into its normal position, and review the boundaries and subdivisions of the anterior triangle (page 49). Reidentify the major contents of each subdivision, as previously identified.

The following dissection should be performed on the attached side. If you have not done so previously on this side, cut across the clavicle with the bone saw at the junction of the lateral one third with the medial two thirds of the bone, and reflect the medial two thirds medially. With the access you have now obtained, reidentify the scalenus anterior muscle (pages 122 and 227), and trace it down to its insertion into the first rib. Now study the important relationships of this muscle. The muscle separates the subclavian vein, which lies anterior to it, from the subclavian artery, which lies posterior to it. Reidentify the subclavian vein and trace it medially to the site where it unites with the internal jugular vein, posterior to the position of the sternoclavicular joint, to form the brachiocephalic vein. Review the position and course of the latter vein (page 100). Clear connective tissue and fascia from the anterior surface of the scalenus anterior muscle and identify its important anterior relationships. These include the **phrenic nerve** (page 123), which slopes across the anterior surface from the lateral edge of the muscle, passing inferomedially to leave the medial edge of the muscle and proceed behind the subclavian vein into the thorax, where you previously studied it (pages 106 and 107). Close to the medial edge of the muscle, identify the **inferior thyroid artery,** which ascends along the muscle, to curve medially and course toward the posterior aspect of the inferior pole of the thyroid gland. Trace the artery peripherally to the gland, and centrally to its origin from the **thyrocervical trunk,** a branch of the first part of the subclavian artery.

The left subclavian artery leaves the arch of the aorta posterior to the upper half of the manubrium sterni (page 99) to curve upwards and laterally, passing behind the sternoclavicular joint, and then crossing the upper surface of the first rib. On the right the artery commences behind the joint (page 99). On both sides, the portion of the subclavian artery medial to the scalenus anterior muscle is the **first part** of the artery, the portion of the artery posterior to the muscle is the **second part,** and the portion of the artery lateral to the muscle is the **third part.** At the lateral edge of the first rib the artery changes its name to the **axillary artery.**

The thyrocervical trunk gives off, in addition to the inferior thyroid artery, two branches that cross the anterior surface of the scalenus anterior muscle, passing from medial to lateral. Superiorly, identify the **transverse cervical artery,** and inferiorly, the **suprascapular artery,** both of which pass deep to the sternocleidomastoid muscle and enter the posterior triangle of the neck. To facilitate exposure of structures surrounding the scalenus anterior muscle, mobilize the internal jugular vein, ensuring that you have freed it from related structures such as nerves and arteries, and, using a pair of scissors, excise a section of the vein about 3 cm in length, a little above the site of fusion with the subclavian vein. Using careful scissor dissection in the connective tissue bed thus exposed, dissect to clear remnants of the carotid sheath and surrounding fat away, and expose deep-lying structures.

Careful dissection should be performed at the site of junction of the internal jugular and subclavian veins, which you have left intact, to expose the termination of the thoracic duct. This important lymphatic vessel, after ascending on the right side of the thoracic esophagus and crossing to the left, posterior to the esophagus, at the level of T5, continues upwards into the neck region (pages 106 to 107 and 123). It passes posteromedial to the arch of the aorta, then passes behind the left carotid sheath and curves anterolaterally to end in the junction of the large veins or in one of the veins close to the junction. The duct, which resembles a vein, may be up to several millimeters in diameter, and may even contain blood through reflux from the veins. Follow the thoracic duct behind the common carotid artery, and downwards to the medial aspect of the arch of the aorta. Establish continuity of it with that part of the duct that you previously dissected in the thorax (page 123).

In your scissor dissection of the thoracic duct posterior to the common carotid artery you will encounter the **sympathetic trunk** (page 106), which extends up from the thorax into the neck. Note that although the sympathetic trunk continues into the neck and downwards into the pelvis to end on the anterior surface of the sacrum, it derives preganglionic fibers only from segments T1 to L2 (page 106). The extensions of the trunk superiorly and inferiorly comprise fibers which have passed through sympathetic ganglia in the T1 and L2 regions, without synapsing, and are yet to synapse in ganglia situated outside of these regions. These extensions of the trunk also contain postganglionic fibers on their way to supply distant organs (fig. 52, page 104). For example, the cervical sympathetic trunk, which you are now about to dissect, contains cervical sympathetic ganglia, and postganglionic fibers that supply structures in the neck and in the head. The trunk continues superiorly as the **internal carotid nerve,** to form a plexus around the internal carotid artery, which enters with that artery into the intracranial region (page 200).

The thoracic duct passes in front of the sympathetic trunk as it arches laterally toward its termination. To find the trunk you should dissect posteromedially to the common carotid artery. Having removed remnants of carotid sheath fascia, you have to dissect through another layer of the deep cervical fascia, the **prevertebral fascia** (see also the **investing layer,** page 55). The prevertebral fascia is a tough layer of fascia which lies anterior to the vertebrae and some muscles which cover the vertebral column on its anterior surface. The sympathetic trunk lies deep to the prevertebral fascia. Using scissor dissection open the prevertebral fascia and identify the nerve trunk and trace it inferiorly, posteromedial to the common carotid artery and, further down, posterior to the left subclavian artery. Just inferior to where it passes behind the subclavian artery, establish continuity between the cervical and thoracic portions of the trunk at the site where you previously dissected the superior end of the thoracic portion of the structure. Superiorly you should trace the trunk upwards to identify all of the cervical sympathetic ganglia situated respectively at the levels of C2 to C3 (**superior cervical ganglion**), C6 (**middle cervical ganglion**), and C7 to T1 (the **inferior cervical ganglion,** which may be fused with the superior thoracic sympathetic ganglion, to form the **stellate ganglion**). The trunk should be followed up as far as possible on the attached half of the cadaver. It will subsequently be further dissected on the detached half head and neck specimen.

Study the branches of the subclavian artery. Returning to the medial aspect of the scalenus anterior muscle, trace the **thyrocervical trunk** and its branches. The **transverse cervical** and **suprascapular arteries** may originate close to the site at which the **inferior thyroid** artery also comes off, or they may originate from a common trunk. The transverse cervical artery divides into a superficial branch, the **superficial cervical artery,** and a deep branch, the **dorsal scapular artery.** The inferior thyroid artery curves medially, to pass either posterior (usually) or anterior to the cervical sympathetic trunk, to then recurve inferiorly and gain the inferior pole of the thyroid gland into which it passes on the posterior aspect of the gland. Identify also the **internal thoracic artery,** which passes inferiorly, and which you previously studied in the anterior chest wall (page 92), as well as the **vertebral artery.** This branch ascends very close to the vertebral column for the first part of its course, then enters the **foramina transversaria** (holes in the transverse processes) of the sixth to first cervical vertebrae. Identify now, on a skeleton, the foramina transversaria of cervical transverse processes (fig. 131, page 244). The vertebral artery should be traced to the site where it disappears into the sixth foramen transversarium. (The companion **vertebral vein** passes through all seven foramina.) Superiorly, the vertebral artery arches around the lateral mass of the atlas (the first cervical vertebra), and enters the skull through the **foramen magnum** where you have previously seen it in relationship to the last cranial nerves (page 199) and where it attains the inferior surface of the brain to participate in the circle of Willis.

The remaining branch of the subclavian artery, the costocervical trunk, will be studied later.

Just posterosuperior to the lateral aspect of the scalenus anterior muscle you will observe several large nerve bundles belonging to the **brachial plexus** (page 230 and fig. 124, page 228), passing posterolaterally towards the upper limb. These will be studied subsequently. Superiorly, in series with the nerves that enter the brachial plexus, identify other nerves, posterior to the internal jugular vein and internal and external carotid arteries. These are nerves of the **cervical plexus,** which pass to the lateral edge of the sternocleidomastoid muscle, and then curve around that edge to become subcutaneous. You previously identified some of the nerves of this plexus (page 55). For example, the great auricular nerve ascends vertically, on the lateral surface of the sternocleidomastoid muscle (fig. 8, page 16) to supply the skin of the ear and adjacent regions.

Reverting now to the detached specimen, identify the pharynx and dissect in the connective tissue-filled space posterior to the pharynx and anterior to the vertebral column. Gently separate the loose connective tissue by blunt dissection, to identify the carotid sheath and its contents: the common, internal, and external carotid arteries, the internal jugular vein, and vagus nerve. Inspection of the horizontally cut surface through the neck, which should be approximately at the level of C5, will aid in this identification. Study the prevertebral fascia, and deep to it, the sympathetic trunk, which you should free from the deep fascia by careful scissor dissection. Dissecting upwards, you should identify the large superior cervical ganglion, situated approximately at the level of C2. Small branches from the superior cervical ganglion to surrounding structures, including the pharynx, should be identified and dissected. Elevate the prevertebral fascia, and identify cervical plexus nerves passing from intervertebral foramina inferolaterally toward the posterior edge of the sternocleidomastoid muscle. Separate tissues at this posterior edge, reidentifying the investing layer of deep cervical fascia which surrounds the sternocleidomastoid, and splits at its anterior and posterior edges (page 55). Previously, you followed the fascia anteriorly in the anterior triangle. At this site, you can follow it posteriorly in the posterior triangle of the neck and trace it to the trapezius muscle at the back of the neck.

Separate the posterior surface of the pharynx from the retropharyngeal fascia, the layer of connective tissue connecting the pharynx with the anterior surface of the **prevertebral muscles.** Follow the pharynx superiorly, to its attachments to the pharyngeal tubercle (page 61). To facilitate this and the ensuing dissection, extend, if necessary, the bony incision you previously made in your study of the inner ear, using a bone saw if needed. Clear the posterior surface of the pharynx of fascia, using scissor and scalpel dissection and tracing in an anteroinferior and medial direction in your dissection, thus following the direction that nerves and muscles take as they approach the posterolateral aspect of the pharynx. By stroking your scalpel in the direction mentioned you will gradually expose through connective tissue and fat the component fibers of the pharyngeal plexus of nerves, reaching the pharynx from the sympathetic trunk (pages 233 to 234), the glossopharyngeal nerve, and the vagus nerve. Trace as many of the branches as you can, both to the pharyngeal wall and to their trunks of origin. In this way you should be able to reidentify the glossopharyngeal nerve and the stylopharyngeus muscle (page 72) as they slope downwards, forwards, and medially toward the pharynx. The glossopharyngeal nerve lies medial to the muscle as they both approach the pharynx but more superiorly, the nerve lies posterior to the muscle. That is, the nerve swerves down on the posterior edge of the muscle and then sweeps across the lateral surface and around the anterior edge to become medial to it. The muscle can be traced superiorly to its origin from the styloid process and this will also enable you to identify and review other muscles which have similar origin (pages 52, 56, 58, and 24). Branches of the sympathetic trunk to the pharyngeal plexus should be identified, and branches from the vagus should be traced back to the trunk of that nerve, reconfirming the situation of the vagus nerve in the carotid sheath. The rich **pharyngeal venous plexus** should also be identified on the posterior surface of the pharynx. Note any lymphatic glands that you may encounter.

Dissecting more superiorly in the retropharyngeal space, reidentify the **levator veli palatini** muscle (pages 60 and 66). In the present dissection you will more readily be able to fully appreciate in three dimensions the relationship of this muscle to the pharyngeal tube.

Study the region of the styloid process by a combined medial and lateral approach (i.e., gaining access medially in the region posterior to the pharynx where you dissected the stylopharyngeus muscle and glossopharyngeal nerve, and from the lateral aspect, where you previously identified the styloid process when dissecting the ear [page 218]). Identify and clear, by means of careful scissor and scalpel dissection, the muscles taking origin from the styloid process. Follow the stylohyoid to the styloid process, carefully preserving and studying the relationships of this muscle to the common carotid artery, which it passes laterally. Close to its insertion into the hyoid bone the stylohyoid splits to enclose the intermediate tendon of the digastric muscle (page 52).

Reidentify and trace the paths of the last four cranial nerves. Identify the glossopharyngeal nerve on the medial aspect of the stylopharyngeus muscle, and follow its path superiorly, as it winds around the stylopharyngeus muscle. Identify it from the lateral approach, lateral to the muscle. With a skull at your side to assist you in getting your bearings, and making use of reference to atlas illustrations, trace the glossopharyngeal nerve up to its site of exit from the skull, at the jugular foramen (page 193). In its characteristic position in relationship to the internal carotid artery and the internal jugular vein identify the vagus nerve, reconfirming the branches it gives off to the pharynx, and trace it similarly up to the jugular foramen. Note that the vagus nerve lies slightly posterior to cranial nerve IX. Superiorly, on the posterior aspect of the vagus, identify the spinal accessory nerve, which shortly after existing from the jugular foramen, swerves posteriorly. It passes lateral to the internal jugular vein to gain the medial surface of the sternocleidomastoid muscle. Finally, in the lateral approach to the cranial nerves, identify the hypoglossal nerve as it passes lateral to the vagus and accessory nerves. Reidentify (pages 198 to 199), on your dissection specimen and on a skull, its exit through the anterior condylar foramen, above and in front of the occipital condyle. The nerve is initially posterior to the other cranial nerves at exit. It curves forwards and downwards, passing, as mentioned, lateral to the other nerves, to ultimately emerge below the inferior border of the posterior belly of the digastric muscle. It crosses lateral to the internal and external carotid arteries to attain its characteristic position in relation to the upward loop of the lingual artery (pages 56 to 57, and fig. 24, page 50). You have previously dissected its subsequent course, which you should now review (page 24 and fig. 12, page 22).

Laboratory Session 22B

Your objective for this dissection session is to study the posterior triangle of the neck. Review the borders of this triangle (page 225) on the attached half head and neck specimen of your cadaver. It will facilitate access to the posterior triangle to rotate the half head specimen to the opposite side. You should, however, repeatedly rotate the head anteriorly again, to obtain an image of the way the triangle and its contents appear in the unrotated position.

To skin the posterior triangle, continue the skin incision you made previously in dissecting the anterior triangle (page 55 and page 232) upwards toward the mastoid process and then beyond it, skirting posterior to the auricle, to attain the superior nuchal line (fig. 94, page 184), to which the sternocleidomastoid muscle attaches. Your incision should be approximately in the middle of the belly of the sternocleidomastoid muscle, so that its posterior edge can be displayed and studied. Make an incision along the superior nuchal line about 5 cm in length, and then incise the skin parallel to and slightly behind the border of the trapezius muscle so that, again, the border of the muscle that bounds the triangle will be accessible to study. Carry this incision down to the **acromion process** of the scapula (page 226 and fig. 122, page 228), to which the trapezius muscle attaches and which you should reidentify on a skeleton. The trapezius attachment continues forwards and then medially onto the clavicle, and your incision should follow this attachment, thus continuing onto the third side of the triangle. Complete the incision by carrying it along the clavicle until it meets the original incision, medial to the lateral free edge of sternocleidomastoid.

Elevate skin carefully, identifying the platysma muscle, the fibers of which lie in the subcutaneous connective tissue and run in an inferolateral direction, splaying out both anteriorly and posteriorly. Also in the subcutaneous tissue, identify branches of the cervical plexus of nerves, emerging at the posterior border of the sternocleidomastoid muscle. Some of these were previously encountered in your dissection of the anterior triangle of the neck (pages 55 and 233). You are likely to find the supraclavicular branches, passing downwards and laterally to cross the clavicle. The great auricular nerve (fig. 8, page 16) ascends on the superficial surface of the sternocleidomastoid muscle, toward the auricle, supplying sensory fibers to both sides of the auricle and to a region anterior to it. The lesser occipital nerve runs superiorly along the posterior edge of the sternocleidomastoid, before then similarly ascending vertically on the superficial surface of the muscle, to the region behind the auricle. The transverse cervical nerve will probably have been dissected previously.

Carefully dissect the investing deep cervical fascia (page 55) from the anterior surface of the sternocleidomastoid muscle, toward the posterior edge. Recall (page 55) that this fascia encloses sternocleidomastoid in a sheath. Anteriorly, the two layers fuse and can be followed into the anterior triangle. The strap muscles are embedded in the posterior layer of this fascia (page 55). At the posterior border of sternocleidomastoid, the two layers of fascia also fuse. This double layer can be traced posteriorly to the anterior edge of trapezius where the two layers again separate to enclose the latter muscle. This should be verified on your cadaver, but you should dissect carefully because just deep to the fused layers in the posterior triangle, a very important nerve, the **spinal accessory** (cranial nerve XI) will be found. This nerve enters the posterior triangle at about the midpoint of the posterior border of the muscle, having curved inferoposteriorly from its exit through the jugular foramen (page 193), deep to the sternocleidomastoid muscle (but occasionally passing **through** the belly of the muscle). The nerve passes from the posterior edge of the muscle posterolaterally toward the free edge of the trapezius muscle, deep to which it passes to enter the back. (Recall that this nerve supplies motor fibers to the sternocleidomastoid and trapezius muscles, and the integrity of this cranial nerve is in part tested by activating these muscles.)

The posterior edge of the sternocleidomastoid muscle and the anterior edge of trapezius should be carefully dissected free of the investing deep fascia and traced superiorly, where they converge at the superior angle of the posterior triangle. This dissection requires care, because of fusion of fascial layers. As previously studied on a living model (page 231), within the triangle forming its deep surface, several muscles are found (figs. 124, pages 228 and 125 and 126, page 238) covered by the prevertebral layer of deep cervical fascia (page 233). These muscles include, inferiorly, the scalenus anterior and scalenus medius muscles (page 231). These take their origins from anterior and posterior tubercles of cervical transverse processes respectively, and insert, as mentioned previously, into the upper surface of the first rib. Above the scalenus muscles identify the **levator scapulae** muscle (page 231 and fig. 125, page 238) that also arises from cervical transverse processes (though from posterior tubercles). The muscle passes posteroinferiorly to the superomedial angle of the scapula. As a landmark, note that the accessory nerve usually lies just superficial to the levator scapulae muscle. Superior to the levator scapulae, dissect the **splenius capitis** muscle (figs. 125 and 126, page 238). This muscle arises in the posterior midline from vertebral spines (cervical and thoracic) to pass superolaterally to the mastoid process and the adjacent region of the skull just inferior to the superior nuchal line. The fibers of this muscle are fine, highly parallel, and closely packed, giving it a very characteristic appearance of which the name splenius (bandage) is descriptive.

Toward the superior angle of the posterior triangle of the neck, the posterior surface of the investing layer of deep cervical fascia, on the posterior surfaces of sternocleidomastoid and trapezius muscles and spanning the gap between the two, may be somewhat fused with the deep-lying prevertebral layer of cervical fascia. This makes careful scalpel dissection of the fascia layers necessary, to expose splenius capitis and, even more superiorly, the one further muscle found in the deep surface of the triangle, the **semispinalis capitis**. The fibers of this muscle will be found running almost vertically, close to the midline above and behind the posterosuperior border of splenius (figs. 126, 127, page 238 and fig. 129, page 244).

In the course of dissecting deep fascia posterior to the posterior border of sternocleidomastoid, identify the **occipital branch** of the **external carotid artery** (page 58). Preserve this artery, trace it superiorly toward the occipital region, and trace it inferiorly toward its origin (deep to the sternocleidomastoid muscle).

To allow access to the inferomedial reaches of the triangle, carefully detach the inferior attachment of the sternocleidomastoid muscle from the clavicle and sternum, and reflect the muscle upwards and medially. As you did previously, in other dissections, you should repeatedly replace the muscle in order to visualize the natural relationships. As you reflect the muscle, you will detach its connective tissue sheath from the underlying sling of the omohyoid muscle. Pursue the course of the inferior belly of the omohyoid, and review its role in dividing the posterior triangle of the neck into two subdivisions (pages 226 to 227). Study the course of the anterior jugular vein as it passes posterior to the sternocleidomastoid to end in the external jugular vein (page 230) just before the latter terminates in the **subclavian vein.**

The subclavian is the large vein of the upper limb that accompanies the artery of the same name and unites with the internal jugular vein to form the brachiocephalic vein (page 100). Although elsewhere in the body veins accompany arteries and nerves in the neurovascular plane (page 86), the subclavian vein is an exception and lies anterior to the scalenus anterior muscle, separated by the latter from the artery. However, the neurovascular plane between scalenus anterior and scalenus medius does indeed contain nerves, which you should now identify. These are the nerves of the **brachial plexus** (plexus meaning network) that you previously palpated (page 230). The brachial plexus, like the cervical plexus and other limb plexuses of the body (lumbar, sacral, and coccygeal plexuses) is derived entirely from ventral rami of spinal nerves (page 86). This is because in embryologic development the limb buds, which develop into the limbs, grow out from the body anterior to the coronal morphological plane which divides the ventral part of the body from the dorsal (pages 86 and 230).

Before proceeding to a more detailed study of the brachial plexus, reidentify superficial structures that may become damaged in the ensuing dissection. These include the branches of the cervical plexus and superficial veins such as the suprascapular and transverse cervical veins, which drain into the external jugular vein. The latter vein should be traced to the subclavian vein, reidentifying, in the process, the termination of the anterior jugular vein in the external jugular (or the subclavian).

Carefully remove prevertebral fascia to expose the brachial plexus, in the neurovascular plane between scalenus anterior and scalenus medius. The **roots** of the plexus are derived from the ventral rami of C5, C6, C7, C8, and T1. Identify these as they descend to form the **trunks** of the plexus, C5 and C6 combining to form an **upper trunk;** C7 continuing on its own as the **middle trunk;** and C8 and T1 joining to form the **lower trunk** of the plexus. Confirm the rope-like nature of the trunks as you palpated them in the living body (page 230). With the aid of an atlas illustration, identify the branches of the roots and trunks of the brachial plexus, and trace these branches as far peripherally as you can. The suprascapular nerve, passing posteriorly in the general direction of the inferior belly of the omohyoid muscle, is usually easy to identify. It passes through the suprascapular notch (page 227) to supply muscles on the back of the scapula. Similarly, the medial pectoral nerve (page 87) may arise from the lower trunk of the chord and is also easily found (although this nerve usually arises from the brachial plexus more peripherally).

By means of careful scalpel dissection, reidentify the phrenic nerve (pages 106, 107 and 123) and trace it proximally to its origin from the cervical plexus.

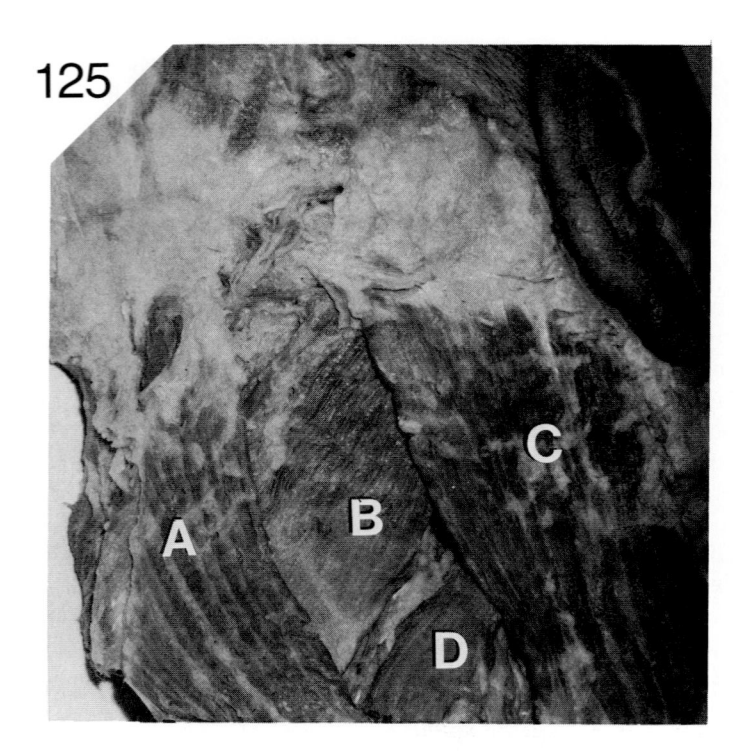

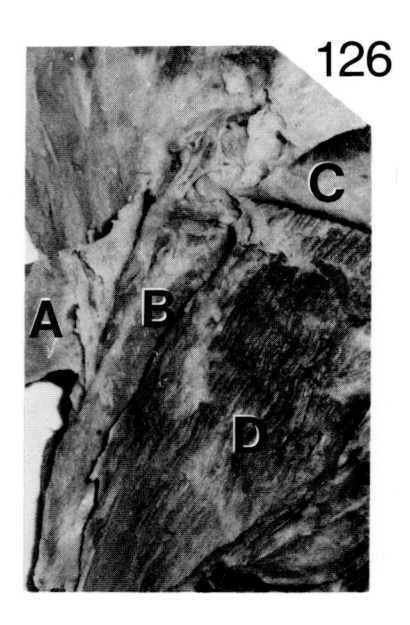

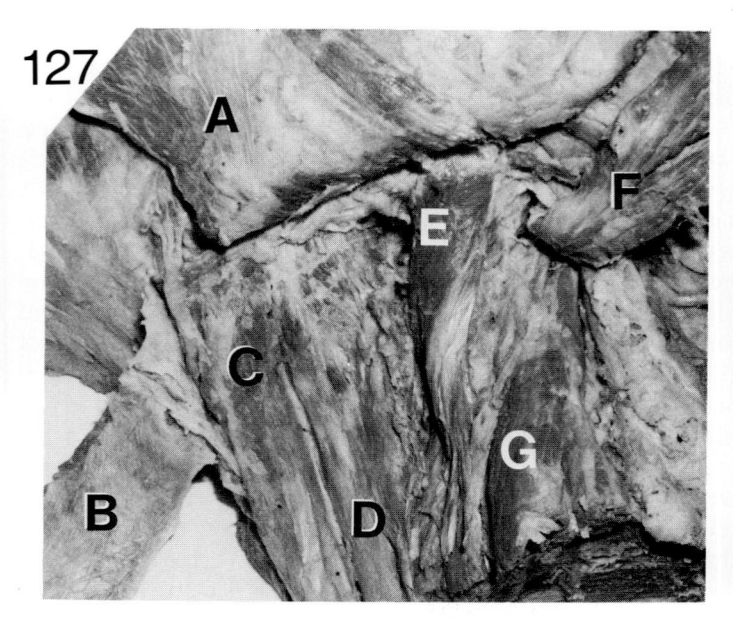

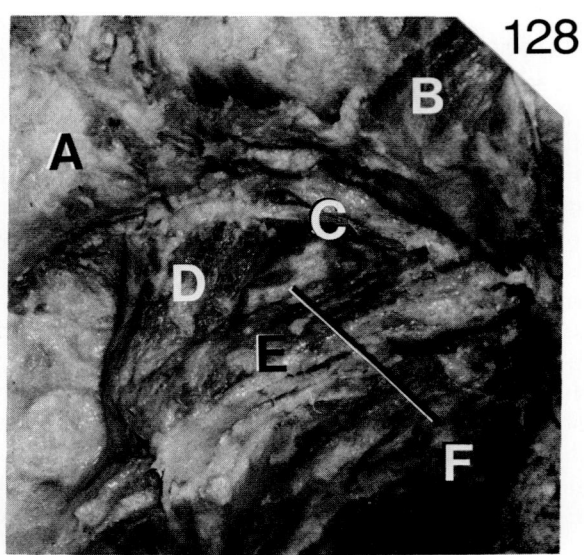

238 YOUR PATIENT'S ANATOMY: A CLINICAL VIEW OF HUMAN MORPHOLOGY

Figure 125. Posterolateral view of upper part of right posterior triangle of neck, and back of neck.
A. Trapezius
B. Splenius capitis
C. Sternocleidomastoid
D. Levator scapulae

Figure 126. Posterolateral view of upper part of right posterior triangle of neck, and back of neck; deeper aspect than in Figure 125
A. Trapezius, reflected medially
B. Semispinalis capitis
C. Sternocleidomastoid, reflected upwards
D. Splenius capitis

Figure 127. Posterolateral view of upper part of right posterior triangle of neck, and back of neck; deeper aspect than in Figures 125 and 126.
A. Splenius, reflected upwards
B. Trapezius, reflected medially
C. Semispinalis capitis
D. Semispinalis capitis (with semispinalis cervicis lying deep)
E. Longissimus capitis
F. Levator scapulae, reflected upwards
G. Longissimus cervicis

Figure 128. Posterolateral view of upper part of right posterior triangle of neck, and back of neck; deepest level (suboccipital triangle)
A. Semispinalis capitis and cervicis, reflected upwards
B. Longissimus capitis, reflected upwards
C. Obliquus capitis superior
D. Rectus capitis posterior major
E. Obliquus capitis inferior
F. Third part of vertebral artery within triangle

Review also its distal course (pages 106, 107, 146 and 232).

Using careful scissor dissection deep to the sternocleidomastoid muscle reidentify the contents of the carotid sheath (page 56). Establish the communication of the common carotid artery with the subclavian artery. Now carefully complete your dissection of the branches of the subclavian artery. Recall (page 233) that only one branch descends into the thorax. The **vertebral artery** should be traced superiorly until it disappears into the foramen transversarium of the sixth cervical transverse process. The **thyrocervical trunk** should be dissected to display all its branches including the inferior thyroid artery (page 232), the ascending cervical artery, and the transverse cervical and suprascapular branches. Finally, identify and trace the **costocervical trunk,** which arches posteriorly over the cervical pleura (page 226) toward the neck of the first rib. Use atlas illustrations to aid you in identifying and following these branches of the subclavian artery.

With the lung in place, palpate bimanually the relationship of the apex of the lung to the supraclavicular region, to establish the proximity of the lung apex to the supraclavicular fossa of the living individual, in order to reinforce your visualization of these relationships, and the reason why apical lung sounds can be auscultated in this vicinity (page 95 and page 226).

To complete your study of this region, review the adjacent structures of the anterior triangle and the deep prevertebral region previously dissected (page 233). This review should include the cervical sympathetic trunk, the thoracic duct, the prevertebral muscles, and the deep cervical lymph glands.

Clinical Session 23

■ **TOPICS**

Posterior Aspect of Neck
The Back
Bony Prominences
Muscles of Neck and Back
Movements of Neck and Back

The objective of the exercises described below is to familiarize yourself with the visible and palpable bony and muscular landmarks of the neck and back, and the movements and other functions of these structures. To perform these examinations your model should be stripped to the waist. You should correlate your clinical examination with study of a skeleton and isolated bones, and consult atlas illustrations. While studying structures of the back you should also bear in mind that their clinical relevance, though often not obvious to the student, is nevertheless considerable. Lower back pain and pathology are among the most common symptoms and signs encountered in medical practice. As another example, the muscles of the upper back and neck stabilize the head in normal stance and gait, and act as fixators for mastication, among other functions.

Inspection of General Outline and Spinal Curvatures

With your model in the anatomical position examine the neck and back from the posterior aspect (figs. 130 and 132, page 244) and note the superolateral border of the trapezius muscle sloping downwards and laterally from the superior nuchal line to the shoulder; the latissimus dorsi muscle (latissimus means the broadest; dorsi means of the back) sloping from just below the axilla downwards and medially toward the waist; the median spinal furrow (page 125 and fig. 64, page 128); the vertebra prominens (page 94 and fig. 131, page 244); the sacrospinalis muscle column, extending upwards on either side of the median spinal furrow (page 158 and fig. 64, page 128); and the medial border of the scapula. Note, by inspection of the median spinal furrow and the adjacent sacrospinalis muscles, whether the vertical axis of your model's back is entirely linear, or whether, as is usual, a slight degree of **scoliosis** (lateral curvature) is detectable. Minor degrees of scoliosis are often associated with "handedness," and a slight degree of scoliosis to the one side in the thoracic region will most often be compensated by a slight degree to the opposite side in the lumbar region.

Observe your model from the lateral aspect to ascertain the degree of anteroposterior curvature in each region. In the cervical region there is a slight normal **lordosis** (anterior convexity). The thoracic part of the vertebral column presents a physiological **kyphosis** (posterior convexity); the lumbar region has a lordosis and the sacrum and coccyx show kyphosis. Exaggerated degrees of any of these curvatures may be pathological. The opposite curvature to that normally observed may also be a presenting sign of vertebral abnormality.

Palpation of Bony Landmarks

Identify on the posterior aspect of the skull the inion (page 186) and the superior nuchal line extending out laterally on either side of it. Palpate the mastoid process and the vertebra prominens. Attempt to palpate the spines of the cervical vertebrae. Although it may occasionally be possible to palpate some of them, this examination is rendered difficult by the presence of a tough elastic ligament, the **ligamentum nuchae.** This is a triangular structure lying in the sagittal plane. The long side of the triangle is a free edge extending from the inion to the vertebra prominens. A short side of the triangle attaches to the external occipital crest (fig. 94, page 184), and the remaining side attaches to the tips of all the cervical spinous processes. Visualize this structure by examining a skeleton and Figure 131, page 244. From vertebra prominens (C7) count vertebral spinous processes downwards, to identify all the thoracic and lumbar spines. Correlate the following vertebral levels by palpating the appropriate structure with one hand and placing the index finger of the other hand on the appropriate vertebral spinous process: body of hyoid bone (C2); upper edge of thyroid cartilage (C3) (page 53); sternal angle (T4–T5); inferior angle of scapula (T8) (fig. 41, page 90); subcostal plane (L3) (page 125); intercristal plane (L4) (page 125); intertubercular plane (L5) (page 125); anterior superior iliac spine (S2); and posterior superior iliac spine (S2) (page 126 and fig. 64, page 128).

Identify the palpable portions of the scapula. You have previously located the inferior angle of the scapula at the level of the spinous process of T8. Recall that the scapular line passes through this landmark (page 94 and fig. 41, page 90). The medial border of the scapula is more or less parallel to the median vertebral furrow, and when your model pulls his or her shoulders backwards, this border of the scapula becomes very obvious, an appearance known as "winging" of the scapula. Palpate this medial border superiorly, to the **superior angle**. From the medial border, about 2 cm inferior to the superior angle, palpate the **spine of the scapula** extending out laterally in an approximately horizontal direction. The spine lies quite superficially and can be palpated toward the shoulder, where it bend forwards at approximately a right angle to form the acromion process (page 226 and fig. 122, page 228). The flattened surface and the lateral edge of the acromion are usually easily palpable in most subjects. At its anterior end, the acromion process articulates with the lateral end of the clavicle, and this joint and its cavity are usually also easily identifiable.

Starting again at the inferior angle of the scapula, identify the lateral border and attempt to palpate it as far superolaterally as possible. This palpation is somewhat hampered by the presence of the teres major muscle, which takes its origin from this border (fig. 123, page 228 and fig. 132, page 244) and participates in formation of the posterior wall of the axilla. For completeness, reidentify the **coracoid process** of the scapula in the deltopectoral triangle (page 85).

Finally palpate in the space between the posterior border of the ramus of the mandible and the mastoid process to identify the **transverse process** of the **atlas** (the first cervical vertebra), about 1 cm anteroinferior to the mastoid process (fig. 131, page 244).

Palpation of Muscles of the Posterior Aspect of the Neck and Back

The muscles of the posterior aspect of the neck are, in part, muscles that belong to the so-called back muscles, and in part smaller, deeper-lying muscles intrinsic to the neck and occipital region of the skull. The latter group cannot readily be seen or palpated in the living, and will be studied by dissection. Of the former, the only muscle easily accessible to examination is trapezius.

As previously studied (page 194) trapezius is capable of elevating the shoulders when its superior attachment (to the superior nuchal line) is fixed. With your patient holding his or her head stabilized and elevating a shoulder against the resistance of one of your hands, palpate the free margin of the trapezius from its superior attachment on the superior nuchal line, inferolaterally to its inferior attachment to the scapula spine, the acromion process and the lateral portion of the posterior aspect of the clavicle (figs. 121 and 124, page 228). This margin has previously been identified as the posterior border of the posterior triangle of the neck (page 225).

The trapezius muscles are so named because right and left sides, when viewed together from the posterior aspect, create the approximate shape of a trapezoid (figs. 130 and 132, page 244 and 133, page 252), with the superior angle at inion (and extending somewhat out laterally along the superior nuchal lines); the lateral angles at the medial borders of the acromion processes (fig. 122, page 228); and the inferior angle at the spine of T12. The muscle of each side attaches **medially** to the inion (and the medial third of the superior nuchal lines), the entire length of the posterior edge of the ligamentum nuchae, and the tips of the thoracic spines and the **supraspinous ligaments** (ligaments extending between the tips of vertebral spines, to which ligamentum nuchae is homologous in the cervical region). The **lateral** attachments of the muscle are such that it can be regarded as having a superior and an inferior portion. The fibers taking origin from the superior nuchal line are those that insert furthest anteriorly (i.e., on the clavicle). Those originating from the ligamentum nuchae and the uppermost of the thoracic vertebrae insert into the acromion process and the upper border of the spine of the scapula. Fibers originating from the remaining thoracic vertebral spines pass upwards and laterally to insert into the tubercle and inferior border of the spine of the scapula (fig. 122, page 228), creating an inferolateral border of the muscle (fig. 41, page 90 and fig. 84, page 160). It will be evident that it is the superior of these fibers that are capable of elevating the scapula. The inferior fibers depress the scapula. Palpate the two groups of fibers separately, as your model performs elevation and depression of the scapula against resistance. Visualize by consideration of the attachments and position of the trapezius muscles what other functions they may have, and confirm these on your model by palpation during activation against resistance.

By activating the sternocleidomastoid muscle and trapezius muscle in appropriate movements against resistance (page 49 and as above), delineate the posterior triangle of the neck and examine the muscles that are situated within it. Review the palpation of the inferior belly of the omohyoid muscle (page 227). Palpate the scalenus anterior and scalenus medius muscles as your model flexes the head (page 231) against resistance. The scalenus muscles, acting bilaterally, are cervical flexors. Acting unilaterally, they will bend the head and neck to the corresponding side. Levator scapulae muscle (page 231 and fig. 125, page 238) can be palpated as the subject elevates the shoulder or with the shoulder fixed, bends the neck laterally. Review the attachments of the muscle. Splenius capitis (page 237 and fig. 125, page 238) may also be palpable in the upper reaches of the occipital portion of the posterior triangle of the neck. Your model should rotate his or her head against resistance; the muscle will be palpable on the side toward which the head is rotating, since the origin of the fibers is from the lower half of the ligamentum nuchae and the insertion is into the lateral portion of the superior nuchal line.

Muscles of the Back

The muscles of the back are disposed in six layers.

The muscles of the **first layer** are trapezius and latissimus dorsi, and both are readily palpable. Review the actions of **trapezius** in elevating and depressing the scapula. In addition, palpate the superior part of trapezius as your model elevates his or her arm laterally. Palpate the lower portion of the muscle when the scapulae are being retracted and winged. Finally, get your model to extend (page 225) his or her head against resistance, and palpate the upper parts of his or her right and left trapezius muscles simultaneously.

The **latissimus dorsi** muscle is responsible for the tapering of the lateral borders of the back, as viewed from the posterior aspect. This muscle has been studied previously in various contexts (e.g., as the posterior border of the axilla [page 85 and fig. 35, page 82], and in forming the above-mentioned tapering border of the back [figs. 130 and 132, page 244]). Palpate the muscle, including its origins and insertion, while your model activates it by powerful extension at the shoulder joint (i.e., by pulling something toward him or herself against resistance, as in a rowing motion. This can be achieved, for example, by your model standing in a doorway, and "pulling" the doorframe toward him or herself). The lateral border of the muscle, as illustrated in Figures 61 and 63, page 128, should be palpated along the side of the back. Follow the border superiorly into the axilla, visualizing that the muscle twists around the side margin of the body to attain its insertion into the humerus (the long bone of the arm). As the muscle passes toward its insertion, it forms the posterior wall of the axilla. Trace the inferolateral border of the muscle (figs. 61 and 63) down to its origin from the posterior portion of the iliac crest. The muscle also takes origin, in part, from the lower four ribs, these origins interdigitating with serratus anterior. (The latter muscle also interdigitates with the external oblique muscle of the abdomen [fig. 34, page 82].) The latissimus dorsi also takes origin indirectly (i.e., through a very tough layer of fascia) from the tips of vertebral spines: the lower six thoracic, the lumbar, and the upper sacral. The tough fascia constitutes part of the posterior layer of the **thoracolumbar fascia.** You previously encountered the anterior layer of this fascia on the anterior surface of quadratus lumborum muscle, and from the fused anterior and middle layers you observed the origin of the transversus abdominis muscle (page 167). The origin of the latissimus dorsi from these midline attachments can be palpated while your model activates the muscle. Finally, trace the upper border of the muscle from the region of the sixth thoracic vertebral spine, laterally in a gentle slope upwards, across the inferior angle of the scapula (to which the muscle usually attaches), to converge with the inferolateral border of the muscle where the muscle turns around the body wall to attain its insertion into the humerus.

A muscle that is intimately related to latissimus dorsi, and that is also visible and palpable in the back but constitutes a muscle of the upper extremity rather than a true back muscle, is the **teres major muscle,** which should now be examined (figs. 123, page 228, 134, page 252, and 35, page 82). As mentioned previously (page 227) teres major takes its origin from the lateral border of the scapula. It is seen as a swelling on the lateral border of the back, just above the site where the latissimus dorsi twists onto the anterior aspect of the trunk (figs. 130 and 132, page 244). The teres major performs a similar twist, and thus participates also in the formation of the posterior wall of the axilla, and furthermore inserts into the humerus close to the latissimus dorsi (fig. 35, page 82).

The **second layer** of muscles of the back comprises the levator scapulae, which you examined in the living model (page 231), and the **rhomboid major** (figs. 130 and 132, page 244) and **rhomboid minor** muscles. The latter two take origin from tips of vertebral spines (C7–T5) and insert into the medial border of the scapula, below the site of insertion of levator scapulae (i.e., inferior to the spine of the scapula). The muscles can be palpated as your subject retracts the scapula. Note that since the scapula belongs to the upper limb, all the back muscles of the first and second layers attach the upper limb to the back.

The muscles of the **third layer,** the **serratus posterior superior** (fig. 129, page 244) and **serratus posterior inferior,** are neither visible nor palpable in the living. They lie in the same plane as, and are morphologically associated with, the serratus anterior muscle (page 85), which you should review and palpate now. Your model should activate the latter muscle by pushing against a wall. Observe the serrations of the muscle (figs. 35 and 34, page 82). These represent the origins of serratus anterior from the upper eight ribs. The insertion of the muscle is into the **anterior** surface of the medial border of the scapula. Visualize from this information the situation of the muscle, and formulate, by considering what action your model is executing in order to acti-

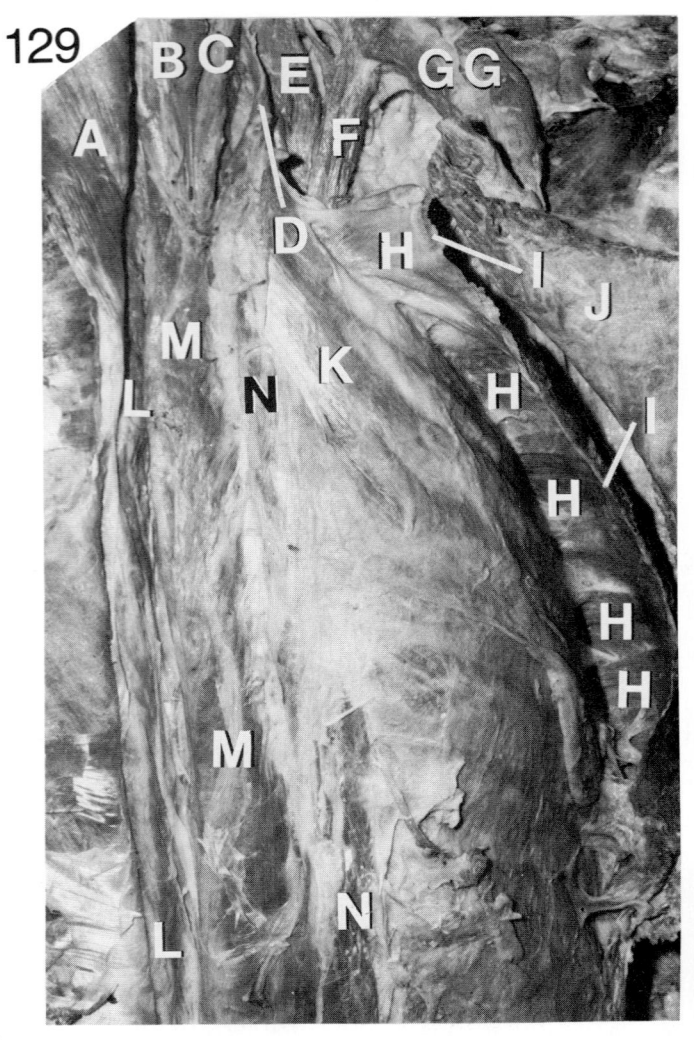

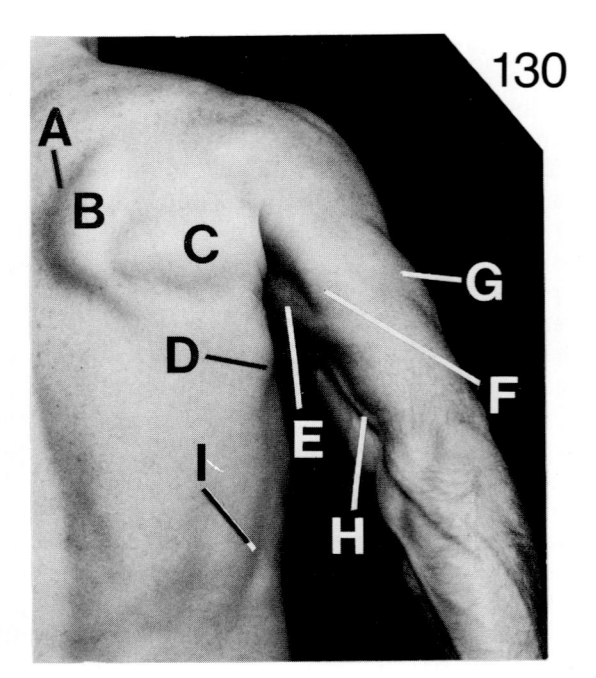

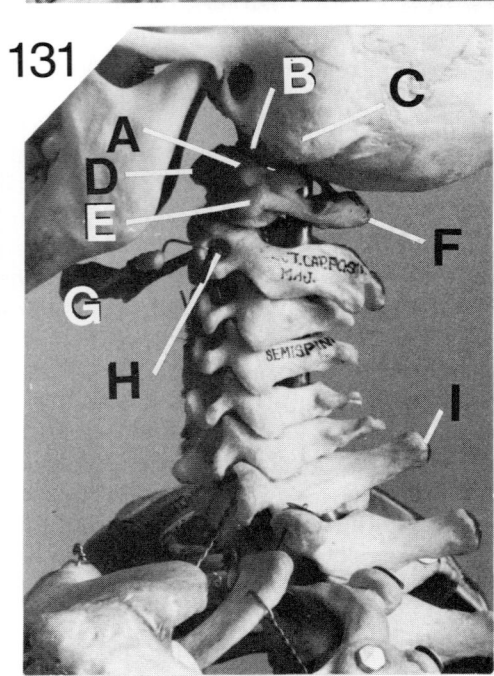

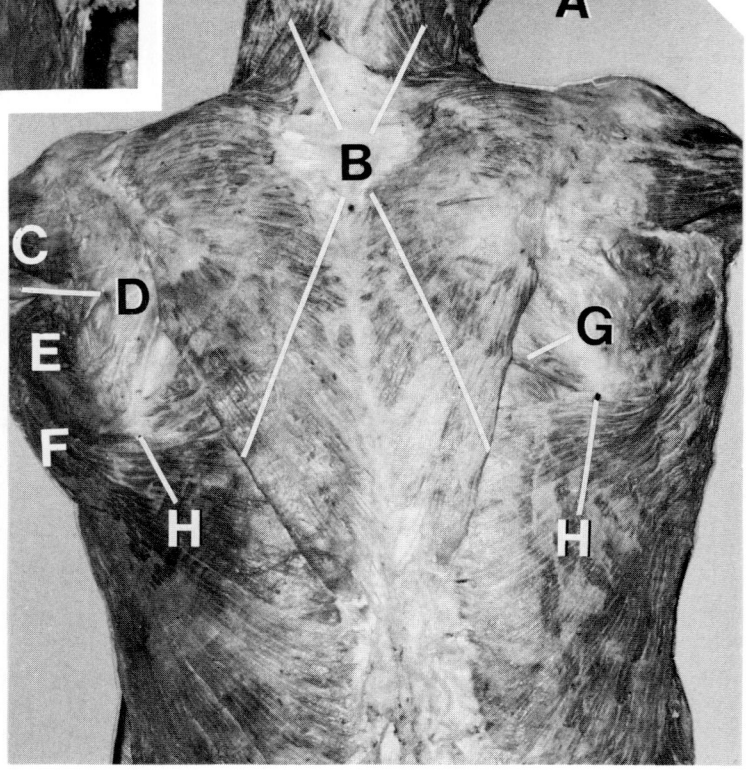

244 YOUR PATIENT'S ANATOMY: A CLINICAL VIEW OF HUMAN MORPHOLOGY

Figure 129. Muscles of right side of back, right scapula displaced laterally.
A. Splenius capitis and cervicis, reflected to left
B. Semispinalis capitis and cervicis
C. Longissimus capitis and cervicis (part of middle column of sacrospinalis)
D. Costocervicalis (part of lateral column of sacrospinalis)
E. Scalenus medius
F. Scalenus posterior
G. Levator scapulae
H. Digitations of serratus anterior (pulled laterally, medial aspect visible)
I. Cut edge of serratus anterior
J. Rhomboids, reflected laterally
K. Serratus posterior superior
L. Spinalis (middle column of sacrospinalis)
M. Longissimus (middle column of sacrospinalis)
N. Ileocostalis (lateral column of sacrospinalis)

Figure 130. Posterior aspect of right shoulder and arm.
A. Inferolateral border of trapezius
B. Rhomboid major
C. Teres major
D. Serratus anterior, covered by latissimus dorsi
E. Medial head of triceps
F. Long head of triceps
G. Lateral head of triceps
H. Basilic vein
I. Latissimus dorsi

Figure 131. Lateral view of cervical vertebrae and relations.
A. Superior articular facet of lateral mass of atlas (for articulation with occipital condyle)
B. Occipital condyle
C. Mastoid process
D. Anterior tubercle of atlas
E. Transverse process of atlas
F. Posterior tubercle of atlas
G. Hyoid bone
H. Foramen transversarium of axis
I. Spinous process of vertebra prominens

Figure 132. Muscles of back (reconstructed by replacement of dissected right half head and neck).
A. Splenius (see also fig. 125, page 238).
B. Edges of trapezius
C. Deltoid
D. Long head of triceps
E. Teres major
F. Converging fibers of latissimus dorsi
G. Rhomboid major
H. Superior edge of latissimus dorsi

vate the muscle, a description of the actions of this muscle.

The **fourth layer** of back muscles is constituted entirely by the splenius capitis and **splenius cervicis** muscles. Review the splenius capitis (page 237). The splenius cervicis similarly takes origin from vertebral spines (T1–T6) and is inserted into cervical transverse processes.

The **fifth layer** of back muscles comprises the **sacrospinalis (erector spinae)** muscle group (figs. 64, page 128, 84, page 160, and 129, page 244). This massive muscle complex comprises three vertical columns on either side of the median spinal furrow, extending from a very broad, common tendinous origin from the back of the sacrum, the iliac crests, and the sacral and lower lumbar vertebral spines, all the way up the back to the cervical vertebrae and the occipital region of the skull. The three columns are difficult to distinguish in the living, and will be studied more fully in dissection. The actions of the muscle columns can, however, readily be observed. Ask your model to extend his or her head against resistance, and palpate the upper portion of the muscle mass just lateral to the median spinal furrow. Then ask your model to extend the lower portion of the back against resistance, and similarly palpate the sacrospinalis muscles in the thoracic and lumbar regions. Now repeat the exercise previously described for study of these muscles (page 158), (i.e., to palpate the muscle mass as your model repeatedly shifts the weight of his or her body from one to the other lower limb by standing on one foot at a time). Extend this examination by palpating as your model walks slowly forwards. The muscle tenses on one side as the model places weight on the lower limb of the opposite side. This exercise demonstrates one of the very important functions of this group of muscles; that is the elevation, through the muscle's insertion into the pelvis, of the lower limb of one side as weight is transferred to the other side in walking.

The **sixth and last layer** of back muscles is the **transversospinalis** or oblique (spiral) group. These muscles have, as the names imply, a spiral course of direction as compared to those of the previous layer which have a more or less vertical course. They consist mainly of small muscles passing between transverse processes of vertebrae at one level to spinous processes of vertebrae at a higher level; or spinous processes of vertebrae at a lower level to transverse processes of vertebrae at a higher level, thus creating the spiral course. One of the larger muscles of this group has previously been encountered, the semispinalis capitis, which you observed in the uppermost reaches of the occipital portion of the posterior triangle of the neck (page 237 and figs. 126 and 127, page 238, and fig. 129, page 244). The portion of the muscle which you previously dissected is close to its insertion just below the superior nuchal line, deep to trapezius. This insertion, although it extends out laterally from the midline, is homologous with the insertions, at lower levels, into spinous processes. The origin of the muscle is from transverse processes.

Also visible in the back, but not comprising a true back muscle, is **quadratus lumborum** (figs. 84, page 160 and 87, page 164; and pages 158, 159, and 163). This muscle receives its innervation from ventral rami, whereas most of the true back muscles receive innervation from dorsal rami. Note however, that the serratus muscles and latissimus dorsi, which are muscles of the upper limb and which receive their innervation from ventral rami, are nevertheless regarded as also belonging to the back musculature.

MOVEMENTS OF THE HEAD, NECK, AND BACK

Review by palpation, and enumerate, the muscles involved as your model performs each of the following movements of the head and neck: flexion; extension; lateral bending; and rotation about a vertical axis. In considering muscle actions, recall that these will differ depending on whether bilateral muscles are acting bilaterally or unilaterally, and on which attachment of the muscle is fixed (stabilized) and which is mobile.

Movements of the head and neck involve movements between individual cervical vertebrae, and movements between the head and the atlas. The atlantooccipital joint involves the occipital condyle, which articulates with a corresponding articular facet on the atlas. The main movements are flexion and extension. Rotation of the head occurs largely through rotation of the skull and atlas about the vertical axis formed by the **dens,** a peg-like upward projection of the second cervical vertebra. This vertebra is called the **axis.** The dens rotates within a ring formed by the anterior arch of the atlas and the transverse ligament of the atlas. Most rotational movement of the neck occurs at this joint, whereas the other cervical vertebrae have little rotation between them, but considerable flexion and extension.

Ask your model to rotate (twist) his or her back, and observe the angle of maximal rotation. Similarly observe the maximal angles of flexion, extension, and lateral bending. Which muscles are involved in each of these movements?

Movements in the back take place at intervertebral joints. These include the **synovial** joints between **articulating facets** of vertebrae, and the joints between vertebral bodies, in which **intervertebral disks** are interposed. These are fibrous disks of tissue which intervene between the **hyaline cartilage** which covers the superior and inferior surface of each vertebral body. Because cartilage is interposed between the bones participating in these joints they are **cartilaginous joints** (page 39). A cartilaginous joint in which a layer of fibrous tissue intervenes, as here, is called a **secondary cartilaginous joint** or **symphysis** (see also symphysis pubis, page 120). A cartilaginous joint without fibrous tissue is a **primary cartilaginous joint,** an example of which is the **sphenooccipital synchondrosis** (pages 39 and 61).

Laboratory Session 23

In this dissection your objectives are to study the functional anatomy of the posterior aspect of the neck and of the back. Commence by removing skin from the regions to be studied. This can be performed by three teams, one skinning the posterior aspect of the detached half head and neck specimen, and one working on either side of the back and attached half head and neck. To remove skin from the head and neck specimen, the incision previously made for the study of the posterior triangle of the neck, along the superior nuchal line, should be extended to the midline. On the back make a midline incision down to the lumbosacral joint, then out laterally, skirting the iliac crest on each side, to the midaxillary line. Now reflect the skin of the back out laterally on each side, dissecting superficially on one side, to study nerves (mainly **dorsal rami:** see pages 86, and 230) and vessels, and more deeply on the opposite side, to attain the muscular level more rapidly. Your skin reflection should include the shoulder region and should expose the muscles of the posterior wall of the axilla (pages 85 and 243). Instructions for the deep dissection continue on page 249.

On the detached half head, demonstrate the superior bony attachments of trapezius and sternocleidomastoid, as previously done on the opposite side. Review the manner in which the investing layer of deep cervical fascia splits to enclose sternocleidomastoid anteriorly and trapezius posteriorly. Then clear this fascia to expose the contents of the posterior triangle, including splenius capitis muscle, and, in the superior angle of the triangle, semispinalis capitis (pages 236 to 237 and figs. 125, 126, and 127, page 238). On the superficial aspect of the latter muscle, preserve the **greater occipital nerve** (a dorsal ramus), and the accessory nerve as it crosses the posterior triangle of the neck, superficial to the levator scapulae muscle, to pass deep to the superolateral edge of trapezius. The latter nerve should be traced proximally, deep to (or through) sternocleidomastoid.

Reflect sternocleidomastoid upwards and forwards, and trapezius upwards and backwards, to expose splenius capitis (page 237 and fig. 126, page 238). Clear connective tissue from splenius, and trace the muscle superiorly to its attachment just below the superior nuchal line. To do so, it will be necessary to dissect and reflect sternocleidomastoid extensively. Medial to the medial edge of splenius, identify and clear of connective tissue semispinalis capitis, and study the greater occipital nerve, which passes through the substance of semispinalis capitis, may similarly pierce trapezius, and passes superiorly to supply sensory fibers to the occipital region of the skull.

Splenius capitis should be mobilized, and connective tissue on its deep surface dissected, to allow you to reflect the muscle superolaterally. Preserve for subsequent study the blood vessels and nerves that you encounter on the deep surface of splenius. On the lateral edge of splenius, separate this muscle from levator scapulae. As previously mentioned (pages 231 and 236), levator scapulae takes origin from the posterior tubercles of cervical vertebrae. More specifically, it arises from the posterior aspect of the transverse processes of the atlas and axis, and the posterior tubercles of C3 and C4. The muscle is peculiar in that the portion of the muscle which arises most superiorly (i.e., from the atlas) inserts furthest inferiorly on the scapula, and vice versa. As a result, the four slips of the muscle are twisted in relationship to each other. Attempt to trace the portions arising from the atlas and axis to their respective origins. To do this palpate the transverse process of the atlas as previously performed on the living (page 242). Carefully clear connective tissue from the uppermost head of origin of the levator scapulae, and trace it up to the transverse process of the atlas. Then reflect the muscle anteriorly, and in the space between the anterior edge of splenius and the posterior edge of levator scapulae, identify **longissimus capitis,** which inserts onto the mastoid process, deep to splenius; and **longissimus cervicis,** which lies lateral to longissimus capitis, deep to levator scapulae, and attaches to cervical transverse processes.

Medial to splenius, on the superficial surface of semispinalis capitis, identify the terminal portion of the **occipital artery,** which ascends in company with the **greater occipital nerve** to supply the occiput. Trace the artery proximally, following it medially deep to splenius. Carefully reflect the latter muscle as far superiorly as is necessary to display the course of the artery, dissecting muscle and artery free from each other and from connective tissue as you do so. This dissection should demonstrate the path of the artery between splenius capitis and the semispinalis capitis. Trace the artery further anteromedially, to the site where it emerges from having passed deep to longissimus capitis. This muscle should now be fully mobilized, to further trace the course of the occipital artery anteriorly. Careful scissor dissection at the anterolateral edge of longissimus capitis will reveal the artery, lying just deep to the inferior border of the posterior belly of the digastric muscle (pages 58 and 237), which itself should now be carefully cleared and exposed. Continuity with the intermediate tendon, and through this, with the anterior belly, can now be established. Furthermore, the occipital artery can now be traced from the portion previously studied, to that part that has been dissected in this session.

Return to longissimus capitis and reflect the muscle upwards. In doing this, you will expose, just deep to the site where longissimus capitis attaches to the superior nuchal line, the small **inferior oblique muscle** of the head (**obliquus capitis inferior**). The muscle extends from the transverse process of the atlas downwards and medially toward the bifid (two-pronged) spinous process of the axis. Clear this muscle and trace it medialwards as it passes deep to semispinalis capitis. To facilitate adequate exposure of the inferior oblique muscle, detach the origin of semispinalis capitis from cervical transverse processes, to allow you to reflect this muscle upwards and medially. Try to avoid damage to deep lying nerves and vessels.

After reflecting the semispinalis capitis muscle, study the important nerves lying deep to it. Inferior to the obliquus capitis inferior, identify the dorsal ramus of C2, a branch of which continues up as the greater occipital nerve (page 248). Other branches supply neighbouring muscles, namely the splenius capitis, the semispinalis capitis and the longissimus capitis. Superior to the obliquus capitis inferior, the **suboccipital nerve** (C1) should now be dissected carefully. In so doing, and clearing connective tissue away, identify another small muscle, passing from the spinous process of the axis upwards and slightly laterally. This is the **rectus capitis posterior major,** which inserts in the area below the inferior nuchal line. Identify the site of insertion of this muscle on a skull now. Proceed with your dissection in this region to also identify the **obliquus capitis superior,** which originates on the transverse process of the atlas and passes upwards and medially to insert in the region between the superior and inferior nuchal lines. Thus the obliquus capitis inferior, rectus capitis posterior major and obliquus capitis superior enclose a triangle. This triangle is known as the **suboccipital triangle** (not to be confused with the occipital triangle, see page 226). The suboccipital nerve is thus an important content of the suboccipital triangle. Its branches to the muscles bordering the triangle, and to the **rectus capitis posterior minor,** which lies medial to the rectus capitis posterior major, should now be traced. The rectus capitis posterior minor originates from the vertebral spine of C1, and inserts below the inferior nuchal line, medial to its "major." Other important contents of the suboccipital triangle that should be identified in the course of your dissection are the suboccipital plexus of veins, and, deep in the triangle, the third part of the vertebral artery. Recall that you previously identified the vertebral artery elsewhere. The first part of the artery rises as a branch of the subclavian (page 233). The second part of the artery is that which traverses the foramina transversaria (page 233). The third part is that which you are now able to dissect, and is defined as the portion of the artery from where the artery emerges through the foramen transversarium of the atlas, and passes on the upper surface of the posterior arch of the atlas to disappear into the foramen magnum. The fourth part of the vertebral artery is seen within the cranium (page 199), and ascends to participate in the formation of the basilar artery.

Also observe the tough membrane here, which connects the posterior arch of the atlas to the posterior edge of the foramen magnum. This is called the **posterior atlantooccipital membrane,** which you should identify in relation to the medial portion of the posterior arch. The vertebral artery passes anterior, at the lateral edge of the latter, to enter the foramen magnum, and this relationship should be demonstrated in your dissection.

The last muscle which you should identify at this time is the **semispinalis cervicis,** the most superior attachment of which can be observed at the vertebral spine of C2. The muscle is therefore immediately inferior to obliquus capitis inferior, and has a characteristic tapering insertion into the C2 spinous process.

Also on the **detached side** of the back of the neck dissect in the subcutaneous fatty tissue to demonstrate as many of the cutaneous branches of the dorsal rami of spinal nerves as possible. Use atlas illustrations to help you locate the distribution of these nerves, which emerge close to the midline superiorly, and somewhat more laterally in the inferior segmental regions.

First Layer of Back Muscles

On the side on which you are performing a deep dissection, remove subcutaneous tissue, and identify the deep fascia covering the back muscles. Dissect this fascia free to demonstrate the superficial layer of back muscles: trapezius and latissimus dorsi (figs. 130 and 132, page 244, and pages 242 to 243). Dissect to demonstrate all the attachments of both trapezius (pages 242 to 243) and latissimus dorsi (page 243). Review the functions of both muscles (pages 243–243).

Inferolateral to the inferolateral border of latissimus dorsi, display the fibers of the external abdominal oblique muscle, and demonstrate the interdigitations of the slips of these two muscles on the lower ribs. Superiorly, demonstrate the attachment of the upper border of latissimus dorsi to the inferior angle of the scapula. Then dissect carefully on the posterior aspect of the scapula, and along the borders of this bone, to demonstrate the following muscles.

Second Layer of Back Muscles

Along the medial border of the scapula, inferior to the inferolateral border of trapezius, identify the rhomboid major muscle (page 243 and figs. 130 and 132, page 244). Rhomboid major is seen to insert along most of the medial border of the scapula, having taken its origin from the cervical vertebral spines of T1 to T5. Elevate the lateral border of trapezius, detaching it from its attachment to the spine of the scapula. Continue the detachment of the trapezius insertions along the medial edge of the acromion process and the posterior edge of the lateral third of the clavicle, avoiding damage to deep-lying structures by elevating the muscle from below in advance of dissecting its attachment free. In this way, preserve for subsequent study the **supraspinatus muscle** (a muscle of the upper limb, originating above the scapular spine) and the structures around the shoulder joint. In addition, the spinal accessory nerve, having crossed the posterior triangle of the neck and passed deep to the anterior edge of trapezius (page 236), will be found, in this dissection, on the deep surface of the muscle.

As you reflect the muscle medially, toward the midline, you will expose the full extent of the rhomboid major muscle, and the rhomboid minor muscle, superior to the major, originating from the spinous processes of C7 and T1, and inserting into the medial border of the scapula above the spine of the scapula. Also exposed by this reflection of trapezius is the lower portion of levator scapulae (pages 231, 236 to 237 and 243). The component parts of the levator scapulae should be carefully dissected from deep-lying structures, to expose the dorsal scapular nerve (a branch of the brachial plexus), and the deep branch of the transverse cervical artery, which you previously saw in the posterior triangle of the neck (pages 232 and 240).

Elevate the superior border of latissimus dorsi and mobilize the muscle from deep-lying tissues. Identify, on the deep surface of the muscle, the nerve to latissimus dorsi, called the **thoracodorsal nerve** (derived from the brachial plexus). The nerve reaches this site by passing anterior to the scapula and the **subscapularis muscle** (fig. 135, page 252), an upper limb muscle which can be palpated on the anterior surface of the scapula by inserting a finger deep to the inferior angle of the scapula. Progressively elevate and mobilize the latissimus dorsi, and then make a vertical incision through the belly of the muscle from its superior border to its inferolateral border. Elevate the two cut halves of the muscle in their respective directions. In reflecting the lower portion of the muscle toward the midline, you will detach, and thus be able to study, the attachments of the muscle to the lower ribs. Follow the lateral border of the muscle inferiorly, identifying the **lumbar triangle,** formed by the border of latissimus dorsi, the posterior border of the external abdominal oblique muscle, and the iliac crest. As you approach the midline in this reflection, you will discover that the tendonous origin of the latissimus dorsi merges with the tough, posterior layer of the thoracolumbar fascia (page 243).

Superiorly, make a vertical scissor incision through the rhomboid major and minor muscles, and reflect medial and lateral halves of these muscles toward their respective bony attachments.

Study the components of the levator scapulae muscle, confirming that the slip of the muscle originating from the first cervical vertebral spine has the lowest insertion on the scapula, and the slip with the lowest origin (C4) has the highest insertion, the four digitations therefore making a twist about each other. Elevate levator scapulae from deep-lying tissues.

Third Layer of Back Muscles

With the reflection of levator scapulae and the rhomboid muscles superiorly, and latissimus dorsi inferiorly, the serratus posterior superior and serratus posterior inferior muscles are exposed. Serratus posterior superior has a fiber direction similar to that of the rhomboids, and the muscle may inadvertently be elevated with the rhomboid muscles, because the serratus posterior muscles are vestigial. Similarly, the inferior of these muscles runs parallel with latissimus dorsi, and may similarly be elevated with the latter.

Pull the medial border of the scapula laterally, and examine, on the deep surface of the scapula, the origins, course, and insertions of the digitations of serratus anterior muscle (figs. 35 and 34, page 82 and pages 85 and 243). Observe the interdigitations of the insertions, into the lower ribs, with digitations of the external oblique muscle. Study the origins of serratus anterior muscle from the anterior surface of the medial border of the scapula. Note that the insertions, into the upper eight ribs, lie in the same plane as the insertion of the serratus posterior superior muscle, thus confirming the point previously made (page 243) that, in the morphological sense, the serratus muscles all belong in the same plane and are all muscles of the upper limb. Serratus anterior retains this relationship in the human, whereas serratus posterior superior and inferior have lost their connections with the limbs. Nevertheless, all these muscles retain their limb-type innervation (i.e., from ventral rami [page 230]). Using scissors, and where necessary a scalpel, cut the serratus anterior muscle close to its origin from the medial border of the scapula, in the entire length of the muscle, thus allowing further reflection of the scapula laterally and further access to the deep surface of the scapula.

Fourth Layer of Back Muscles

The fourth layer of muscles (i.e., splenius capitis and splenius cervicis) can now be studied. As previously seen on the detached half specimen of head and neck, splenius capitis inserts on the mastoid process, deep to sternocleidomastoid, and in the line extending posteriorly from that site. Splenius cervicis attaches to the first, second, and third cervical transverse processes. The origins of the muscle are from the lower half of the ligamentum nuchae and the upper thoracic vertebral spines. Review the functions of splenius (pages 246 and 237). To gain full exposure of splenius, and subsequent layers of muscle, cut through serratus posterior superior close to its midline attachment.

Fifth Layer of Back Muscles

The fifth layer of muscles of the back comprises the muscles of the three columns of the sacrospinalis group on each side of the midline (page 246).

Follow the slips of levator scapulae upward from their insertion in the scapula toward their origins from posterior tubercles of cervical transverse processes. Intimately related to levator scapulae is the uppermost portion of the lateral of the three columns, **costocervicalis** (fig. 129, page 244), which takes origin from ribs, and inserts into cervical transverse processes, close to the origin of levator scapulae. Just medial to costocervicalis, and anterior to it, lie **longissimus cervicis** and, more medially, **longissimus capitis.** The latter has previously been dissected at its insertion on the mastoid process (page 248 and fig. 127, page 238). Both longissimus muscles can be identified deep to serratus posterior superior (page 250), which crosses them superficially.

Deep to splenius, the semispinalis cervicis and semispinalis capitis, which have also been dissected at their superior attachments (pages 248 and 249 and fig. 127, page 238) can now be seen.

Inferiorly, incise serratus posterior inferior with a vertical scissor incision, and expose the common origin of the three columns of sacrospinalis. Trace the three columns superiorly, and identify the medial **spinalis,** the middle **longissimus,** and the lateral **iliocostalis** components. Trace all three components as far as possible on your cadaver, establishing continuity, where possible, with those components of the columns which you dissected superiorly in the back. Note that the longissimus is the only one of the three columns that reaches the cranium.

Sixth Layer of Back Muscles

The sixth layer is known as the **transversospinalis** or **oblique** layer. The fibers pass obliquely upwards and medially from transverse processes to spines of vertebrae and are segmentally arranged (i.e., span one interspace, or else are the result of fusion from several segments, and span several interspaces).

The **semispinalis** muscles (**semispinalis capitis, cervicis,** and **thoracis**) are the most superficial of the transversospinalis group of muscles and are also the largest, spanning five or more interspaces. Review semispinalis capitis (page 248 and Fig. 127, page 238). It lies deep to splenius and medial to the longissimus muscles. It arises from the tips of transverse processes of the upper sixth thoracic vertebrae and from the backs (articular processes) of the lower cervicle vertebrae, and inserts into the medial part of the area between the superior and inferior nuchal lines. Semispinalis cervicis also arises from the transverse processes of the upper thoracic vertebrae and inserts into the spines of cervicle vertebrae. The portion inserting into the axis is large, and gives the characteristic tapered appearance of the muscle. Semispinalis thoracis arises from lower thoracic transverse processes and is inserted into upper thoracic vertebral spines.

In a selected region on one side, extending a distance of about four vertebral segments, remove the sacrospinalis muscles to expose the small muscles of the transversospinalis layer. With the aid of atlas illustrations, identify **multifidus** fibers, which span several vertebral segments, passing from transverse processes to spinous processes.

The smallest intervertebral muscles (i.e., those spanning only one segment) are also the deepest of this group. They are the **rotators,** the **intertransverse** muscles, the **interspinal** muscles, and the **levator costarum muscles.** The rotators run from transverse processes to the vertebral lamina just above and are best developed in the thoracic region. The intertransverse muscles are small slips between the transverse processes of consecutive vertebrae. The interspinals are similarly small slips between spinous processes, and the levatores costarum pass from the tips of transverse processes to the ribs just below.

Upper Limb Muscles in the Back

The following muscles belong morphologically and

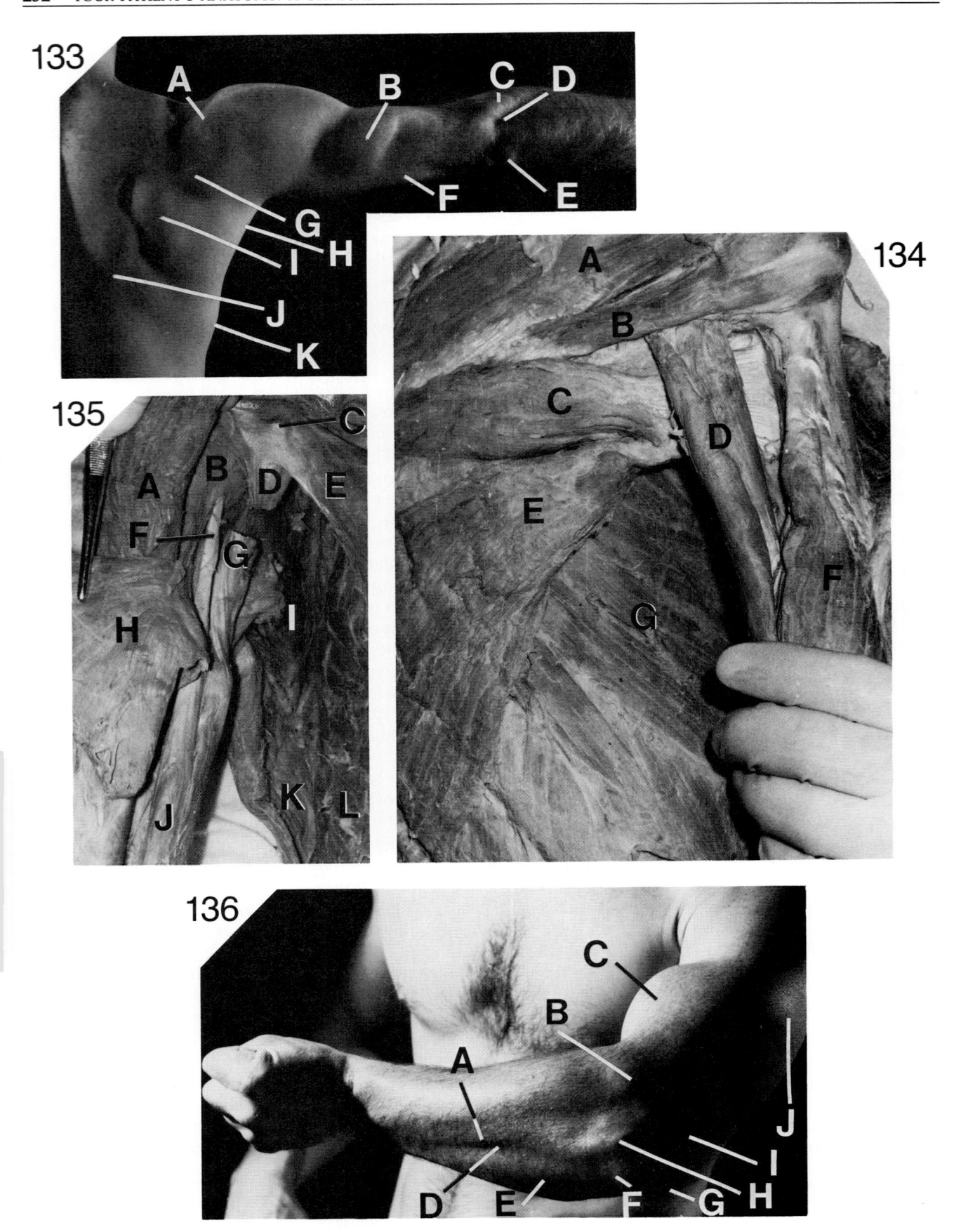

252   YOUR PATIENT'S ANATOMY: A CLINICAL VIEW OF HUMAN MORPHOLOGY

Figure 133. Posterior aspect of right shoulder, arm, and elbow regions.
A. Lateral (acromial) part of deltoid muscle
B. Lateral head of triceps muscle
C. Mobile muscle wad: brachioradialis, extensors carpi radialis longus and brevis
D. Depression in which head of radius and capitulum of humerus are palpable
E. Tip of olecranon process
F. Long head of triceps
G. Posterior part of deltoid (from spine of scapula)
H. Teres major
I. Infraspinatus
J. Inferolateral edge of trapezius, crossing rhomboid major
K. Latissimus dorsi

Figure 134. Posterior aspect of right shoulder and upper part of arm.
A. Infraspinatus muscle
B. Teres minor
C. Teres major
D. Long head of triceps
E. Latissimus dorsi
F. Lateral head of triceps
G. Serratus anterior muscle

Figure 135. Anterior aspect of right shoulder and arm.
A. Anterior (clavicular) part of deltoid
B. Capsule of shoulder joint
C. Coracoid process
D. Upper part of cut short head of biceps
E. Pectoralis minor
F. Tendon of long head of biceps in bicipital groove
G. Lower part of cut short head of biceps
H. Portion of pectoralis major, reflected laterally (in forceps) to display insertion
I. Subscapularis muscle
J. Belly of biceps brachii
K. Latissimus dorsi
L. Serratus anterior

Figure 136. Left arm and forearm.
A. Extensor carpi ulnaris
B. Brachioradialis
C. Biceps brachii
D. Posterior border of ulna
E. Flexor carpi ulnaris
F. Anconeus
G. Tip of olecranon process
H. Lateral epicondyle of humerus
I. Triceps
J. Lateral head of triceps

functionally to the upper limb, rather than the back, but, being in part situated in the back, their medial (proximal) attachments are appropriately studied together with the back muscles.

Review the positions, origins, insertions, actions, and nerve supplies of latissimus dorsi, trapezius, rhomboid major, rhomboid minor, and serratus anterior.

Identify **supraspinatus** (page 250) and mobilize it from the deep-lying surface of the scapula. Follow the muscle fibers out laterally as far as possible, until they disappear deep to the acromion process. Attempt to ascertain where they insert. With finger dissection, identify the superior border of the scapula, and, using scissors and scalpel dissection demonstrate the suprascapular nerve and vessels, the suprascapular ligament and notch, and the attachment of the inferior belly of the omohyoid muscle (page 227).

Mobilize the posterior part of **deltoid,** and incise its belly below and parallel to the spine of the scapula. Reflect the two portions of the muscle upwards and downwards respectively, to display deep-lying structures. Preserve the **axillary nerve and accompanying vessels,** which you will encounter on the deep surface of the inferior part of the deltoid muscle as you reflect this down toward the humerus.

These will be important landmarks for your further study of the upper limb.

Infraspinatus muscle is covered by a tough fascia, which you should dissect away to demonstrate the muscle. Follow its fibers out laterally toward their insertion on the humerus. Carefully separate the lower border of infraspinatus from the closely adherent, and often partly fused **teres minor** muscle. This muscle takes origin from the lateral border of the scapula and passes laterally to also insert on the humerus. It will subsequently be seen that supraspinatus, infraspinatus, and teres minor insert in sequence on the humerus to form a "lateral rotator cuff." Review the concept of lateral and medial rotation of the arm (page 85).

Teres major, which takes its origin from the inferior angle of the scapula, just below **teres minor,** is separated from the latter by the **long head of triceps.** Thus teres major passes anterior to the long head, to its insertion on the anterior surface of the humerus, while teres minor passes posterior to the long head of the triceps, to the back of the humerus at its superior end. Follow the tendons of insertion of latissimus dorsi and teres major toward the humerus, noting the manner in which the two tendons twist about each other to reach their destination (figs. 130 and 132, page 244, and fig. 134, page 252).

Clinical Session 24

■ **TOPICS**

Shoulder, Arm, and Elbow
Bony Landmarks
Muscles of Shoulder and Arm
Vessels and Nerves
Functional Anatomy and Movements

The limbs are basically somatic structures (pages 5 to 6). Their development is initiated as limb buds which appear to grow out from the body wall, ventral to the coronal morphological plane (page 230). Cells originating from somites migrate into the limb buds, to form limb muscles (and bones). The musculature is thus comparable in its origin to that of the body wall, and it similarly receives innervation from the somatic nervous system (page 6).

Many features of the muscles and bones of the limbs, and some arteries and nerves, are readily studied by inspection and palpation. Furthermore, muscle action can, in many cases, be observed or deduced by clinical examination of specific movements performed by a living model.

Bony Landmarks

With your model stripped to the waist and standing in the anatomical position review palpation of the scapula (pages 241 to 242). Having reidentified the general features of the scapula, repalpate the **spine of the scapula** and follow it out laterally to where it bends forward as the **acromion process** (fig. 122, page 228 and fig. 138, page 258). Palpate the articulation of the acromion process with the lateral end of the clavicle, and follow the sinuous course of the clavicle medially to its articulation with the sternum.

Review the palpation of the coracoid process of the scapula in the deltopectoral triangle (pages 85 and 242; fig. 33, page 82 and fig. 121, page 228).

You will recall that when you initially palpated the corocoid process it was pointed out that you should distinguish between it and the **lesser tubercle of the humerus** (page 85). This was done by asking the model to perform medial and later **rotation** (about a vertical axis). Since the lesser tubercle of the humerus will move with rotation of the upper limb, and the corocoid process, being part of the scapula, will not, the latter should remain stationary during this exercise. Repeat this palpation, this time to identify the lesser tubercle of the humerus.

The humerus is the long bone which constitutes the skeleton of the arm (that part of the upper limb which extends from the shoulder to the elbow). Like most long bones it consists of an elongated shaft, the **diaphysis** and, at each end, a somewhat expanded portion called the **epiphysis**. During growth and development of the long bones, the epiphyses are each separated from the diaphysis by a disk of cartilage, the **epiphyseal disk.** On the diaphyseal side of the epiphyseal cartilage, a region called the **metaphysis** is found, in which conversion of cartilaginous material to bone takes place. Growth in length of the bone is brought about by the cartilage of the disk growing, by cell division of cartilage cells, on the epiphyseal side of the disk. Growth in length ceases when the process of conversion of cartilage to bone (the metaphysis) "overtakes" the process of cell division in the epiphyseal cartilage (i.e., the epiphyseal disk becomes entirely converted to bone). At this time no further cartilaginous growth can take place (because there is no more unconverted cartilage left). This process is entirely similar to one you previously have been aware of in respect of growth in length of the base of the skull, at the sphenooccipital synchondrosis (page 61). Such synchondroses constitute **primary cartilaginous joints** (page 39).

In most long bones, the epiphyses are themselves converted to bone by several **centers of ossification** (bone formation), at the same time as the diaphysis is ossified. This usually results in there being several bony prominences at the expanded ends of the long bones. Of major importance, at the proximal (upper) end of the humerus, is the rounded **head** of the humerus (fig. 138, page 258), which faces medially to articulate with the slightly concave **glenoid surface** of the scapula. This creates a very mobile "ball and socket" synovial joint. Another such discrete prominence is the lesser tubercle of the humerus. In addition, palpate now the **greater tubercle of the humerus,** which lies lateral to the lesser tubercle. Commence by reidentifying the acromion process, and palpate its lateral edge. Inferior to this edge, the greater tubercle of the humerus constitutes an easily palpable bony prominence, which, like the lesser tubercle, will move on rotation of the limb. Consult Figure 138 and atlas illustrations to assist you in identifying this and the subsequent features of the humerus. Inferior to the greater tubercle, most of the lateral aspect of the humerus is difficult to palpate because of covering muscles. At the elbow, however, a distinct prominence, the **lateral epicondyle** (epi indicates above), is easily identified, and, above it, the sharp **supracondylar ridge.** Recall that condyle (knuckle) refers to the smooth, rounded bone ends which participate in articulations of the synovial type (pages 37 and 39). The lateral epicondyle, in accordance with this terminology, is situated just above a rounded component of the bone end called the **capitulum,** which participates in articulation with the **head of the radius.** The radius is the lateral of the two long bones of the forearm.

With the elbow joint extended in the anatomical position, identify on the posterior surface of the elbow a depression in which bony structures can be palpated (fig. 133, page 252). At the lateral edge, identify the lateral epicondyle and the lateral supracondylar ridge. Medial to the lateral epicondyle, the rounded capitulum can be palpated. Immediately inferior to the capitulum the joint space between the latter and the head of the radius can be felt as a groove. Keeping the upper limb in extension and the **arm** in the anatomical position, your model should turn the palm of the hand to face posteriorly by swinging the thumb across from a lateral to a medial position. The term **pronation** is used to describe this movement, in respect of the forearm only. The opposite movement, in which the thumb is moved back to a lateral position and the palm once again brought to face anteriorly, is called **supination.** Compare this to rotation of the arm (page 85). While your model repeatedly pronates and supinates his or her forearm, palpate the head of the radius, which you will be able to feel rotating about a vertical axis passing through it and the capitulum.

Medial to the lateral epicondyle, the head of the radius and the joint space between them, palpate on the posterior aspect (with the elbow in extension) the **olecranon process** of the **ulna.** The ulna is the medial of the two long bones of the forearm, and the olecranon process is situated at its proximal end. Palpate the olecranon process as your model flexes the elbow, and confirm that the process constitutes the point of the elbow in flexion. With the arm once more in extension, delimit the superior edge of the olecranon, and confirm that a joint space can be palpated superior to it. With the elbow in slight flexion, a depression in the posterior surface of the humerus, just above the olecranon process, can be palpated, into which the olecranon process fits in full extension. Medial to the olecranon process, the **medial epicondyle** of the humerus and its **medial supracondylar ridge** can also be palpated. (Anteriorly, the capitulum and the head of the radius cannot be palpated because of overlying muscles. Similarly, the trochlea [pulley], which is that part of the inferior end of the humerus which articulates with the ulna, is also covered by soft tissues.)

Above the medial supracondylar ridge, the humerus can be palpated on most of the medial aspect of the arm. Perform this examination on your model, reviewing at the same time the role of the humerus and related soft tissues of the arm in forming the lateral wall of the axilla (page 85).

Muscles of Shoulder and Arm

On the posterior aspect of the shoulder, reidentify by inspection and palpation the muscles of the scapula. (The scapula and clavicle together form the **upper limb girdle** [page 122], so named because it tethers the limb to the trunk. In the lower limb, homologously, the pelvis forms the limb girdle.)

Above the spine of the scapula, identify the **supraspinatus** muscle, which takes its origin from the scapula. Its tendon passes laterally, inferior to the **acromioclavicular joint,** to insert into the top of the greater tubercle of the humerus. Palpate the muscle as your model abducts (raises laterally) the upper limb, first to a horizontal and then into a vertical position. Is the supraspinatus activated in the first part or the latter part of this movement?

The **infraspinatus** muscle is palpable inferior to the spine of the scapula. It takes origin from the posterior surface of the scapula, and its tendon passes laterally, behind the shoulder joint, to insert a little below that of the previous muscle. Palpate both the supraspinatus and the infraspinatus while your model performs medial and lateral rotation of the upper limb (page 85). Which of these functions is executed by these two muscles?

With your model performing an appropriate exercise against appropriate resistance, palpate the activated levator scapulae muscle (page 231). With the model retracting and "winging" the scapula, palpate the **rhomboid major** and **rhomboid minor** muscles, medial to the medial border of the scapula, and review their muscle attachments (pages 243 and 250).

The **teres major** muscle (figs. 130 and 132, page 244 and fig. 134, page 252) should be inspected and palpated at the lateral border of the scapula. Ask your model to abduct the arm to a horizontal position and then horizontally adduct it (i.e., pull it in toward the chest wall, against resistance), and palpate the teres major as it passes out toward its insertion in the humerus. Recall that teres major and **latissimus dorsi** muscles both participate in the posterior wall of the axilla. In passing from the back to the arm (anterior surface of the humerus), these two muscles perform a twist such that the latissimus dorsi, which is posterior and inferior on the back, comes to lie anterior and superior as they approach the humerus. Review the entire extent, including origins, of latissimus dorsi (page 243 and figs. 130

and 132, page 244). (The **teres minor** muscle, which also takes its origin from the lateral border of the scapula, above teres major and inserts on the greater tubercle of the humerus, below infraspinatus, is not visible or palpable in the living.)

Review the origin, insertion, and action of serratus anterior (page 243). One other muscle of the scapula that cannot be studied in the living is the **subscapularis,** which takes origin from the anterior surface of the scapula and inserts into the lesser tubercle of the humerus.

Review the attachments of trapezius (pages 242 to 243). The superior portion of the muscle inserts on the superior border of the spine of the scapula; the medial border of the acromion process; and the posterior border of the lateral one third of the clavicle. Now identify the deltoid muscle. Previously (page 85) you identified the deltopectoral triangle (in which you palpated the coracoid process) as a space that separates muscle fibers of the pectoralis major and deltoid muscles (fig. 33, page 82). Referring to Figure 38, confirm that the muscle participates in forming the rounded prominence of the shoulder, by virtue of fibers which originate from the clavicle; fibers which are attached to the lateral edge of the acromion process; and, posteriorly, fibers which originate from the inferior border of the spine of the scapula. Thus the origins of the deltoid approximately circumscribe the insertions of the upper portion of the trapezius. Palpate the three separate origins of the deltoid muscle, and their convergence into a common insertion, on the lateral aspect of the humerus. Your model should perform the following three movements, while you palpate the muscle. In flexion of the arm (i.e., lifting the upper limb anteriorly) the clavicular fibers are mainly active. In lateral elevation of the arm (i.e., lifting the upper limb to the side) the acromial fibers of the muscle are mainly functional. The fibers which originate from the spine of the scapula are activated in extension of the arm (i.e., when your model raises the arm backwards from the anatomical position).

Examine the muscle group on the anterior surface of the arm. These muscles can be grasped by the examiner between thumb and fingers or, in the case of a larger model, between the fingers of two hands. The muscle mass is more readily mobilized from the humerus on the medial aspect than on the lateral. Three major flexor muscles constitute this group, two of which participate in flexion of the arm at the shoulder joint, and two of which flex the forearm at the elbow. Thus one muscle, the **biceps brachii,** participates in both actions (bi means two; ceps means heads; brachii means of the arm). The biceps brachii is the muscle which lies anteriorly, and is the one which is demonstrable on flexion of the forearm (i.e., is the muscle in which spinach induced hypertrophy in Popeye). As the name implies, this muscle has two heads. The **short head** takes its origin from the coracoid process and is palpable from that process to the site at which this head joins with the other on the anterior surface of the arm. The long head takes its tendinous origin from the **supraglenoid tubercle,** a small projection just above the **glenoid surface** of the scapula, onto which the large, rounded **head of the humerus** articulates. The tendon of the long head of biceps passes through the shoulder joint, and down onto the anterior surface of the humerus between the greater and lesser tubercles, in a groove called the **bicipital groove.** With the model flexing the forearm slightly against resistance, attempt to palpate the short head in its extent from the arm to the coracoid process, and the long head upwards as far as possible into the bicipital groove.

The bicipital groove is also the site of attachment of three other muscles you have previously studied. Review now the insertions of the pectoralis major muscle (page 84) into the lateral lip of the bicipital groove; of the latissimus dorsi (page 243) into the central area of the bicipital groove; and of teres major (page 254) into the medial lip of the groove.

Inferiorly, the biceps brachii muscle passes anterior to the elbow joint, to insert into a tubercle on the medial aspect of the radius. Because it passes in front of the joint, it acts as a flexor of the elbow joint. Because the tendon of the long head of biceps passes anterior to the shoulder joint, and because the short head also passes onto the anterior surface of the humerus, the biceps is a flexor of the shoulder joint. Finally, because it inserts on the medial aspect of the radius, the biceps also acts as a **supinator** of the forearm. Ask your model to supinate (page 256), while you palpate the biceps brachii muscle, and, especially, its tendinous insertion.

Coracobrachialis is a small flexor muscle of the arm. It takes origin from the coracoid process and passes medially to attach to the medial aspect of the humerus. With the upper limb abducted (raised laterally from the body wall) observe the longitudinal ridge, both visible and palpable, that extends from the coracoid process to this medial side of the arm. As it approaches the humerus this ridge converges with that formed by the short head of the biceps. A neurovascular bundle is palpable just posterior to the coracobrachialis.

The last and deepest of the muscles of the flexor group on the anterior surface of the arm is the **brachialis.** On the lateral aspect of the arm it is usually possible to observe a groove between the biceps brachii and the brachialis muscle, which takes it origin on the anterior surface of the humerus and in-

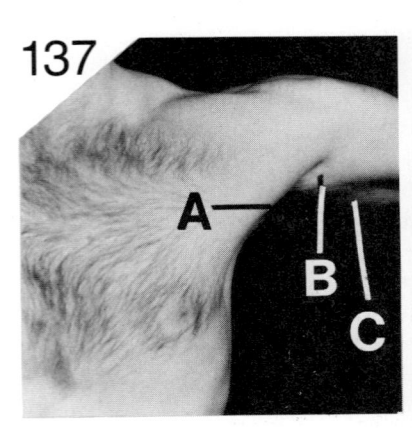

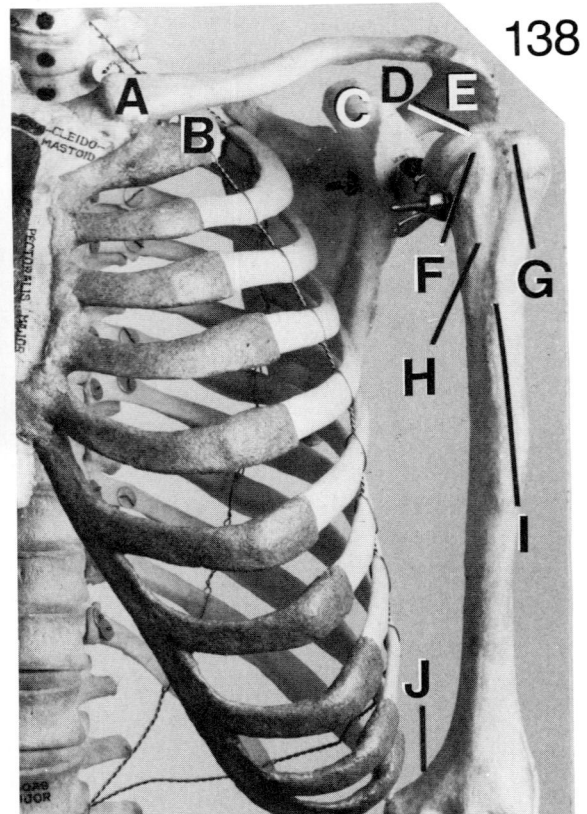

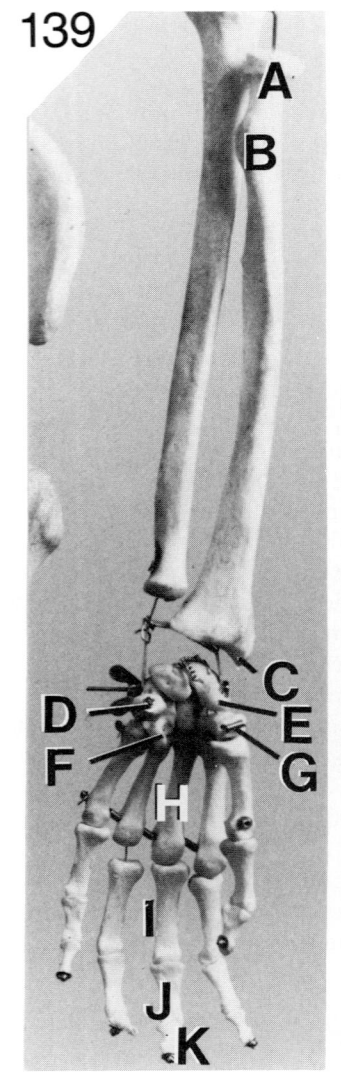

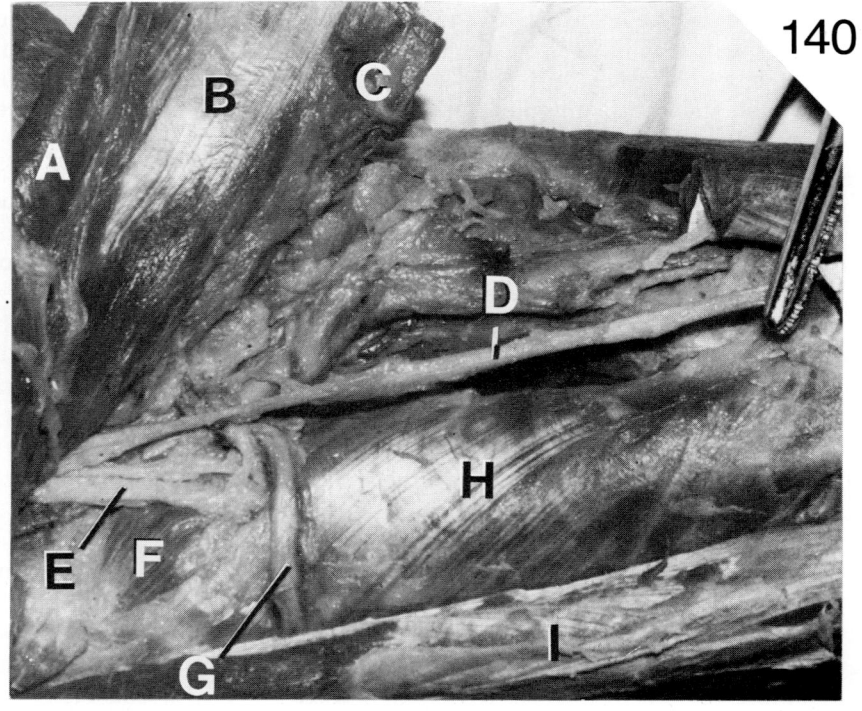

258 YOUR PATIENT'S ANATOMY: A CLINICAL VIEW OF HUMAN MORPHOLOGY

Figure 137. Anterior aspect of right axilla and arm.
A. Lateral edge of pectoralis major
B. Neurovascular bundle and coracobrachialis
C. Triceps

Figure 138. Anterior view of skeletal structures of right shoulder and arm.
A. Clavicle (medial end)
B. First rib
C. Coracoid process
D. Head of humerus
E. Acromion process
F. Lesser tubercle of humerus
G. Greater tubercle of humerus
H. Medial lip of bicipital groove
I. Lateral lip of bicipital groove
J. Medial supracondylar ridge

Figure 139. Skeleton of right forearm, wrist, and hand, seen from front, radius displaced distally.
A. Head of radius
B. Radial tuberosity
C. Styloid process of radius
D. Pisiform
E. Tubercle of scaphoid
F. Hook of hamate
G. Trapezium
H. Middle (third) metacarpal
I. Proximal phalanx of third digit
J. Middle phalanx of third digit
K. Distal phalanx of third digit

Figure 140. Right supinator muscle and relations, exposed and viewed from the lateral aspect.
A. Extensor carpi radialis longus, cut and reflected superomedially
B. Extensor carpi radialis brevis, cut and reflected
C. Brachioradialis, cut and reflected
D. Superficial branch of radial nerve (held in forceps)
E. Deep branch of radial nerve
F. Humeral head of supinator
G. Radial recurrent artery
H. Ulnar head of supinator
I. Extensor digitorum

serts on the ulna, just below the elbow joint. With the model flexing slightly against resistance, palpate the brachialis muscle on the lateral aspect of the arm, and distinguish it from the biceps brachii at this site. It is also palpable medially, just above the elbow (fig. 141, page 264).

The muscles on the extensor (posterior) aspect of the arm mainly comprise the three heads of **triceps**. With the upper limb extended in the anatomical position, identify the **long head of triceps** passing from the **infraglenoid tubercle** (just inferior to the glenoid facet of the scapula) on the posterior aspect of the humerus, down to the elbow. It inserts, in a common tendon for all three heads of triceps, into the posterior aspect of the olecranon process of the ulna. The **lateral head** (fig. 133, page 252) forms a ridge sloping downwards and laterally on the posterolateral aspect of the arm. The **medial head** creates a fullness medially. Previously you confirmed that on the medial aspect of the arm, the ventral musculature can more readily be displaced from the humerus than on the lateral aspect. Reidentify the soft tissue groove on the medial aspect, to confirm that the medial head of triceps lies posterior to this groove (fig. 141, page 264). Ask your model to flex the arm at the elbow and then extend the forearm against resistance, while you palpate the three heads of triceps.

Vessels and Nerves

Blood Vessels

The subclavian artery was previously palpated in the subclavian triangle, the lower and smaller subdivision of the posterior triangle of the neck (page 227). Reidentify the supraclavicular fossa (fig. 33, page 82 and pages 94, 226, and 227). The fossa extends beyond the subclavian triangle into the occipital subdivision of the posterior triangle and contains the subclavian artery, which you should palpate again now. After the artery leaves the lateral border of the first rib it changes its name to the **axillary artery** and, as previously indicated (page 227), enters the axilla. The artery passes through the axilla (i.e., posterior to the anterior axillary wall, consisting of the pectoralis muscles). As the artery leaves the axilla it again changes its name, this time to the **brachial artery**. The brachial artery is easily palpable in the groove on the medial aspect of the arm, which separates the anterior flexor muscles from the posterior extensor (triceps). Palpate the brachial artery at this site, and simultaneously palpate the wrist pulse. Demonstrate that by pressure of the brachial artery against the deep-lying humerus, the pulse at the wrist can be made to disappear (i.e., in a severe hemorrhage pressure at the site of the brachial artery could satisfactorily stem the flow).

Examine the surface of the arm and identify the two large superficial veins on the anterior aspect. Medially, the **basilic vein** extends halfway up the arm, to then pass to a deeper situation where it accompanies the axillary artery (fig. 130, page 244). On the lateral side of the anterior aspect of the arm, identify the **cephalic vein**, which passes upwards to the shoulder, then veers across the anterior aspect of the shoulder joint to attain the deltopectoral triangle, where it dives deeply (to subsequently pass through the clavipectoral fascia [page 122], to end in the axillary vein). (The latter continues upwards to become the subclavian vein.) Anterior to the elbow (in the space known as the **antecubital fossa**) the cephalic and basilic veins are usually connected by the **median cubital vein**, which slopes upwards and medially from cephalic to basilic.

Nerves

The brachial plexus was previously palpated in the posterior triangle of the neck (page 230). At that site, the roots of the plexus, comprising the ventral rami of C5 to T1, combine to form the upper trunk (C5 and C6), the middle trunk (C7), and the lower trunk (C8 and T1).

Posterior to the clavicle, as the nerves of the brachial plexus pass downwards and laterally to enter the axilla (page 230), each trunk divides into **ventral (anterior) and dorsal (posterior) divisions**. These are concerned with sensory supply to ventral and dorsal surfaces of the limb respectively, and motor supply to ventral (flexor) and dorsal (extensor) muscles, respectively. All three posterior divisions unite to form the **posterior cord** of the brachial plexus. The anterior divisions of the upper and middle trunks unite to form the **lateral cord**, while the anterior division of the lower trunk continues down on its own as the **medial cord** of the brachial plexus. These cords lie below the clavicle (i.e., in the axilla, deep to the pectoralis minor muscle, which, with pectoralis major, lies in the anterior wall of the axilla). The cords give off nerve branches which proceed down into the upper limb. The main branches of the brachial plexus take characteristic positions in relation to the axillary artery, and therefore are also palpable in the neurovascular bundle that lies just posterior to the coracobrachialis and the short head of the biceps (page 257 and fig. 137, page 258). Palpate this bundle on your model, and attempt to identify nerves, by their stringy consistency, surrounding the axillary artery in this site. One nerve, the **musculocutaneous nerve**, passes laterally to enter the substance of the coracobrachialis muscle, and may be palpable as it does so. This nerve is the one that supplies the three flexor muscles in the anterior compartment of the arm. The posterior cord may be palpable, again as a stringy structure, from the posterior aspect of the arm. Its major nerve branch, the **radial nerve,** curves around the humerus on its posterior aspect. An important nerve, the **ulnar nerve,** lying immediately medial to the axillary artery in the axilla passes posterior to the medial epicondyle of the humerus, where it can be palpated as a mobile, stringy structure. Gently "roll" the ulnar nerve on the posterior aspect of the medial epicondyle. In this site, the ulnar nerve is especially susceptible to trauma, and is the anatomical basis for the phenomenon known as the "funny bone."

Movements of the Shoulder Girdle

The shoulder joint is the most mobile of all the synovial joints of the body. The arm can perform flexion, extension, **abduction** (lateral movement, away from the median plane), **adduction** (medial movement, towards the median plane), medial and lateral rotation, and **circumduction** (in which the upper limb is moved in such a way that the hand circumscribes an approximate circle that forms the base of a cone of which the head of the humerus is the apex). In addition, movements of the upper limb also involve the upper limb girdle. The scapula is capable of being elevated, depressed, protracted, retracted, and rotated. Movement of the scapula necessarily also involves the clavicle, by virtue of the acromioclavicular joint. Therefore movement at the sternoclavicular joint also takes place in upper limb movement.

Ask your model to **abduct** the upper limb (i.e., raise it laterally). Palpate the muscles involved in this movement, and observe carefully the changes in position of the muscle masses and bony landmarks, including, among the latter, changes in position of the clavicle and scapula. In the initial stages of abduction, movement is mainly between the humerus and the scapula at the shoulder joint. The muscle primarily involved in the early stages is supraspinatus, and thereafter deltoid. Beyond the first 90 degrees of abduction, further movement involves **rotation of the scapula.** Aided by appropriate palpation, and applied resistance to the movement, deduce which muscles are further involved in the movement of elevation of the limb. **Adduction,** the opposite movement in which the limb is brought back to the anatomical position, is powerfully effected by pectoralis major and latissimus dorsi. If the arm is slightly flexed or extended in order to bring it in front of or behind the trunk, further adduction can take place.

Elevation of the scapula occurs when one shrugs the shoulders. Observe the changes in position of both scapula and clavicle as your model performs this movement, and, by palpation, determine which muscles are involved, as well as which muscles produce **depression.**

Protraction of the scapula is the movement caused by pushing. With your model pushing against a wall, study this action, which is primarily executed by serratus anterior muscle (pages 243 and 250). Retraction of the shoulder produces "winging" of the scapula. Which muscles perform this action?

Review the movements of flexion and extension at the shoulder joint, determining which muscles are involved in each. Review the movements of rotation of the upper limb about its vertical axis (page 255), and enumerate the muscles which might be involved in medial and lateral rotation respectively. Finally, observe and palpate your model's upper limb and shoulder girdle muscles during circumduction of the upper limb, and enumerate the muscles sequentially employed. (Pronation and supination of the forearm will be studied with that region of the upper limb.)

Laboratory Session 24

Remove the skin from the anterior and posterior surfaces of the arm and the elbow region and the antecubital fossa. Superiorly, the pectoralis major, deltoid, and shoulder muscles should also be exposed. When skinning the anterior surface of the arm identify the two large superficial veins previously observed in the living (page 260). Follow the cephalic vein proximally, to the site where it passes deep in the deltopectoral triangle (page 260 and fig. 33, page 82). Follow the vein into the triangle by a blunt finger dissection, elevating and mobilizing the superior edge of pectoralis major as you do so. Similarly elevate the inferior edge of pectoralis major, and, after establishing communication posterior to the muscle belly several centimeters medial to the arm, cut through the muscle with a vertical scalpel incision. This will allow you to abduct the upper limb, and approach the medial aspect of the arm. Elevate the pectoralis major and reidentify pectoralis minor. Review the clavipectoral fascia, which ensheaths the pectoralis minor, and which superiorly is pierced by the cephalic vein. The fascia also encloses the subclavius muscle superiorly, and is attached medially to the costoclavicular ligament (situated between the clavicle and first rib) and laterally to the coracoclavicular ligament (situated between the coracoid process and the clavicle).

Identify and palpate the coracobrachialis muscle, extending from the coracoid process to the medial aspect of the humerus (page 365). In the space between the coracobrachialis and pectoralis minor muscles, expose by finger dissection and gentle removal of connective tissue, the neurovascular structures that pass from the trunk to the upper limb. By definition, you are dissecting in the axilla (because the structures you are dissecting are posterior to the anterior wall of the axilla (i.e., the pectoral muscles) and are between the lateral and medial walls of the axilla. Review the boundaries of the axilla (page 227).

Study of the brachial plexus can be simplified by identifying three key nerves, which create a landmark by forming a configuration shaped like the letter "M." Commence by identifying, by careful scissor and scalpel dissection, a large nerve sloping downwards and laterally into the substance of the coracobrachialis muscle. This is the **musculocutaneous nerve** (page 260), and this nerve is the first limb of the "M" (the first vertical or left-sided downstroke, if you are dissecting a right upper limb, or the last and right-sided downstroke if you are dissecting a left upper limb). Trace the musculocutaneous nerve proximally until you encounter another nerve which appears to be branching off the same stem from which the musculocutaneous nerve has come (i.e., the first half downstroke of the "M" in a right upper limb or the second half downstroke in a left limb). This is the **lateral root** of the **median nerve.** Follow this root downwards, to identify the large median nerve and find the **medial root** of the median nerve (which, of course, forms the second half downstroke of the "M") in the right axilla and the first half stroke of a left-sided "M". By tracing the medial root of the median nerve proximally, one can, in the same fashion, identify the remaining limb of the "M" (the final downstroke of the right axilla or the first downstroke in the left axilla). This is also a larger nerve known as the **ulnar nerve,** and should now be traced down distally to its typical relationship with the medial epicondyle (page 261). Note that in tracing the ulnar nerve you will ascertain its passage through a layer of deep fascia, which inferiorly attaches to the medial supracondylar ridge (page 256). This layer of fascia, which lies in a frontal plane, divides the muscles of the arm into a ventral or **flexor compartment,** and a dorsal or **extensor compartment.** A similar sheet of fascia is found on the lateral side of the arm, attached to the lateral supracondylar ridge. Distinguish carefully between the ulnar nerve, which lies at a more posterior level than a slightly smaller nerve which has a similar course (the **medial cutaneous nerve of the forearm**). An even smaller nerve, the **medial cutaneous nerve of the arm,** also runs parallel with the above two.

Posterior to the ulnar nerve and the intermuscular septum that the ulnar nerve pierces inferiorly, identify the **medial head of the triceps** (page 260). Follow the medial head distally by finger dissection to its junction with the **long head,** and then trace the latter proximally to its origin.

Returning to the coracobrachialis muscle, define it from its origin at the coracoid process to its insertion into the humerus. Carefully separate it, by finger dissection, from the **short head of biceps,** which has a similar origin from the coracoid process. By following the short head of the biceps distally, identify the **long head,** with which the short head fuses, and mobilize the two-headed belly of the muscle from the deeper lying **brachialis** muscle (page 257). Follow the tendon of the long head of the biceps proximally, into the **bicipital groove** (fig. 135, page 252). To do so you will have to reflect the lateral half of the pectoralis major laterally and free it from surrounding tissues all the way to its attachment on the lateral lip of the bicipital groove. Now use finger dissection to elevate and mobilize the anterior edge of the deltoid. Separate the deltoid from deep-lying structures which include the capsule of the shoulder joint. When the muscle is sufficiently freed and elevated, use a scalpel to cut through the deltoid muscle about a centimeter away from and parallel to its clavicular and acromial attachments. Reflect deltoid laterally to expose the shoulder joint. You may encounter a small out-pocketing or out-pouching of **synovial membrane** lining the joint (the fluid-secreting membrane that lubricates the internal surface of the synovial joints [page 39]). The out-pouching is known as a **bursa,** and in this case the bursa lies deep to the deltoid muscle or deep to the acromion process (separating the latter from the supraspinatus muscle). The bursa is accordingly known as the **deltoid** or **subacromial bursa.**

Lift the tendon of the long head of biceps in one hand and incise the joint capsule with a scalpel. Trace the tendon proximally and observe the glistening synovial surface on the interior of the joint and the synovial fluid. Identify the tough **coracoacromial ligament,** which forms a superior reinforcement of the capsule of the joint. Insert a closed pair of forceps into the bicipital groove, anterior to the long tendon of biceps, as far up as you can, and then use a scalpel to incise between the limbs of the forceps, to cut through the capsule of the joint. Use the same technique to cut through the coracoacromial ligament, to enable you trace the tendon of the long head of biceps to its origin at the **supraglenoid tubercle** (page 257).

Returning to the axilla, identify the axillary artery and study its relationships to the nerves of the brachial plexus which you so far have identified: the musculocutaneous, median, and ulnar nerves, as well the medial cutaneous nerves of forearm and arm. Using finger dissection, dissect deep to the axillary artery to define the posterior wall of the axilla: the latissimus dorsi and teres major muscle insertions. Establish continuity with those portions of these muscles you have previously seen on the back (page 254).

The musculocutaneous nerve and the lateral root of the median nerve, which you used as your initial landmarks for studying the brachial plexus (page 262), are both branches of the **lateral cord** of the brachial plexus (page 260). The medial root of the median nerve comes from the **medial cord,** as does the ulnar nerve. You will recall from the previous description given (page 260) that medial and lateral cords are derived from anterior divisions of the trunks of the brachial plexus. These anterior divisions are concerned with flexion, and accordingly supply muscles on the ventral aspect of the upper limb. All the posterior divisions unite to form a **posterior cord,** responsible for innervating the dorsal muscles or extensors of the upper limb.

The major nerve of the posterior cord is the **radial nerve.** Identify this nerve posterior to the axillary artery. It is large, and slopes inferolaterally anterior to the teres major and latissimus dorsi. Reidentify the medial and long heads of the triceps, as previously, by following the ulnar nerve downwards and finding the medial head posterior to the ulnar nerve. By following the medial head inferiorly, its fusion with the long head should be reestablished. The radial nerve passes posterior to the upper fibers of the medial head and anterior to the long head and it curves around the posterior aspect of the humerus in the **radial groove,** which you should now identify on a skeleton. Follow the nerve through to the lateral aspect of the arm, where it pierces the intermuscular septum to become, briefly, anterior. (It returns to the dorsal surface in the forearm.) On the lateral aspect of the arm it can be found in the groove between the brachialis and the **lateral head of the triceps.**

Follow the radial nerve proximally, starting at its course behind the axillary artery, and anterior to the teres major and the latissimus dorsi. By tracing it proximally, you will find that, superiorly, its posterior relation is the **subscapularis muscle** (pages 250 and 257, and fig. 135, page 252). This muscle arises from the anterior surface of the scapula and its fibers pass laterally and converge to insert in the lesser tubercle of the humerus. Expose this portion of the muscle, and identify the **quadrangular space,** bounded superiorly by the lower border of the subscapularis, inferiorly by the upper border of the teres major muscle, laterally by the humerus, and medially by the long head of the triceps. Also emerging from the posterior cord is the thoracodorsal nerve, which should be identified and traced to the site where you previously studied it (page 353).

Another branch of the posterior cord of the brachial plexus, arising higher up than the radial nerve, passes through the quadrangular space. This is the **axillary nerve,** which winds around onto the posterior surface of the humerus, and supplies, among other structures, the deltoid muscle, where you previously found it (page 254). It also gives sensory branches to the shoulder. These may mediate referred diaphragmatic pain (page 146). A branch of the axillary artery (i.e., the **posterior circumflex humeral artery**) accompanies the axillary nerve.

The axillary artery has three parts, separated by the pectoralis minor muscle. Thus the first part is proximal to the medial border of the muscle. The second part lies posterior to the muscle and the third part lies distal to the lateral border of the muscle. The first part gives off one branch, the second part

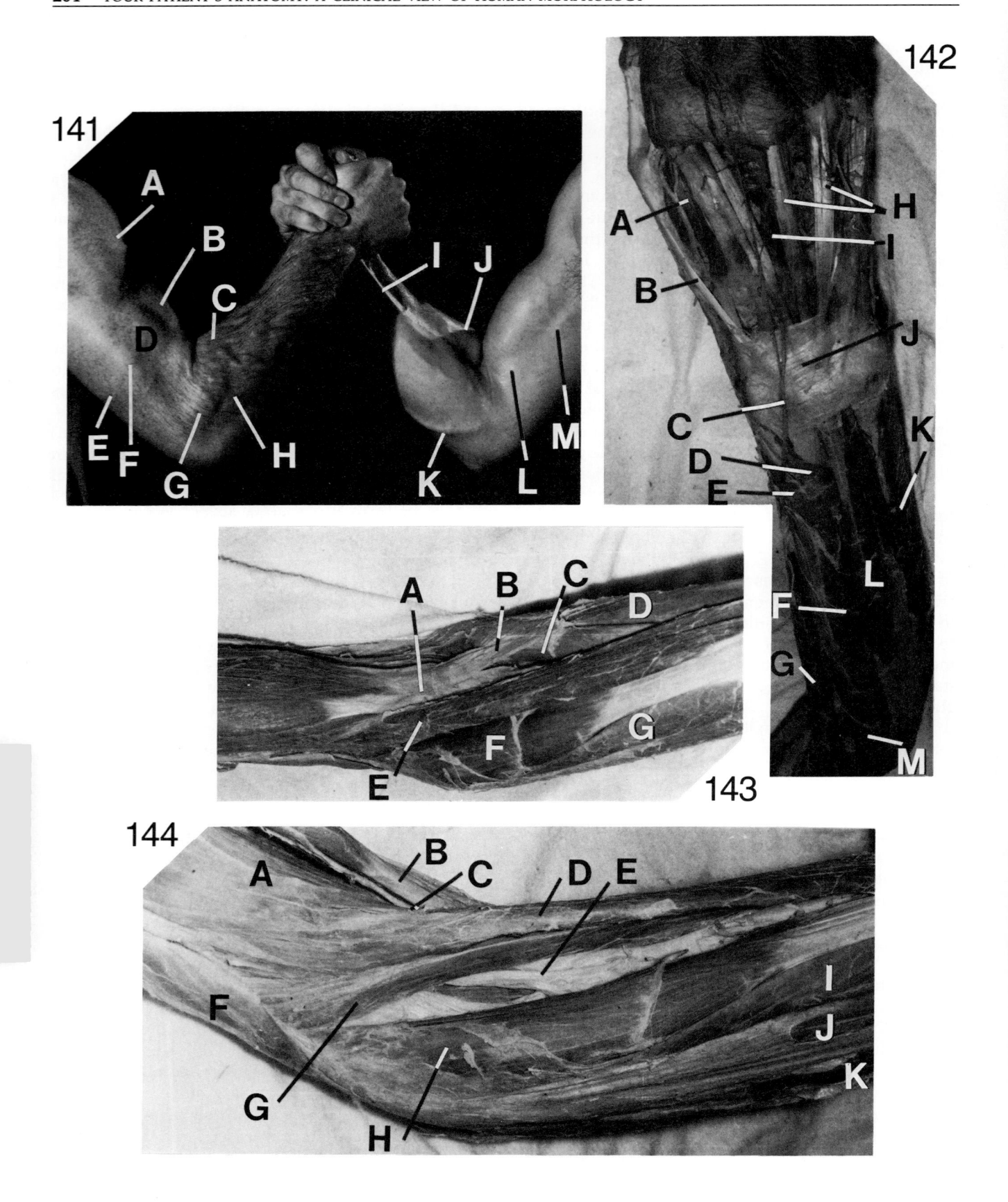

264 YOUR PATIENT'S ANATOMY: A CLINICAL VIEW OF HUMAN MORPHOLOGY

Figure 141. Lateral and medial aspects of right upper limb.
A. Deltoid
B. Biceps
C. Brachioradialis
D. Brachialis
E. Long head of triceps
F. Lateral head of triceps
G. Extensor carpi radialis longus
H. Extensor digitorum
I. Tendon of palmaris longus
J. Brachioradialis
K. Basilic vein
L. Brachialis
M. Medial head of triceps

Figure 142. Dorsal aspect of right hand, wrist, and forearm in pronation.
A. First dorsal interosseous muscle
B. Extensor pollicis longus
C. Superficial branch of radial nerve
D. Extensor pollicis brevis
E. Abductor pollicis longus
F. Extensor carpi radialis brevis
G. Brachioradialis
H. Tendons of extensor digitorum
I. Second dorsal interosseous
J. Extensor retinaculum
K. Extensor carpi ulnaris
L. Extensor digitorum
M. Extensor carpi radialis longus

Figure 143. Anterior aspect of right arm, antecubital fossa, and forearm.
A. Biceps tendon
B. Bicipital aponeurosis
C. Pronator teres muscle
D. Flexor muscles
E. Brachioradialis
F. Extensor carpi radialis longus
G. Extensor carpi radialis brevis

Figure 144. Lateral aspect of right arm and forearm.
A. Brachialis
B. Biceps tendon
C. Musculocutaneous nerve (cut at site of continuation with lateral cutaneous nerve of forearm)
D. Brachioradialis
E. Superficial branch of radial nerve
F. Triceps
G. Brachioradialis
H. Extensor carpi radialis longus
I. Extensor carpi radialis brevis
J. Extensor digitorum
K. Extensor carpi ulnaris

two branches, and the third part three branches, of which the posterior circumflex humeral artery is the first. The others are an **anterior circumflex humeral artery** (which anastomoses with the posterior circumflex humeral) and a **subscapular artery.** The latter anastomoses, around the scapula, with the dorsal scapular artery (page 233). This anastomosis may ensure blood supply to the limb in blockage of the subclavian-axillary trunk between these two branches. The branches of the second part are the **lateral thoracic artery,** which runs parallel to the lateral border of the pectoralis minor, (page 87) and a **thoracoacromial branch.** The branch of the first part is the **superior thoracic artery.**

Lymph glands, of clinical importance in such conditions as cancer of the breast, are found in relation to the axillary artery and its branches (see also pages 86 to 87 and 92). Identify these structures, which vary in size from a few millimeters to a centimeter or more in diameter. The lymph vessels connecting to them are usually small and fragile.

To complete your study of the axillary artery, note its relationship to the median nerve (page 262). Trace the artery (continuing as the **brachial artery**) and the median nerve down to the antecubital fossa.

Now turn the cadaver into the **prone** position (i.e., ventral surface facing downwards. When the cadaver is lying on the posterior surface, it is said to be in the **supine** position). Reidentify the quadrangular space which you dissected on the anterior surface. On the posterior surface the subscapularis muscle is not evident as a boundary of the space; instead the teres minor muscle forms this boundary. Furthermore, medial to the long head of triceps, a **triangular space** can also be identified, bounded by the medial border of the long head of triceps, the inferior border of teres minor and the superior border of teres major. The **circumflex scapular** branch of the subscapular branch of the axillary artery passes through the triangular space. Review the relationships of the long head of the triceps to teres minor, which passes posterior to the long head and the teres major, which passes anterior (fig. 134, page 252). Study the insertions of the supraspinatus, infraspinatus, and teres minor into the greater tubercle of the humerus in sequence from above and anteriorly to below and posteriorly, to form the lateral rotator cuff (pages 254 and 256).

Use finger dissection along the lateral border of the long head of the triceps to separate it from the lateral head of this muscle. Deep in the space so exposed and inferior to the teres major, reidentify the radial nerve which you previously saw in this situation on the anterior surface of the long head of the triceps. In the present situation you can study its curving course around the posterior surface of the humerus. This approach also gives you access to study the origins of the lateral and medial heads, and their respective relations to the radial groove, because the nerve curves around the humerus between the origins of these two heads. The nerve is accompanied by a large artery, the **profunda brachii** branch of the brachial artery (the continuation of the axillary artery).

Review the course of the ulnar nerve, this time as seen from the posterior aspect. Reidentify the medial head of triceps in the same way as previously (i.e., as a posterior relation of the lower part of the ulnar nerve in the arm). Now study the way the three heads of triceps merge to form a common tendon of insertion. Dissect this tendon down as far as its attachment to the olecranon process of the ulna. Mobilize the elbow joint by bending the forearm into flexion and back into extension a few times, and then test the function of the triceps muscle by tugging on the tendon of insertion, with the arm in slight flexion. Correlate the appearance of the lateral head of triceps, as seen in this dissection, with the appearance of this head on the posterior aspect of the arm in the living body (fig. 133, page 252).

Clinical Session 25

■ **TOPICS**

Forearm, Wrist, and Hand
Bony Landmarks
Muscles and Muscle Actions
Arteries and Nerves

The anatomy of the forearm is complicated by the fact that lateral and medial relationships change, and in some cases are reversed, between **pronation** and **supination** (page 256). Also, although the forearm is in supination in the anatomical position, it is mainly in pronation during most normal activity. Because supination relationships on one side resemble pronation relationships on the other, it is essential, throughout clinical and dissectional study of this region, to examine both right and left sides (e.g., on different models).

BONY LANDMARKS

Review the bony features of the upper ends of the **radius** and **ulna**, as palpated previously (page 256 and fig. 139, page 258). In the anatomical position, the radius is lateral and the ulna medial, and both participate in the elbow joint. Repalpate the movement of the head of the radius in repeated pronation and supination of the forearm. Recall that the head of the radius articulates with the capitulum of the lower end the humerus. In addition, palpate deeply on the medial aspect of the head of the radius, to identify its articulation with the ulna. A notch on the lateral aspect of the latter bone accommodates the edge of the head of the radius in this movement.

Palpate the posterior border of the ulna throughout the entire length of the forearm. At the lower end of the ulna, identify the somewhat bulky **head** at the lateral aspect of the ulna, and the **styloid process**, which is medial. The head makes a prominent elevation when the wrist is pronated. In supination, a groove can be palpated between the head and styloid process, and on extending the hand, cord-like tendons can be palpated in this groove. These are **extensor tendons** to the wrist and little finger.

The radius is not palpable through its entire extent in the forearm, but the lower portion of it is quite superficial and easily felt. At its lower end, on the lateral aspect, the radius also has a **styloid process**, which is somewhat smaller and sharper than that of the ulna and can be easily palpated on the lateral aspect of the wrist. By turning the wrist into a position of half pronation (i.e., with the thumb facing anteriorly), and by moving the thumb upwards and forewards away from the index finger; (a movement which constitutes extension of the thumb), the styloid process of the radius can be palpated in a groove or depression which is created between powerful tendons on the radial aspect of the wrist. The depression is known as the "anatomical snuffbox." With your model's wrist again in the anatomical position, and the thumb extended, repalpate the anatomical snuffbox and the styloid process, and identify the boundaries of the snuffbox. The lateral tendinous boundary consists of two tendons: most laterally the **abductor pollicis longus** and, just medial to it, the **extensor pollicis brevis.** The medial boundary of the depression is formed by the tendon of **extensor pollicies longus** (figs. 145 and 146, page 272).

The skeleton of the wrist is comprised of eight small, irregular bones, arranged in two rows, a proximal and a distal, each comprised of four bones (fig. 139, page 258). On the anterior aspect of the wrist (i.e., in the superior part of the **palm**), the most easily palpable of these eight **carpal bones** is the **pisiform,** situated just distal to the inferior end of the ulna, in the **hypothenar eminence** (the swelling on the medial or ulnar side of the proximal part of the palm). On the lateral side, just distal to the radius, in the corresponding **thenar eminence,** the **tubercle** of the **scaphoid** bone can also be palpated. Similarly, just distal to the pisiform and scaphoid, the **hook** of the **hamate** (on the medial side) and the **trapezium** (on the lateral side can be identified by deep palpation. A tough fibrous band, the **flexor retinaculum,** attaches to these four palpable carpal bony structures. It serves as a tether, deep to which flexor tendons of the hand and wrist pass. In certain pathological conditions excess pressure may occur on tendons, nerves, and arteries deep to the retinaculum, producing the painful "carpal tunnel syndrome," which may necessitate surgical incision of the retinaculum. The remaining four carpal bones are not usually easily felt on the anterior aspect of the wrist. Posteriorly, individual carpal bones are difficult to identify, but collectively the **carpus** (wrist skeleton) can be felt. An **extensor retinaculum** is found on this surface (fig. 142, page 264). Identify the remaining carpals (the capitate, trapezoid, lunate, and triquetral) on a skeleton, with the aid of an atlas illustration.

The hand skeleton comprises five **metacarpal** bones, all of which are quite easily felt on the dorsal aspect. Their distal ends are best identified by flexing the fingers (i.e., clenching the fist). In this situation, the inferior extremities or **heads** of the metacarpals form the **knuckles**. With the fingers lightly flexed, palpate just distal to the heads of the metacarpals to identify the joint cavities between them and the **bases** of the **proximal phalanges**, the first of the small bones of the fingers. The four medial **digits** (fingers) each have three phalanges, a **proximal, middle,** and **distal,** whereas the lateral digit (the thumb) has only a proximal and distal phalanx. Palpate the individual phalanges and their joint spaces.

Flexor Muscles of the Forearm

The flexor surface of the upper limb is ventral, and muscles of the ventral aspect of the forearm are mainly flexors of the wrist and fingers. In addition, flexors of the elbow joint, the muscle bellies of which lie in the arm, insert on the ventral aspect of the forearm. Reconfirm the latter fact by flexing the forearm at the elbow, and palpating the major **agonist** (page 225) of this movement, the biceps brachii. Now palpate the muscle mass of the forearm while your model flexes the hand at the wrist, and, with an extended wrist, flexes the fingers.

With your subject flexing strongly at the wrist and with a clenched fist (i.e., flexed fingers), palpate alternately at the medial and lateral sides of the elbow joint. Confirm that the flexor muscles are situated on the medial aspect of the forearm. In fact, they largely originate from the **medial epicondyle** of the **humerus** (page 256). With the fingers in strong flexion and the wrist in partial flexion, identify the following structures by inspection and by palpation, on the ventral aspect of the lower end of the forearm (figs. 141, page 264 and 148, page 272). (Tendons are felt as tough, cord-like structures. They can be distinguished from other, usually softer cord-like structures, such as nerves, by the fact that tendons can be made to tighten and harden by muscle contraction.) Medially, a tendon can be seen and palpated in its course toward the pisiform (page 267). This is the tendon of the **flexor carpi ulnaris** muscle (fig. 148, page 272). By repeated flexion of the wrist against slight resistance, make your model activate this muscle, while you palpate its entire extent. This muscle has two heads. One takes origin from the anterior surface of the medial epicondyle of the humerus, and the other arises from the medial aspect of the olecranon and the posterior border of the ulna. You will recall that the **ulna nerve** passes posterior to the medial epicondyle of the humerus (page 261). As it proceeds down into the forearm, the ulna nerve passes into the space between these two heads.

Lateral to the tendon of flexor carpi ulnaris at the wrist, there is a slight depression, in which movements of flexor tendons to the fingers can be felt as the fingers are clenched and opened. There are two main flexor muscles to the fingers: the **flexor digitorum superficialis** (fig. 148, page 272) and the deeper lying **flexor digitorum profundus.** By flexing and extending the fingers, movement in the superficial flexor can be palpated to its origin, from the anterior aspect of the medial epicondyle of the humerus. A portion of the muscle also takes origin from the anterior surface of the radius. The deep flexor of the fingers takes origin from the anterior surface of the ulna bone, including its proximal extremity. The muscle also folds around the ulna and takes origin, by aponeurosis, from the posterior border of the bone.

More laterally, two superficial and prominent tendons are usually evident at about the middle of the wrist. The medial of these two is that of **palmaris longus** (fig. 141, page 264 and fig. 148, page 272). (The palmaris longus is absent in more than 10 percent of forearms, more frequently on the left than on the right, and more frequently in females.) When present, it takes origin from the medial epicondyle of the humerus, and ends as an aponeurosis (the **palmar aponeurosis**) in the superficial region of the palm of the hand. Lateral to the palmaris longus tendon, at the wrist, is the powerful tendon of **flexor carpi radialis** (figs. 146 and 148, page 272). With your model flexing the wrist powerfully against resistance, palpate this muscle, which takes its origin from the anterior surface of the medial epicondyle of the wrist and passes down to insert into the bases (proximal ends) of the second and third metacarpals (i.e., those corresponding to the index and middle fingers).

Laterally, the remaining tendon which is prominently visible at the wrist is that of the **abductor pollicis longus,** which you identified earlier (page 267) as the lateral boundary of the anatomical snuffbox. The abductor pollicis longus, however, originates on the dorsal aspect of the forearm and is not a flexor muscle.

Note that the thumb is morphologically set at right angles to the other four fingers. As a result, while flexion and extension of the other fingers entails a movement in a parasagittal plane, flexion and extension of the thumb takes place in a frontal plane. That is, from the anatomical position, flexion of the thumb carries the thumb medially, across the palm of the hand and extension brings it out laterally. Abduction of the thumb pulls it anteriorly, and adduction brings it back posteriorly. Palpate the deep-lying **flexor pollicis longus** which takes its origin mainly from the anterior aspect of the radius, as your model flexes the thumb against resistance.

Important relations of the tendons you have just identified on the ventral aspect of the wrist follow. Laterally, between the tendons of flexor carpi radialis and abductor pollicis longus, the radial pulse can be felt. This is the **radial artery,** which becomes superficial at this site. The radial artery is one of the two terminal branches of the brachial artery, which divides into the radial and ulnar arteries in the **antecubital fossa.** Palpate the radial pulse in your model. From this site, the radial artery passes dorsally, deep to the tendon of the abductor pollicis longus, to enter the anatomical snuffbox, where you should now also palpate it again. Returning to the anterior aspect of the wrist, note that the **median nerve,** which passes downwards from the site at which you previously identified it (page 266) to the wrist, is situated between the tendons of palmaris longus and flexor carpi radialis.

In lean and muscular individuals, the **pronator teres muscle** (fig. 143, page 264) may be palpable high up on the anterior aspect of the forearm. This muscle has two heads, the superficial of which takes its origin from the anterior aspect of the medial epicondyle of the humerus and passes inferolaterally to insert on the lateral aspect of the radius. (A deep head arises on the anterior aspect of the olecranon process of the ulna, and joins the other head.) With your model pronating the wrist strongly against resistance, palpate deeply at the site indicated, and attempt to identify this muscle. (Another pronator muscle, **pronator quadratus,** is situated distally at the wrist, but is not palpable in the living.)

Supination is in part performed, as mentioned previously, by the biceps brachii muscle (page 257). Palpate the biceps, and in particular its tendon of insertion into the tuberosity of the radius (on the medial aspect of the radius, just distal to the radial head), as your model supinates strongly against resistance.

Extensor Muscles of the Forearm

With your model's forearm in the anatomical position (i.e., in extension) reidentify the depression on the posterior aspect of the elbow, in which you previously palpated the head of the radius rotating on the capitulum of the humerus, in pronation and supination of the forearm. The lateral boundary of this depression is formed by a muscle mass, which you can grasp between the index finger and thumb of an examining hand (fig. 133, page 252). Three muscles, all of which take their origin from the supracondylar ridge, constitute this pad of muscles. When your model extends the wrist strongly, the lower part of this pad may be felt to tense: this is due to the extensor muscles of the radial side of the wrist, and they can be palpated from their origins at this ridge, along the length of the forearm to their insertions at the bases of the second and third metacarpals. These muscles are called **extensor carpi radialis longus** and **extensor carpi radialis brevis** (figs. 141 and 142, page 264). The longus takes it origin from the supracondylar ridge a little higher up than the brevis, and the longus inserts into the base of the second metacarpal and the brevis into the base of the third. Taking off even higher on the supracondylar ridge is a muscle called **brachioradialis,** which inserts on the lateral side of the lower end of the radius. Palpate this muscle as your model flexes the forearm at the elbow, against resistance (see figs. 133 and 136, page 252 and figs. 141 and 142, page 264). Note that this muscle, as with the previous two, receives its innervation from the radial nerve. These muscles are therefore extensors. However, brachioradialis, although morphologically an extensor, has, through evolutionary change, modified its insertion and accordingly acts as a flexor.

Most of the other extensor muscles of the forearm take their origin from the anterior surface of the lateral epicondyle of the humerus (compare with the flexor muscles, which originate from the medial epicondyle). Palpate deeply on the anterior surface of the lateral epicondyle (by displacing the mobile wad of the previously mentioned three muscles anteriorly), while your model extends the hand and fingers strongly, to confirm the origin of extensor muscles from this site.

The long extensors of the fingers thus originate under cover of the mobile wad, and then emerge on the posterior surface of the forearm. Palpate the tendons and their movements there as the model extends the fingers. These long finger extensors are called **extensor digitorum** (figs. 141 and 142, page 264), and the tendons run to the fingers to be inserted into the bases of the proximal, middle, and distal phalanges. In addition to extensor digitorum, a separate **extensor digiti minimi** passes to the little finger, but can rarely be distinguished from the extensor digitorum in the living. The next muscle, however, is easily palpable. The **extensor carpi ulnaris** (figs. 142 and 144, page 264) has two heads, one of which originates, in common with the other extensors, on the anterior surface of the lateral epicondyle of the humerus. The second head is attached by aponeurosis to the posterior margin of the ulnar bone. Palpate this posterior margin in the anatomical position, and identify the muscle, just lateral to the margin, as your model extends the wrist strongly. The tendon of insertion of extensor carpi ulnaris should be palpated as far as possible toward its distal attachment on the base of the fifth metacarpal.

Note that the aponeurosis by which this muscle takes its origin from the posterior border of the ulna is continuous, at that posterior border, with the aponeurosis through which flexor digitorum profundus attaches (page 268). In addition, flexor carpi ulnaris (page 268) also attaches at this site. Confirm by palpation on your model that both flexor carpi ulnaris and extensor carpi ulnaris can be felt to contract in close relationship to the posterior border of the ulna, in flexion and extension respectively.

To complete your examination of the superficial extensor muscles of the forearm, palpate the small triangular **anconeous** (fig. 136, page 252), from the lateral epicondyle to the lateral edge of the olecranon process, in powerful extension at the elbow.

There is a deep layer of muscles on the posterior surface of the forearm, the bellies of which cannot be palpated because they are covered by those you already have identified. However, the tendons of three of them have already been studied by you, at the sites where they emerge from under this cover of the superficial muscles. These three are the tendons that bound the anatomical snuffbox. Review abductor pollicis longus, extensor pollicis longus, and extensor pollicis brevis (page 267 and figs. 145 and 146, page 272). A fourth deep extensor muscle, also not identifiable on the living subject, is **extensor indicis,** to the index finger. Finally **supinator** (fig. 140, page 258) can be palpated, although covered by superficial muscles. It has a head that originates with the extensor muscles from the anterior surface of the lateral epicondyle, and which passes down to the lateral aspect of the upper part of the radius. A second head of the supinator should be palpated in the groove between the superior parts of the radius and ulna. Commence by reidentifying the head of the radius, in the depression posterior to the elbow joint. Inferior to the head, the first part of the shaft of the radius can be palpated, and the finger can be placed between it and the lateral aspect of the olecranon. Your model should supinate against strong resistance, and you will feel muscle fibers of supinator tense. These fibers take origin from the lateral edge of the olecranon process and wind around posterior to the shaft of the radius, to then insert on the lateral and anterior aspect close to where the other head also inserts. Review the fact (pages 257 to 269) that biceps brachii also participates in supination.

The various extensor muscles that originate from the anterior aspect of the lateral epicondyle do so largely through a common tendon, known as the **common extensor tendon.** As a result, this tendinous origin is subjected to strain through the action of many different muscles, and it is accordingly susceptible to injury. "Tennis elbow" is the colloquial term used for sprain and inflammation at this site.

Hand and Wrist

Examine the palmar surface of the wrist, hand, and fingers. The skin of this surface is characterized by various markings. On the skin covering the distal phalanges, characteristic minute ridges and intervening furrows create the dermatoglyphics, or fingerprints. These are genetically determined, and create specific patterns. Using a magnifying glass, examine the ridges of the pattern on your subject, and identify the minute openings of sweat glands, situated at intervals along the ridges. Similar dermatoglyphic patterns are also found on the palm, at the bases of the fingers as well as on the hypothenar eminence.

A second system of markings of the skin is the creases formed at specific sites that correspond to regions of flexion and extension (fig. 147, page 272). At the wrist, identify the **distal crease of the wrist,** which demarcates the proximal extent of the thenar and hypothenar eminences from the anterior surface of the forearm. Proximal to this, one and sometimes two other creases are found. These are known as the **middle wrist crease** and **proximal wrist crease.** The middle wrist crease, when present, usually indicates the approximate site of the distal extremity of the radius and ulna, and the articulation between these and the carpus. The distal wrist crease is accordingly situated within the carpus.

At the base of the fingers, identify the **proximal digital crease.** The metacarpophalangeal joint is situated proximal to the **crease,** a fact that you should confirm by flexing the fingers at this joint. The **middle digital crease** is, on the other hand (so to speak), situated proximal to the interphalangeal joint between the proximal and middle phalanges, and the **distal digital crease** is similarly proximal to the joint between the middle and distal phalanges.

In the palm itself, identify a distal transverse crease of the palm, which extends from the space between the index and middle fingers and curves toward the medial border of the palm. A proximal transverse palmar crease passes from the lateral border of the palm medially, at about the middle of the palm. A longitudinal palmar crease is usually present between the thenar and hypothenar eminences, and curves distally and laterally to meet the proximal transverse palmar crease. The transverse palmar creases do **not** usually extend across from medial to lateral borders of the hand. The presence of such a complete tranverse palmar crease is anomalous, and is often associated with certain syndromes of multiple congenital abnormality (particularly the Down syndrome).

You previously established (pages 267 to 268) that the metacarpals are easily palpable on the dorsal surface of the hand. That they are less superficial on the palmar surface is due to the fact that several small muscles are found anteriorly. Palpate the thenar eminence as your model **flexes** the thumb, as your model **abducts** the thumb, and as your model **opposes** the thumb to the little finger (i.e., rolls the thumb across to touch the tip of the little finger). The thenar eminence is made up of small muscles extending from the region of the wrist bones (carpals) to the thumb, and which perform **flexion, opposition,** and **abduction.** The small abductor in the thenar eminence, **abductor pollicis brevis** (fig. 148, page 272) assists the abductor pollicis longus which you previously studied as a relation to the anatomical snuffbox. Now palpate distally and deep to the thenar muscles as your model adducts the thumb: the **adductor pollicis** can be felt.

In the hypothenar eminence, palpate as your model **flexes** the little finger, **opposes** it toward the thumb and **abducts** it away from the hand. Note that in the hand the terms abduction and adduction refer to movement away from or toward the midline of the hand itself (i.e., a line running vertically through the middle finger). This differs from the use of the terms abduction and adduction in respect to the limbs themselves, where the midline being considered is that of the body.

In addition to the small muscles of the hypothenar and thenar eminences and the adductor pollicis, described above, three other groups of small muscles of the hand should now be examined. With your model extending the fingers as strongly as possible, identify in the palm, very close to the spaces between the fingers, small swellings due to the **lumbrical muscles** (figs. 147 and 148, page 272). Identify all four lumbricals: the first lumbrical between the thumb and index fingers; the second between the index finger and middle finger; the third between the middle finger and the ring finger; and the fourth between the ring and fifth fingers. These small muscles are unusual, in that they arise not from a skeletal structure, but from the tendons of the deep flexor muscle (flexor digitorum profundus). The small tendons of the lumbricals pass through the spaces between the fingers to insert with the tendons of extensor digitorum. Because of these unusual relationships, the lumbricals can act both as flexors at the metacarpophalangeal joint but can also, when the fingers are already extended, reinforce the action of the extensor digitorum on the interphalangeal joints.

The remaining small muscles of the palm are the interosseous muscles (inter means between; osseous means of bones. These muscles are so named because they are situated between the metacarpal bones). There are two groups, the **dorsal interosseous** (figs. 145 and 146, page 272 and fig. 142, page 264) and the **ventral (palmar) interosseous** muscles.

Your model should place his or her hand palm downwards on a flat surface, such as a desk top, and abduct the index finger away from the middle finger against resistance, while you palpate in the space between the metacarpals of the index finger and thumb. You are palpating the first dorsal interosseous, which is situated between these two metacarpal bones, and inserts at the base of the lateral aspect of the index finger so that it causes this abduction. Raise your model's hand, and palpate between your own index finger and thumb the muscle mass of the first dorsal interosseous. Within this muscle mass you may be able to also palpate the pulsation of the radial artery. You previously felt the radial pulse at the wrist, between the tendons of flexor carpi radialis longus and abductor pollicis longus. Subsequently, you followed the course of the artery deep to the tendon of the latter, to enter the anatomical snuffbox. From there, the artery passes deep to the tendons of extensor pollicis longus and brevis and pierces the muscle mass of the first dorsal interosseous muscle (fig. 146, page 272) where you can now palpate it. From this point, the radial artery enters the palm of the hand.

Replacing your model's hand palm downwards, similarly palpate the fourth dorsal interosseous muscle between the metacarpals of the ring and fifth finger, as the ring finger is abducted from the middle finger. Thus dorsal interosseous muscles abduct. The second and third dorsal interosseii are situated between the second and third, and third and fourth metacarpals respectively, and insert on either side of the base of the first phalanx of the middle finger.

With the model's hand now supinated, the first, second, fourth, and fifth digits should be adducted toward the middle finger, as you palpate on the appropriate side of the relevant metacarpal in each case (i.e., on the medial sides of the first and second metacarpals and the lateral sides of the fourth and fifth metacarpals). The palmar interosseous muscles which you may be able to palpate in this action insert on the appropriate side of the base of the first phalanx of the respective fingers. However, the palmar interosseii are smaller than the dorsal, and are not as easily palpable as the latter.

ARTERIES AND NERVES

The brachial artery divides into the radial and ulnar arteries in the antecubital fossa, but is not usually palpable there because of overlying muscles. The radial artery can be felt at the situations previously indicated above. The ulnar pulse is usually also

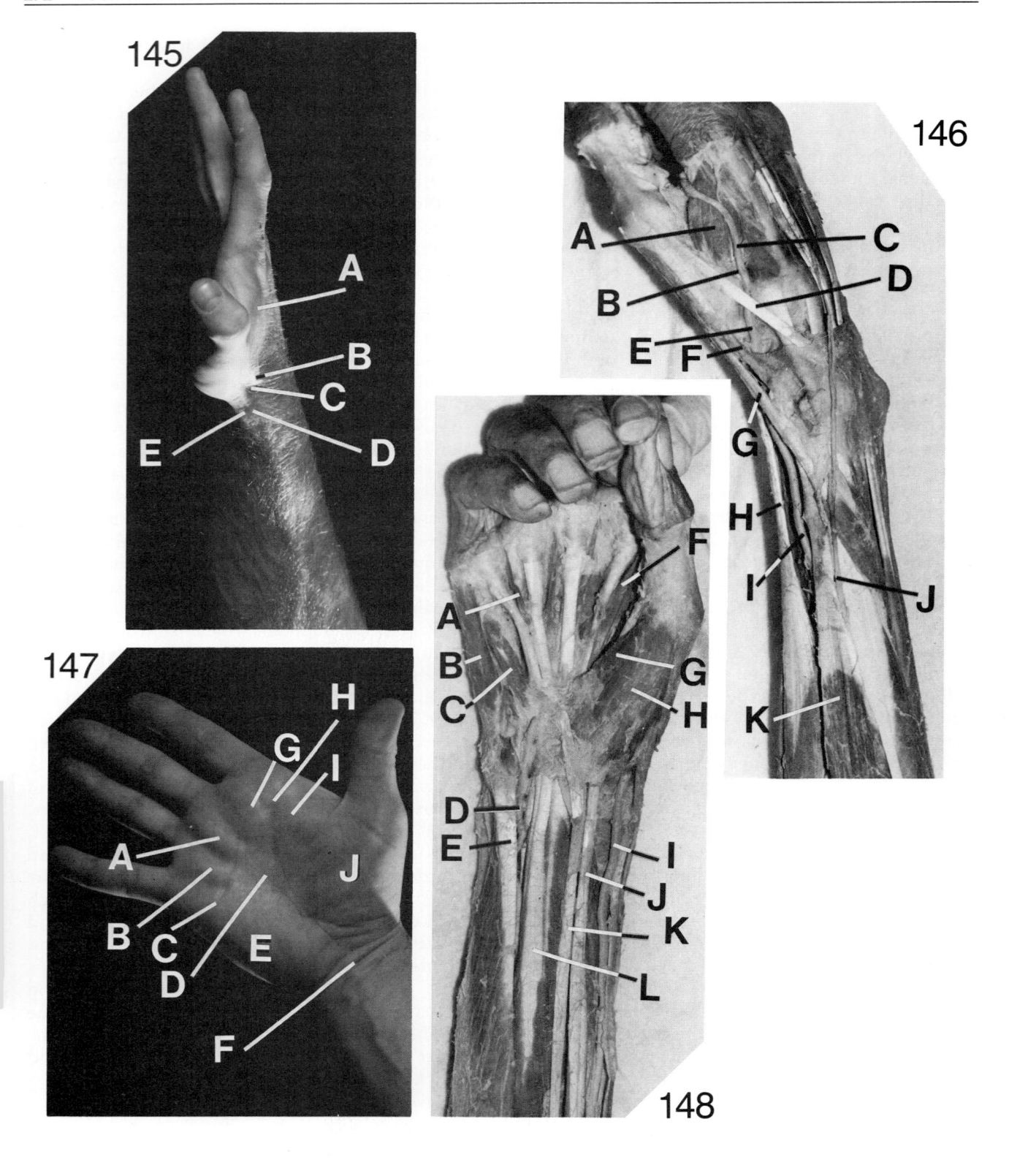

272 YOUR PATIENT'S ANATOMY: A CLINICAL VIEW OF HUMAN MORPHOLOGY

Figure 145. Lateral aspect of right hand and forearm.
A. First dorsal interosseous muscle
B. Tendon of extensor pollicis longus
C. Anatomical snuffbox
D. Tendon of extensor pollicis brevis
E. Tendon of abductor pollicis longus

Figure 146. Lateral aspect of right hand and forearm.
A. First dorsal interosseous muscle
B. Radial artery (passing through first dorsal interosseous muscle and giving off dorsalis indicis artery at this site)
C. Dorsalis indicis artery (branch of radial artery to index finger)
D. Tendon of extensor pollicis longus
E. Radial artery in anatomical snuffbox
F. Tendon of extensor pollicis brevis
G. Tendon of abductor pollicis longus
H. Flexor carpi radialis
I. Radial artery in wrist
J. Superficial branch of radial nerve
K. Brachioradialis muscle

Figure 147. Right palm and wrist.
A. Third lumbrical
B. Fourth lumbrical
C. Distal transverse palmar crease
D. Longitudinal palmar crease
E. Hypothenar eminence
F. Middle wrist crease
G. Second lumbrical
H. Proximal transverse palmar crease
I. First lumbrical
J. Thenar eminence

Figure 148. Ventral aspect of hand, wrist, and forearm.
A. Fourth lumbrical
B. Abductor digiti minimi
C. Flexor digiti minimi
D. Ulnar artery
E. Flexor carpi ulnaris
F. First lumbrical
G. Flexor pollicis brevis
H. Abductor pollicis brevis
I. Radial artery
J. Flexor carpi radialis
K. Plamaris longus (palmar aponeurosis has been removed)
L. Flexor digitorum superficialis

easily palpable in the wrist, just lateral to the tendon of the flexor carpi ulnaris. It was mentioned previously that the radial artery enters the palm. So too does the ulnar artery, and together they form, by anastomosis, a superficial and a deep palmar arterial arch. The arches give off branches to the hand and fingers.

The median and ulnar nerves are derived from ventral divisions of the brachial plexus (page 260), and supply the ventral aspect of the upper limb. Therefore, they are the nerves responsible for the innervation of the flexor muscles, as well as supplying the ventral skin surface with sensory fibers. Review the relationship of the ulnar nerve to the medial epicondyle of the humerus (page 261). The nerve proceeds, as mentioned previously (page 268) by passing between the heads of the flexor carpi ulnaris. As this course, and indeed the name of the nerve, imply, this nerve lies on the medial side of the forearm. Accordingly, it supplies the flexor muscles as well as sensory innervation on that side. In the hand, the palmar surface of the little finger and the medial half of the ring finger receive their supply from the ulnar nerve. The median nerve correspondingly supplies the flexor muscles and sensory innervation of the lateral part of the forearm, hand, and fingers.

The radial nerve is derived from posterior divisions of the brachial plexus, and is responsible for activating the extensor muscles on the dorsal surface of the limb. It also carries sensory fibers to most of this surface. In the hand, however, the ulnar nerve extends its sensory supply onto the dorsal aspect of the hand, in respect of the medial one-and-a-half fingers and the corresponding part of the hand. In addition, the median nerve also supplies a very small part of the dorsal surface of each finger, specifically the skin over the distal phalanges of the lateral three-and-a-half fingers.

The integrity of the sensory nerve supply to the hand and the rest of the upper limb should be tested, using the techniques described on page 191. In addition, with your subject's eyes closed, passively flex and extend his or her index finger repeatedly, finally placing it in an extended or flexed position, and ask your subject to state whether the finger is flexed (bent) or extended (straight). This tests the status of the sensory nerve tracts that convey position sense or **proprioception.** These pathways are specifically disrupted in certain neurological conditions.

Finally, test the integrity of the upper limb reflexes, as follows. First, review the concepts of the **reflex arc** (pages 5 to 6, 26 and 86) and the **spinal segment** (page 86 and fig. 51, page 104). Next, reidentify the tendons of insertion of the biceps brachii (page 257) and triceps (page 260). For the clinical procedure of testing the reflexes, you will need a **reflex hammer,** an instrument comprising a metal or wooden shaft and a rubber head. Holding it at the end of the shaft, the examiner strikes the target gently with the head of the hammer. With your model seated, place his or her upper limb at rest (for example on a table), and in partial passive flexion at the elbow. To test the **flexor reflex,** place the thumb of your palpating hand (the left, if you are right-handed) on the model's biceps brachii tendon, close to the insertion, and tap on your thumb with the hammer, thus transmitting the force of the strike to the tendon. This results in a stretching of the muscle. From sensory receptors that detect this stretch, an impulse is initiated and then conveyed in afferent nerves to the spinal cord. A reflex efferent impulse returns to the muscle, which then contracts. To test the **extensor reflex,** with your model positioned as above, strike gently, directly over the site of the triceps tendon, close to its insertion into the olecranon process.

Laboratory Session 25

Remove the skin of the forearm and hand, taking care to study superficial veins and nerves when encountered on the left side, and dissecting rapidly down to the layer of deep fascia and muscles on the right. The basilic, cephalic, and median cubital veins should be identified in the forearm, and their continuity into the arm studied (page 260). The basilic and cephalic veins commence from a venous plexus on the dorsal surface of the hand, and curve around onto the ventral aspect at the wrist. Cutaneous branches of the medial and lateral cutaneous nerves of the forearm will be encountered, as well as the superfical branches of the radial, ulna, and median nerves. Review the distribution of these (page 274), consulting atlas illustrations to assist you.

Now examine the muscles on the skinned ventral surface of the forearm.

Define the **antecubital fossa** (also sometimes known as the cubital fossa). It is the triangular area bounded superiorly by the imaginary line connecting the humeral epicondyles, laterally by the medial edge of brachioradialis muscle (page 269) and medially by the pronator teres muscle (page 269). Identify the latter two muscles from the descriptions given on page 269 recalling the exercises you carried out on the living body, in order to palpate them. The biceps brachii tendon enters the fossa, whereas the bicepetal aponeurosis fans out medially, to cover the surface of the pronator teres muscles (fig. 143, page 264). Elevate the lower part of the belly of the biceps brachii muscle, and cut across the belly with a scalpel. Reflect the lower portion of the muscle anteriorly to allow you to follow the tendon down to its insertion, on the radial tuberosity, and visualize the functions of the biceps brachii muscle in flexion and in supination.

The posterior aspect of the antecubital fossa is formed by the brachialis muscle. Expose the brachialis by freeing the medial edge of the pronator teres, and trace the brachialis to its insertion on the **coronoid process of the ulna.** Identify this site of attachment on the skeleton.

Incise the bicipetal aponeurosis to expose the pronator teres fully. Elevate the medial edge of the muscle, and examine the other important content of the antecubital fossa: the brachial artery and the median nerve. The brachial artery divides into the radial artery which passes laterally, and the ulna artery which passes medially. Follow the radial artery distally, as it veers out laterally to come to lie posterior to the medial edge of the brachioradialis muscle. Also deep to the brachioradialis muscle, identify the superficial branch of the radial nerve (figs. 140, page 258; 142, page 264, and 146, page 272) which lies lateral to the artery and, close to the wrist, escapes from behind the muscle to pass onto the side of the wrist, and subsequently onto the dorsal surface of the hand.

By loosening muscles gently from their deep fascial covering, study the muscles of the flexor surface of the forearm, referring to the description given on pages 267 to 269 and the illustrations given in Figures 141 and 143, page 264 and Figure 148, page 272. You should thus identify the brachioradialis, flexor carpi radialis, palmaris longus, flexor digitorum sublimis and profundus, and flexor carpi ulnaris. Just medial to the flexor carpi ulnaris identify the ulnar artery and lateral to this, the ulnar nerve. Gently dissect the ulnar nerve away from the artery, and trace the nerve proximally, under cover of the flexor carpi ulnaris muscle, back up to the medial epicondyle of the humerus, where you should be able to establish the relationship of the nerve posterior to the epicondyle, and its passage through the two heads of the flexor carpi ulnaris into the forearm (pages 261, 268 and 274). Following the descriptions given on pages 267–269 study the origins of the muscles of the forearm, and trace their tendons down to the level of the wrist.

Place the forearm in pronation, and identify the muscles on the dorsal surface of the forearm (figs. 141, 142, and 144, page 264). Commence with brachioradialis and the two muscles closely associated with it at their common origin from the supracondylar ridge, above the lateral epicondyle of the humerus: the extensor carpi radialis longus and extensor carpi radialis brevis. In these muscles and others you will be dissecting in the subsequent stages of this session, you will encounter considerable fusion of muscle bellies as you approach muscle origins. Carefully mobilize the brachioradialis muscle, freeing it from the deep-lying connective tissue, and taking care to preserve the superficial branch of the radial nerve lying posterior to the muscle. Under visualization, to avoid damaging the nerve, cut the belly of brachioradialis across with a scalpel, and reflect the proximal part of the muscle upwards, toward its origin from the supracondylar ridge. Now similarly mobilize extensor carpi radialis longus, cut it, and reflect it upwards and then do the same with the extensor carpi radialis brevis. This dissection will allow you to expose and study the supinator muscle (page 270, and fig. 140, page 258). Review the origin, attachments, insertion, and function of supinator (page 270). Also observe the relationships of the radial nerve to supinator: the nerve divides somewhat above this site, into its deep and superficial branches (fig. 140, page 258). The deep branch passes between the two heads of supinator, to the dorsal aspect of the forearm, where it supplies extensor muscles. The superficial branch continues downwards deep to the brachioradialis, where you have studied its further course.

To study the **elbow joint,** begin by detaching supinator from the **radial collateral ligament** and the **anular ligament.** The former extends from the lateral epicondyle (pages 255 to 256) to the anular ligament, which in turn encircles the head and neck of the radius (page 256) and attaches medially to the ulna. Mobilize the common extensor tendon from the lateral epicondyle and the flexor tendon from the medial epicondyle. Expose the triangular **ulnar collateral ligament,** the three components of which connect the medial epicondyle and the medial aspects of the coronoid process and the olecronon. Divide the biceps and brachialis tendons, and expose the **joint capsule.** Then make an incision into the capsule to display the **synovial membrane** and to study the joint.

Medial to the extensor carpi radialis brevis identify the extensor digitorum, and trace its bellies and tendons downwards, toward the wrist. Medial to this muscle, identify extensor carpi ulnaris. Trace its aponeurosis to the posterior border of the ulna, which is easily palpable on the cadaver as it is on the living body (pages 267 and 269). Note that medial to the posterior border of the ulna, the flexor carpi ulnaris is also visible and palpable, and it also attaches to this border by aponeurosis. Reestablish that the muscle you now are palpating medial to the posterior border of the ulna is the same one that you previously dissected on the ventral surface of the forearm, when studying the course of the ulnar nerve.

You should now attempt to display the extensor retinaculum (page 267 and fig. 142, page 264), on the dorsal aspect of the wrist. Follow the tendons as they pass deep to the retinaculum.

On the lateral aspect of the wrist, identify the tendons, and the related muscle bellies, of the boundaries of the anatomical snuffbox (page 267 and figs. 145 and 146, page 272). Define, by removing connective tissue surroundings, the abductor pollicis longus, and extensor pollicis longus and brevis muscles. Dissect in the snuffbox, to reidentify the radial artery, and palpate deeply in the space to identify the styloid process of the radius (page 267). The superficial branch of the radial nerve also usually crosses in the vicinity of the anatomical snuffbox.

Dissect the radial artery distally, from the anatomical snuffbox toward the space between the thumb and index finger. Remove connective tissue to identify, in this region, the first dorsal interosseous muscle. Follow the course of the artery through the two heads of the muscle (fig. 146, page 272).

On the ventral aspect of the wrist, identify the flexor retinaculum (page 267), by dissection in the deep fascia. As in the case of the extensor retinaculum, the flexor tendons pass deep to the retinaculum, in order to reach the hand.

In the palm, trace the palmaris longus tendon (if present) into the **palmar aponeurosis,** which lies quite superficially. After defining the aponeurosis, and following its extensions toward the fingers, dissect in the spaces between these to identify the **lumbrical** muscles (figs. 147 and 148, page 272) and the small arteries and nerves that run toward the fingers.

Remove connective tissue from the thenar and hypothenar eminences and identify the thenar and hypothenar muscles, making reference to the descriptions previously given (page 271) and to atlas illustrations. In dissecting the thenar eminence, take care to attempt to identify a small nerve entering the muscles from the distal aspect. This is the **recurrent branch** of the **median nerve.** By holding the thumb in the abducted position, dissect the adductor pollicis muscle, deep to the thenar eminence.

Reidentify the ulnar artery at the wrist, and trace its course distally, deep to a slip of the flexor retinaculum, into the palm. Follow its deep branch between the two superficial muscles of the hypothenar eminence, to where this branch participates in the **deep palmar arch.** (The radial artery completes the arch on the lateral aspect [page 274].)

Follow the flexor tendons of both flexor digitorum superficialis and flexor digitorum profundus forwards toward the fingers. In the fingers, note how the flexor superficialis tendon "splits" to allow the profundus tendon to become superficial. Accordingly, the profundus tendon continues down to the distal phalanges, whereas the superficialis tendon inserts into the ventral surface of the middle phalanges. Note that the flexor tendons are enclosed in tough fibrous **flexor sheaths,** which, like the retinacula, restrain the tendons from bowing. The tendons are also enclosed in a very thin synovial sheath, which maintains a lubricated surface between the outer surface of the tendons and the inner surface of the fibrous flexor sheath.

Complete your dissection of the palmar surface of the hand by identifying the small palmar interosseous muscles (page 271).

Turn again to the dorsal surface of the hand. Review your dissection of the first dorsal interosseous muscle and proceed to find the remaining dorsal interosseous muscles. Trace the tendons of the extensor digitorum beyond the level of the extensor retinaculum, in their pathway across the dorsal surface of the palm, toward the dorsal surface of the fingers. Trace the tendon of the middle finger all the way down to its insertion on the dorsal aspects of the proximal, middle, and distal phalanges. The broad, flattened aponeurosis which the tendon forms is known as the **extensor expansion.** The lumbrical and interosseous muscles insert into the extensor expansion close to the base of the proximal phalanx. Recall that the different interosseous muscles insert differently, and that the common function of dorsal interosseous muscles is abduction toward the middle finger while that of the palmar interosseous muscles is adduction.

Review the muscles and tendons of the thumb. These constitute a special case. Reidentify the borders of the anatomical snuffbox: the extensor pollicis longus medially and the extensor pollicis brevis laterally (figs. 145 and 146, page 272). Follow the extensor longus to its insertion on the distal phalanx. This is homologous with a tendon of the extensor digitorum. Extensor pollicis brevis inserts mainly on the base of the proximal phalanx of the thumb, though it usually also has a slip to the distal phalanx. Now study abductor pollicis longus, which reinforces the lateral boundary of the snuffbox (i.e., the extensor pollicis brevis), laterally. Trace this tendon to the base of the thumb metacarpal. On the ventral aspect of the thumb, examine the muscles of the thenar eminence (page 271). All of these muscles take the origin from the radial half of the flexor retinaculum. Flexor pollicis brevis passes to the radial side of the base of the proximal phalanx of the thumb. It inserts there via a minute **sesamoid bone** (a bone that develops within a tendon). Abductor pollicis brevis originates on the radial side of the flexor retinaculum, and inserts in common with flexor pollicis brevis on the radial side of the base of the proximal phalanx. The opponens pollicis has its origin from the front of the flexor retinaculum and inserts on the radial part of the palmar surface of the first metacarpal. The adductor pollicis muscle (page 271) has two heads, one originating from the anterior margin of the middle metacarpal, and the other from the middle of the carpus. The two heads converge to insert on the medial side of the base of the proximal phalanx of the thumb, also through a sesamoid bone. The radial artery passes between these two heads, as it emerges in the palm, having reached this site after having passed through the first dorsal interosseous muscle (page 271 and fig. 146, page 272). This first dorsal interosseous muscle also inserts on the medial aspect of the base of the proximal phalanx of the index finger. Finally, identify the flexor pollicis longus muscle, the actions of which you discerned on the living body (page 268). This muscle takes its origin from the anterior aspect of the radius (and a small slip from the coronoid process of the ulna) and its tendon inserts on the palmar surface of the base of the distal phalanx of the thumb. Being a flexor muscle on the lateral aspect of the forearm, it is innervated by a branch of the median nerve (page 274).

Identify the median nerve at its characteristic position in the wrist (page 269). Follow it distally into the palm of the hand, and trace its branches toward the lateral three-and-a-half fingers and the corresponding region of the palm. In accordance with the "division of labor" between the median nerve and the ulnar nerve, the lateral two lumbrical muscles are supplied by the median as are the muscles of the thenar eminence. The ulnar nerve correspondingly supplies the two medial lumbricals and the muscles of the hypothenar eminence. However, all the interosseous muscles are supplied by the ulnar nerve, and so is the adductor pollicis. Attempt to trace branches of the ulnar nerve to these muscles, and to its superficial supply of the medial one-and-a-half fingers and corresponding region of the hand. Attempt to dissect the deep palmar arterial arch and its branches to the fingers.

On the dorsal aspect of the hand, dissect the superficial branches of the radial and ulnar nerves, to their supply of sensory structures in their respective regions of the hand and of the fingers.

Study the wrist joint, which has three separate synovial joint compartments. Displacing tendons where necessary, incise the capsule anteriorly, from the radial styloid process to the ulnar styloid process (page 267). This exposes the **radiocarpal joint.** Identify the **articular disk,** which intervenes between the ulna and carpal bones. A second incision, 1 cm distal to the first will expose the **midcarpal joint,** which is situated between the two rows of carpal bones (page 267). Finally, study the third component of the wrist joint, the **carpometacarpal joint.**

The interphalangeal joints of the fingers are also synovial. Dissect open the capsule of one such joint. Note the **palmar ligaments,** which are thickenings on the ventral aspect of the joint capsules. Also study the **collateral ligaments** on each side of the joint, and determine what mechanism causes them to tighten in flexion and what effect this has.

Contents of Section 5

Lower Limb

26. **Homology of the Upper and Lower Limbs; Lower Limb Girdle and Bony Landmarks of the Thigh; Muscles and Movements of the Thigh; Vessels and Nerves of the Thigh**
    Clinical Session 26 281
    Laboratory Session 26A 289
    Laboratory Session 26B. 301

27. **Knee Joint; Popliteal Fossa; Bony Landmarks of Leg and Foot; Muscles of Leg and Foot; Vessels and Nerves of Leg and Foot**
    Clinical Session 27 308
    Laboratory Session 27A 315
    Laboratory Session 27B. 324

Clinical Session 26

■ **TOPICS**

Homology of the Upper and Lower Limbs
Lower Limb Girdle and Bony Landmarks of the Thigh
Muscles and Movements of the Thigh
Vessels and Nerves of the Thigh

HOMOLOGY OF THE UPPER AND LOWER LIMBS

It is a matter of common, everyday observation that there is a certain similarity between the upper and lower limbs in the human body. Consider, as you now make a preliminary examination of your model's lower limb, the correspondence between fingers and toes; hand and foot; wrist and ankle; forearm and leg; and arm and thigh (review definitions on pages 4 to 5). In morphological terms, the upper and lower limbs exhibit a degree of **homology** (structural similarity due to common genetic and developmental origin). In contrast, **analogy** means similarity of structure and/or function with different genetic and developmental origins. For example, the upper limbs of a human and the wings of a bird are **homologous,** in the sense of evolution from a common genetical origin (see also page 130). The wings of the bird and the bat are **analogous,** because they have similar structure and function, but they do not correspond in genetic and developmental origin.

An awareness of the morphological similarity between the upper and lower limbs will simplify understanding and learning of anatomical detail. In addition, by attempting to understand the contrasting evolutionary pathways that the limbs nevertheless have traveled to meet their remarkably different functional demands, you will acquire insight into the differences in normal function and susceptibility to pathology of the respective upper and lower limb components.

Both limbs originate embryologically as limb buds that grow out from the body wall, ventral to the coronal morphological plane (pages 86, 230, and 255). The musculature is accordingly innervated by the somatic nervous system (page 6), more specifically by ventral rami of mixed spinal nerves (page 230).

In addition to the regional homologies of the upper and lower limbs mentioned above, there is also a **lower limb girdle** that corresponds to the upper limb girdle defined on pages 122 and 256. The lower limb girdle comprises the **bony pelvis,** which, through its attachments to both the thigh and the lower portion of the vertebral column, is responsible for girdling or binding the lower limb to the trunk. In addition, the bony pelvis also houses the visceral structures belonging to the pelvis and perineum and has accordingly been studied as a part of the abdominal region. It was stressed early on in your study of the human body that subdivision of the body into organ systems and regions is arbitrary. There is necessarily overlap and continuity between regions. It is important to emphasize this continuity and to link your knowledge of one region with that of the others, thus integrating the information into an understandable body of knowledge that you can apply in practice. In the following, the components of the lower limb girdle will be identified. Attempt to correlate each part with its homologue in the upper limb girdle.

BONY LANDMARKS

For this clinical session your model should have the lower limbs and as much as possible of the pelvic region exposed, wearing at most a small bikini or undergarment. Examine your subject in the normal anatomical position where possible, or in the supine or prone position when necessary. Have an articulated skeleton available for reference. Begin by reviewing the identification and palpation of the bony landmarks of the pelvis, including the pelvic outlet. Relevant instructions for these exercises will be found as follows: crest of the ilium, including the intercristal plane (page 125; point 3 to 181, point 4); anterior superior iliac spine (page 126); posterior superior iliac spine (page 126, point 11); pubic symphysis (pages 120 and 125); pubic crest and pubic tubercle (page 126); sacrum (page 126); coccyx (page 168); ischial tuberosity (page 168); and the ischiopubic rami that form the pubic arch (page 168).

Review the following points on your model and the skeleton.

The posterior surface of the sacrum is continuous laterally with the posterior portion of the curving blade of the ilium. The broad surface gives rise, on either side, to the powerful **gluteal muscles** of the buttock (fig. 64, page 128). Identify the largest of these, the **gluteus maximus** (fig. 149, page 294), as your model tenses the buttock. The gluteal fold (page 168) is due to the inferior edge of this muscle (fig. 150, page 294) and the gluteal cleft (page 158) to its medial border.

Review the attachment of the anterior abdominal wall to the pubis and to the inguinal ligament, which extends from the pubic tubercle to the anterior superior iliac spine (page 126, points 10 to 13). The inguinal ligament demarcates the trunk from the thigh; for example the external iliac artery changes its name to the femoral artery as it passes deep to this ligament (page 169, point 6; and page 172). The ischiopubic ramus passes posterolaterally from the pubic symphysis toward the ischial tuberosity and has a surface that faces anterolaterally and inferiorly toward the thigh. This surface gives rise to the important adducter muscles (page 261) on the medial aspect of the thigh, and you should palpate this region briefly to identify this group of muscles. Now return to the ischial tuberosity, and note that several powerful muscles attach to it and pass inferiorly into the posterior surface of the thigh. What function will these muscles have? From the lateral aspect of the crest of the ilium, a muscular band arises and passes down onto the lateral aspect of the thigh, to act in abduction (page 261). Finally, to complete this preliminary palpation, with your model in the supine position, identify the anterior superior iliac spine, and palpate about 3 cm inferior to it, as your model slightly flexes (raises) the thigh. Here you will be able to identify another powerful muscle passing into the anterior aspect of the thigh.

In summary of this overview, note that you have identified muscles that connect the thigh to the pelvic girdle, attaching to the following surfaces: the posterior aspect of the sacrum and ilium; the lateral aspect of the crest of the ilium; the anterior superior iliac spine; and the anterolateral aspect of the ischiopubic ramus.

Certain parts of the **femur,** the long bone of the thigh, are palpable. Begin by identifying the **greater trochanter,** a bony prominence on the lateral aspect of the superior end of the femur. To locate its position, first reidentify the highest (most superior) point of the iliac crest (page 125, point 4). About 12 to 13 cm vertically below (inferior to) this point (more in taller subjects and less in shorter subjects), the greater trochanter of the femur can be identified by deep palpation on the lateral aspect of the thigh. What upper limb structure might be homologous to this prominence (page 255)? Having located the greater trochanter, palpate the ischial tuberosity with one hand and the anterior superior iliac spine with the other. In the supine position, the imaginary line between the latter two points, known as **Nelaton's line,** passes through the upper end of the greater trochanter (fig. 92, page 176). In certain pathological conditions, for example fracture of the **neck of the femur** (fig. 159, page 302), the position of the greater trochanter is displaced; its relationship to Nelaton's line provides a simple clinical test of whether the trochanter is in its normal position.

Much of the lateral aspect of the femur is palpable, from the greater trochanter to the inferior end of the femur. There the femur expands, on the lateral aspect, into a rounded **lateral condyle,** which articulates with the corresponding **lateral condyle of the tibia,** the larger of the two long bones of the leg. Review the meaning of the word condyle (page 37), and, referring to a skeleton for orientation, palpate the lateral condyles of the femur and tibia, as your subject flexes (bends) and extends (straightens) the knee joint. Review the meanings of the words flexion and extension (page 225). In fetal life, the lower limb undergoes a **rotation** that brings the morphological dorsal surface of the limb onto the anterior aspect of the body in the anatomical position. As a result, the lower part of the anterior aspect of the thigh, the anterior aspect of the leg, and the superior aspect of the foot are morphologically dorsal surfaces, and the posterior aspect of the thigh and leg and the inferior aspect of the foot are ventral surfaces. This explains why bending the lower limb at the knee joint constitutes flexion.

On the medial aspect of the knee, palpate the **medial condyle of the femur.** As in the case of the lateral condyle, it presents a rounded surface that faces inferiorly and somewhat posteriorly. This surface articulates with the **medial condyle of the tibia.** Palpate the medial femoral and tibial condyles during flexion and extension of the knee. In addition, a bony prominence called the **adductor tubercle** can be palpated, on the medial aspect, just above the medial condyle. This tubercle can be identified by asking your subject to **adduct** his or her thigh (i.e., move it toward the medial plane [page 261]) against resistance. Several powerful tendons will become visible and palpable on the medial aspect, just above the knee. The most anterior of these, due to a muscle called the **adductor magnus,** should be followed down inferiorly to where it inserts into the adductor tubercle, and upwards to where it takes origin from the ischial tuberosity.

One further bony prominence of the femur may be palpable in some subjects. On the posterior aspect, a few centimeters lateral to the ischial tuberosity, the **lesser trochanter** (fig. 159, page 302) may be discerned on deep palpation, on the posteromedial aspect of the upper end of the femur.

Other than those parts and structures mentioned above, the rest of the femur is by and large not easily palpable. Note that this is due to the abundance of large and strong muscles that cover the femur on all aspects. These will be studied in a later section. Note in passing that, in the anatomical position, the lower ends of the femurs are in contact (at the knee) whereas their upper ends are divergent (at the hips). Consider the possible functional consequences of this fact.

Articulation Between the Lower Limb and the Trunk

You previously reviewed palpation of specific parts of each of the three paired bones of the pelvis: the ilium, ischium, and pubis (pages 281–282). These parts were the iliac crest and its tubercle, the posterior superior iliac spine and anterior superior iliac spine, the ischial tuberosity and the ischial ramus, and the body of the pubic bone, inferior pubic ramus, pubic tubercle, and crest. With the aid of atlas illustrations and a skeletal specimen of the bony pelvis, identify the extent of each of these three bones of the pelvis, and visualize their relationships to each other. Note that although the three bones each have an independent existence in fetal life, in the adult they are fused together to form the **hip bone** of each side, and the three components display no mobility between each other (although the three bones can still be identified in adults).

The sacrum, situated between the hip bones, is triangular, with the base facing superiorly and the apex inferiorly (page 158), and as such the weight of the trunk, which rests on the superior surface of the sacrum, tends to force the sacrum down as a wedge between the two hip bones. Stability is created by very strong ligamentous bands between the two hip bones and the lower lumbar and sacral regions of the vertebral column (see the following paragraph). In this manner, the weight of the trunk is distributed to the two lower limbs, and the sacrum functions as an integrated and stabilized part of the **bony pelvis.** Indeed its anterior surface constitutes a wall of the true pelvis (page 169), and the **promontory** of the sacrum forms part of the **brim of the pelvis** (page 169).

The joints between the hip bones and the sacrum are, anteriorly, the **symphysis pubis** between each half pelvis (pages 120, 125, and 168), and, posteriorly, the **sacroiliac joints** between the ilium of each side and the sacrum. A certain degree of mobility is possible at these joints. In particular, slight increases in the dimensions of the pelvis are possible, due to such movement, in the female pelvis during birth. Furthermore, the sacroiliac joints constitute the main connection between the trunk of the body and the pelvis, and through it, the lower limbs. Finally, on each side the pelvis articulates with the femur.

Sacroiliac Joints

The importance of the sacroiliac joints is that, for the most part, they sacrifice mobility for stability, because they subserve the function of transmission of the weight of the body to the lower limbs. In the adult male there is very limited mobility; in the female, after puberty, the range is greater and is somewhat increased in the later months of pregnancy.

On both the isolated ilium and sacrum an ear-shaped, rough articular facet may be seen. In view of the uneven surface, it may appear surprising that the joint between these surfaces is of the synovial type (page 39). Powerful **ventral, dorsal,** and **interosseous sacroiliac ligaments** stabilize this joint. Furthermore, three extremely important ligaments, which extend from the lower vertebrae to the pelvis, reinforce the sacroiliac joint. The situations of these three ligaments should be visualized now on your model, referring to a pelvic skeleton and an atlas. For general orientation, repalpate the posterior superior iliac spine (PSIS), which you can identify by the dimple in which it is situated (page 126 and fig. 64, page 128). On each side palpate a more or less vertical ridge that extends downwards from the PSIS to the **posterior inferior iliac spine,** usually not easily palpable. The ridge represents the posterior edge of the ilium, at the sacroiliac joints. Medially, the posterior surface of the sacrum can be palpated. The bony points of attachment and, at least in some lean subjects, the ligaments themselves, can be discerned by deep palpation, as follows.

1. The **iliolumbar ligament** extends from the transverse process of the fifth lumbar vertebra to the crest of the ilium. It gives partial origin to the quadratus lumborum muscle (page 246) and is continuous with the thoracolumbar fascia (page 167).

2. The **sacrotuberous ligament** is approximately triangular in shape. The attachment of the base of the triangle extends vertically, from the posterior superior and inferior iliac spines, down the posterior surface of the sacrum and coccyx. The fibers converge laterally to an apex that inserts into the medial margin of the ischial tuberosity. The lower margin can usually be palpated between the coccyx and the tuberosity. This ligament is an extremely important landmark, playing a major role in numerous structural and functional relationships. For example, the gluteus maximus muscle (page 281) takes its origin from this ligament, as well as from adjacent bony structures.

3. The **sacrospinous ligament** lies anterior and medial to the sacrotuberous ligament. It, too, is more or less triangular in shape. Its apex is attached to the **spine of the ischium** (page 170), a somewhat sharp protrusion that points in medially, 1 to 2 cm superior to the ischial tuberosity, where it is palpable in some subjects. The base of the sacrospinous ligament is attached to the lateral edge of the sacrum and coccyx. Note that because the ischial spine is superior to the tuberosity, and also projects in further medially, the two

ligaments are quite widely separated at their ischial attachments.

Identify, first on the skeleton and, in approximate position on your subject, the **greater** and **lesser sciatic** notches, the spaces respectively above and below and medial to the ischial spine. Each is converted into the corresponding **sciatic foramen** by the attachment of the ligaments mentioned in numbers 2 and 3 above.

As an example of the functional stability of the pelvis, including the sacrum, review and repeat the exercise previously performed (pages 158 and 246) that demonstrated the way in which the **sacrospinalis** muscles play a role in tilting the pelvis in walking, and in so doing, in elevating the lower limb of one side as weight is transferred to the other.

Pubic Symphysis

The pubic symphysis (page 120) is a true symphysis or secondary cartilaginous joint. Review the concepts of primary and secondary cartilaginous joints (page 247). As in the case of the sacroiliac joints, the symphysis pubis normally has relatively little mobility, but, in postpubertal females, and in particular late in pregnancy and at delivery, some considerable mobility and expansion is possible. Palpate the symphysis pubis on your model, and locate the adjacent body of the pubic bone, the pubic tubercle and crest, and the inferior pubic ramus (page 168). These landmarks are of importance, because several muscles of the lower limb take their attachment in this vicinity.

Hip Joint

The joint between the femur and the bony pelvis is the hip joint. It is a ball and socket joint, the ball being the **head of the femur** and the socket being the **acetabulum,** a shallow depression on the side of the hip bone.

At its superior end the femur has a **neck** that passes superomedially, from the greater and lesser trochanters (page 282) at an angle of about 125 degrees (male) or less (female) to the shaft of the femur (fig. 159, page 302). Palpate the ischial tuberosity on your model with your right hand and the greater trochanter with your left hand, the latter to be used as a landmark. Moving your right hand lateral to the ischial tuberosity palpate superiorly, as deeply into the tissue as you can. Your hand will be able to approach the level of the greater trochanter, at which level it will encounter resistance. When your model flexes his or her hip joint (i.e., raises the upper portion of the anterior surface of the thigh toward the anterior abdominal wall), you will (in most subjects) be able to get an impression of the neck of the femur. The head of the femur is at the end of the neck and is not normally palpable.

Explore the movements possible at the hip joint, by asking your subject to flex the hip. Note that because of the **developmental rotation** of the lower limb, discussed on page 282, from about mid-thigh and inferiorly, the anterior aspect of the lower limb is the morphological dorsal surface. In the superior portion of the thigh, the anterior aspect is morphologically ventral, which accounts for the fact that the action of flexion of the thigh involves bringing that part of the thigh toward the ventral aspect of the abdomen. Extension of the hip joint is performed when the subject brings the thigh backwards from the anatomical position. **Abduction** takes place when the model raises his or her lower limb out laterally from the anatomical position, and **adduction** (page 282) is the movement in which the subject brings the lower limb back toward the midline. The thigh can also undergo **medial** and **lateral rotation,** about a vertical axis that passes through the head of the femur superiorly and the **intercondylar notch** between the condyles, inferiorly (compare with medial and lateral rotation of the upper limb, pages 85 and 255). Finally, **circumduction** can also be performed at the hip joint, though to a more limited degree than at the shoulder joint (page 261). This is because the socket of this ball and socket joint is considerably deeper than is the somewhat flattened glenoid fossa of the shoulder joint.

MUSCLES AND MOVEMENTS OF THE THIGH

Muscles Acting on the Hip Joint

Commence by studying the muscles involved in flexion at the hip joint. Palpate the anterior superior iliac spine (ASIS) (page 135), and the anterior inferior iliac spine, which can be identified by deep palpation 3 to 5 cm inferior to and slightly medial to ASIS. From each of these two spines an important and powerful flexor of the thigh originates, and each can be palpated as a discrete muscle bundle, several centimeters broad, as your model slowly flexes his or her thigh. The muscle originating from the ASIS is the **sartorius** (tailor), so called because it is active when the subject sits "tailor fashion." With your subject on an examination couch or on the floor in a sitting position with the knees extended (page 282), ask him or her to bend the lower limbs in to the tailor position as you palpate the muscle from its origin at the ASIS, downwards and medially as it crosses the thigh and passes inferior and posterior to the knee joint, to insert on the medial aspect of the medial condyle of the tibia (figs. 157, and 160, page 302; fig. 155, page 298). Note that the actions of the muscle are thus flexion of the thigh at the hip joint, flexion of the knee joint, and lateral rotation of the thigh. Because the muscle inserts quite far posteriorly, it has the action of flexing the knee joint, and by virtue of it crossing from a superior, anterior, and lateral origin to an inferior, posterior, and medial insertion, the direction of pull (superiorly and laterally toward the ASIS) will cause lateral rotation of the thigh.

From the anterior inferior iliac spine, the **straight head** of the **rectus femoris** muscle arises, and should also be palpated (fig. 160, page 302) as the subject flexes the thigh at the hip joint. (Rectus femoris has a second, **reflected head** of origin, from a groove superior to the acetabulum [page 284]. This origin cannot be palpated in the living and will be studied later.) As one palpates the rectus femoris inferiorly, it loses its discrete identity and merges into the muscle mass on the anterior aspect of the thigh, known as the **quadriceps femoris** (quadriceps means four heads, femoris means of [related to] the femur). This muscle has, as the name implies, four components, of which the rectus femoris is one. Of these, only the rectus femoris is a thigh flexor. By palpating the muscle mass on the anterior aspect at the thigh and following it inferiorly, confirm that the four muscles converge on the **quadriceps tendon,** which inserts into the superior edge of the **patella** (kneecap). Inferiorly the patella, in turn, is attached to the anterior aspect of the upper end of the tibia, by the **patellar tendon.** Indeed, developmentally, quadriceps femoris inserts into the tibia, and the patella develops secondarily within the tendon, as a **sesamoid bone** (page 277). Thus rectus femoris also has, in addition to its function as a flexor of the thigh, an action at the knee joint in common with the rest of quadriceps femoris. What main action do you think this is? In light of this action, what is the relevance of the point mentioned on page 282 concerning divergence of the upper ends of the femurs, to the potential movement of the patella when quadriceps femoris contracts? These questions will be returned to subsequently.

An additional important flexor of the thigh is the conjoined muscle formed by the union of the **psoas major** and **iliacus muscles** (page 163 and figs. 87 and 88, page 164). Review the situation and bony origins of these two muscles. The iliacus inserts mainly into the tendon of the psoas major, although some fibers continue independently to their common destination, which is the lesser trochanter of the femur (page 282). The muscle can be palpated by deep examination in the inguinal region, as follows. Reidentify the straight head of rectus femoris, originating on the anterior inferior iliac spine. Just medial to the rectus femoris, deep palpation allows one to feel tension in iliopsoas, as the subject flexes the thigh and laterally rotates it. A site where you are able to palpate the muscle is thus just medial to the anterior inferior iliac spine. In fact there is a groove for the muscle, which you should identify now on a skeleton, just anterior to the brim of the pelvis, between the anterior inferior iliac spine and the iliopubic (iliopectineal) eminence (page 169). Just inferiorly, the muscle passes anterior to the **hip joint** itself. Reidentify the lesser trochanter (page 286) and again palpate it while your model flexes the hip joint. To understand the action of iliopsoas and other muscles on rotation of the thigh, you should picture, with the aid of atlas illustrations, the vertical axis of rotation of the thigh as passing, as mentioned on page 284, through the head of the femur superiorly and through the intercondylar notch inferiorly. Muscles such as iliopsoas and the adductor group (to be discussed in the following paragraphs), that pass downwards and laterally from the pelvis, to insert on the posteromedial aspect of the femur, will tend to rotate the femur laterally when it is raised off the ground. However, when the foot is on the ground, these same muscles will tend to rotate the trunk of the body laterally.

The adductor muscles should now be palpated on the medial aspect of the thigh. Reidentify the pubic crest and pubic tubercle (page 126). With your model's thigh slightly abducted, palpate the origin of the powerful and very prominent **adductor longus,** immediately lateral to the pubic tubercle. As your model adducts the thigh, follow the muscle inferiorly and posteriorly (figs. 157 and 160, page 302). It inserts on the posterior aspect of the femur into the ridge known as the **medial lip** of the **linea aspera,** which should be identified on a femur with the aid of atlas illustrations. Two other muscles have their origins very close to that of the adductor longus. **Pectineus** is just lateral, taking origin from the pectineal line (pages 132 and 169), which extends from the pubic tubercle to the iliopectineal eminence (page 169). The muscle can also be palpated very deeply, as it, too, passes to its insertion high up on the medial lip of the linea aspera. The **gracilis** originates from the pubis, posterior to the adductor longus (fig. 158, page 302). Palpate it from its origin and down on the medial aspect of the thigh. It is a long, thin muscle (gracilis means graceful), that inserts on the medial condyle of the tibia, just posterior to sartorius (page 284 and fig. 155, page 298). With the model's thigh strongly abducted (i.e., with the adductor muscles on the stretch), the tendinous origin of gracilis may be seen and palpated one to two cm posterior to that of adductor longus.

Adductor magnus, which has two heads of origin, should now be palpated. The **adductor origin** of adductor magnus can be palpated on the posterior portion of the ramus, in continuation of the origin of the gracilis. Posteriorly, gracilis overlaps the adductor magnus, which is thus more extensive than may at first be appreciated. Indeed the bulk of the "adductor mass" of muscle that can be palpated on the medial aspect of the thigh as your model adducts the lower limb, is provided by the adductor magnus, as is implied by its name. The muscle inserts on the posterior aspect of the thigh, into the linea aspera, lateral to the insertion of the other adductor muscles.

In addition to the adductor origin of adductor magnus, there is also a **hamstring origin,** from the lateral aspect of the ischial tuberosity (page 282). The adductor and hamstring origins are continuous, curving from the lateral edge of the ischial ramus posterosuperiorly onto the tuberosity to form a U-shaped bony attachment to the ischium. Palpate the hamstring origin as your model extends the thigh, to appreciate the continuity between the two origins of the muscle. In general, the hamstring muscles are those that take origin from the ischial tuberosity and pass along the posterior aspect of the thigh, to reach the knee joint, which they flex. They also have the additional function of extending the hip joint. Bear in mind that the upper portion of the posterior aspect of the thigh is morphologically dorsal, which accounts for this movement being extension (pages 282 and 284). The morphological origin of the hamstring muscles is mainly ventral; their role in extension of the hip joint is a result of a late evolutionary modification. The hamstring component of the adductor magnus inserts in the adductor tubercle, which you previously palpated (page 282). Reidentify this tubercle on the medial aspect of the thigh close to the medial condyle, and palpate the tendon of the adductor magnus down to the adductor tubercle, just anterior to the sartorius and gracilis muscles, as your model extends the thigh.

Adductor brevis also originates from the ischiopubic ramus, deep to gracilis. It is not possible to palpate it as a separate muscle, but only as part of the adductor mass. It inserts into the upper part of the linea aspera, lateral to the insertion of the pectineus. Finally, with respect to the adductor muscles originating from the lateral aspect of the ishiopubic ramus, the **obturator externus** takes origin from the lower border of the **obturator foramen** of the hip bone, and the adjacent inferolateral surface of the pubis and ischium. It passes posteriorly and laterally, to wind round behind the neck of the femur and insert into the medial aspect of the greater trochanter.

The flexors and adductors of the hip joint are developmentally derived from a common muscle mass, and share innervation from ventral divisions of the nerves of the lumbar plexus. The entire plexus is, of course, as with other limb plexuses, derived from ventral rami (pages 86, 166–167, and 230). As in the case of the upper limb (page 260), these ventral **rami** then divide into **ventral divisions** that supply, in general, flexor and adductor muscles on the ventral surfaces, and **dorsal divisions** that supply extensor muscles on dorsal surfaces (pages 260 and 263). Thus the fact that the ventral divisions supply the true flexor pectineus as well as adductors, corresponds, in terms of upper and lower limb homology, to the fact that ventral divisions of the brachial plexus supply biceps brachii, brachialis, and coracobrachialis, the flexors at the shoulder joint, as well as upper limb adductors. Sartorius, rectus femoris and iliopsoas, supplied by dorsal divisions, become flexors because of the evolutionary modification of limb rotation.

For completeness you should, at this stage, get a general impression of the situation of the other components of the **quadriceps femoris** group of muscles, in addition to rectus femoris that you already studied (page 285). Using atlas illustrations, palpate **vastus medialis,** on the medial and anterior aspect of the thigh (figs. 157 and 160, page 302) and **vastus lateralis** on the lateral and anterior aspect (fig. 160, page 302). **Vastus intermedius** lies mainly on the anterior aspect, deep to the other heads, and is therefore not palpable as a separate structure. The four heads unite to insert in the superior border of the patella (page 285). These muscles act on the knee joint rather than on the hip joint. You have previously deduced (page 285) what action these muscles have on the knee.

Just as flexors and adductors of the hip joint derive from a common muscle mass, which is morphologically ventral, so are the major extensors and the abductors of the hip joint (the gluteal muscles) morphologically dorsal structures. They are innervated by dorsal divisions of the lumbar and sacral plexuses of nerves (see above). The **abductor muscles** of the hip joint should now be studied. **Gluteus maximus** (fig. 64, page 128; page 281; and figs. 149 and 150, page 294) should be palpated as the main muscle mass of the buttock, as your model extends the lower limb, rises from a sitting position and rises from touching his or her toes. This muscle takes origin from the posterior portion of the iliac crest, from the back of the sacrum, and from the sacrotuberous ligament (page 283). The fibers pass inferiorly and laterally, over the greater trochanter of the femur, and some insert into the **gluteal tuberosity,** a rough region on the posterior aspect of the femur that you should identify on the skeleton, inferior to the greatest trochanter, at the upper end of the lateral lip of the linea aspera. Most fibers of gluteus maximus proceed inferiorly to insert in a very strong fibrous band, known as the **iliotibial tract,** which can be palpated down to the lateral aspect of the knee joint. At this point it forms the more anterior of the two easily palpable and visible cord-like bands on the lateral side of the knee (fig. 149, page 294). Palpate the tract to its insertion on the lateral condyle of the tibia. By reviewing your findings in this examination, summarize the functions of gluteus maximus, including any evolutionary significance it may have.

Superiorly, the iliotibial tract is also continuous with another abductor muscle, the **tensor fasciae latae,** which takes origin from the lateral aspect of the iliac crest, and may be palpated, in active abduction, as a discrete round muscle head, 2 to 3 cm in diameter, slightly superior to the ischial tuberosity (fig. 149, page 294 and figs 157 and 160, page 302).

Deep to gluteus maximus are **gluteus medius** and **gluteus minimus,** which attach to the lateral and anterior parts of the greater trochanter, respectively; both abduct and laterally rotate the thigh. Gluteus medius extends superiorly to gluteus maximus and can be palpated as a discrete structure (figs. 149, 150, and 151, page 294).

Study the hamstring muscles of the thigh, reviewing first the hamstring component of the adductor magnus (page 286). The other hamstring muscles similarly take origin from the ischial tuberosity, and they can be palpated on the posterior aspect of the thigh. Recall, however (page 286), that the situation of the hamstrings, and therefore their role in hip extension, are evolutionarily secondary phenomena. Ask your model to extend the lower limb against resistance while you palpate from the region of the ischial tuberosity inferiorly. Muscle bands can be identified that pass from the ischial tuberosity inferiorly and laterally, as well as other bands that pass inferiorly and medially. Identify the band that passes inferiorly and laterally formed by the **long head** of the **biceps femoris.** This takes its origin from the upper, medial part of the ischial tuberosity (fig. 152, page 294). About half way down the thigh it is joined by the **short head** of the muscle, which takes its origin from the lateral lip of the linea aspera (and is the only hamstring component which is truly dorsal in origin). The two heads unite and the common tendon passes the knee joint on its lateral aspect to insert into the **head** of the **fibula,** which should now be palpated. This is a prominent, rounded bony structure on the lateral side of the leg, just below the knee joint and about 1 cm inferior to the lower border of the patella (fig. 149, page 294 and fig. 153, page 298). The tendon is felt as a powerful cord, the most posterior of those palpable on the lateral aspect of the knee, and is activated not only when your model extends the lower limb at the hip joint, but also when he or she flexes the knee against resistance.

The remaining hamstring muscles are the **semitendinosus,** which also arises from the medial portion of the ischial tuberosity (fig. 152, page 294), and the **semimembranosus,** which arises from the lateral aspect of the ischial tuberosity. Both these muscles can be palpated, passing downwards and medially, and both cross the knee joint on its medial aspect to insert into the **medial condyle** of the **tibia.** The two tendons of insertion can be felt as distant cords on the posteromedial aspect of the knee joint, the semitendinosus being the most posterior structure palpable and the semimembranosus tendon being a slightly less distinct band, just anterior to the tendon of semitendinosus. At this time review the insertions, on the medial aspect of the knee, of adductor magnus (page 282); sartorius (pages 284 to 285); gracilis (page 285); and, now, semimembranosus and semitendinosus (figs. 155 and 156, page 298). Note that on the posterior aspect of the knee a shallow depression is seen, bounded superolaterally by the biceps femoris tendon and superomedially by the semitendinosis and semimembranosis tendons, as each of these pass toward their respective insertions. This depression is called the **popliteal fossa.** In the center of this depression attempt to palpate the pulse of the **popliteal artery** using the three middle fingers of one hand while you hold the subject's knee in slight passive flexion with your other hand.

Other muscles that are not palpable in the living and that will be studied in the dissection laboratory, include **quadratus femoris,** which passes from the lateral edge of the ischial tuberosity to the posterior aspect of the femur (fig. 152, page 294); **obturator externus** (page 286); **obturator internus,** which originates from the medial aspect of the obturator membrane, covering the obturator foramen, and winds around the posterior aspect of the ischium, just superior to the ischial tuberosity, to insert into the medial aspect of the greater trochanter; and **piriformis,** which originates from the anterior aspect of the sacrum and passes out laterally, also to insert in the greater trochanter of the femur (fig. 151, page 294). What functions will these four muscles have in common?

VESSELS AND NERVES OF THE THIGH

Review the femoral point (page 126, point 14), and recall that the external iliac artery changes its name to the femoral artery as the vessel passes deep to the inguinal ligament, to enter the thigh at this point (page 169, point 6). Palpate the pulse of the femoral artery on your model. Just lateral to the situation of the femoral artery, a cord-like, fibrous structure may be palpated by rolling the finger lateromedially. This is the **femoral nerve** that reaches this position by passing from the lumbar plexus in the pelvis (page 166), lateral to the psoas muscle, also deep to the inguinal ligament. Previously, by deep palpation, you identified at this site, posterior to the femoral nerve, the iliopsoas tendon, which in turn lay just anterior to the hip joint (page 285). About 2 cm medial to the femoral artery and not palpable as a discrete structure, the **femoral vein** is situated. Superficially, an important tributary of the femoral vein may be visible in some subjects. The **long saphenous vein** runs up on the medial aspect of the lower limb. It passes the medial aspect of the knee joint, and ascends on the medial side of the thigh to a position about 2 cm medial to and 2 cm inferior to the femoral point, where the vein passes deep, through a gap in the deep fascia known as the saphenous opening, to enter the femoral vein.

In a subject in which the vein is prominently visible, perform the following clinical examination, used by the physician to determine the competence of the valves of the veins (folds of the inner lining of the veins, which prevent reverse [distal] flow of blood). Place the index finger of your left hand firmly across the vein. This will obstruct the normal (proximal) venous flow. Stroke the index finger of your right hand firmly along the vein in the proximal direction, from your left index finger toward the femoral point, lifting your right finger away from the skin surface at the end of the stroke. The vein should empty, and momentarily remain empty between the position of your left index finger and the site of the nearest valve, which may be several centimeters along the course of the vein. Normally the valves should maintain their competence for several seconds. Incompetence of the valves is demonstrated if the vein does not stay empty at all, but refills instantaneously from above. Valvular incompetence may be associated with the condition of varicose veins. Why should this condition be more common in humans than in lower primates and other mammals? Why should it be more common in women than in men?

To complete this session, review the emphasised key words, paying special attention to the homologies, between the upper and lower limbs, of ventral flexor–adductor and dorsal extensor–abductor muscle groups and their corresponding nerve supplies.

Laboratory Session 26A

Your lower limb dissection is to be performed on a cadaver in which one lower limb and its hemipelvis has been detached from the rest of the body by a sagittal section through the pelvic region and a transverse section on one side at about L5 or S1 (page 171). You will, of course, need to study both the detached and the attached lower limbs.

Begin your dissectional study of the lower limb by reviewing the lumbar plexus (pages 166–167). Reidentify, on the attached lower limb, the branches emerging from the lateral or medial aspects or on the anterior surface of the psoas major muscle. Complete your dissectional review of the plexus by tracing the branches proximally, in through the substance of the psoas major muscle, removing parts of the muscle piecemeal as necessary. Recall that the plexus, like other limb plexuses, is formed from ventral rami of spinal nerves (page 167); specifically, it is derived from ventral rami of the segments L1 to L5. The ventral rami of L2, L3, and L4 divide into ventral and dorsal **divisions,** respectively destined to supply ventral (flexor) and dorsal (extensor) surfaces of the lower limb. As noted (page 286), this situation is homologous to that of the brachial plexus of the upper limb, which comprises ventral and dorsal divisions of ventral rami that respectively supply ventral and dorsal surfaces of the upper limb (page 260). In particular, reidentify the main dorsal component, the femoral nerve (page 288), which is derived from L2, L3, and L4 and supplies the large extensors of the knee joint (page 286), and the major ventral component, the obturator nerve (also from L2, L3, and L4), which supplies the flexor–adductor group of muscles of the thigh (page 286). To find these nerves refer to page 167 and fig. 88, page 164. The obturator nerve accompanies the obturator artery (page 173) through the **obturator foramen** to enter the thigh. Identify also the nerve supply to psoas major itself, coming from dorsal divisions of L2, L3, and L4. This dorsal origin of a flexor muscle of the hip joint is explained on page 286.

Also on the attached lower limb side of your dissection, review now the hypogastric autonomic plexus (page 173), including the S2 and S3 contributions to it (page 106), and the pelvic splanchnic nerve. Then complete removal of the pelvic peritoneum, preserving the deep-lying structures including nerves and vessels. Reidentify the internal iliac artery and its branches and the accompanying veins. On the lateral surface of the lower lumbar vertebrae medial to psoas major and on the anterior surface of the first sacral vertebrae, identify the sympathetic trunk, readily visible because of its ganglia and the numerous small, thread-like branches that emanate from it to join the hypogastric plexus and to pass to other destinations. Review the morphological significance of the sympathetic trunk (page 106). Trace the trunk downwards, onto the anterior aspect of the sacrum, to its termination in the ganglion impar (unpaired). The trunk is firmly bound down by connective tissue where it is met by the trunk of the opposite side at the level of the coccyx.

In the following paragraphs you will study in greater detail, on both the attached and detached lower limbs of your cadaver, the sacral plexus. Review your previous encounter with this plexus on page 173, where one of its important branches, the pudendal nerve, was studied. As is the case for other limb plexuses (pages 260, 286, and this page), the ventral rami participating in the plexus divide into ventral and dorsal **divisions,** which supply flexor–adductor and extensor–abductor muscle groups respectively.

A good starting point for this dissection, and an extremely valuable landmark, of which you should obtain a clear visual image, is the **lumbosacral trunk,** derived from L4 and L5, to be found at the medial margin of the psoas major, posterior to the obturator nerve. The trunk descends from the lumbar plexus, crosses the sacral portion of the pelvic brim (page 169), and enters the true pelvis to join with sacral nerves to form the sacral plexus. The branches of the sacral plexus are somewhat intimately interwoven with the arterial branches of the internal iliac artery, and the accompanying venous tributaries of the internal iliac vein (page 173). As a result, it can, at times, be difficult to get one's bearings, with respect to which sacral segment one is observing. In this connection, the lumbosacral trunk can be very helpful as a landmark, because it is easy to identify as it crosses the pelvic brim. The first ventral ramus that joins it from the medial aspect, inferior to the pelvic brim, can be identified as S1, from which you can, thereafter, count downward serially, to identify S2, S3, and so forth.

Returning to the junction of the lumbosacral trunk and S1, use careful scissor dissection to trace branches of the internal iliac artery as they weave between the sacral nerves, emerging through sacral foramina on the anterior aspect of the sacrum. Study the relationship of the arterial branches to the nerve components of the sacral plexus. When you have removed fat and surrounding connective tissue by scissor dissection, ensure that you have identified the S1, S2, S3, and S4 components of the sacral plexus. (The S5 contribution is small and is mainly associated with the minute **coccygeal plexus.**) Also note and study the intimate relationship of the emerging sacral nerves with the **piriformis** muscle (page 287), which takes its origin from the anterior aspect of the sacrum. Elevate the fibers of piriformis and follow them medially to their origin from the sacrum. Establish from which sacral segments the piriformis takes its origin, by counting sacral segments in the sagittal section through the sacrum (using the promontory [page 169] as your landmark for identifying the upper edge of the first sacral vertebra). Inferior to piriformis, identify the **coccygeus** muscle which originates from the lateral aspect of the lower sacral vertebrae and from the coccyx, and inserts by converging fibers on the ischial spine (pages 170 and 283). The coccygeus muscle fibers are intermingled heavily with tendinous fibers. Further inferoanteriorly, coccygeus is in continuity with the most posterior portion of the pelvic diaphragm (page 168), the **iliococcygeus** muscle (page 173), which also takes its origin from the ischial spine. Palpate the spine, an important landmark (as done in the living subject, pages 170 and 283–284). Trace the fibers of the iliococcygeus medially, to their insertion into the tough, fibrous **anococcygeal** ligament, which stretches from the tip of the coccyx to the coccygeal body, a midline fibrous structure situated just posterior to the anal canal. The iliococcygeus muscle is so named because evolutionarily its lateral attachment was originally to the ilium. In lower primates, this muscle still takes that origin, and is functional in tail-wagging. In the human, because of the upright stance, and the conversion of the pelvic diaphragm into a support structure for pelvic organs (pages 168 to 169 and 173 to 174), rather than a tail-wagging muscle, the muscle takes on a more horizontal situation. In keeping with this, its origin has migrated down from the ilium to the ischium, and so strictly speaking the muscle in the human would more correctly be called the ischiococcygeus.

After reidentifying the iliolumbar branch of the internal iliac artery (page 173), and tracing it laterally, posterior to the psoas muscle and into the greater pelvis, your dissection of the sacral plexus can now be facilitated by cutting this branch and reflecting the iliac vessels and the arterial branches medially. Reidentify, once again, the deep-lying structures, including the sympathetic trunk and its branches. Note that in this region, as elsewhere, the sympathetic trunk is not always an entirely discrete structure, but rather is composed of numerous, fine threads, which are easily separable, and may give the impression of a very loose network of nerve fibers. In performing this reflection of the iliac vessels, take care to avoid tearing or destroying them, in particular those that pass deeply and posteriorly. Those serving the lower limb (the superior and inferior iliac branches, and the obturator branch, as well as the pudendal artery), will be studied on the posterior aspect of this dissection shortly, and it will greatly enhance your understanding of the three-dimensional relationships, to be able to trace the same structures from the internal, pelvic aspect of your dissection to the external, gluteal aspect.

By palpating the ischial spine (pages 170 and 284), and the region around it, and with the aid of a pelvic skeleton to observe while you perform this dissection, reidentify (page 284) the **greater sciatic notch,** the curved bony edge of which is in continuity with the superior edge of the ischial spine, and the **lesser sciatic notch,** the corresponding curved edge of which is continuous with the inferior edge of the spine. Each of these notches is converted into a foramen—the **greater sciatic foramen** and the **lesser sciatic foramen,** respectively—by the surrounding tough ligamentous structures. Mobilize the superior edge of the coccygeus muscle (see above), palpate posterior to it, and identify the **sacrospinous ligament** (page 283) posterior to the coccygeus, extending from the ischial spine to the lower sacral vertebrae. Also identify the space above the ligament, the greater sciatic foramen. Note that the piriformis leaves the pelvis through this foramen, as do branches of the sacral plexus and accompanying branches of the internal iliac vessels, some inferior to **piriformis** and others superior to the upper edge of the muscle. Create an artificial separation, by finger dissection, between the lowermost fibers of coccygeus and the upper fibers of iliococcygeus (see above). Through this opening palpate laterally to reidentify the ischial spine, and follow its curved inferior border (lateral boundary of the lesser sciatic foramen) down to the ischial tuberosity. This can be palpated, as well as the **sacrotuberous ligament** (page 283), which completes the boundary of the lesser sciatic foramen and should be followed medially to the sacrum and coccyx.

Review (page 173) the branches of the internal iliac artery and their relationships to the sacral plexus as a prelude to finding sacral plexus branches that accompany some of the iliac artery branches. The internal iliac artery sometimes appears to divide into an anterior and posterior division, before giving off its branches, although the pattern is somewhat variable. The posterior division branches are usually more superior. They comprise the following, which should be reviewed now: the iliolumbar artery; the lateral sacral arteries (one or two); and the superior gluteal artery, which passes between the lumbosacral trunk and S1, or between S1 and S2, to leave the pelvis superior to the upper border of piriformis.

The anterior division gives off the umbilical artery, which in turn gives off vesical and vaginal branches, and a branch to the vas deferens, proximal to where it becomes obliterated (page 133); the inferior vesical artery, the middle rectal artery; the obturator artery, which runs with the obturator nerve to the obturator foramen (page 289); the uterine artery of the female; and the two terminal branches, the inferior gluteal artery and the internal pudendal artery, which pass between S1 and S2, or S2 and S3, to leave the pelvis inferior to the piriformis muscle. In order to gain better access to display the course of the arteries it may be necessary to cut some of the veins and dissect them away. Thrombosis (clotting of blood within the lumen) may frequently be encountered in these veins. Recall that tributaries of this system of veins establish anastomoses between the portal and systemic venous systems. Consider while you dissect these veins what possible consequences thrombosis within the system may produce clinically (pages 173 and 288).

In separating the iliococcygeus from coccygeus you will be entering, from the anterior aspect, the ischioanal fossa, previously dissected from an inferoposterior approach (page 179). Its main content, fat, will have been partly removed in your previous dissection.

Disconnect entirely the attachment of the coccygeus to the lateral edge of the sacrum and coccyx, and reflect the muscle, which is made up of muscular and tendinous fibers, laterally, to give access to the posterior aspect of the muscle. Trace the muscle fibers to the spine of the ischium. Posteriorly, the muscle may also be seen to take origin, in part, from the sacrospinous ligament. On its posterior surface, you may be able to see it receiving its nerve supply from a branch of S4 and/or S5. Now trace the two terminal branches of the anterior division of the internal iliac artery. The internal pudendal artery is the smaller of the two. At its origin, it lies lateral to the other terminal branch, the inferior gluteal artery. However, the internal pudendal artery crosses somewhat medially as it passes posterior to the upper edge of coccygeus and then behind the ischial spine. At this site it is joined by the pudendal nerve (page 173), which emerges from the sacral plexus, having picked up ventral division branches from S2, S3, and S4 (page 289). The pudendal nerve is the most medial structure passing posterior to the ischial spine, and the artery and vein lie just lateral to it. These, and the inferior gluteal artery, leave the pelvis below the inferior edge of piriformis. Note, however, that the inferior gluteal artery swerves out laterally and posteriorly, to pass toward the gluteal region. Thus only the pudendal structures reenter the perineum, passing through the lesser sciatic foramen. Establish this passage by identifying both the sacrospinous and sacrotuberous ligaments and observing the vessels and nerve pass between them. The vessels and nerve then pass onto the medial aspect of the ischium, to enter a fascial "tunnel," the **pudendal canal,** on the obturator internus muscle.

Reflect iliococcygeus inferiorly, to expose the pudendal canal on the surface of the obturator internus muscle. Trace it inferomedially, and review the branches of the pudendal artery and nerve, tracing as many as possible to their destinations. This dissectional approach provides a good opportunity for reviewing the relationship of the levator ani (iliococcygeus) muscle (pages 173 and 290) to the rectum and anal canal, and the nervous and vascular structures of the perineum. Also, you should, at this time, palpate the obturator internus muscle and visualize its origin from the medial aspect of the obturator membrane and adjacent bony structures. The fibers of the muscle pass posteriorly to converge in the groove between the ischial spine and the ischial tuberosity (the lesser sciatic notch). You should palpate the muscle here as the fibers turn sharply laterally to exit the pelvis. Their destination is the greater trochanter of the femur (page 282).

Dissection of the Gluteal Region

Turn both lower limb specimens (the detached half pelvis with lower limb and the rest of the cadaver with attached lower limb) into the prone position. Make a skin incision on the lateral side of the thigh, from the level of the transverse section of the detached limb or corresponding position on the attached side, down to the knee joint. Remove skin from the level of the transverse section to the knee joint. Use the rapid method on the detached side, and the method for preservation of superficial structures on the attached side (page xv). On both sides, correlate frequently the structures dissected on the dorsal surface with landmarks previously identified within the pelvis as dissected from the anterior aspect of your specimen.

In the gluteal (buttock) region, skin removal will reveal abundant fat deposition, more marked in females than in males. Superficial blood vessels and nerve branches will be encountered. As your dissection proceeds to deeper levels, you will display, in the inferomedial quadrant of the buttock, the important vessels and nerves seen previously to leave the pelvis and pass toward the gluteal region, emerging either above or below the piriformis muscle. Branches and tributaries splay out from this site, for which reason intramuscular injections into the buttock should be given into the superolateral quadrant. Repeated injections are bound to damage branches of the gluteal vein, and in your dissection you are likely to encounter such thrombosed veins caused by injections in individuals who prior to death had been subjected to prolonged medical treatment with numerous injections. The veins so damaged are small tributaries of the superior and inferior gluteal veins, which accompany the corresponding arteries superior and inferior to the piriformis respectively (pages 290 to 291). Nerve branches that will be encountered in the subcutaneous tissue, and that should be demonstrated in the superficial skinning, include the following, which should be identified with the aid of an atlas:

1. The lateral cutaneous branch of the iliohypogastric nerve (pages 131 and 167), which passes downwards and laterally over the posterior portion of the iliac crest to supply the upper, outermost part of the buttock. This part of the iliac crest is frequently used as a source of a bone transplant in a spinal fusion operation for intervertebral disk degeneration ("slipped disks"). A frequent complication of the operation is that this nerve may be cut, with resulting "numb bum syndrome."
2. Cutaneous branches of upper lumbar spinal nerves, at the posterior part of the iliac crest and the sacrum, medial to 1. above.
3. Cutaneous branches of dorsal rami from sacral and coccygeal nerves, emerging on the medial aspect of the buttock.
4. The **perforating cutaneous nerve,** a branch of the sacral plexus, which winds around the medial part of the gluteal fold (pages 168 and 281), to pass upwards and supply the skin of the buttock.
5. Just lateral to 4, gluteal branches of the **posterior cutaneous nerve of the thigh,** a branch of the sacral plexus, which similarly curve around the gluteal fold.
6. The **perineal branch of the posterior cutaneous nerve of the thigh,** which emerges just inferior to the gluteal fold, and passes anteromedially, to contribute to the sensory supply of the perineum.
7. Other cutaneous branches of the posterior cutaneous nerve of the thigh, emerging just inferior to the gluteal fold and at varying distances down the posterior aspect of the thigh.
8. **Posterior branches of the lateral cutaneous nerve of the thigh** (a branch of the femoral nerve), which emerge on the lateral aspect of the buttock and thigh to curve around the lower limb onto its posterior aspect.

Remove the fatty tissue of the buttock, exercising care as you approach the layer of the deep fascia and deep-lying muscles. You can get an impression of approximately where this level is, by observing the cut surface at the transverse section that severed the detached limb from the trunk. Depending on the degree of obesity of the subject you are dissecting, from one to several centimeters deep to the skin layer you will encounter the fascia enclosing the **gluteus maximus** muscle, fibers of which characteristically run inferolaterally. Carefully remove fat to display the muscle, the origins of which you should now review (page 286). Identify, on a pelvic skeleton, the **posterior gluteal line,** which will be found on the posterior part of the lateral aspect of the ilium, running more or less vertically from the highest point of the ilium down to the back of the sciatic notch. Gluteus maximus takes origin from the ilium posterior to this line. The fibers pass downwards and laterally and converge to the lateral aspect of the thigh, where they insert mainly into the iliotibial tract (page 287), a very powerful thickening of the deep fascia of the thigh. Review your palpation of this structure on the living person and recall that the tract passes inferiorly on the lateral aspect of the thigh, to the lateral condyle of the tibia. The iliotibial tract also has other attachments. At the knee, it attaches to the lateral aspect of the patella, and adjacent structures just above and below the patella (i.e., the lateral aspect of the quadriceps tendon and the lateral aspect of the patellar tendon respectively).

As in the arm (page 262), in the thigh ventral flexor muscles and dorsal extensor muscles are separated by an intermuscular septum lying in a frontal plane. The lateral intermuscular septum of the thigh binds the iliotibial tract to the lateral lip of the linea aspera and the downward continuation of this lip into the **lateral supracondylar line** of the femur. Superiorly, this intermuscular septum continues up to the gluteal tuberosity (page 286) (which you should reidentify on the bony skeleton) and thus some fibers of the gluteus maximus also insert into this tuberosity. Dissect fat away to demonstrate the entire gluteus maximus muscle. At its anterior edge, trace the deep fascia forwards, across the deep-lying **gluteus medius** and to the **tensor fasciae latae** (page 287). The **fascia lata** (deep fascia of the thigh) splits to ensheath the gluteus maximus muscle posteriorly and tensor fascia latae anteriorly. This arrangement is somewhat reminiscent of that of the investing layer of cervical fascia and its relationship to the trapezius and sternocleidomastoid muscles (pages 55, 236, and 248). Demonstrate tensor fascia latae and show that its fibers also insert into the iliotibial tract. From its origins and in sections you can deduce that its actions and therefore nerve supply are likely to be similar to those of Gluteus medius and minimus, and indeed they are.

Extend the skinning of the thigh onto the lateral aspect, and anteriorly as necessary, to expose the iliotibial tract. Dissecting at the sloping, anterior free edge of the gluteus maximus, separate its covering fascia to expose gluteus medius (figs. 149, 150, and 151, page 294), which lies in part deep to the gluteus maximus, and extends beyond the anterior edge of the latter to emerge with the posterior edge of tensor fasciae latae. Gluteus maximus can be separated from the deep-lying gluteus medius by blunt finger dissection. Gluteus medius is more difficult to separate from the tensor fasciae latae, because the muscle fibers merge, and because tensor fasciae latae takes origin not only from the anterior portion of the iliac crest, but also from its own fascial sheath, most importantly that part which is continuous with the iliotibial tract. Separate these three muscles by dissection, and also dissect the iliotibial tract inferiorly to display its attachment as described previously (pages 287 and 292).

Complete the skinning of the gluteal region and posterior aspect of the thigh down to a level 5 cm inferior to the lower border of the patella, and laterally to your incision line down the middle of the lateral side of the thigh.

Trace by finger dissection the outline of the upper and lower borders of the gluteus maximus, which slope downward, laterally, and anteriorly as the muscle passes in this direction from the back of the sacrum to the lateral aspect of the thigh. With finger dissection, mobilize the upper and lower borders of the muscle, and insert your fingers deep to the muscle to free it from deep-lying structures. This dissection, although difficult, should be performed with care, to avoid damaging the important nerves and vessels lying deep to the muscle, including those that pass into the substance of the muscle from the deep surface. These are the inferior gluteal nerves and vessels. When the muscle has been adequately mobilized from the deep-lying structures, cut across its belly approximately at right angles to the direction of its fibers, starting inferiorly just medial to the ischial tuberosity and continuing upwards and laterally toward the position of the greater trochanter (fig. 150, page 294). Reflect the two halves of the muscle medially and laterally respectively, carefully dissecting away fat and connective tissue to expose the inferior gluteal vessels entering the muscle. Trace the nerve and vessels proximally and in this way identify the superior and inferior borders of the piriformis muscle, where the nerves and vessels emerge (page 290). Clear the entire region deep to the gluteus maximus, and study the following structures (figs. 149, 150, 151, and 152, page 294).

Superior to the piriformis identify gluteus medius, the origin of which can be seen to commence just anterior to that of gluteus maximus (i.e., at the posterior gluteal line [page 292]). The attachment extends from this line to the **middle gluteal line,** which you should identify now on a pelvic skeleton. It commences approximately at the position of the iliac tubercle (page 125, point 5), and curves downwards and posteriorly to the greater sciatic notch. Anteriorly, the fibers of gluteus medius merge with those of the tensor fasciae latae, but the muscle can be distinguished by the fact that the latter fibers are more or less vertical whereas those of gluteus medius are more or less parallel with those of gluteus maximus. Mobilize these two muscles and elevate the gluteus medius from the deep-lying gluteus minimus.

The fibers of gluteus medius are arranged in a fan-shaped pattern, with the apex of the fan on the greater trochanter of the femur. Laterally, the muscle is separated from tensor fasciae latae by the tough fibrous sheath of the latter muscle.

Gluteus medius should be elevated in much the same way as you previously did for the gluteus maximus. Insert your fingers deep to the muscle from its inferior edge, just superior to piriformis. The muscle belly should similarly be cut with an incision parallel to that of the gluteus maximus. In order to protect deep-lying structures, after mobilizing the muscle with your fingers, insert a pair of forceps deep to the muscle and incise the muscle, using a scalpel. Reflect the two halves of gluteus medius laterally and medially respectively, using scalpel dissection to separate the muscle from its attachment to the ilial crest and iliac surface where necessary. You will thus expose the superior gluteal nerves and vessels, emerging from above the superior edge of piriformis. Branches of these enter the gluteus medius on its deep surface and other branches pass deeply to enter the last of the gluteus muscles now exposed, gluteus minimus. This muscle, which can be recognized by the much finer fiber structure it has as compared to the gluteus medius and maximus, takes origin from the iliac surface between the middle and anterior gluteal lines. The latter should be identified on the pelvic skeleton, curving from the anterior superior iliac spine, past the posterior edge of the acetabulum (page 284), toward the greater sciatic notch. Carefully clear connective tissue from the superior gluteal nerves and vessels by scissor dissection, and trace branches to their destinations, clearing gluteus minimus as you do so. Review the main functions of gluteus maximus, medius, and minimus (pages 285 to 286. Do you expect the nerve supply of these muscles to be from ventral or from dorsal divisions of ventral rami?

By combined scissor and scalpel dissection, sep-

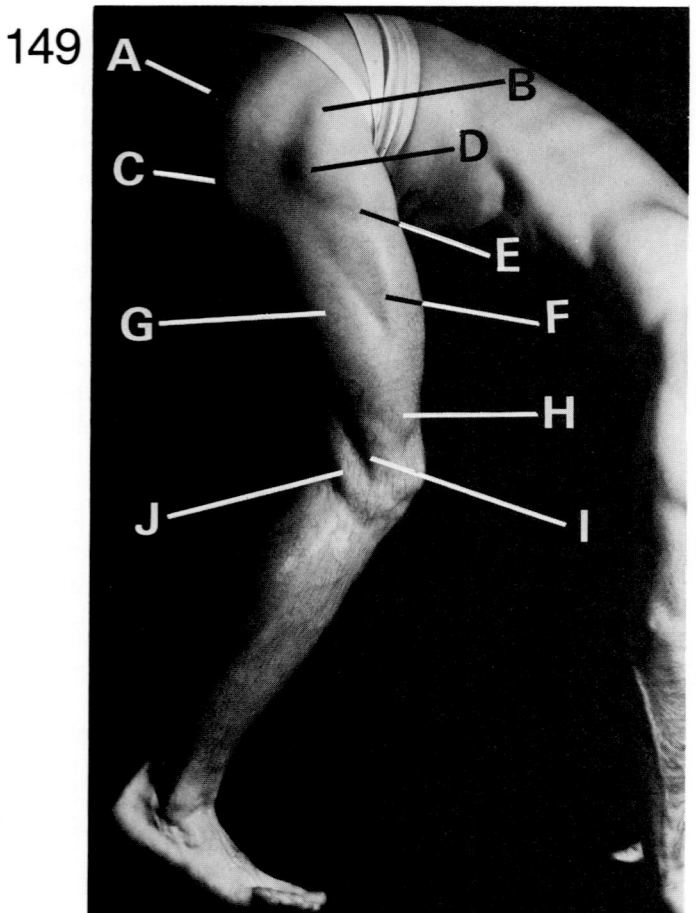

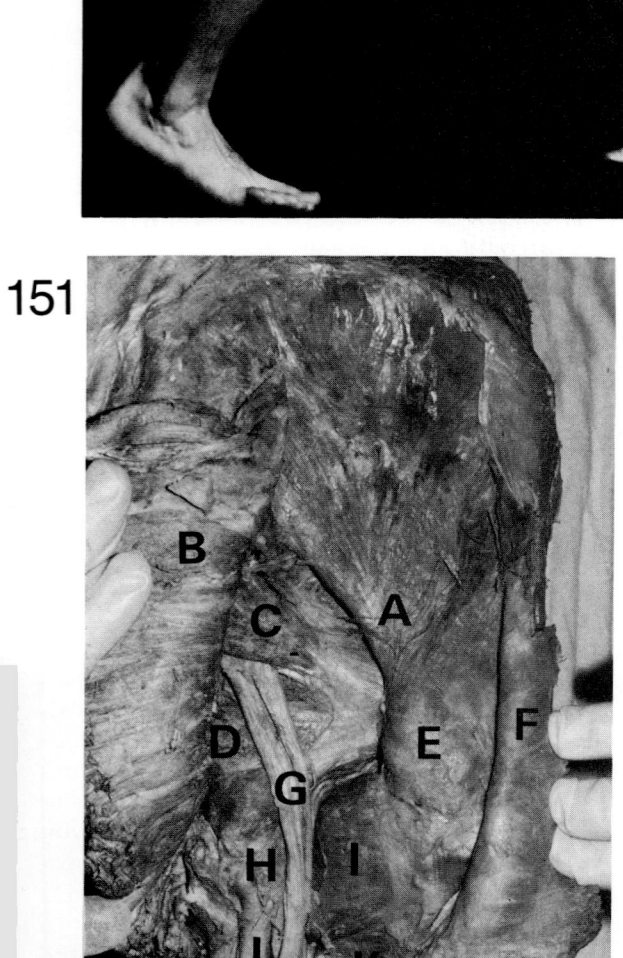

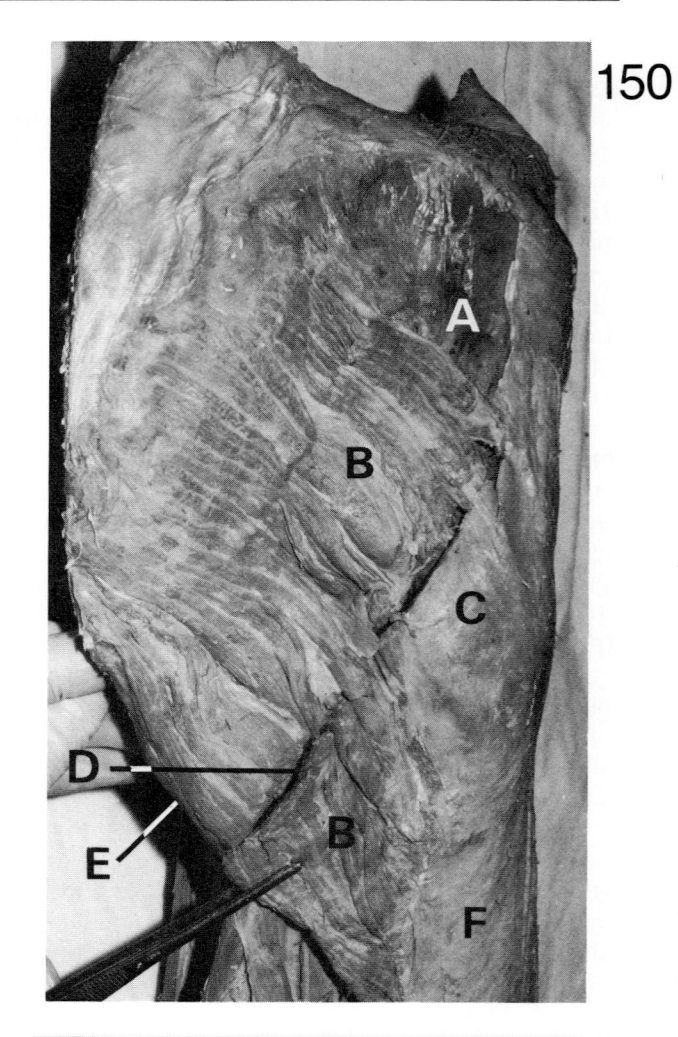

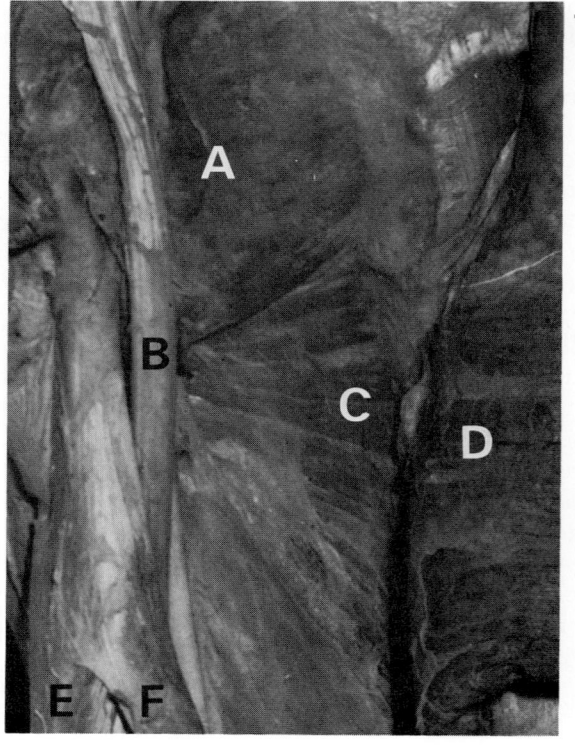

294 YOUR PATIENT'S ANATOMY: A CLINICAL VIEW OF HUMAN MORPHOLOGY

Figure 149. Lateral aspect of right lower limb.
A. Gluteus maximus
B. Gluteus medius
C. Ischial tuberosity
D. Greater trochanter of the femur
E. Tensor fasciae latae
F. Anterior edge of iliotibial tract
G. Iliotibial tract
H. Vastus lateralis
I. Tibial insertion of iliotibial tract
J. Biceps femoris tendon of insertion

Figure 150. Posterior view of gluteus maximus and related structures.
A. Gluteus medius
B. Gluteus maximus
C. Greater trochanter of the femur
D. Dissectional incision at right angles to fibers of gluteus maximus
E. Inferomedial edge of the gluteus maximus, cause of the gluteal fold
F. Iliotibial tract

Figure 151. Posterior view of dissection of gluteal region.
A. Gluteus medius
B. Superomedial part of gluteus medius, reflected medially
C. Piriformis
D. Obturator internus and gemelli superior and inferior
E. Greater trochanter of femur
F. Inferolateral part of gluteus maximus reflected laterally
G. Sciatic nerve
H. Ischial tuberosity
I. Quadratus femoris
J. Tendon of origin of biceps femoris and semitendinosus
K. Superior fibers of adductor magnus

Figure 152. Detail of posterior view of dissection of gluteal region.
A. Quadratis femoris
B. Sciatic nerve
C. Adductor magnus
D. Inferolateral part of gluteus maximus reflected laterally
E. Semitendinosus
F. Biceps femoris

arate gluteus maximus further medially from its attachment to the ischial tuberosity, the dorsal surface of the sacrum, and the sacrotuberous ligament, which you will, in this dissection, now be able to expose from the dorsal surface. When adequately exposed, you should, by placing one finger on the ligament's dorsal surface and another on its ventral surface as exposed within the pelvis (page 290) correlate these two different approaches to the structure. In the process of this dissection, you should be mentally reviewing the origins, insertions, and actions of the gluteus maximus—the best way to learn muscle attachments is to think about the function of the muscles as you examine them clinically and then dissect them. Also during this dissection, you may encounter a nerve piercing the sacrotuberous ligament. This is the perforating cutaneous nerve, previously encountered in your superficial dissection (page 292).

The next major structure to identify at this time is the large **sciatic nerve,** emerging at the inferior border of piriformis, and passing inferiorly. On its posterior aspect, identify the much smaller posterior cutaneous nerve of the thigh, branches of which you also encountered in your superficial dissection (page 292).

An important relation of the sciatic nerve is the **quadratus femoris muscle,** which passes from the lateral edge of the ischial tuberosity to the medial edge of the greater trochanter of the femur. As its name implies, it is quadrangular in shape; the fibers run transversely. The sciatic nerve crosses it posteriorly. From the direction of its fibers, its position, and its attachment, deduce what functions this muscle may have.

Relationships at Inferior Border of Piriformis

Piriformis is a notable landmark for the region now being dissected, and it is worth spending a little time studying the structures inferior to the inferior border of piriformis, in order to get a thorough grasp of the three-dimensional relationships (page 290). With a pelvic skeleton at hand, place a finger at the inferior border of piriformis, and displace the muscle slightly upwards and the sciatic nerve slightly medially. The palpating finger will encounter the posterior surface of the ischial spine (pages 170, 283, 290). Palpate the spine and follow its superior edge upwards into the curvature of the greater sciatic notch (page 290). Your finger will be able to discern the junction of the ilium and the sacrum, and, by running your finger upwards on the ventral aspect of these bones, the anterior edge of the sacroiliac joint (pages 283 to 284). Explore the entire extent of the sciatic notch, confirming the passage of the sciatic nerve, piriformis, and other nerves and vessels, through the notch, and also palpating the sacrospinous ligament (pages 283 and 290). Careful scissor dissection should now be used to further expose the nerves and vessels that emerge below the piriformis. The inferior gluteal vessels should be further defined. At the extreme medial end of the piriformis' lower border, dissect out the internal pudendal artery and pudendal nerve as they pass over the posterior surface of the ischial spine (page 291). One way of facilitating this type of dissection is to put a clamp onto the structures from the dorsal aspect, and then turn the specimen over to reexamine it from the pelvic aspect, or vice versa. Trace the internal pudendal vessels and the pudendal nerve across the ischial spine, and then follow them as they curve forwards to reenter the perineum by passing inferior to the spine and the sacrospinous ligament, in the narrow space between the latter and the sacrotuberous ligament.

In the space inferior to the ischial spine and superior to the quadratus femoris muscle, displace the sciatic nerve medially and identify the obturator internus muscle. Trace its fibers medially to the groove between the ischial spine and the ischial tuberosity, at which site the muscle fibers emerge from the pelvis to make a sharp right-angle bend and cross toward the greater trochanter (page 287). This dissection will be facilitated by achieving full detachment of gluteus maximus from the ischial tuberosity. Having done this, insert a finger deep to gluteus maximus and palpate the medial aspect of the obturator internus above and anterior to the ischial tuberosity, and then follow its fibers by palpating posteriorly and around the ischium. Scissor dissection anterior to the sacrotuberous ligament will allow you to follow fibers as they pass anterior to that ligament. In this dissection you should also encounter the nerve to the obturator internus, from L5, S1, and S2 of the sacral plexus, as it passes the ischial spine just lateral to the pudendal nerve and vessels, and courses toward the obturator internus muscle. On the dorsal aspect, using scissor dissection, separate the inferior border of obturator internus from the quadratous femoris and the superior border of the obturator internus from piriformis. Trace the three muscles laterally to their insertions on the femur. Just above and just below the obturator internus tendon respectively, identify the **gemellus superior** and **gemellus inferior,** small muscles that take their origin from adjacent bone and insert into the obturator internus tendon.

To reinforce your understanding of the muscles of this region and their actions on the hip joint, grasp the foot of your lower limb specimen while an assistant firmly stabilizes the pelvis. Observe carefully, by inspection and palpation, the muscles that you have dissected, while you perform the following movements of the lower limb: medial and lateral rotation; flexion at the hip joint; extension at the hip joint; adduction; and abduction.

Return now to the inferolateral half of gluteus maximus and clear fat and connective tissue from both its superficial and deep surfaces. On the superficial aspect, follow the fibers inferolaterally and trace them into the iliotibial tract. The tract should be followed downwards, as a distinct thickening in the deep fascia of the thigh, the fascia lata. Flex your specimen at the knee, and confirm that this puts stretch on the fascia lata; conversely tug on the gluteus maximus with the knee in the flexed position and observe the action of the gluteus maximus fibers in extending the knee through stretch on the iliotibial tract. Now reflect the lower half of the gluteus maximus laterally, and dissect on its deep surface. Follow the fibers inferiorly, to their attachment to the gluteal tuberosity on the femur, situated at the upper end of the linea aspera. Palpate the tuberosity and the linea, and further trace some of the fibers downwards into the tough fibrous band, the lateral intermuscular septum, which attaches medially to the lateral lip of the linea aspera. Laterally, the lateral intermuscular septum is attached to the iliotibial tract. The superior fibers of gluteus maximus converge from the posterior aspect on muscle fibers approaching the linea aspera from the medial aspect. These are fibers of the adductor magnus. The uppermost of these lie just inferior and parallel to the lower fibers of quadratus femoris (fig. 152, page 294). Dissect these fibers clear of connective tissue, and, sliding a finger across the posterior surface of these muscle fibers, trace them medially to their origin (page 287). Medial to the uppermost, horizontal fibers, identify the vertical fibers of the other component of the adductor magnus, which should also be traced upwards to their origin (pages 285 to 286). Review the two components of the adductor magnus, and their functions.

Hamstring Muscles

Trace the lateral intermuscular septum downwards, and identify a muscle belly lying just posterior to the septum, and attaching to the lateral lip of the linea aspera. This is the **short head of biceps femoris,** the only developmentally dorsal hamstring component. Trace it downwards to its junction with the tendon of the **long head** of that muscle (page 287), and then follow the long head up superiorly to its origin from the ischial tuberosity (fig. 152, page 294). Study the relationship of the sciatic nerve to the long head of the biceps femoris. Returning to the ischial tuberosity, study the remaining two hamstring muscles that take their origin from it (page 287). Unlike biceps femoris that passes downwards and laterally, these two diverge from biceps to pass downwards and medially. Trace the muscle mass containing these two hamstring muscles downwards to the medial aspect of the knee joint, removing connective tissue and fat to expose the muscle bellies. Inferiorly, the more superficial of the two becomes clearly tendinous: this is known as the **semitendinosus.** Review the relationships of the important muscle insertions on the medial aspect of the knee (page 287 and figs. 155 and 156, page 298). Separate the semitendinosus tendon from the deeper lying **semimembranosus,** all the way upwards to the ischial tuberosity, to establish the separate origins of the two muscles: semitendinosus from the upper lateral portion of the tuberosity, in common with the long head of the biceps femoris; semimembranosus from the medial aspect of the ischial tuberosity.

Fully mobilize the hamstring bundle of muscles by finger dissection, separating them from more anterior structures. With the muscles firmly grasped in one hand, flex the knee of the specimen with assistance from a colleague. By tugging on the hamstrings while the knee is gently extended, review the function of the hamstrings at the knee joint (page 287). Also, with an assistant firmly stabilizing the hip, attempt to flex the thigh on the hip, and confirm the role of the hamstrings in this movement. Having clearly established in your mind the actions of the hamstring muscles, reidentify the sciatic nerve (page 296) and separate it from the deep surface of the upper portion of the hamstring bundle, and then cut the hamstrings, avoiding damage to the sciatic nerve. Reflect the proximal half of the cut hamstring muscles proximally, and clear away connective tissue to demonstrate the deep-lying adductor magnus muscle. You can now more fully demonstrate the two components of adductor magnus and their origins from the ischium.

Trace the sciatic nerve inferiorly, clearing connective tissue as you do so, and follow and observe carefully the branches that it gives off to the related muscles. The nerve may clearly be seen to be composed of two components, a lateral and a medial. They separate from each other at a variable distance down the thigh, but always by the time they reach the **popliteal fossa** (page 287). The two components are often separated at higher levels, but in any event are usually visible as distinct structures because of a slight color difference. Observe carefully which component, the lateral **(common peroneal)** or the medial **(tibial)** nerve, gives branches to the different hamstring muscles. The **common peroneal nerve** is comprised of dorsal divisions (i.e., divisions nor-

298 YOUR PATIENT'S ANATOMY: A CLINICAL VIEW OF HUMAN MORPHOLOGY

153

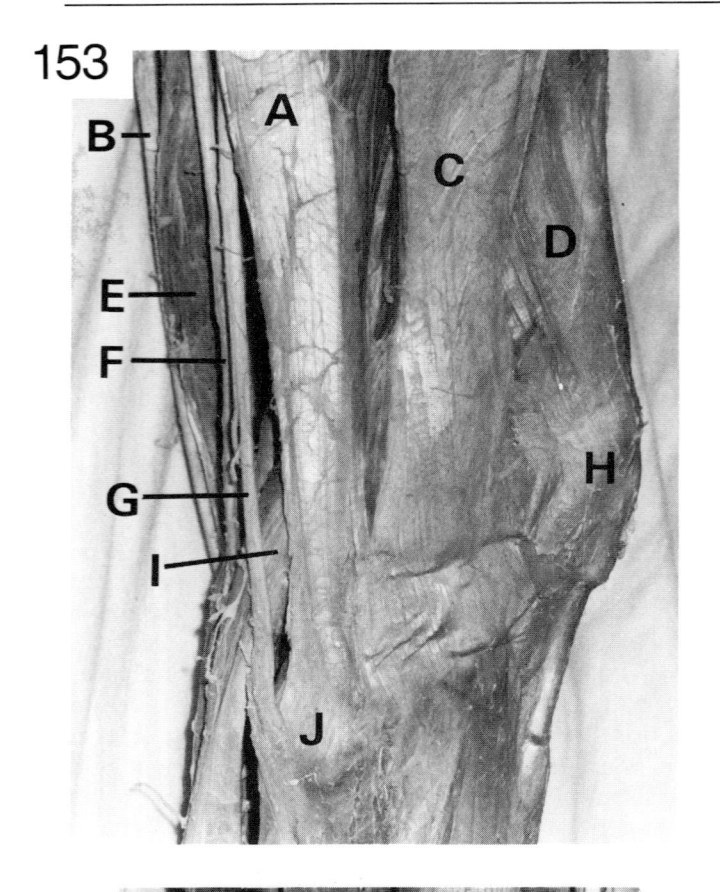

154

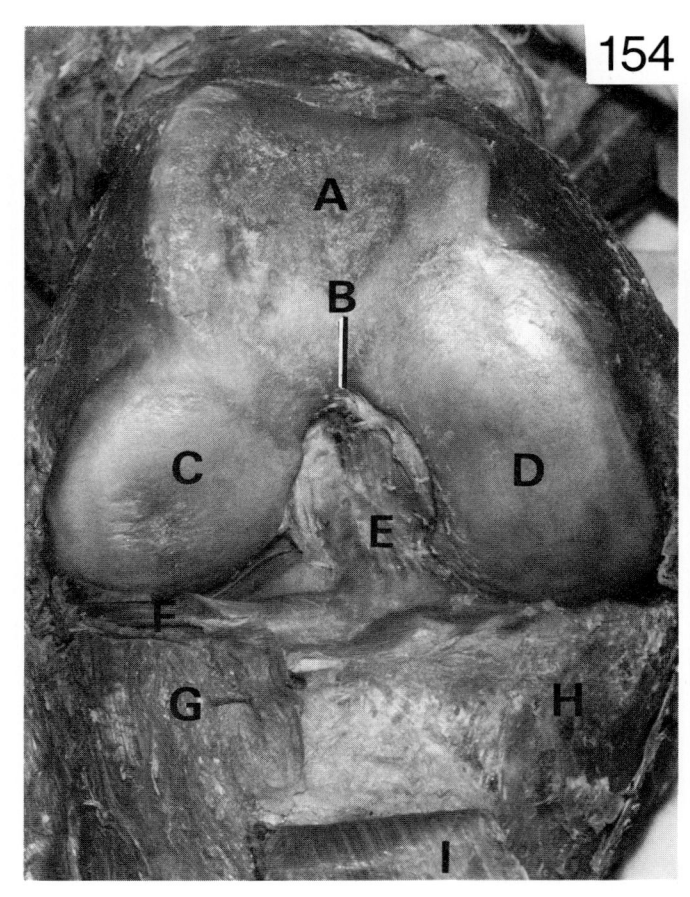

155

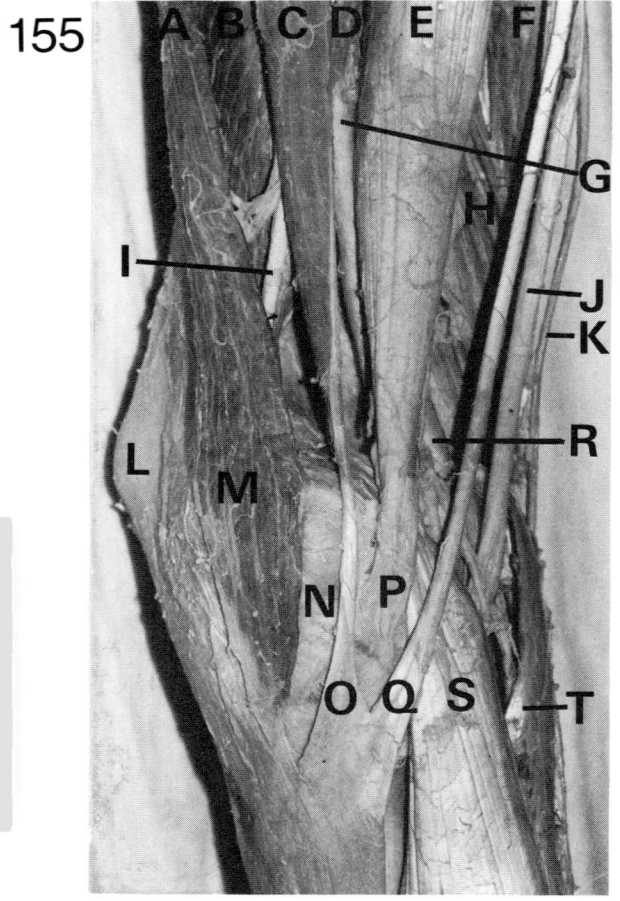

156

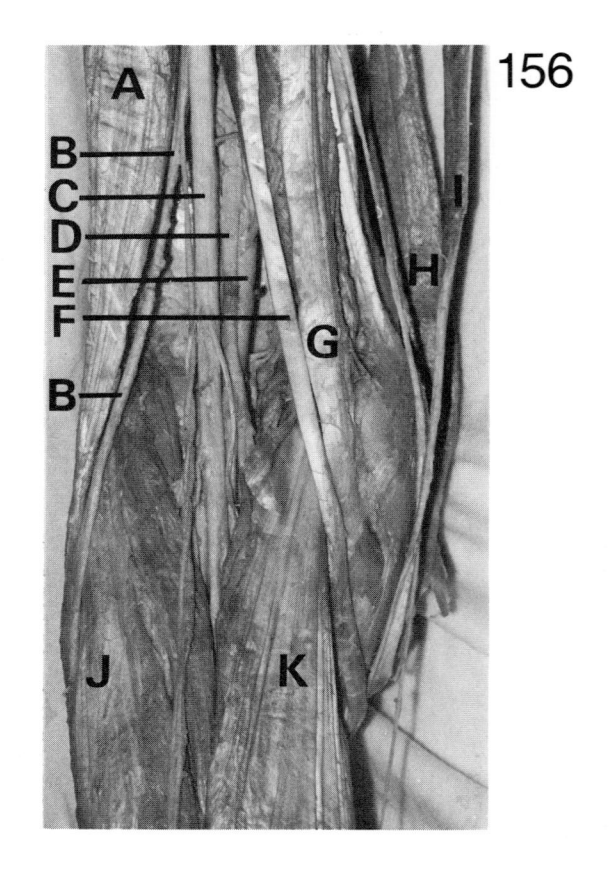

Figure 153. Lateral and posterolateral aspects of right knee.
A. Biceps femoris tendon of insertion
B. Semitendinosus
C. Iliotibial tract
D. Vastus medialis
E. Semimembranosius
F. Tibial nerve
G. Common peroneal nerve
H. Patella
I. Lateral head of gastrocnemius
J. Head of fibula

Figure 154. Anterior view of knee joint, with tibia in anatomical position and femur flexed at the knee.
A. Patellar surface of the femur
B. Intercondylar notch
C. Lateral femoral condyle
D. Medial femoral condyle
E. Anterior cruciate ligament
F. Lateral meniscus, anterior edge
G. Lateral tibial condyle
H. Medial tibial condyle
I. Inferior portion of patellar ligament, reflected inferiorly

Figure 155. Medial and posteromedial aspects of right knee.
A. Sartorius
B. Vastus medialis
C. Gracilis
D. Adductor magnus
E. Semimembranosus
F. Long head of biceps femoris
G. Popliteal vessels emerging through adductor magnus
H. Short head of biceps femoris
I. Tendon of insertion of adductor magnus
J. Tibial nerve
K. Common peroneal nerve
L. Patella
M. Insertion of sartorius
N. Medial condyle of tibia
O. Insertion of gracilis
P. Insertion of semimembranosus
Q. Insertion of semitendinosus
R. Popliteal vessels in popliteal fossa
S. Medial head of gastrocnemius
T. Lateral head of gastrocnemius

Figure 156. Left popliteal fossa.
A. Biceps femoris
B. Common peroneal nerve
C. Tibial nerve
D. Popliteal vein
E. Popliteal artery
F. Semitendinosus
G. Semimembranosus
H. Sartorius
I. Gracilis
J. Lateral head of gastrocnemius
K. Medial head of gastrocnemius

mally destined to supply dorsally situated extensor and abductor muscles and dorsal sensory surfaces), while the tibial nerve is derived from ventral divisions, and would be expected to supply ventrally situated flexor and adductor muscles and ventral sensory areas. Review (pages 286 and 287) the facts that the hamstrings are, with the exception of the short head of biceps femoris, morphologically ventral, in keeping with their role as flexors of the knee. Their positioning and consequent extensor function at the hip joint is a late evolutionary development, and their ventral nerve supply (with the above exception) is thus paradoxical in respect of this function.

Laboratory Session 26B

Anterior Aspect of the Thigh

Turn your specimen into the supine position and remove the skin from the remainder of the anterior aspect of the thigh not previously skinned. On the side in which you are doing the superficial dissection identify as many as possible of the cutaneous, sensory nerves of this region, using atlas illustrations to assist you. Among those you should identify are, superiorly (close to the inguinal sulcus [page 126, point 10]), the femoral branch of the genitofemoral nerve, and the ilioinguinal nerve (pages 131 and 167), which curves downwards onto the thigh after passing through the superficial inguinal ring (page 127). Further inferiorly one finds branches of the lateral cutaneous nerve of the thigh laterally, and branches of the anterior cutaneous branch of the femoral nerve on the anterior aspect of the thigh.

In both superficial and deep skinning dissections, the long saphenous vein should be identified (page 288). Recall that this vein is of clinical importance as it is frequently involved in the condition of varicose (distended) veins. Trace the vein up on the medial side of the thigh and dissect it to the point where it passes through the saphenous opening. Identify the sharp falciform (sickle-shaped) edge of the opening, and dissect the vein to the site where it joins the **femoral vein** (page 288).

Study the numerous tributaries that converge on the long saphenous vein just before it passes through the saphenous opening. In the process of this dissection, you might encounter several lymph glands, situated around the vein. Make a longitudinal incision into the vein, and examine, in its interior, the valves studied previously (page 288). The valves are seen as small folds projecting into the lumen.

The femoral vein lies within a distinct **femoral sheath** of connective tissue that encloses, in addition to the vein, the **femoral artery** (pages 287 to 288) laterally and a lymph gland-filled space, the **femoral canal,** medially. You should carefully dissect the femoral canal and identify its superior opening, the **femoral ring,** situated in the subcutaneous tissue at about the inguinal ligament. Insert a finger into the canal and superiorly through the ring, to confirm that your finger tip will reach the outer surface of the parietal peritoneum. This fact is of clinical importance: abdominal organs may herniate (i.e., be forced by pressure in the abdominal cavity) through the femoral ring and into the femoral canal, causing a **femoral hernia.** A femoral hernia of a loop of bowel may become obstructed or strangulated (i.e., have its blood supply compromised), and this constitutes a surgical emergency.

On the lateral aspect of the vein identify the femoral artery, and dissect it and its branches carefully from surrounding tissue, as follows. After identifying the artery lateral to the vein, insert a finger deep to the deep fascia and pass it inferiorly to elevate the fascia lata (page 292) from the femoral artery. Using the usual techniques of dissection, cut through the fascia and expose the artery, the femoral nerve (page 288), and the boundaries of the **femoral triangle** (figs. 157 and 158, page 302), within which the nerve, artery, vein, and canal are all situated. This triangle is bounded superiorly by the inguinal ligament; laterally by the medial border of the sartorius muscle (page 284); and medially by the medial border of the adductor longus muscle (page 285). Carefully dissect connective tissue away to reveal the entire surfaces of the two muscles and the branches of the femoral nerve and artery that leave their parent trunks within the triangle. Also display the muscles forming the deep surface of the triangle. These are pectineus, medially (page 285) and iliopsoas, laterally (page 285). Confirm the relationship, previously described (page 288), of the femoral nerve to the iliopsoas tendon. Trace the sartorius superiorly to its origin (page 284).

The pectineus muscle should now be cleared. Review its site of origin, the pectineal line (pages 285, 169, and 132) on a pelvic skeleton, curving laterally on the superior aspect of the superior pubic ramus, lateral to the pubic crest, forming part of the brim of the pelvis (page 169). Follow the muscle bellies of the iliacus and psoas superiorly, and establish continuity with the respective muscle bellies in the abdomen. Inferiorly, palpate the common iliopsoas tendon to its insertion in the lesser trochanter (page 285).

Follow the branches of the femoral nerve to the muscles that they innervate. In the femoral triangle, the branch to the sartorius should be identified.

Having mobilized sartorius, now study the rectus femoris (page 285). Sartorius crosses anterior to the upper portion of rectus femoris, just below the anterior superior iliac spine. Reidentify the latter on your cadaver and correlate it with the respective origins of sartorius and the straight head of rectus femoris. Cut sartorius across its belly near to its superior attachment and reflect the two cut portions of the muscle to gain better access to rectus femoris. Follow both the straight and reflected heads of rectus femoris to their respective origins (page 285), and observe the close relationship of the reflected head, and of the iliopsoas, to the hip joint, which you

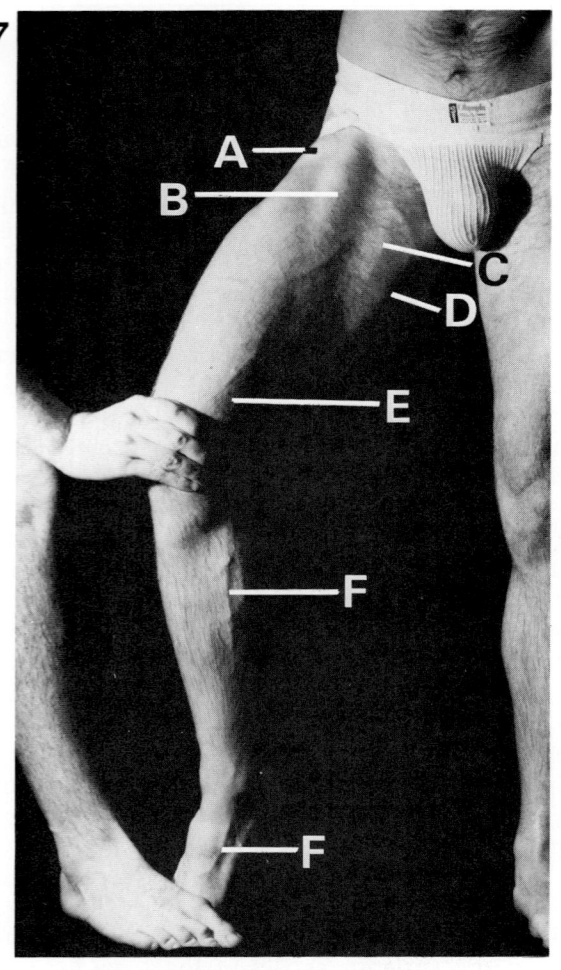

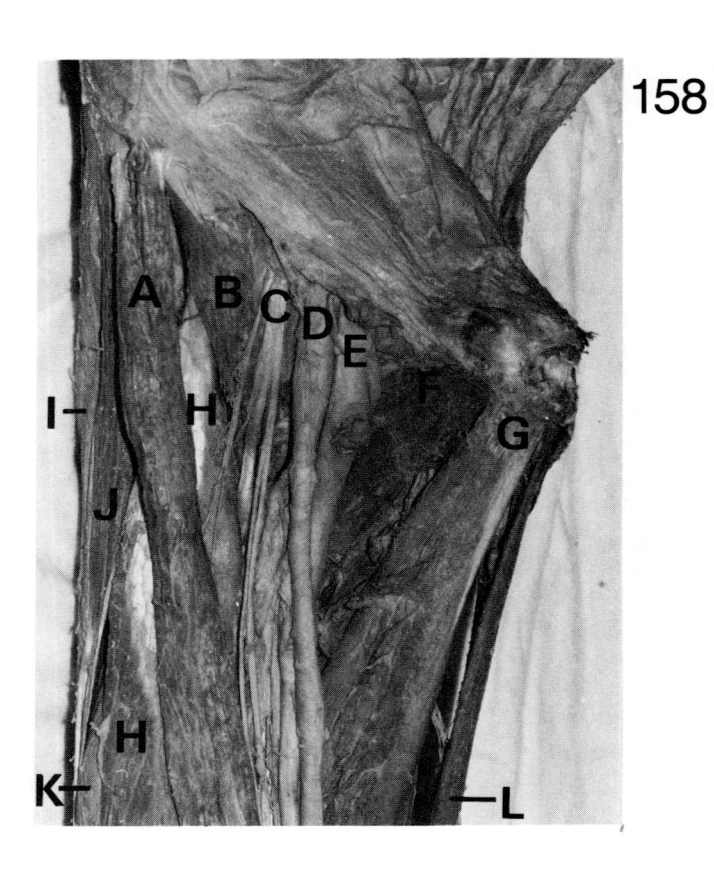

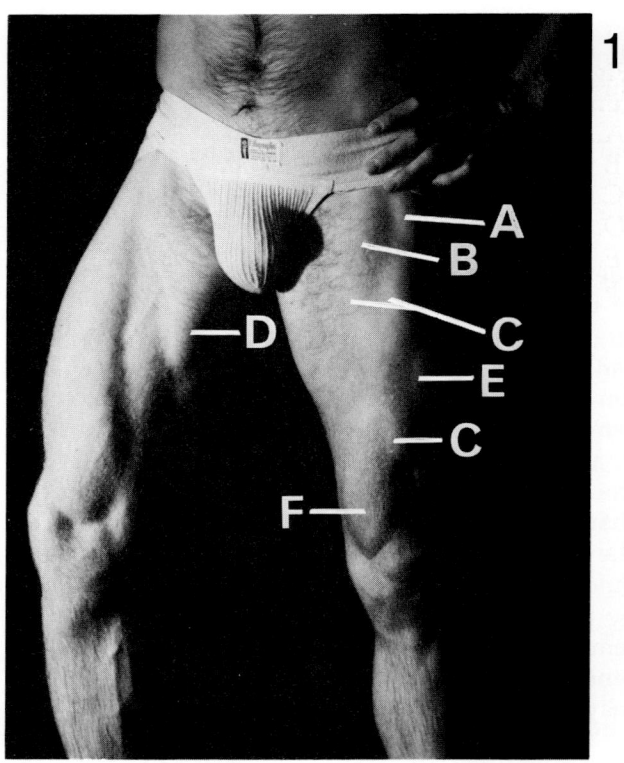

302 YOUR PATIENT'S ANATOMY: A CLINICAL VIEW OF HUMAN MORPHOLOGY

Figure 157. Anteromedial aspect of right lower limb.
A. Tensor fasciae latae
B. Sartorius
C. Femoral triangle
D. Adductor longus
E. Vastus medialis
F. Long saphenous vein

Figure 158. Dissection of right femoral triangle.
A. Sartorius
B. Iliopsoas
C. Femoral nerve
D. Femoral artery
E. Femoral vein
F. Pectineus
G. Adductor longus
H. Rectus femoris
I. Iliotibial tract
J. Tensor fasciae latae
K. Vastus lateralis
L. Gracilis

Figure 159. Dissection of superior end of left femur.
A. Capsule of hip joint
B. Neck of femur
C. Greater trochanter
D. Lesser trochanter
E. Femoral artery
F. Medial circumflex artery

Figure 160. Anterior aspect of left lower limb and anteromedial aspect of right lower limb.
A. Tensor fasciae latae
B. Sartorius
C. Rectus femoris
D. Adductor longus
E. Vastus lateralis
F. Vastus medialis

should identify by mobilizing the lower limb while an assistant steadies the pelvis. Posterolateral to the origin of the sartorius muscle, reidentify the origin of tensor fasciae latae, and follow the muscle downwards to observe its fibers merging into the iliotibial tract. Review the other important muscle insertion into this tract.

Clear the medial aspect of the thigh of connective tissue and fat, and study the adductor muscles on this aspect of the limb. Using finger dissection, trace the adductus magnus fibers, from their origin as described on pages 285 to 286, to their insertion, which you should also reidentify on a skeleton. Medial to adductor magnus, identify gracilis (page 285). Review the origin of gracilis, as described previously, adjacent to that of adductor longus, just lateral to the symphysis pubis. Trace the muscle down to its insertion, clearing connective tissue away to display the muscle belly, and review the relationships of the insertion of this muscle's tendon and those of sartorius, semitendinosus, and semimembranosus (page 287). Flex your specimen's knee to review the last two muscles. Also, review the position, in the living body, of the tendon of insertion of adductor magnus on the adductor tubercle (page 282).

Mobilize adductor longus, and study **adductor brevis,** which is situated posterior to adductor longus and pectineus. Trace the fibers downwards to the insertion, medial to those of the adductor longus, in the linea aspera. Then trace fibers superiorly to the origin of the muscle, from the lateral surface of the pubic ramus. Identify the site of origin on the skeleton.

At the lateral border of adductor brevis, mobilize this muscle from the bulky adductor magnus lying posterior to it. Display the anterior surface of the muscle belly of adductor magnus, and, turning your specimen to study the posterior aspect of the limb, review adductor magnus as seen from that approach and correlate your observations of this muscle from anterior and posterior aspects. In particular, review the proximity of the insertion of the upper fibers of adductor magnus to the insertion of fibers of gluteus maximus in the gluteal tuberosity, at the upper end of the lateral lip of linea aspera.

On the anterior aspect of adductor brevis, identify the anterior branch of the obturator nerve, crossing that muscle and passing to supply adductor brevis, adductor longus, gracilis, and, sometimes, pectineus. Review the course of the first portion of the obturator nerve, which you previously studied in the pelvis (pages 167 and 289). It leaves the pelvis through the obturator foramen, together with the obturator artery. Its passage from that point to the site where you now have identified branches supplying adductor muscles will be studied in the subsequent dissection.

Quadriceps Femoris

Review the situation and functions of the components of the quadriceps femoris muscle complex, as you studied it on the living body (pages 285 and 286). Finger and scalpel dissection should now be used to study the origins of these muscles from the femur, and the convergence of their fibers inferiorly, to form the common quadriceps tendon inserting into the superior edge of the patella. The origin of the **straight head of rectus femoris** could be palpated by deep examination, attaching to the anterior inferior iliac spine. The **reflected head** of the muscle cannot be palpated in the living, but will be studied in its relationship to the capsule of the hip joint. Confirm those relationships by dissection. **Vastus lateralis** takes its origin mainly from the lateral lip of the linea aspera on the posterior aspect of the femur. When traced upwards this lip diverges laterally, and the origin of the vastus lateralis follows this line and also curves around onto the anterior aspect of the femur, at the lateral end of the **intertrochanteric** line. Inferiorly, the origin of the muscle follows the natural divergence of the linea aspera toward the **lateral supracondylar ridge. Vastus medialis** takes origin from the medial lip of the linea aspera and also follows the superior and inferior divergences of this lip toward the medial margin of the femur. Superiorly the origin of vastus medialis also encroaches onto the anterior aspect of the femur, inferior to the lesser trochanter. Inferiorly the origin approaches the **medial supracondylar ridge,** close to the adductor tubercle.

Vastus intermedius takes its origin from a large area on the anterior surface of the femur, extending from just below the intertrochanteric line superiorly to a little more than half way down the anterior surface. Furthermore, the origin curves around on the lateral side of the femur and continues onto the posterior surface, in the middle of the shaft of the femur, reaching the lateral lip of the linea aspera, just lateral to the attachment of vastus lateralis. The vastus intermedius is largely covered by vastus lateralis and vastus medialis, as well as rectus femoris. In order to further study the muscle attachments to the thigh, at this stage the detached lower limb specimen should be further severed transversely across the thigh, about 5 cm inferior to the symphysis pubis. This bisection of the specimen must be done with a saw, under guidance.

On the upper specimen, first reflect sartorius and rectus femoris superiorly, and follow rectus femoris to its origins. Separate tensor fasciae latae and retract it laterally, and then identify and reflect laterally the upper portion of vastus lateralis. This will expose vastus intermedius, attaching just below the intertrochanteric line. Medially, identify the upper fibers of vastus medialis just below the lesser trochanter. Turning to the inferior half of your specimen, reidentify the lower half of rectus femoris, and trace it inferiorly. On its medial edge, dissect free the fibers of vastus medialis, which curve medially to insert into rectus femoris, and, further down, in common with it, into the patella. This dissection is important, to demonstrate clearly the direction of the lower fibers, in keeping with the function of vastus medialis in respect of movement of the patella. You previously considered the action of quadriceps femoris on the patella, and the effect of the angulation of the femora (page 285). Because of the angulated direction of pull of quadriceps in relation to the tibia, the patella would tend to be dislocated laterally, were it not for the direction of the lower fibers of vastus medialis demonstrated here.

Turning now to the posterolateral aspect of the thigh, reidentify the short head of biceps femoris, that takes origin from the lateral lip of linea aspera. Just anterior to it identify the vastus lateralis, also taking its origin from the lateral lip. Study the direction of its fibers as you trace them inferiorly, and contrast this with the direction of the fibers of vastus medialis. On lateral and medial aspects respectively, vastus lateralis and vastus medialis muscle fibers continue into a tough, fibrous aponeurosis (a flattened tendon), which merges with the lower end of the main tendon of the rectus femoris muscle. Deep to rectus femoris and vastus lateralis, identify vastus intermedius, which merges with vastus lateralis. Using finger dissection, separate vastus lateralis from vastus intermedius, which it overlaps. Note that vastus intermedius fibers join the common tendinous insertion into the superior border of the patella from its posterior aspect. Flex the knee of your specimen passively and, by tugging on the components of quadriceps femoris, confirm the major action of the muscle complex (page 286). Finally, trace branches of the femoral nerve to each of the components of the quadriceps.

Returning to the femoral triangle, reidentify the femoral artery and trace it inferiorly. At the upper edge of adductor longus, find and study a major branch of the femoral artery that passes posterior to adductor longus, as the main trunk of the femoral artery passes onto the anterior surface of the muscle. The posterior branch is called the **profunda femoris artery.** The subsequent course of the femoral artery and the profunda femoris will be studied in the following paragraph. Reexamine the entire extent of adductor magnus, including the insertion of the adductor magnus tendon into the adductor tubercle (page 282). The femoral artery may be followed down to its disappearance through a hiatus in the adductor magnus, close to the insertion of the tendon. The artery thus leaves the anterior aspect of the thigh to pass posteriorly, and into the popliteal fossa (pages 287 and 297). Here the artery changes its name to the **popliteal artery,** which you previously palpated (page 287) in the fossa, where it may now be seen with its accompanying **popliteal vein** and tibial nerve (page 297).

Study the profunda femoris branch of the femoral artery and its accompanying vein. Shortly after leaving the parent trunk, the profunda femoris gives off a **lateral circumflex femoral artery** and a **medial circumflex femoral artery.** (Both of these branches may occasionally arise directly from the femoral artery [fig. 159, page 302.] Trace the profunda femoris artery and the femoral artery inferiorly, noting that they diverge from each other at the upper border of the adductor longus, where the femoral artery passes onto the anterior surface of the muscle and profunda femoris onto the posterior surface. Study the perforating branches of the profunda femoris. These are three branches that pass posteriorly, through the adductor magnus close to its attachment to the femur, in serial sequence inferiorly. In series with these, the profunda femoris artery ends by also passing through adductor magnus to the posterior surface.

Study the upper portion of the bisected limb specimen. Reidentify the pectineus muscle, detach its femoral insertion, and reflect it laterally, retaining its superior attachment, to expose the deep-lying structures. Using scissor dissection, identify the medial circumflex branch of the profunda femoris artery (or of the femoral artery), and trace it medially to the site where it passes posterior to the posterior edge of adductor brevis. The dissection on the anterior surface of adductor brevis should also reveal the anterior division of the obturator nerve, crossing that muscle and sending branches to supply the adductor muscles of the thigh (pages 285 to 286). Detach the insertion of adductor brevis and reflect the muscle laterally, leaving its superior attachment intact, as you did in the case of pectineus. This will expose the posterior branch of the obturator nerve, sloping downwards and medially on the anterior aspect of adductor magnus. It will also give you further access to the medial circumflex artery of the thigh. Similarly reflect adductor longus and gracilis superiorly; connective tissue surrounding the origins of the adductors should be cleared away to demonstrate these attachments (pages 285–286). Study these, using a pelvic skeleton to reinforce the lines of attachment. Reconfirm the U-shaped origin of adductor magnus (page 286) in continuation of the curve of origins of the other muscles. Careful scissor dissection between the origins of pectineus and adductor brevis will allow further examination of the course of the obturator nerve and the medial circumflex femoral artery. The obturator artery will be seen to anastomose with the latter.

At the upper edge of the site of the femoral attachment of pectineus, close to the lesser trochanter and between pectineus and the iliopsoas tendon insertion into the lesser trochanter, the medial circumflex artery passes posteriorly. Having reflected adductor brevis upwards and medially, the artery can be further followed posteriorly, deep to adductor brevis. Observe the intimate relationship with the obturator externus muscle, which can now be identified and displayed from this approach. Review adductor magnus and its insertion into the femur as seen from the anterior aspect. Next, turn your specimen over and review again adductor magnus as you previously saw it on the posterior aspect. Reidentify quadratus femoris muscle, and the space between this and the adductor magnus. Now, reverting to the anterior aspect of the specimen, detach the adductor component of adductor magnus from the femur.

Reflect the adductor fibers of adductor magnus superiorly and medially, expose quadratus femoris from the anterior aspect, then invert the specimen again and reconfirm the relationship between adductor magnus and quadratus femoris. Define the superior border of quadratus femoris on its anterior aspect, using the border as defined on the posterior aspect to guide you. In this way you will be separating quadratus femoris from the obturator externus muscle, just superior to quadratus femoris. With the aid of a pelvic skeleton, study the origin of this muscle from the borders of the obturator foramen and define its fibers as they curve superoposteriorly to pass behind the femur on their way to their insertion into the medial aspect of the greater trochanter. The entire course of the muscle should be studied on both anterior and posterior aspects of your specimen. Returning to the anterior aspect, follow the medial circumflex artery as it passes between adductor brevis and obturator externus, and then between adductor magnus and quadratus femoris. Find its transverse branch, which passes laterally, posterior to the femur.

Follow the anterior and posterior branches of the obturator nerve, which you previously found on the anterior surfaces of adductor brevis and adductor magnus respectively, and trace them proximally to their respective points of passage above and through obturator externus.

On the anterior aspect of your specimen, follow now the fibers of adductor magnus proximally. In the approach used in this dissection, you can now clearly see the separate origins of the two components of adductor magnus: the adductor component, the upper portion of which you recently detached from the femur; and the hamstring component, which takes its origin from the ischial tuberosity. Study these two origins and the way in which the two sets of fibers form a curve.

Reidentify the iliopsoas muscle and tendon, and cut them across transversely, displacing the femoral nerve medially. Recall that these structures lie anterior to the hip joint (pages 285, 288, and 301), and this transection of the iliopsoas will give you access to that joint.

While on the anterior aspect of the specimen, trace the lateral circumflex femoral artery laterally, deep to sartorius and rectus femoris. The artery usually passes between the branches of the femoral nerve. Its important transverse branch passes into the muscle mass of vastus lateralis, to wind around the femur. This branch is important because it participates, with the corresponding transverse branch of the medial circumflex femoral artery, in forming the **cruciate anastomosis** (cruciate means cross-shaped). The vertical limbs of the cruciate anastomosis come from a descending branch of the inferior gluteal artery and an ascending branch from the first perforating branch of the profunda femoris artery (page 305). This anastomosis is capable of maintaining circulation to the lower limb, in the event of blockage in the external iliac or femoral arteries. What homologous anastomosis plays a similar role in the upper limb (page 266)?

On the posterior aspect, after reidentifying the quadratus femoris and obturator internus, bisect quadratus femoris vertically, leaving its origin intact on the ischial tuberosity and its insertion intact on the femur. Using careful scissor and scalpel dissection deep to the quadratus femoris, identify the fibers of obturator externus, as they wind around the femur toward the greater trochanter. Revert to the anterior aspect of your specimen to reidentify obturator externus where you previously saw it (see above), and, with finger dissection, follow its fibers from its origin to its insertion, turning the specimen as necessary.

It will now be possible to mobilize the upper, cut portion of the femur, and study the movements of the hip joint by moving the femur while the bony pelvis is stabilized. Review the actions of all the muscles around the hip joint by this means. Note also the relationship of the reflected head of rectus femoris to the capsule of the joint (page 285). Then bisect the obturator internus, gemelli, and piriformis muscles, to further expose the hip joint capsule. Study the external reinforcing ligaments of the capsule, the **ischiofemoral, iliofemoral,** and **pubofemoral ligaments,** using atlas illustrations to assist you. Then incise the capsule of the hip joint, and, by moving the cut femur, displace the head of the femur to allow you access to the interior of the hip joint. Study the internal structure of the joint, including the acetabulum (page 284), the **acetabular labrum** (a lip of fibrocartilage attached to the edge of the acetabulum), the **transverse acetabular ligament,** which bridges the acetabular notch at the inferior part of the acetabulum, and the **ligament of the head of the femur.** Observe the smooth, moist synovial surfaces, and the synovial membrane. To complete your study of the joint, examine its blood supply, derived from the gluteal, circumflex femoral, and obturator arteries, and its nerve supply, derived mainly from the femoral and obturator nerves.

Clinical Session 27

■ **TOPICS**

Knee Joint
Popliteal Fossa
Bony Landmarks of Leg and Foot
Muscles of Leg and Foot
Vessels and Nerves of Leg and Foot

In this session you will study the inferior (distal) portion of the lower limb, including the knee, leg (as defined on page 4), ankle, and foot. To begin with, some superfically visible and palpable landmarks will be identified, after which a structural and functional framework will be developed for subsequent dissectional study.

KNEE JOINT

Commence as usual, in a suitably warm and comfortable environment with your subject's lower limbs disrobed. You should also have a skeleton available for repeated identification of bony structures. Place your subject in the normal anatomical position and review the fundamental morphology of the lower limb. Reidentify homologous regions of upper and lower limbs, and note the respective positions of the thumb (pollex) of the upper limb, at the **lateral** aspect of the hand, and its lower limb homologue, the great toe (hallux), at the **medial** aspect of the foot. This discrepancy is one immediately obvious consequence of the developmental rotation of the lower limb (page 282), which, as you can see, brings the morphological lateral digit to a medial position. Previously, it was noted that this rotation also brings the morphological dorsal surface of the limb to the anterior aspect of the body in the anatomical position, whereas the dorsal surface is posterior in most regions. Thus, as explained previously, most of the anterior surface of the thigh and leg, and the superior surface of the foot, are morphologically dorsal. With this in mind, and your memory refreshed with respect to the definitions of the terms **flexion** and **extension** (page 225), review what constitutes flexion and extension at the knee joint (page 282). Determine what constitutes flexion and extension at the ankle joint, as well as at the joints between the toes and the small long bones of the foot, called metatarsals, which you should identify on a skeleton (and compare with the metacarpals of the hand, page 267), and the joints between the component small long bones (the phalanges) of the toes.

With your model in the supine position, reidentify the following landmarks around the knee joint. The patella (knee cap) is easily recognized as a rounded, bony structure. As your subject flexes the hip joint, keeping the knee joint extended, palpate at the inferior edge of the patella, and reidentify the **patellar ligament,** which represents the developmental insertion of the quadriceps femoris. The ligament inserts into the **tibial tubercle.** Recall that the patella develops as a sesamoid bone, with the result that the muscles appear to insert into its upper edge (page 285).

On either side of the patellar ligament reidentify (page 282) the tibial condyles, and, with your subject flexing the knee joint, palpate the articulation of these with the femoral condyles. It is possible to palpate quite deeply into the space between the tibial and femoral condyles, which constitutes the knee joint cavity. Note how far inferior to the joint cavity the inferior edge of the patella extends.

Superior to the patella, the inferior part of the quadriceps femoris muscle mass can be seen and palpated at its insertion. On the medial aspect, the quadriceps femoris muscle component is vastus medialis. Just posteromedial to it is the **adductor depression,** which is bounded posteriorly by the tendons of several muscles, which you should now reidentify with the knee slightly flexed and slightly raised off the examination table by hip flexion. The muscles are semimembranosus and semitendinosus (page 287), gracilis (page 285), and sartorius (page 284).

On the lateral aspect of the knee joint, the quadriceps femoris component is vastus lateralis, reinforced posteriorly by the iliotibial tract (fig. 149, page 294). Follow the tract inferiorly, to its insertion onto the lateral condyle of the femur, the lateral condyle of the tibia, and to the head of the fibula. Posterolateral to the iliotibial tract is an indentation, bounded posteriorly by the biceps femoris muscle, which passes inferolaterally to insert into the head of the fibula (page 287; fig. 149, page 294). Review the functions of biceps femoris.

Study the movements of the knee. These are flexion and extension as well as rotation. Rotation at the knee joint consists of movement of the tibia about a vertical axis. Starting in the anatomical position, your subject should slightly raise the front part of one foot from the floor, keeping the heel on the floor. By pivoting the foot on the heel, the great toe should now be moved in such a way that it is brought from its original position (in which it points anteriorly) to a position in which it points anterolaterally. This movement produces lateral rotation of the leg (or tibia) in relationship to the thigh (or femur).

With your subject now maintaining balance by holding on to a chair, one knee should be flexed. Palpate the major flexor muscles: semitendinosus, semimembranosus, gracilis, biceps femoris, and sartorius. Note that most of these flexors of the knee joint are also extensors at the hip joint. Which of the above-mentioned muscles are exceptions to this rule, and what are the features of their attachments that make them exceptions? Now ask your subject to extend the knee fully, and palpate the extensor muscle group, known as quadriceps femoris. Identify and name the components of this group. Note that as extension is brought to completeness, there is a slight lateral rotation of the leg; similarly, as flexion commences, there is a slight medial rotation. These rotational movements are determined by the configuration of the articulating surfaces and the ligaments of the knee joint. To facilitate your understanding of this note, on a skeleton, that the articulating surface of the medial femoral condyle is more extensive than that of the lateral condyle. How does this correlate with the observed rotation?

POPLITEAL FOSSA

With your subject now in the prone position identify on the posterior aspect of the knee, the diamond-shaped **popliteal fossa.** Superolaterally it is bounded by the biceps femoris, and superomedially by semimembranosus and semitendinosus (page 287). Inferolaterally and inferomedially the space is bounded by the **lateral** and **medial heads** of the **gastrocnemius** muscle (figs. 163 and 162, page 316; and fig. 156, page 298). Gastrocnemius constitutes the major bulk of the calf, and both heads mainly take their origins superiorly from the posterior aspects of the respective femoral condyles. Attempt to grasp each head separately, to appreciate its bulkiness and its independence of origin from the other head. Inferiorly, the two heads do merge, and form the major part of a very tough tendon, the **tendo calcaneus** (figs. 163 and 162, page 316), which should now be palpated as it inserts into the tuberosity of the **calcaneum** (heel bone). Identify this bone on your subject and on the skeleton. The tendo calcaneus is also known as the Achilles tendon. In Greek mythology, Achilles was dipped into the river Styx, to afford him mortal protection. However, he was held by the ankle in the dipping process, and this part of his body accordingly did not receive the protective anointing. As a result the Trojan warrior Paris was subsequently able to kill him with an arrow that hit him at this site. You will be responsible for this information in your final examination.

Palpate the gastrocnemius heads as your subject flexes and extends first the ankle joint and then the knee joint. What are the actions of gastrocnemius? Which divisions of the sacral plexus nerves will supply gastrocnemius?

To complete this preliminary examination of the knee region, palpate again the **popliteal artery** (pages 287 and 305) in the popliteal fossa. To do this, passively flex the knee slightly and palpate deeply in the middle of the fossa. The artery has an approximately vertical course through the deeper part of the fossa. Its accompanying **popliteal vein** and **tibial nerve** (the tibial branch of the sciatic nerve, pages 297 and 300) lie superficial to it (fig. 156, page 298). This relationship (of the nerve being superficial) is unusual. What practical (clinical) consequence may this relationship have? Note that the popliteal artery is a segment of the main arterial channel to the lower limb, being continuous with the femoral artery of the thigh and continuing inferiorly to supply the leg and foot.

BONY LANDMARKS OF LEG AND FOOT

Palpate the tibia through its entire length, from the medial tibial condyle superiorly to the inferior end of the bone, which forms the **medial malleolus.** Since the bone is superficial and easily palpable throughout its length, the medial malleolus is an easily identified landmark. Examine a skeleton to assist you in visualising this structure.

The broad, flat, and very superficial surface of the tibia that you have identified (known also as the shin) is the medial surface of this bone. The anterior border of the bone is also easily palpable; superiorly it continues into the tibial tubercle. Lateral to the anterior border, a muscle group is palpable, which can be felt to be activated when your subject extends the foot at the ankle joint. Recall that extension is the movement that brings the dorsal (superior) surface of the foot toward the dorsal (anterior) surface of the leg. This movement is often popularly called "dorsiflexion;" however, this term is oxymoronic. It should be remembered that in true flexion dorsal surfaces are moved **away** from each other. Now palpate lateral to the tibial tubercle, across the mass of the extensor muscles just identified, until you encounter the lateral tibial condyle and, just posterolateral to it, the prominent **head of the fibula,** which you should also identify on the skeleton. Starting at the head palpate inferiorly and identify as much of the fibula as you are able to, ending your palpation at its lower extremity, where it expands into the **lateral malleolus.** You will note that the shaft of the fibula, extending between the head and the lateral malleolus, is not quite as superficial as the shaft of the tibia. Your identification of the position of the fibula will establish, however, that the extensor muscle group you have identified is situated between the lateral surface of the tibia and the medial surface of the fibula. A tough interosseous membrane (inter means between; osseous means of bones) extends between the interosseous border of the fibula, which faces anteriorly, and a similar such border of the tibia, which faces laterally. These surfaces and borders of the fibula and tibia should be identified on the skeleton, and the position of the interosseous membrane between the interosseous borders visualized. The extensor muscle compartment is therefore enclosed by the interosseous membrane, the respective surfaces of the fibula and tibia described above, and, anterolaterally, by the deep fascia covered by skin and subcutaneous tissue.

Superiorly, in the slightly flexed knee, a tough **lateral collateral ligament** of the knee joint can be palpated between the head of the fibula and the lateral femoral condyle. A corresponding **medial collateral ligament** between the medial condyles of the tibia and femur can also be palpated medially. The medial ligament is, phylogenetically, a downward extension of adductor magnus.

In preparation for further study of the muscles of the extensor compartment and other leg muscles that insert in the foot, turn now to identifying some bony and other landmarks in the foot. As usual, identify all bony structures mentioned both on your living model and on the skeleton.

The bones of the foot are, in principal, homologous with those of the hand, although, as mentioned, the rotation of the lower limb during development brings the great toe to lie medial whereas the thumb is lateral. In accordance with this homology, the foot bones comprise the bones of the **digits** (toes), homologous to the bones of the upper limb digits (fingers); the bones of the ankle, called **tarsal** bones, homologous with those of the wrist (carpals, page 267); and, between the digits and the ankle bones, the **metatarsals** (meta means next to or after), or bones of the sole of the foot, homologous to the metacarpals, or bones of the palm of the hand. As in the hand, the bones of the digits are called **phalanges.** The phalanges and the metacarpals are, in their shape and manner of development, similar to the long bones of the upper and lower limbs. They are accordingly called "small long bones."

Reidentify the medial malleolus (page 309) and palpate a small bony projection about 2 cm inferior to the lower edge of the malleolus. This projection is the **sustentaculum tali,** which projects medially from the calcaneum, or heel bone that you previously identified (page 309) (sustentaculum means sustainer or supporter of; and tali means of the **talus**). This ledge supports the talus bone, another of the ankle bones that you should now observe on the skeleton. Note that the talus articulates medially with the medial malleolus of the tibia, and laterally with the lateral malleolus of the fibula. The rounded **head of the talus** is the portion of that bone that is sustained by the sustentaculum tali.

Returning to palpation of your living subject, palpate on the medial aspect of the ankle 3 cm anterior to the sustentaculum tali, and identify a rounded projection, the **tuberosity** of the **navicular** (navicular means boat-shaped: confirm the reason for this name on the skeleton). On the skeleton you will see that this tuberosity is just anteroinferior to the head of the talus.

On the medial aspect of the foot notice the arch of the foot, and palpate the **base** (the expanded proximal part) of the first metatarsal, which articulates with tarsal bones, and the rounded, expanded distal portion (the **head**) of this first metatarsal, which articulates with the first phalanx of the big toe. On the lateral aspect of the foot, the remaining characteristic bony landmark palpable is the **tubercle** of the base of the fifth metatarsal. This metatarsal is longer than the first, and projects further posteriorly. Accordingly, the tubercle is palpated further posteriorly, on the lateral aspect of the foot, than the base of the first metatarsal is on the medial aspect of the foot. Seen from the dorsal aspect (i.e., from above), study the clinically important triangle that is formed by the tubercle of the fifth metatarsal, the base of the first metatarsal, and the tuberosity of the navicular. Another triangle to review at this stage is that formed by the tuberosity of the navicular, the sustentaculum tali, and the lower edge of the medial malleolus. The characteristic relationships of these bony points are used to determine whether or not there is any displacement of bony structures in the event of trauma to the foot.

Other bony structures of the foot are less easily palpable but their approximate positions can be ascertained by reference to the skeleton of the foot. Making use of this method, palpate the approximate positions of the remaining tarsal bones. Between the navicular and the first, second, and third metatarsals (named from medial to lateral) the first, second, and third **cuneiforms** (cuneus means wedge) are found. Finally, lateral to the third cuneiform and the navicular the **cuboid** is found. This bone articulates posteriorly with the calcaneum and anteriorly with the fourth and fifth metatarsals. The metatarsals can be palpated for much of their length on the dorsal surface and less easily on the ventral surface. The heads of the metatarsals collectively constitute the "ball" of the foot. Individual phalanges of the digits can be palpated, as can the interphalangeal joints.

MUSCLES OF LEG AND FOOT

Now return to the extensor compartment of the leg (pages 309 to 310) and identify the individual muscles, as follows. The fleshy muscle mass situated just lateral to the superior end of the anterior border of the tibia comprises the **tibialis anterior** muscle medially and the **extensor digitorum longus** just lateral to the previous muscle and separated from the head of the fibula by a groove. To distinguish between these two muscles, palpate tibialis anterior while your subject extends the foot against resistance placed on the medial half of the dorsal surface of the foot. Palpate extensor digitorum longus while your subject extends the foot against resistance placed on the lateral part of the dorsal surface of the foot, or on the lateral toes. Inferiorly, at the front of the ankle just lateral to the anterior border of the tibia and close to the medial malleolus, the powerful tendon of tibialis anterior can be palpated as the foot is extended. About 1 cm lateral to this tendon, a second powerful tendon, that of extensor digitorum longus can also be palpated. Finally, a third tendon between these two can be identified. This is the tendon of **extensor hallucis longus**. It can be specifically palpated while your model extends the great toe against resistance. The tendons of these three extensor muscles can be palpated down to their insertions, as your subject extends the foot and toes appropriately (figs. 169, 170, 171, and 172, page 326). The tibialis anterior tendon inserts into the base of the first metatarsal and the adjacent 1st cuneiform. Extensor digitorum longus tendon splits into four slips that each insert into the base of the first phalanx of the second, third, fourth, and fifth digits (toes). The extensor hallucis longus tendon inserts into the base of the second (distal) phalanx of the great toe. The origin of the extensor hallucis longus is less easy to identify than that of the other two extensors because it lies deep to them, taking its origin from the fibula. To complete this description of the extensor muscles it should be mentioned that a small muscle belly, called **peroneus tertius,** takes its origin from the inferior part of the fibula, in common with extensor digitorum longus. (Tertius means third: the first and second peroneus muscles will be studied in the ensuing paragraphs.) Peroneus tertius separates from extensor digitorum longus, and its tendon inserts into the base of the fifth metatarsal, where it can be palpated as your subject extends the foot against resistance (figs. 170 and 171, page 326). Identify the metatarsal and phalangeal attachments, mentioned earlier, on the skeleton as well as on your subject.

Reidentify the head of the fibula and the insertion into it of the biceps femoris tendon. About 1 cm inferior to this, on the lateral aspect of the fibula, the **common peroneal nerve** can be palpated as a cord, emerging from the popliteal fossa (fig. 156, page 298) to curve across the lateral aspect of the neck of the fibula. As it winds onto the anterior surface it divides into deep and superficial branches, which will be followed to their destinations in due course. The nerve can be rolled upon the fibula, eliciting a mild sensory response in your subject that will confirm the identity of the structure. This nerve contains the dorsal divisions of those nerve components of the sacral plexus destined to supply the leg. That is, this nerve will supply the morphological extensors of the leg, including those muscles you have just identified, as well as sensory fibers to the dorsal surface. Morphologically, in the lower limb the extensor group also includes the abductors (pages 286 and 300), and these will now be studied.

Peroneal Compartment

With your subject in the supine position, ask him or her to move one foot in such a way that the ventral (inferior) surface of the foot (the sole) is turned to face out laterally. This movement is called **eversion** and is the equivalent of abduction. The muscles that perform this movement are **peroneus longus** and **peroneus brevis,** the bellies of which can be palpated on the anterolateral aspect of the leg, starting at the head of the fibula and working downwards as your subject everts the foot. Peroneus longus takes its origin from the upper part of the lateral aspect of the shaft of the fibula and peroneus brevis, deep to the longus from the lower part (fig. 161, page 316). By placing some resistance against your subject's foot as it is everted, and palpating with the other hand inferior to the lateral malleolus, the two tendons can also be identified (figs. 169 and 171, page 326).

Superiorly the origin of the peroneus longus is continuous with the lateral (fibular) collateral ligament of the knee joint (page 310). This is an evolutionary relic of the fact that this muscle, in earlier phylogenetic stages, took its origin from the femur.

In the foot the peroneus brevis tendon is the more anterior of the two, and can be palpated to its insertion on the lateral side of the base of the fifth metatarsal (figs. 169 and 171, page 326). The peroneus longus tendon passes more posteriorly, having crossed the peroneus brevis tendon superficially, just posterior to the malleolus (fig. 171, page 326), and the longus tendon then curves into the sole of the foot and passes on the inferior surface to reach the base of the first metatarsal and the adjacent medial cuneiform bone (page 310; see fig. 175, page 330). Both bones, and the insertion of the peroneus longus, can be palpated on the inferomedial aspect of the foot, in the arch of the sole of foot. The shaft of the first metatarsal is easy to identify on the medial side of the foot, and the medial cuneiform is situated just posterior to this metatarsal. Both bones should be identified on the skeleton to aid you in finding them on your model. Palpate both peroneus longus and peroneus brevis tendons to their insertions while your subject everts the foot.

Flexor Muscles of the Leg (Calf Muscles)

The true flexors of the foot are the muscles found on the posterior aspect of the leg. They are supplied by nerves derived from ventral divisions of the sacral plexus. Reidentify the medial and lateral heads of gastrocnemius as your subject flexes the foot.

Elicit strong true flexion of the foot in your subject by asking him or her to stand on his or her toes. On the lateral aspect of the calf identify another muscle belly, visible and palpable anterior and inferior to the lateral head of gastrocnemius. This is the lateral edge of **soleus**, which, as indicated, lies deep (anterior) to gastrocnemius (fig. 161, page 316). In flexion against resistance, as in standing on tiptoes, these two muscles are clearly demarcated in lean subjects, as is a groove between the anterior edge of the soleus and peroneal muscles. Soleus is also palpable, though not quite as obviously visible, on the medial aspect of the calf, anterior to the medial head of gastrocnemius. Inferiorly, soleus inserts, in common with gastrocnemius, into the tendo calcaneus (fig. 166, page 320).

A third superficial muscle of the calf, the **plantaris,** cannot be examined in the living and will be seen subsequently in dissection.

Deep Flexor Muscles

In addition to the three superficial flexor muscles mentioned above, three other flexors are deeply situated in the calf, anterior to those previously mentioned. Their tendons can be identified on the medial aspect of the ankle. Palpate the medial malleolus. Now ask your subjects to **invert** the foot (i.e., turn the sole of the foot medially) (fig. 172, page 326). Clearly, the **prime mover** muscles (page 225) of the movement of eversion (page 311) are **antagonists** in the movement of inversion. With your subject performing inversion against resistance of your left hand, palpate with your right hand immediately posteroinferiorly to the medial malleolus, and identify the tendon of **tibialis posterior.** This tendon can be followed inferoanteriorly to its insertion onto the inferior surface of the tuberosity of the navicular (page 310) and neighboring tarsal bones. The tendon of tibialis posterior can also be palpated proximally, passing posterosuperiorly immediately behind the medial malleolus and posterior to the tibia.

In addition to inversion, tibialis posterior is, of course, by virtue of its insertion onto the inferior surface of the foot, a flexor at the ankle joint. Note that tibialis anterior, which you previously studied as an extensor of the foot (page 311), is also, like the tibialis posterior, an inverter of the foot. Thus while tibialis anterior and tibialis posterior are antagonistic to each other in respect of their functions as extensors and flexors respectively, they cooperate as prime movers in inversion.

Immediately posterior to the tendon of tibialis posterior at the level of the medial malleolus, also curving inferiorly and then anteriorly into the sole of the foot, palpate **flexor digitorum longus** (figs. 166, 167, and 168, page 320). Its tendon can be palpated as your subject flexes his or her toes against resistance. The slips of insertion into the 2nd to 5th toes can be palpated at the bases of the distal phalanges. The muscle belly can also be palpated 1 cm posterior to the tibia, immediately posterior and lateral to the tibialis posterior. Finally, slightly posterolateral to flexor digitorum longus muscle is **flexor hallucis longus** (hallucis means of the hallux), best palpated as your subject flexes the great toe against resistance. This tendon passes anteriorly in a groove on the inferior surface of the sustentaculum tali (page 310) to reach the base of the distal phalanx of the great toe.

Two small muscles of the foot are quite easily palpable in the living subject. On the medial aspect of the arch of the instep, a small muscle belly, the **abductor hallucis,** can be palpated, extending from a small bony elevation called the **medial tubercle** of the calcaneum to the medial side of the base of the proximal phalanx of the great toe (fig. 165, page 320 and figs. 173, 175 and 176, page 330). It can be palpated in most of its length from its origin on the calcaneum to its insertion on the proximal phalanx, passing inferior to the tuberosity of the navicular. To assist in this palpation, it can be passively stretched by gently pushing your model's great toe laterally. Despite its name (an evolutionary relic) the muscle's function is flexion.

On the dorsal aspect of the foot palpate a slight depression just anterior to the medial malleolus. Then, just anterior to this depression, identify the muscle belly of **extensor digitorum brevis** as a fleshy mass that tenses as your subject extends the toes (figs. 169, 170, 171, and 172, page 326).

Palpate the relevant bones and joints as your model performs the following movements of the foot, and in doing so review previous descriptions of the relevant muscles, tendons, and bones. True flexion and extension (the latter also, but inaccurately, called dorsiflexion) occur mainly at the joint between the talus and the lateral and medial malleolae, and to a lesser extent at the joint between the navicular and talus. Review the muscles mainly responsible for flexion and extension of the foot. Inversion and eversion occur partly in the joint between the talus and the calcaneum and partly in the **midtarsal joint**. This latter joint is formed medially by a ball and socket, formed by the talus and calcaneum (the ball) in the navicular (the socket), and laterally by the joint between the calcaneum and the cuboid. Review the muscles involved in inversion and eversion. Which movement is more extensive?

Study flexion and extension of the toes, paying attention to movements at the metatarsophalangeal joints as well as the interphalangeal joints. Review the actions of those muscles that you have studied thus far, which act in flexion and extension of the digits (extensor digitorum brevis [see above]; extensor digitorum longus [page 311]; extensor hallucis longus [page 311]; flexor digitorum longus [page 312]; flexor hallucis longus [page 312]; and abductor hallucis [page 312]).

The foot serves the dual purposes of supporting the weight of the body when stationary and of participating in locomotion. These functions are, to an extent, in conflict with each other, since ideally, for support, the foot should lie flat on the ground, whereas for it to act in thrust-off in walking, the foot functions more effectively when constructed as an arch. The plantar surface of the foot is in fact arch-shaped in its anteroposterior as well as in its mediolateral dimensions (fig. 165, page 320). The anteroposterior or longitudinal arch has a medial and a lateral portion; the mediolateral arch is also known as the transverse arch of the foot. Study the arch-shape of your model's foot in standing and in walking, and identify the bones involved in the medial portion of the longitudinal arch (the calcaneum, talus, navicular, cuneiforms, and first, second, and third metatarsals), and in the lateral portion of the longitudinal arch (the calcaneum, cuboid, and fourth and fifth metatarsal bones). Also study the bony components of the transverse arch, as seen both anteriorly and posteriorly in the foot. The arches are maintained by strong ligamentous structures, to be studied by dissection later, as well as by several of the muscles and their tendons, some of which you have already studied. Which of the muscles and tendons studied thus far are likely to play a role in the maintenance of the arches of the foot?

VESSELS AND NERVES OF LEG AND FOOT

Now study the veins and arteries of the foot and leg. On the dorsum of the foot, between the visible tendons of extensor hallucis longus and the extensor digitorum longus slip to the second toe, previously identified, the pulse of the **dorsalis pedis artery** is clinically palpated. Place your index, middle, and ring fingers in the groove between and parallel to these two tendons, close to the interdigital cleft between the 1st and 2nd toes. The pulsation of this artery should be palpable. The presence of a detectable pulse at this site is taken as an indicator of the integrity of the arterial supply to the distal part of the lower limb. The dorsalis pedis artery is the continuation of the anterior tibial artery, one of the two terminal branches of the popliteal artery that you previously palpated in the popliteal fossa (page 309). The anterior tibial artery supplies the anterior part of the leg. The dorsalis pedis artery ends by giving off branches which supply the toes.

The other terminal branch of the popliteal artery is the posterior tibial artery, which supplies much of the posterior muscles of the leg. A palpable pulse of this artery is also used clinically and should be identified now. Place your three palpating fingers posterior to the medial malleolus, and identify the pulsation of this artery.

The superficial veins of the lower limb are often clearly visible at their commencement in the foot. On the medial aspect of the dorsum of the foot, identify the start of the **long saphenous vein**. This vein passes anterior to the medial malleolus, up the medial aspect of the calf (fig. 157, page 302), and continues into the thigh to end, as previously identified, close to the inguinal ligament where it dips deeply to empty into the femoral vein (pages 288 and 301). The **short saphenous vein** can also be identified at its commencement on the dorsum of the foot, on the lateral aspect. This vein passes posterior to the lateral malleolus, then continues superiorly onto the lateral aspect of the leg, and then veers posteriorly. It ends by passing deep to the deep fascia covering the popliteal fossa where it terminates by joining the popliteal vein (page 309).

The peripheral nerves that supply the lower limb may be tested as you previously did for the cranial nerves (page 191) and for the nerves of the upper limb (page 274). Using the techniques described previously, test for the integrity of the modalities of light touch, pain, and proprioception (position sense).

The modality of proprioception is clinically tested as follows. With your subject in the supine position, instruct him or her to close his or her eyes. Grasp the great toe between your thumb and index finger and repeatedly move it passively (i.e., without the subject participating by flexion or extension) into the flexed and extended positions. Cease the movement in either the extended or the flexed position, and ask your subject to state whether the toe is pointing "up" or "down." In certain disease conditions of the peripheral nervous system, the sense of proprioception may be lost (i.e., the subject may be unable to state in which direction the toe is pointing).

To complete your clinical examination of lower limb morphology, test the lower limb reflexes. Review the concepts of the **reflex arc** (pages 5 to 6, 26, and 86) and of the **spinal segment** (page 86 and fig. 51, page 104). Also refresh your memory on how you tested the flexor and extensor stretch reflexes of the upper limb (page 274). Now examine the **extensor reflex** at the knee joint. To do this, your model should be in a relaxed, supine position. Passively flex one lower limb slightly at the knee, then place the palm of your left hand posterior to the flexed joint, to support it in the popliteal fossa, leaving the model's heel resting on the examining couch. The weight of the limb should thus be suspended. Alternatively, your model may be seated with crossed lower limbs, one thigh resting on the other so that the leg of the crossed limb is loosely suspended. Palpate inferior to the patella of the suspended limb to identify the patellar ligament, and, using the reflex hammer (page 274), strike across the ligament midway between the inferior edge of the patella and the tibial tubercle. Observe the consequences, which, on the assumption that the test is correctly and gently performed, can be interpreted in terms of the explanation on page 274. To test the **flexor reflex** place the foot in passive extension and strike the Achilles tendon.

A common test of whether the connections between the peripheral nervous system and the **extrapyramidal** nervous pathways of the central nervous system are intact is to elicit the **plantar reflex**. The extrapyramidal pathways carry motor inhibitory impulses, in contrast to those of the **pyramidal tract**, which carry the impulses that initiate voluntary movement (page 130). Using a blunt instrument such as the back of a pen, stimulate the ventral (plantar) surface of the foot by drawing the instrument in a posteroanterior direction, starting near the heel and ending near the great toe. In a normal plantar reflex the toes curl into flexion. Abnormally, they will go into extension, an event known as the "Babinski response," which, as mentioned earlier, is indicative of an extrapyramidal lesion.

Laboratory Session 27A

Remove the skin from below the knee to beyond the ankle on both the dorsal and ventral aspects of both lower limbs. As usual, one team should be dissecting the superficial structures on the attached limb, and a second team should be dissecting down to the deep fascia and muscles on the detached and now severed specimen.

Study the superficial veins and nerves of the leg. The long saphenous vein (page 313) should be dissected, passing anterior to the medial malleolus on to the medial aspect of the leg, and the short saphenous vein (page 313) posterior to the lateral malleolus and on to the lateral aspect. Dissect the valves of both veins as you previously did in the upper portion of the long saphenous vein (page 301), and identify some of the deep communicating branches, which connect these superficial veins to the deep veins of the leg. When the valves are nonpatent, the communicating veins may be the cause of varicose superficial veins. Follow the short saphenous vein as it swerves medially on the posterior aspect of the calf, to dip deeply at the lower border of the popliteal fossa, where it joins the popliteal vein (page 313).

Study the superficial nerves that accompany the long and short saphenous veins respectively. In the leg, the saphenous nerve emerges through deep fascia about 5 cm below the knee, on the medial aspect, and accompanies the long saphenous vein. The saphenous nerve, a branch of the femoral nerve, passes downwards and medially deep to muscles in the thigh before emerging at this superficial site. It passes inferiorly to supply the medial aspect of the ankle. On the posterior aspect of the calf, dissect the sural nerve, which runs most of the length of its course next to the short saphenous vein. It passes down onto the lateral aspect of the foot, which it supplies all the way to the middle toe (fig. 171, page 326).

Other cutaneous nerves should also be dissected. On the anterior aspect of the calf, infrapatellar branches of the saphenous nerve pierce deep fascia just inferior to the patella, and branches of the lateral cutaneous nerve of the calf, a branch of the tibial nerve, curve around the calf and become superficial on the anterior aspect. Inferiorly, on the dorsum of the foot (i.e., the superior aspect) the superficial peroneal nerve, a branch of the common peroneal nerve, pierces the fascia, and supplies branches to each side of the lateral three interdigital (between toes) spaces, as well as a branch that passes to the medial aspect of the big toe. On the posterior aspect of the calf, the lateral cutaneous nerve of the calf supplies cutaneous branches just below the knee, and the posterior cutaneous nerve of the calf just below the popliteal fossa. On the medial aspect of the calf, branches of the saphenous nerve become superficial. On the ventral surface of the foot, cutaneous branches of medial and lateral plantar nerves (terminal branches of the tibial nerve) supply the medial and lateral sides of the foot respectively, and plantar digital nerves supply the interdigital spaces. These superficial nerves are, of course, the ones responsible for conveying centripetally (toward the central nervous system) the sensory impulses for which you tested clinically (pages 313 to 314).

Identify, through its covering layer of deep fascia, the medial head of the gastrocnemius muscle (page 309). Carefully incise the deep fascia to mobilize the muscle. Rotating the limb medially, similarly expose the lateral belly of the muscle. Reflect the deep fascia with care, to preserve nerves lying just deep to it, which send superficial branches through the fascia. For example, by following the sural nerve superiorly, its origin from the tibial nerve can be studied in the popliteal fossa. Between the two heads of gastrocnemius, the medial cutaneous nerve of the calf (branch of the tibial nerve) passes down inferiorly. Laterally, the common peroneal nerve winds around the neck of the fibula (page 311). It can be dissected medially (proximally), across the lateral head of gastrocnemius to the popliteal fossa (fig. 156, page 298). Swerving down from the common peroneal nerve, the communicating peroneal nerve passes medially and joins the medial cutaneous nerve of the calf, to form the sural nerve, which thus has both tibial and peroneal components.

Using scissor dissection, study the relationships of the tibial and common peroneal nerves, and the popliteal vein and artery. Review the important sequence in which these structures lie (page 309). The nerves are most superficial, unlike many other parts of the body. Therefore they are exposed to the risk of trauma in this region. Trace the popliteal vein and artery superiorly, and follow them through the hiatus in the adductor magnus muscle (fig. 155, page 298). Turn the cadaver to reconfirm from the anterior approach that it is at this site that the femoral vessels pass from the anterior to the posterior surface, changing their name as they do so (page 305). Dissect the common peroneal and tibial nerves superiorly, to the point where they diverge from their common stem, the sciatic nerve, which you previously followed down to this site (page 297).

Carefully dissect the medial and lateral heads of gastrocnemius superiorly, to their respective ori-

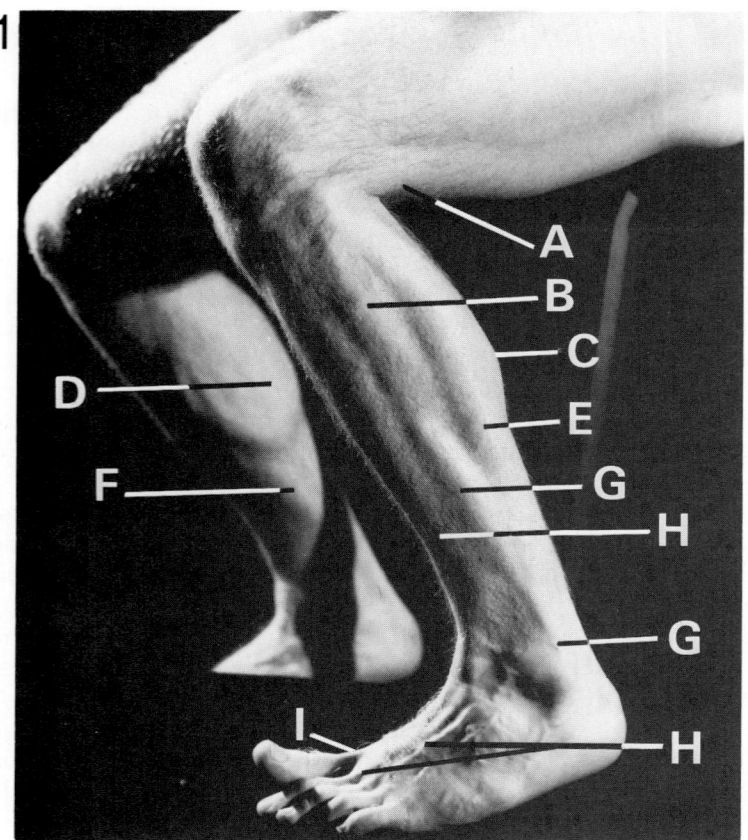

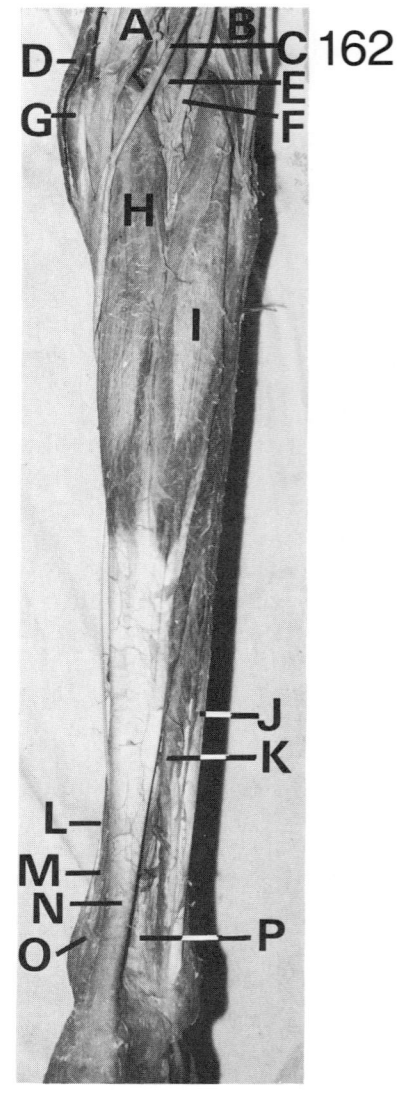

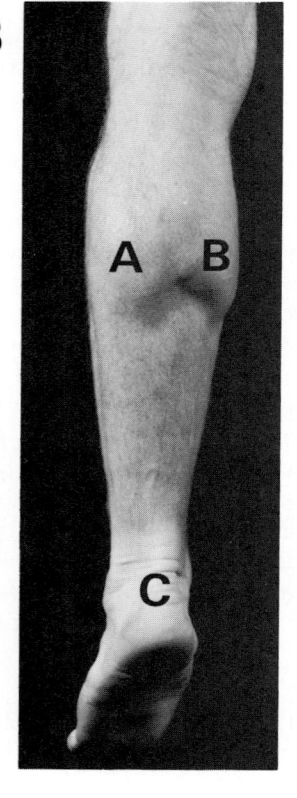

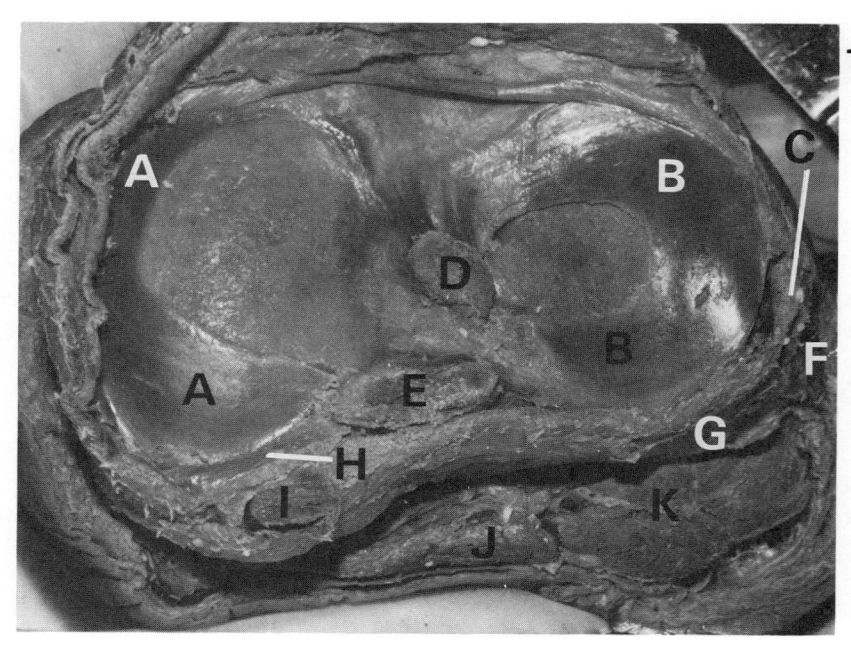

316 YOUR PATIENT'S ANATOMY: A CLINICAL VIEW OF HUMAN MORPHOLOGY

Figure 161. Lateral aspect of left lower limb and posteromedial aspect of right leg.
A. Biceps femoris
B. Peroneus longus
C. Left gastrocnemius
D. Right gastrocnemius
E. Lateral edge of left soleus
F. Medial edge of right soleus
G. Peroneus brevis
H. Extensor digitorum longus
I. Extensor hallucis longus

Figure 162. Dissection of posterior aspect of right leg.
A. Semimembranosus
B. Biceps femoris
C. Semitendinosus
D. Sartorius
E. Popliteal artery
F. Tibial nerve
G. Gracilis
H. Medial head of gastrocnemius
I. Lateral head of gastrocnemius
J. Peroneus longus
K. Peroneus brevis
L. Flexor digitorum longus
M. Tibialis posterior
N. Achilles tendon
O. Flexor retinaculum
P. Flexor hallucis longus

Figure 163. Posterior view of left calf.
A. Lateral head of gastrocnemius
B. Medial head of gastrocnemius
C. Achilles tendon

Figure 164. Superior (articulatory) surface of tibial condyles and related structures.
A. Medial meniscus
B. Lateral meniscus
C. Popliteus
D. Anterior cruciate ligament
E. Posterior cruciate ligament
F. Biceps femoris
G. Plantaris
H. Semimembranosus
I. Semitendinosus
J. Medial head of gastrocnemius
K. Lateral head of gastrocnemius

gins on the posterior aspect of the femur, just above the respective femoral condyles. The two heads of this muscle form the inferolateral and inferomedial boundaries of the popliteal fossa, the superior borders of which you should now review (page 287).

You should perform two exercises, comprising movements of the knee joint and of the ankle joint respectively, to test the function of the gastrocnemius muscle. Flex the knee joint, and note that by tugging on the gastrocnemius tendon, you can simulate flexion at this joint. Extension will conversely place stretch on the muscle. By tugging upwards on the belly of gastrocnemius, confirm also that this will cause flexion at the ankle joint.

Mobilize the lateral head of the gastrocnemius muscle, and, dissecting on its medial aspect, identify and define, and follow to the muscle, the branches of the tibial nerve, and the vein and artery, which supply this muscle. Carefully cut across the muscle belly, about 3 cm inferior to its superior attachment onto the femoral condyle. Reflect the muscle medially to expose deep-lying structures.

You will now be able to study the soleus muscle, which takes its origin from the posterior surface of the tibia. The muscle passes inferiorly deep to gastrocnemius. Inferiorly, the tendons of soleus and gastrocnemius unite, and insert, as the **Achilles tendon,** into the posterior surface of the **calcaneus.** Study the soleus muscle, including its origin, its tendon, and its insertion with the gastrocnemius (fig. 161, page 316, and fig. 166, page 320).

Reflect the superior portion of the gastrocnemius muscle, and follow the lateral head superiorly to its attachment to the lateral femoral condyle. This will display the intimate relationship of this muscle and its tendon attachment to that of **plantaris,** which comes into view just medial to the lateral gastrocnemius origin. The belly of plantaris will be displayed deep in the popliteal fossa. Inferiorly, it becomes continuous with the very thin tendon that runs down, as the name of the muscle implies, to the plantar surface of the foot. Plantaris is a variable muscle, and is present in approximately 40 percent of subjects. (Compare this muscle with palmaris longus, variably present in the upper limb [page 268]). Cut the belly of plantaris close to the point where it becomes continuous with the tendon, and reflect it superiorly, dissecting it away, on the deep surface, from the condyle of the femur, using scalpel dissection. You should repeatedly palpate the bony surrounding structures and flex the knee joint to assist you in maintaining your bearings in respect of the knee joint. Deep to plantaris, find the branch of the tibial nerve passing to the soleus muscle. By reflecting the lateral head of gastrocnemius far over medially and inferiorly, expose the soleus attachment in its entirety.

Turning now to the medial side of the popliteal fossa, remove fat and connective tissue to display structures lying medial to the upper medial boundary of the fossa. These are the tendons of the **gracilis** and **sartorius** muscles, the tendinous insertions of which you previously followed down to the medial aspect of the knee joint (fig. 155, page 298). Dissect these tendons downwards and display their attachments to the medial tibial condyle. Your dissection will also bring you to demonstrate the close relationship of the semitendinosus tendon, just posterolateral to that of gracilis. Finally, yet more medially and deep to semitendinous, trace the semimembranosus tendon, which ends in a complex fashion, attaching partly to the posterior aspect of the medial tibial condyle and reflecting also medially onto the medial head of gastrocnemius, and, deep to that, into a tough fibrous sheet called the **popliteus fascia.** To demonstrate the latter, you should now embark on the dissection of the medial head of gastrocnemius.

Cut through the fascial extension of the semimembranosus tendon over gastrocnemius medial head, to expose the head, then cut this head of gastrocnemius across and reflect the bulk of the belly inferiorly. Finally reflect the superior part of the medial head of gastrocnemius superiorly, and dissect its attachment to the medial condyle of the femur, displaying, deep to the tendon, the **bursa** (see page 263) that lies between the tendon and the bone, and that usually opens into the knee joint.

Use scalpel dissection to get the tendinous portion of the attachment of the medial head of gastrocnemius well separated from the medial portion of the medial femoral condyle. Display the expansion of the semimembranosus tendon into the popliteus fascia, then incise this fascia carefully, and reflect to expose the deep-lying **popliteus muscle.** This muscle has a flat insertion on the posterior surface of the tibia, above the curving soleus line, which you should identify on a skeleton at this time. Observe the inferior lateral and medial genicular arteries and veins (genu means knee), branches and tributaries respectively of the popliteal artery and vein, crossing the popliteus muscle, and then cut these in order to be able to mobilize the muscle superiorly, and trace its tendon out laterally toward the lateral side of the knee joint. The tendon of origin of popliteus is attached to the lateral condyle of the femur, within the knee joint, close to the lateral collateral ligament and biceps femoris tendon. Superior lateral and medial genicular vessels, similarly derived from the popliteal vessels, should be observed completing the vascular anastomosis of the popiteal region.

Proceed now with the following steps. Reidentify the tendon of the biceps femoris, and follow it inferiorly, dissecting on the medial aspect of the short head of biceps. Follow it down to where it inserts into the lateral ligament of the knee joint. Carefully dissect the ligament and display its extent, and its superior and inferior attachments to the lateral condyles of the femur and tibia respectively. Then mobilize the ligament, and the biceps tendon somewhat on its medial aspect, so that the tendon can be displaced slightly posterolaterally, giving access to the lateral side of the knee joint on the deep surface of the ligament. Now use scalpel dissection in the crevice between soleus and popliteus to display the lateral edge of popliteus. Then similarly display the medial edge of popliteus, deep in the popliteal fossa. Finally insert a pair of forceps deep to (anterior to) popliteus, free it from the tibial surface, and then cut across it with a scalpel, using your forceps to cut against. Reflect the superior portion of popliteus upwards, dissecting with a scalpel on its deep aspect, to display a bursa deep to popliteus. This bursa also opens into the knee joint, and on opening the bursa you will enter the knee joint and display its internal structure. Note the synovial fluid that escapes, and study the free edge of the **lateral meniscus** (meniscus means moon-shaped). This structure is also known as the lateral semilunar cartilage of the knee joint, but because it is not made of true cartilage it is more correctly called the lateral meniscus. This structure is a partial **disk,** lying in a more or less horizontal plane, and separating the lateral portion of the knee joint cavity into superior and inferior compartments as does a corresponding **medial meniscus** in the medial portion (compare with the disk of the temporomandibular joint, pages 40 and 48). Follow the popliteus tendon proximally, medial to the lateral ligament of the knee joint, to its origin on the lateral condyle of the femur. The tendon is covered by the synovial membrane, and it grooves the lateral meniscus.

Cut the popliteal artery and vein, and the sciatic nerve, at the upper end of the popliteal fossa, and reflect these structures inferiorly, in order to display the knee joint capsule. Using the site at which you opened into the capsule of the knee joint in your dissection of popliteus as your starting point, extend the opening of the capsule medially and display the lateral meniscus, as well as, on the medial aspect, the corresponding medial meniscus of the knee joint (fig. 164, page 316). Both menisci are surrounded peripherally by coronary ligaments. Expose the knee joint posteriorly through the cut you have made into the capsule, using a pair of scissors to display the tough posterior cruciate ligament (fig. 164, page 316) that becomes visible between the two tibial condyles. It attaches close to the lateral condyle and slopes upwards and medially as well as anteriorly, toward the medial condyle of the femur.

Follow the tendon of the medial head of gastrocnemius superiorly, to its attachment onto the medial condyle of the femur. Just above that, reidentify the hamstring component of the adductor magnus muscle as it inserts into the adductor tubercle close to the gastrocnemius insertion. Follow the adductor tendon inferiorly, using scalpel dissection, and display the fact that fibers from the adductor pass downwards and run into the medial collateral ligament of the knee joint (page 310). Dissect and display this ligament throughout its extent from the medial femoral to the medial tibial condyle. The continuation of adductor magnus fibers into the ligament is of phylogenetic interest: evolutionarily the ligament is part of the adductor magnus tendon, the attachment of which originally was to the tibia. The modification seen in humans is related to the adaption of the knee joint to the upright stance.

Turn your specimen over and commence an anterior approach to the dissection of the knee joint. In the subcutaneous tissue anterior to the knee joint, several superficial bursae may be encountered: the superficial prepatellar bursa; an infrapatellar bursa; and a bursa just anterior to the tibial tubercle, which should be palpated just inferior to the kneecap, on the anterior aspect of the tibia. Recall that it is to this tubercle that the ligamentum patellae attaches. You should dissect down to the ligament, studying the bursae as you encounter them. Define the inferolateral edges of the ligamentum patellae by scalpel dissection, and then gently elevate the ligament from the tibial surface superior to the tubercle. In this space you will encounter yet another bursa, the deep infrapatellar bursa, between the ligament and the tibia. Incise the fascia at the edges of the patella in order to facilitate the following dissection in which you will be exposing the interior of the knee joint. Inspect the joint at each step, because the actual reflection of the patella will damage structures that should be seen before this damage takes place.

Carefully extend the medial incision superiorly into the fascia that binds the patella down medially onto the lateral aspect of the medial condyle of the femur. Insert a pair of forceps deep to the patellar ligament and then make a transverse incision through the ligament. Now place the knee joint across a wooden block, thus flexing the knee as you continue with this dissection. Expand your me-

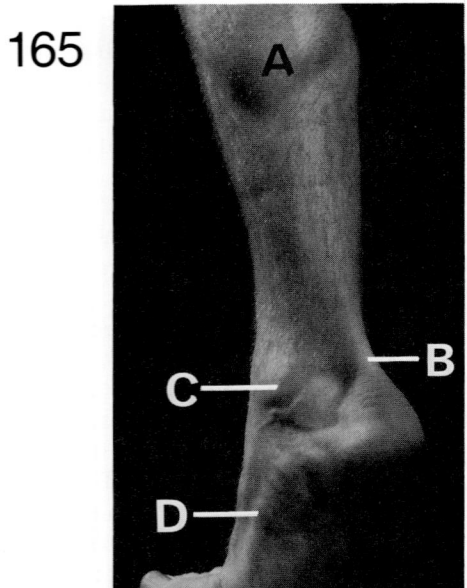

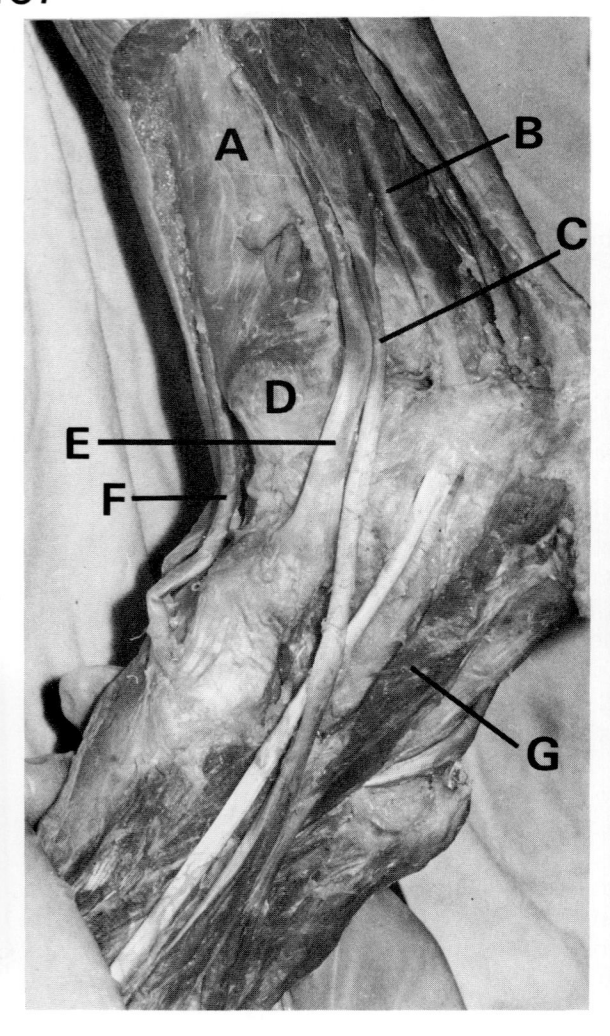

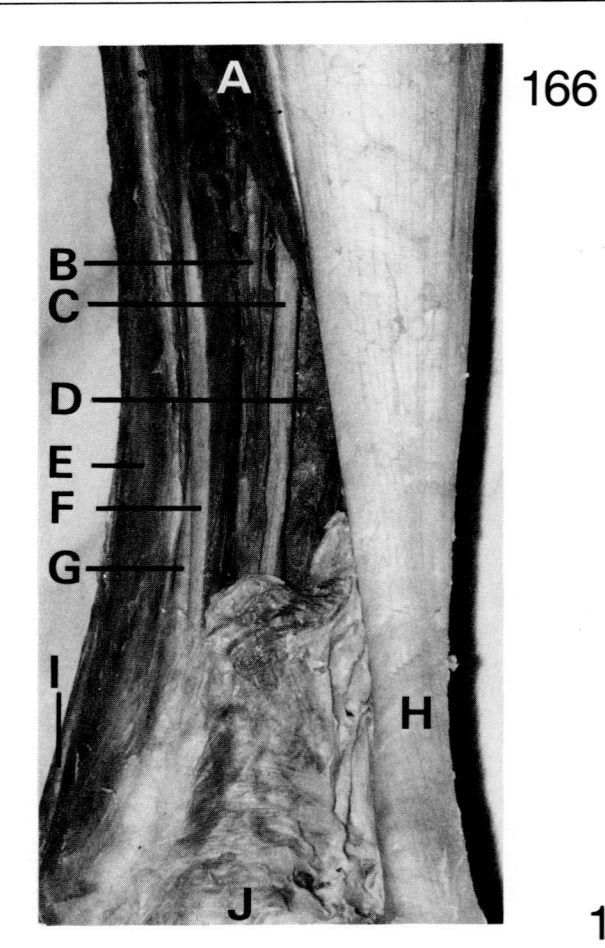

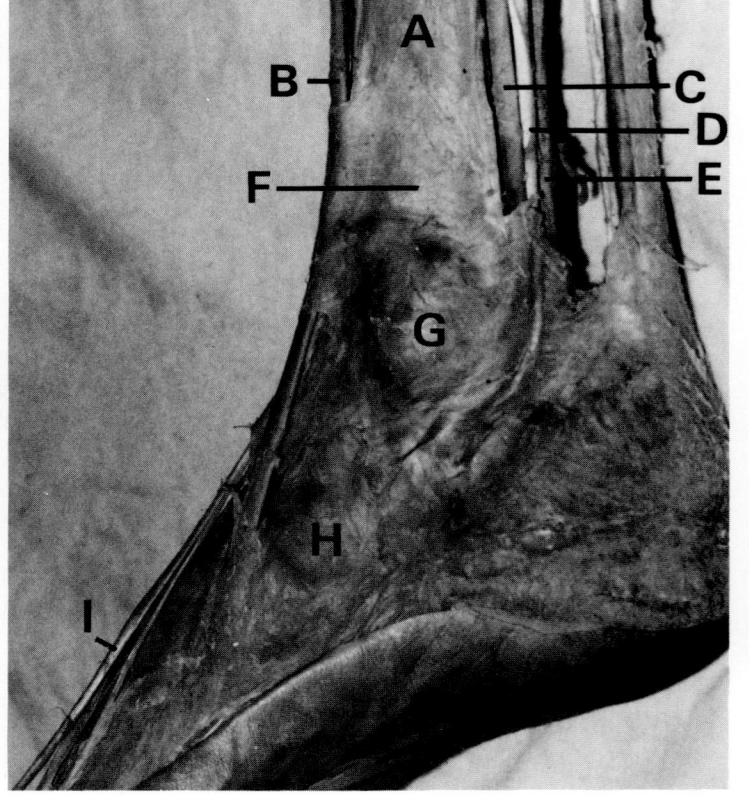

320 YOUR PATIENT'S ANATOMY: A CLINICAL VIEW OF HUMAN MORPHOLOGY

Figure 165. Inferomedial view of right leg and arch of foot.
A. Gastrocnemius
B. Achilles tendon
C. Medial malleolus
D. Abductor hallucis

Figure 166. Posteromedial dissection of right ankle region.
A. Soleus
B. Posterior tibial artery and vein
C. Tibial nerve
D. Flexor hallucis longus
E. Tibia
F. Flexor digitorum longus
G. Tibialis posterior
H. Achilles tendon
I. Tibialis anterior
J. Flexor retinaculum

Figure 167. Inferomedial view of dissection of right ankle and plantar surface of foot.
A. Tibia
B. Flexor hallucis longus
C. Flexor digitorum longus
D. Medial malleolus
E. Tibialis posterior
F. Tibialis anterior, with slack in tendon due to passive extension
G. Quadratus plantae (flexor accessorius)

Figure 168. Medial view of dissection of right ankle.
A. Tibia
B. Tibialis anterior
C. Tibialis posterior
D. Flexor digitorum longus
E. Posterior tibial vessels and tibial nerve
F. Inferior extensor retinaculum
G. Medial malleolus
H. Tuberosity of the navicular
I. Extensor hallucis longus

dial incision in order to allow you to inspect the interior of the knee joint, separating the two sides of your cut gently, in order not to damage intraarticular structures at this stage. Use a pocket torch (flashlight) to observe the **infrapatellar fat pad,** within which the deep **infrapatellar bursa** is situated. The fat pad continues superiorly in a synovial fold that may be seen to attach to the inferior edge of the patella. As you gently pull the patella forwards and displace it laterally, you can observe that the fold of the synovial tissue also attaches to a point between the condyles of the femur. This is the **infrapatellar synovial fold,** and it may be seen to have a lateral and a medial extension, known as the **alar folds** (alar means wing-like). The alar folds extend to the lateral and medial sides of the joint capsule respectively. In order to understand the three-dimensional relationships involved here, it is necessary for you to know that the knee joint, in embryological existence, is subdivided into medial and lateral halves by a synovial membrane passing from a superior attachment between the femoral condyles to an inferior attachment between the tibial condyles. The synovial fold and its alar extensions, and the infrapatellar fat pad are remnants of this partition, as are also, more posteriorly, the **cruciate ligaments,** which are more or less in the same parasagittal plane as the synovial membrane structures.

Separate the lower fibers of vastus medialis carefully from the common quadriceps expansion into the patella, and reflect the patella and the remaining portion of the quadriceps insertion laterally. Study the interior of the knee joint, including the suprapatella bursa thus exposed. Note the small muscle, the **articularis genu,** which attaches the upper end of the bursa to the anterior aspect of the femur, and is responsible for preventing the bursa from falling in deep to the patella and into the knee joint, in movements at the knee.

Incise the knee joint capsule laterally and medially, in the plane of the joint cavity between the femoral and tibial condyles. This will allow a more complete flexion at the knee joint. The synovial fold will be clearly visible anteriorly, and the anterior cruciate ligament will be seen just posterior to it, attached anteriorly between the tibial condyles, on the lateral edge of the medial condyle's articular surface. This ligament passes upwards, posteriorly and laterally to the lateral femoral condyle (fig. 154, page 298). The posterior cruciate ligament has a converse course: its inferior attachment is posterior, medial to the lateral tibial condyle (fig. 164, page 316), and it passes upwards and anteriorly to the medial femoral condyle. Visualize the manner in which the two ligaments cross each other, a relationship from which the name cruciate derives. Study the medial and lateral minisci, observing the differences in shape and position of the two (fig. 164, page 316), and note the manner in which the minisci are attached to the surrounding bony structures.

To complete your study of the knee joint, identify the arteries and nerves that constitute the vascular supply and innervation of the joint.

In fetal life the head of the fibula participates in the knee joint, but with growth of the lateral tibial condyle the fibula loses its contact with the femur and articulates only with the tibia. The **superior tibiofibular joint** is synovial, and the capsule is reinforced by anterior and posterior ligaments, which should be displayed.

Return now to the posterior aspect of the calf. Continue the elevation of gastrocnemius from soleus, partially performed previously. Review the continuity of the gastrocnemius tendon, inferiorly, with that of soleus. In addition, trace the thin tendon of plantaris to the same common insertion, into the Achilles tendon. The plantaris tendon lies, for most of its course, in the plane between gastrocnemius and soleus, and then veers medially before joining the Achilles tendon. Mobilize the Achilles tendon from deep-lying structures, and then dissect in the plane between the Achilles tendon and the deeper muscles of the posterior aspect of the calf. Use blunt finger dissection, assisted by scalpel dissection when necessary, to free the medial edge of soleus from its origin along the medial border of the tibia; then transect the belly of soleus, about 15 cm above the Achilles tendon attachment, and dissect soleus free from its origin from the posterior surface of the upper part of the fibula. As you reflect the upper portion of the soleus belly superiorly, take care to avoid damage to deep-lying vessels and muscles.

Trace the Achilles tendon inferiorly, reflecting it posteriorly and downwards as you use a scalpel to dissect on the deep surface, until you demonstrate its attachment to the posterior surface of the calcaneus bone.

Study the **deep transverse fascia of the leg,** a fascial sheet that separates the soleus muscle from the deeper muscles of the calf. Dissect through this fascia, and identify the tibial nerve and posterior tibial vessels now exposed. Returning to the popliteal fossa, establish, by blunt finger dissection, continuity of the tibial nerve and vessels in the fossa with those branches and derivatives that pass deep to the arch formed by the attachment of soleus to the soleal line, to reach the situation where they are now seen, deep to soleus.

Study the branches of the posterior tibial vessels and the tibial nerve that supply the muscles of the calf. Inferiorly, identify the specialized thickening of the deep transverse fascia of the leg, called the **flexor retinaculum,** extending from the medial malleolus to the medial tubercle of the calcaneum (fig. 166, page 320; compare this to the flexor and extensor retinacula of the wrist [pages 267 and 276]). Observe the tendons of the deep flexors as they pass deep to the retinaculum.

The deep muscles of the posterior aspect of the calf should now be identified in your dissection, reviewing their situation as originally identified on the living body (page 312; figs. 165, 166, 167, and 168, page 320). Recall (page 312) that most medially, close to the medial malleolus and just posterior to it, the powerful tendon of **tibialis posterior** could be palpated when inverting the foot against resistance. Identify this tendon and trace it superiorly to where its muscle belly takes origin from the posterior aspect of the tibia.

Just lateral to the tendon of tibialis posterior, posterior to the medial malleolus (page 312), you previous palpated and should now identify the tendon of **flexor digitorum longus.**

About 3 cm above and posterior to the medial malleolus, the flexor digitorum tendon commences its swerve laterally, crossing posterior to the tibialis posterior tendon and thus leaving the latter as the most immediate posterior relationship of the medial malleolus (page 312). Illustrate the crossing of these tendons by gently elevating the intermuscular septum of deep fascia previously identified, using the forceps and scalpel technique. The dissection should be performed with care, as the posterior tibial vessels and tibial nerve, the course of which you have demonstrated clearly throughout the length of the calf, lie close to where you are dissecting. Just posterior and lateral to the crossing of the tendons of flexor digitorum longus and tibialis posterior, identify the third important tendon you previously examined in the living body at this siste, that of **flexor hallucis longus** (page 312).

Laboratory Session 27B

You should now turn your specimen over in order to study the anterior aspect of the calf. Start by getting your bearings in respect of the lateral malleolus and the calcaneus just inferior to it. Also repalpate the medial malleolus. On the cadaver, identify thickenings (clearly distinct specialized bands) within the deep fascia covering the dorsal surface (superior aspect of the foot and adjacent anterior aspect of the lower part of the leg). These bands are retinacula for the extensor tendons (figs. 170 and 171, page 326), comparable to the flexor retinaculum previously studied (see page 323). Fibers running from the lateral malleolus transversely across to the lower end of the tibia constitute the superior extensor retinaculum. In addition, a band commences on the lateral aspect of the calcaneous and passes across the dorsum of the foot, just inferior to the lateral malleolus, to branch into two bands, one of which passes to the medial malleolus and the other of which passes anteriorly to blend with the fascia on the dorsum of the foot. This Y-shaped thickening is the inferior extensor retinaculum.

Reidentify the tendon of the tibialis anterior muscle (page 311), in its characteristic position in relation to the anterior border of the tibia. Cut the extensor retinaculum using the forceps and scalpel technique, and free tibialis anterior from the retinaculum and elevate it from the tibia. Trace the tibialis anterior tendon proximally, and follow it up into its muscle belly. Dissect the belly free of surrounding connective tissue to clearly display the muscle, and separate it from its adjacent neighbor on the lateral aspect, the extensor hallucis longus (page 311). Slit the ensheathing deep fascia covering the extensor muscles all the way up to the knee joint. Identify the tendon of extensor digitorum longus (page 311), and trace this muscle proximally, separating it from extensor hallucis longus. Finally, identify the last of the extensor muscles, peroneus tertius (page 311), by its expansion on the lateral aspect of the foot. Trace its tendon superiorly and identify its muscle belly (fig. 161, page 316 and figs. 169, 170, and 171, page 326).

Posterolateral to peroneus tertius, identify the intermuscular septum that separates it from peroneus longus and brevis muscles.

Anterior to the ankle joint, in the interval between the tendons of tibialis anterior and extensor hallucis longus, identify the anterior tibial artery and the deep peroneal nerve. The continuation of the artery, the dorsalis pedis artery, is that which you have palpated along the dorsum of the foot in the living subject (page 313). Proximally, the artery can be traced deep to the extensor muscles of the calf accompanied by the deep peroneal nerve. Note that the artery and nerve lie on the plane of the tibial surface. Superiorly, the artery can be traced to an opening in the interosseous membrane between the fibula and the tibia, through which it passes from the popliteal fossa, having arisen as a branch of the popliteal artery. The nerve accompanies the artery along most of its length. Superiorly, the nerve does not enter this compartment of the calf in the same manner as the artery does. Identify the **common peroneal nerve** as it winds around the head of the fibula on the lateral aspect, where you palpated it (page 311) in the living subject. At this site it divides into its **deep** and **superficial** branches. The deep peroneal nerve, after passing deep to the peroneus longus muscle, then enters the substance of, or becomes deep to, extensor hallucis longus, into which you should now trace the nerve superiorly, dissecting muscle tissue from it.

Returning to the course of the anterior tibial artery and deep peroneal nerve in the calf, study all the branches of these two structures, using atlas illustrations to assist you.

Rotate the limb to enable you to dissect peroneus longus and brevis muscles (pages 311 to 312). To do so, expose these two muscles by removing the deep fascia of the calf that covers them superficially, from the head of the fibula superiorly down to the lateral aspect of the foot inferiorly. Note two regions of thickening of the fascia into specialized bands or retinacula. These are the superior peroneal retinaculum, from the back of the lateral malleolus to the lateral surface of the calcaneus, and the inferior peroneal retinaculum, from the upper surface of the calcaneus passing downward to the lateral surface of the calcaneus. The upper end of the inferior peroneal retinaculum is continuous with the inferior extensor retinaculum. The two peroneal muscles are tightly ensheathed in tough deep fascia: the superficial fascia externally; and anterior and posterior intermuscular septa. When you have exposed the muscles by removing the covering deep fascia, identify first the peroneus brevis from its origin from the lower two thirds of the lateral surface of the fibula to its insertion as previously palpated (page 311). Using the forceps and scalpel technique cut through the superior and inferior retinaculae and display the full course of peroneus brevis down to its insertion into bone. Displace the overlapping peroneus longus tendon anteriorly onto the lateral malleolus, to fully expose the tendon of the brevis muscle. Then separate the two muscle bellies by finger dissection, all the way up to the head of the fibula, establishing the origin of the peroneus longus muscle from the upper half of the lateral surface of the fibula and the head of the fibula. Superiorly, trace the superficial peroneal nerve from the point where it branches from the common peroneal nerve (page 324) down to its branches to both muscles now being dissected. The nerve runs for most of its course within the substance of the two muscles. It emerges to become a superficial, cutaneous sensory nerve in the lower third of the calf, supplying the anterior surface of the lower part of the leg and the dorsal (superior) surface of the foot, in the interdigital clefts, with the exception of the first interdigital cleft, which is supplied by the deep peroneal nerve (see below) and the lateral surface of the foot and small toe, which are supplied by a branch of the sural nerve (pages 315, and fig. 171, page 326).

The tendon of the peroneus longus passes onto the plantar surface of the foot to reach its insertion as previously palpated (pages 311 to 312).

Remove the skin from the dorsum of the foot and the interdigital clefts. Reidentify the anterior tibial artery and deep peroneal nerve, where you previously dissected them. Trace the artery forwards as it continues, to become the dosalis pedis artery, to the site where you previously palpated this artery in the living subject. Also trace the accompanying nerve branch of the deep peroneal nerve to its destination to supply the skin of the first interdigital cleft (see above). In tracing the dosalis pedis artery and its accompanying branch of the deep peroneal nerve, they will be seen to pass deep to a small slip of muscle and its inserting tendon passing superficially toward the great toe. This is the **extensor hallucis brevis** muscle, which should be cut across its belly and reflected to allow a better access to dissection of the dosalis pedis artery and the nerve. When you have completed this, trace the extensor hallucis brevis proximally, deep to the tendons of the extensor digitorum longus and peroneus tertius, to the muscle belly situated just anterior to the lateral malleolus. This is a common muscle belly of extensor hallucis brevis and the **extensor digitorum brevis** (page 313; figs. 169, 170, 171, and 172, page 326). The latter sends slips to the remaining toes, and indeed the slip to the great toe is regarded by some authors as part of extensor digitorum brevis. Dissect the common origin of these superficial brevis muscles from the superior surface of the calcaneus. Improved access to this dissection may be obtained by cutting across the tendons of extensor digitorum longus and peroneus tertius, but before making this incision observe the common synovial sheaths around these tendons. Then cut the tendons and reflect the distal ends distally and dissect the extensor digitorum brevis. Follow the tendons of the longus and brevis extensors distally. The extensor hallucis brevis inserts at the base of the proximal phallanx of the hallux. The other brevis tendons insert into the lateral edge of the corresponding tendons of the longus muscle, passing to the second, third, and small toes (regarding the hallux as the first). Note that it is correct anatomical nomenclature to number the toes and the fingers from one to five starting with one at the great toe or thumb. However, it is vitally important not to rely on that numbering system, as confusion can arise because some people do not adhere strictly to the correct terminology, and count, for example, the index finger as number one. Tragic examples of the incorrect finger or toe being surgically removed because of misunderstanding between two surgeons are on record; one should therefore always specify precisely which terminology one is using when communicating with a professional colleague. When you have completed your study of the insertion of the tendons of the extensor brevis muscle, cut them across as you did with the first tendon (that of the hallux), and reflect all the tendons and the muscle belly proximally, dissecting it to its origin from the calcaneus. Correlate your observations with those made on the living subject (page 313).

Dissect the dosalis pedis artery distally as far as you can, identifying its branches: the first dorsal metatarsal artery, passing toward the first interdigital cleft; the deep plantar artery, which passes onto the plantar aspect of the foot to participate in the **(deep) plantar arch** (compare this to the deep palmar arch, page 276); the arcuate artery, which arches laterally, ventral to the metatarsal bones and gives off the remaining dorsal metatarsal arteries; and medial and lateral tarsal arteries.

Position your specimen to provide access to the sole of the foot, and remove the skin from the entire surface. While dissecting the foot and leg, you should repeatedly attempt to establish your bearings in respect of anatomical position. It is especially important in dissections of the upper and lower limbs, that you train yourself in this respect; the symmetry of the limbs, and the ability of the limbs to rotate, evert, invert, supirnate, and pronate can create considerable confusion in the mind of the inexperienced observer who is not in the habit of mentally reorientating to the anatomical position.

In relationship to the dosalis pedis artery and its accompanying branches of the deep peroneal nerve, also identify the vena communitantes (accompanying veins) that unite, and in passing proximally become the long saphenous vein, which passes anterior to the medial malleolus. Similarly on the lateral aspect of the foot, identify the commencement of

326 YOUR PATIENT'S ANATOMY: A CLINICAL VIEW OF HUMAN MORPHOLOGY

Figure 169. Lateral view of right foot and ankle, with toes in extension and foot in partial inversion.
A. Extensor digitorum brevis
B. Tendons of extensor digitorum longus
C. Extensor hallucis longus
D. Lateral malleolus
E. Peroneus longus
F. Peroneus brevis

Figure 170. Dorsolateral dissection of right foot and ankle.
A. Peroneus brevis
B. Peroneus tertius
C. Tibialis anterior
D. Peroneus longus
E. Extensor hallucis longus
F. Superior extensor retinaculum
G. Inferior extensor retinaculum
H. Extensor digitorum brevis
I. Extensor digitorum longus
J. Extensor hallucis brevis

Figure 171. Dissection of lateral aspect of left foot and ankle.
A. Extensor digitorum longus
B. Sural nerve
C. Extensor hallucis longus
D. Peroneus tertius
E. Peroneus longus
F. Superior extensor retinaculum
G. Peroneus brevis
H. Inferior extensor retinaculum
I. Extensor digitorum brevis
J. Abductor digiti minimi

Figure 172. Right and left feet in inversion.
A. Right lateral malleolus
B. Right tibialis anterior
C. Left tibialis anterior
D. Right extensor digitorum brevis

the lesser (short) saphenous vein, passing posterior to the lateral malleolus.

Remove the superficial connective tissue and fat on the sole of the foot, to prepare for study of the deep structures. Bear in mind that the muscles to be dissected here are true flexors, supplied by ventral divisions of the nerves of the lower limb (sacral) plexus.

There are four layers of muscles on the plantar surface of the foot. There are, furthermore, neurovascular planes between the first and second muscle layers, and between the third and fourth layers. Commence your dissection of the first, almost superficial muscle layer, as follows.

On the medial aspect of the foot, identify **abductor hallucis** (page 312; fig. 165, page 320 and figs. 173, 175 and 176, page 330). Recall that the terms abduction and adduction mean, in general, movement away from or movement toward the midline of the body. With respect to the fingers, a special rule applies, the line of reference being through the middle finger, rather than the median plane of the body. In the foot, yet a different rule applies, for reasons of homology that will become apparent in due course. In the foot, abduction indicates movement away from the parasagittal plane through the second toe (i.e., the toe lateral to the great toe); adduction means movement toward that plane.

Carefully separate the tendon of **abductor hallucis** from the **plantar aponeurosis** (compare with the palmar aponeurosis, pages 268 and 276) that covers all three superficial muscles of the foot. Cut the tendon of abductor hallucis, and reflect the two ends proximally and distally, respectively. The distal portion of the muscle should be traced down to its insertion into the medial side of the base of the phalanx of the great toe. In reflecting the proximal portion, expose with care the structures passing deep to it. These are the tibial nerve and posterior tibial vessels, which enter the sole of the foot at this site and divide into lateral and medial plantar nerves and vessels. Observe the branch of the medial plantar nerve that supplies the abductor hallucis. Trace the proximal portion of the muscle to its origin from the medial tubercle of the calcaneus.

Proceed now to study the middle of the three superficial muscles, the **flexor digitorum brevis**, which also takes origin from the calcaneus. Trace its four tendons to the four lateral toes. Each tendon splits at the base of the corresponding digit to allow passage through the tendon of the corresponding tendons of the flexor digitorum longus (page 312). Compare this arrangement with that of the flexor digitorum sublimis and profundus in the upper limb (page 276). To expose these tendons, dissect the plantar aponeurosis away from the tendons. Trace the tendons of flexor digitorum longus and brevis of the middle toe distally, to demonstrate the fibro-osseous canals within which they are contained, and compare the arrangement to that of the fingers (page 276). Incise the canal to demonstrate the tendons' synovial sheath.

The plantar aponeurosis is very firmly attached to the muscle belly of flexor digitorum brevis and can only be separated from it with difficulty. When you have achieved this separation, cut the aponeurosis free and reflect the proximal portion all the way back to the calcaneus. Then mobilize the deep-lying belly of the flexor digitorum brevis, insert a finger deep to it, and run the finger posteriorly to separate by finger dissection the belly of the muscle all the way back to its origin from the medial tubercle of the calcaneus. Cut the belly of the muscle transversely and reflect the proximal portion to the origin and the distal portion anteriorly, to expose the deep-lying tendons of flexor digitorum longus.

Moving to the lateral edge of the sole of the foot, identify and define by scalpel dissection the **abductor digiti minimi** muscle (figs. 173, 175, and 176, page 330). Insert a pair of forceps deep to the muscle on its lateral edge, at the lateral edge of the foot, and elevate the belly of the muscle. Then bisect the belly with your scalpel and reflect proximal and distal parts of the muscle in their respective directions, taking care to avoid damage to the lateral plantar nerve that lies close to the medial edge of the muscle. Trace the muscle proximally to its origin from the inferior surface of the calcaneus, including its medial and lateral tubercles, and the distal portion of the muscle to its insertion onto the lateral aspect of the base of the proximal phalanx of the fifth (little) toe.

Having reflected the muscles of the first layer of the foot, identify the structures of the neurovascular plane. Reidentify the medial and lateral plantar nerves and vessels, tracing them from the division of the tibial nerve and posterior tibial vessels to the termination of the plantar nerves and vessels. Identify the lateral plantar nerve just medial to abductor digiti minimi and the medial plantar nerve and vessels medial to the abductor hallucis.

Proceed now to study the second layer of foot muscles. In the middle of the sole of the foot identify the tough, white fibrous tendon of **flexor digitorum longus**, passing obliquely forwards with a mediolateral slope, with a muscle belly attached to its posterolateral edge. The muscle belly is that of **flexor digitorum accessorius** (accessory flexor of the toes), also known as **quadratus plantae** (fig. 167, page 320 and fig. 174, page 330). Reidentify the tendon of flexor digitorum longus where you previously studied it in the calf (pages 312 and 323), just above and posterior to the medial malleolus. Tug on the tendon at this site, and confirm that this creates tension in the tendon in the sole of the foot and in the small tendons proceeding to each of the toes. Follow these again toward the fibrous flexor sheaths, one of which you previously opened on the middle toe (page 328), and within which the tendons of the flexor digitorum longus pass through the tendons of flexor digitorum brevis to become superficial. Trace quadratus plantae posteriorly to its origin from the calcaneus. Observe that it has two heads of origin, a medial from the medial surface of the calcaneus, and a lateral, from the tuberosity of the calcaneus and an adjacent ligament known as the **long plantar ligament** (fig. 175, page 330). The posterior attachment of this ligament is on the anterior surface of the calcaneus, between the two heads of quadratus plantae. The ligament passes forward, deep to the lateral head of the muscle, to attach to the bases of the 2nd, 3rd, and 4th metatarsals. Between these anterior and posterior attachments the ligament bridges a groove on the cuboid bone, creating a tunnel through which the tendon of peroneus longus passes in the sole of the foot (fig. 175, page 330).

Trace the lateral plantar nerve and vessels proximally to their origin, as they cross the muscle belly of quadratus plantae obliquely.

Also as part of the second muscle layer of the foot, study the **lumbrical** muscle slips (compare with the lumbricals of the hand, pages 271 and 276), which arise from the tendons of flexor digitorum longus (fig. 174, page 330). The medial lumbrical arises from the medial side of the tendon slip of the second toe and passes to the medial side of the base of the first phalanx of that toe. The subsequent lumbricals arise from the adjacent tendinous slips (i.e., the second lumbrical from the first and second slips; the third lumbrical from the third slip; and the fourth from the third and fourth slips). Each lumbrical passes to the medial side of the corresponding first phalanx base. The most medial lumbrical is supplied by a branch of the medial plantar nerve; the remaining three lumbricals are supplied by branches from the lateral plantar nerves.

The final muscle of the second layer of the foot is the tendon of flexor hallucis longus (figs. 166 and 167, page 320 and fig. 174, page 330), which should be now found medial to the medial plantar nerve and the medial tendinous slip of flexor digitorum longus. Identify the muscle belly of flexor hallucis longus in the calf, where you studied it previously (pages 312 and 323), and tug on the belly at that site, and on the tendon in the sole of the foot and the plantar surface of the great toe, to establish continuity. Trace the tendon into the flexor fibrous sheath of the great toe.

Now dissect the muscles of the third layer of the sole. The **flexor hallucis brevis** (figs. 175 and 176, page 330) lies lateral to the abductor hallucis and medial to the tendon of flexor hallucis longus. This muscle takes its origin from the three cuneiform bones and the cuboid. It inserts by two separate tendons into the medial and lateral sides of the base of the proximal phalanx of the toe, via two small sesamoid bones, and is supplied by a branch from the medial plantar nerve that you should now attempt to identify. Mobilize the muscle from its deep relations, and cut across its belly to reflect the proximal and distal ends to their respective attachments. The muscle may be fused to a greater or lesser extent with the abductor hallucis, with which it has a common insertion on the lateral aspect of the base of the first phalanx, through the sesamoid bone.

Mobilization of the flexor hallucis brevis should be performed by elevating it at the medial edge, with a pair of forceps, and cutting in from the medial edge. Scalpel dissection should be performed in the direction of medial to lateral, to establish the edges of the muscle, which are difficult to separate from surrounding muscles.

Dissect the small **flexor digiti minumi brevis** muscle (fig. 175, page 330), which takes its origin from the base of the fifth metatarsal and inserts into the base of the proximal phalanx of the toe adjacent to the insertion of the abductor digiti minumi.

Cut across the tendon of flexor digitorum longus, in the sole of the foot just anterior to the site at which quadratus plantae attaches to it, and reflect the distal slips of these tendons toward the toes, to display the last muscle of the third layer, the **adductor hallucis.** This muscle has two heads: an oblique head, which takes origin from the base of the second, third, and fourth metatarsals and sweeps across anteromedially to the lateral side of the base of the first phalanx of the great toe; and a transverse head, which takes origin from the tough transverse metatarsal ligament that binds the heads of the metatarsals to each other. This head converges on the oblique head and inserts, in common with it, via the medial sesamoid, into the medial aspect of the base of the first phalanx.

Carefully mobilize the lateral edge of the oblique head of the adductor hallucis muscle, separating it from adjacent and deep structures. Then elevate the muscle from the lateral aspect by inserting a pair of forceps, bisect the belly of the muscle across the forceps as usual, and reflect the proximal portion of the muscle proximally, dissecting carefully on the deep surface with your scalpel, to preserve important vessels and nerves deep to the muscle. Return

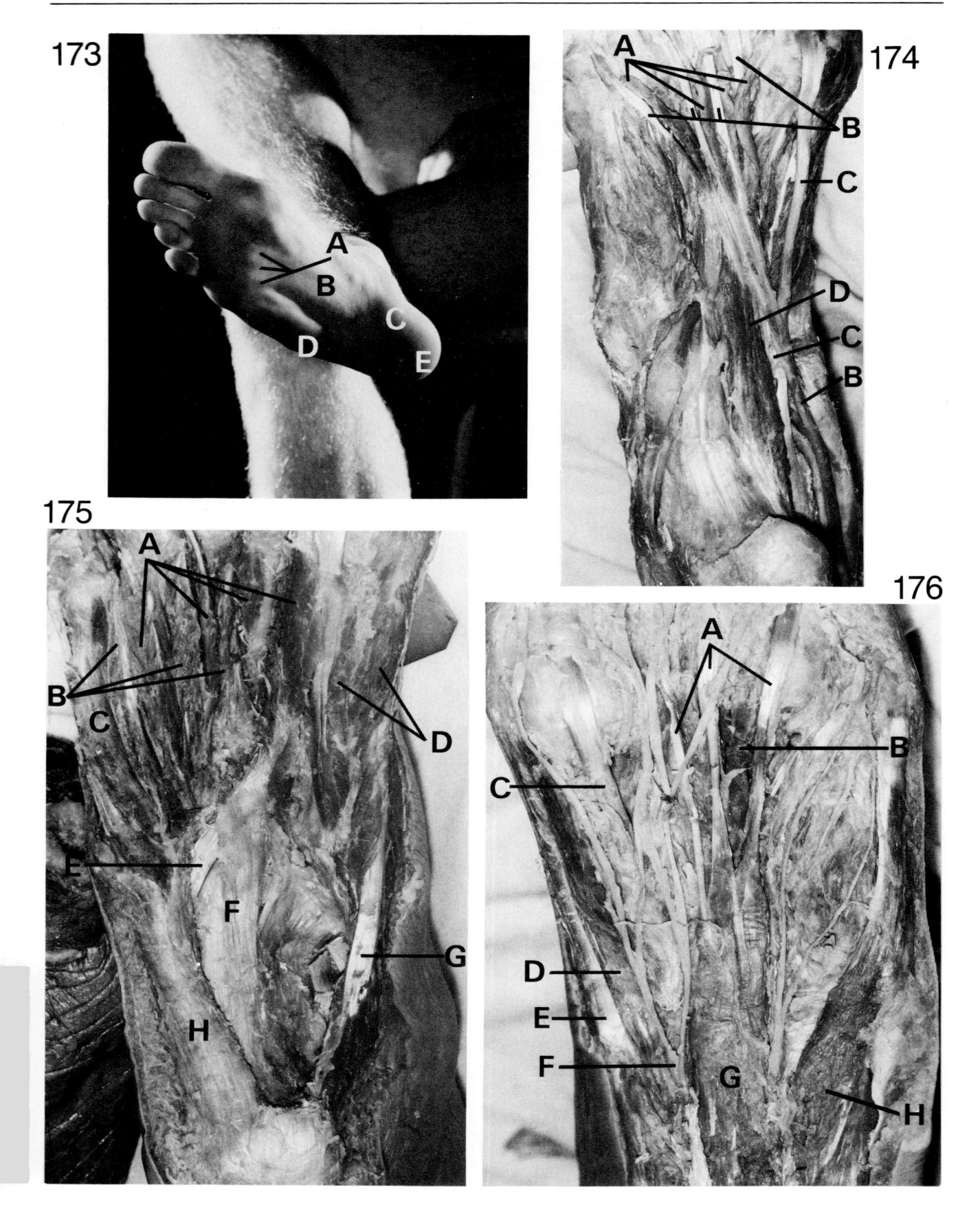

330   YOUR PATIENT'S ANATOMY: A CLINICAL VIEW OF HUMAN MORPHOLOGY

Figure 173. Plantar surface of right foot.
A. Tendons of flexor digitorum brevis
B. Belly of flexor digitorum brevis
C. Abductor hallucis
D. Abductor digiti minimi
E. Calcaneus

Figure 174. Dissection of second muscle layer of right foot.
A. Lumbricals
B. Flexor digitorum longus
C. Flexor hallucis longus
D. Quadratus plantae (flexor accessorius)

Figure 175. Muscles of first, third, and fourth muscle layers of right foot (layer number indicated).
A. Dorsal interossei (4)
B. Plantar interossei (4)
C. Flexor digiti minimi brevis (3)
D. Flexor hallucis brevis (3)
E. Peroneus longus (4)
F. Long plantar ligament
G. Abductor hallucis (1)
H. Abductor digiti minimi (1)

Figure 176. Dissection of left plantar surface, showing superficial nerves, and muscles of layers 1, 2, and 3 (layer number indicated).
A. Tendons of flexor digitorum brevis (1)
B. Lumbrical (2)
C. Flexor hallucis longus (2)
D. Flexor hallucis brevis (3)
E. Abductor hallucis (1)
F. Plantar digital nerves
G. Belly of flexor accessorius (2)
H. Abductor digiti minimi (1)

to the lateral plantar nerves and vessels, and trace the deep branch of the lateral plantar nerve, and the communication from the lateral plantar artery to the deep plantar arch, both of which lie in the plane deep to the oblique head of the adductor muscle. The deep branch of the plantar nerve and the deep plantar arch lie between this muscle and the metatarsal bones. Observe the plantar metatarsal arteries, branches of the arch that proceed between the metatarsals. Anteriorly, they divide into the plantar digital arteries to supply the interdigital clefts. The deep branch of the lateral plantar nerve supplies both heads of the adductor hallucis muscle, and, as mentioned previously, the three lateral lumbrical muscles. Finally, it also supplies the muscles of the fourth and last layer of the sole of the foot, the **interosseous muscles,** which will be studied now.

The **plantar interosseous** muscles (fig. 175, page 330) arise from the medial surfaces of the third, fourth, and fifth metatarsals, and pass to insert into the medial sides of the bases of the first phalanges of the corresponding toes. These are small, thin muscles, and care should be taken in the dissection to avoid damaging them. The **dorsal interossei** (fig. 175, page 330) are situated between the metatarsals: the first dorsal interosseous between the first and second metatarsals, with the tendon of insertion into the medial side of the base of the phalanx of the second toe; the second dorsal interosseous between the second and third metatarsals, with a tendon of insertion into the lateral aspect of the base of the second toe's first phalanx; the third dorsal interosseous from between the third and fourth metatarsals, and inserting into the lateral aspect of the base of the first phalanx of the middle toe; and the fourth dorsal interosseous muscle arising between the fourth and fifth metatarsals, and inserting into the lateral aspect of the base of the first phalanx of the fourth toe. All the interossei are supplied by the lateral plantar nerve. Note that the arrangement of the insertions accounts for the line of adduction and abduction in the foot being in the second toe (page 328).

Reidentify the tendon of peroneus longus muscle, on the lateral aspect of the foot, anterior to the lateral malleolus. Trace the tendon into the sole of the foot, and display the canal that it enters, deep to the long plantar ligament, and in the groove in the cuboid bone. Note that there is a fibrous sheath thus created, and part of the origin of the oblique head of adductor hallucis is from this sheath. Incise this sheath to display the tendon in its course toward its site of insertion on the lateral side of the first metatarsal and the adjacent medial cuneiform bone.

Finally, reidentify tibialis posterior where you previously studied it in the calf (pages 312 and 323), and follow its tendon down to its site of insertion in the sole of the foot, into the tuberosity of the navicular and the adjacent medial cuneiform bone.

Review the superior tibiofibular joint (page 322). In addition to that joint, the tibia and fibula are bound to each other by the tough **interosseous membrane** that extends between the two bones, and by the **inferior tibiofibular joint,** which will now be studied in relation to the ankle joint.

Clear connective tissue away from the surface of the medial malleolus. Displace the tendon of tibialis posterior anteriorly, onto the lateral surface of the malleolus, and cut the tendon of flexor digitorum longus to displace the distal portion of it anteriorly. Displace the posterior tibial vessels and tibial nerve posteriorly. Palpate the sustentaculum tali (page 310) and tuberosity of the navicular (page 310), and then dissect down to the osseous level, using atlas illustrations to guide you, to demonstrate the fan-shaped deltoid (medial) ligament of the ankle joint. Its superior attachment is to the lower edge of the medial malleolus, and the fibers fan out inferiorly, some passing anteriorly to the tuberosity of the navicular, others passing inferiorly to the sustentaculum tali, and some passing posteriorly to the talus.

Displace the tendons of tibialis posterior and flexor digitorum longus, clasping them in a clamp to maintain them out of the dissection area, and demonstrate the powerful **spring ligament** (the **plantar calcaneonavicular ligament**) which extends from the sustentaculum tali anteriorly, to the plantar surface of the navicular. This ligament plays an important role in maintaining the longitudinal arch of the foot (page 313). Some of the fibers of the deltoid ligament insert into the spring ligament.

Turn your specimen and dissect the lateral ligament of the ankle joint, which has three components, passing from the lateral malleolus of the fibula, posteriorly to the talus, inferoposteriorly to the calcaneus, and anteriorly to the talus. Identify these **posterior talofibular, anterior talofibular,** and **calcaneofibular ligaments.** To do this, displace the tendons of the peroneus longus and peroneus brevis anteriorly, as you did with the flexor tendons medially. Now displace peroneus tendons posteriorly, to dissect the anterior talofibular ligament. Move the foot passively into extension and flexion to study the movement of the talus in relationship to the fibula and tibia. Then incise into the ankle joint to further study this movement. Your incision should be made superior to the anterior talofibular ligament. Posteriorly, dissect the posteroinferior tibiofibular ligament.

Incise the long plantar ligament (fig. 175, page 330) after tracing it to its anterior attachments and reviewing its posterior attachment (page 329), and then cut the ligament transversely, to expose and enable you to dissect the **short plantar ligament** (also called the plantar calcaneocuboid ligament).

Review, on your dissection specimen, the movements at the subtalar or talocalcaneal joint, the talocalcaneonavicular joint, and the calcaneocuboid joint, referring back to the description previously given (page 313). Incise the tough ligaments between the tarsal bones where necessary and examine the joint surfaces involved.

Dissect the middle metatarsophalangeal joint, reviewing your previous examination of the fibrous flexor sheaths. Compare the joints in the toes with those of the fingers.

In reviewing the bones, muscles, nerves, and vessels of the lower extremity, attempt to determine the body part in the upper limb to which each major lower limb structure is homologous. Consider the diverse functions they perform and pathologies they may be exposed to as consequences of the different demands made on them by evolution of the upright stance and gait and the related free use of the hands and development of the brain. This will appropriately complete your introductory study of the human body on a morphological, functional, and clinical note.

Index

Abdomen
  components of, 125
  divisions of, 126–127
  lower structures of, examination procedures, 149. *See also* individually named organs; Lower abdominal organs
  palpation of, 125
  pelvic part defined, 168–169
  posterior structures of, *160*
  reflex tests, 130
  and skeletal observations, 4
  upper structures of, examination procedures, 135. *See also* individually named organs; Upper abdominal organs
Abdominal aorta, 159. *See also* Descending aorta, abdominal portion
  pelvis and, 169
Abdominal reflexes, 130
  test for, 130
Abdominal respiration, 94
  male versus female, 127
Abdominal wall
  anterior
    dissection of, *128,* 131–134
    vertebral column and, 159
    sexual differences observable in, 127, 130
  examination by palpation, 126
  posterior
    dissection of, 163–167
    examination of, 158
Abducens nerve
  in cranial cavity dissection, 198
  clinical examination, 187
  in orbit dissection, 213
Abduction
  of lower limb, 284
  of thumb, 268, 271
  of upper limb, 261
Abductor muscles
  in foot dissection
    digiti minimi, 328
    hallucis, 328
  of the lower limb, 286–287
Abductor pollicis longus tendon, 267, 268

Accessory muscles of respiration, 94
Accessory nerve, clinical examination, 193
Accommodation, of the lens, 204
Acetabular labrum, in quadriceps femoris dissection, 307
Acetabular ligament, transverse, in quadriceps femoris dissection, 307
Acetabulum, hip joint and, 284
Achilles tendon, 318, 322
Acromioclavicular joint, 256
Acromion process, 226
  dissection, 236
  shoulder examination and, 255
Acuity, visual, testing of, 186
Adam's apple. *See* Thyroid cartilage
Adduction
  of lower limb, 284
  of thumb, 268
  of upper limb, 261
Adductor depression, in knee examination, 308
Adductor muscle mass, and hip joint flexors, 286
Adductor muscles
  in femur examination, 282
  hallucis, in foot dissection, 329, 332
  and hip joint flexion
    brevis, 286
    longus, 285
    magnus, 285–286
  origins of, 286
  in quadriceps femoris dissection, 305–306
  in thigh dissection, 304
Adrenal gland. *See* Suprarenal gland
Afferent impulses, 6
Agonist, 225
Alae, nasal, 59
Alar folds, in knee joint dissection, 322
Allantois, 133
Alveolar gingiva, examination of, 29
Alveolar nerve, inferior, 28
Alveolar part. *See* Mandible
Alveolar process. *See* Maxilla

Alveolus
  of jaws, 9
  of lungs, 93
Anal canal
  in abdominal wall dissection, 134
  manual examination of, 170
Anal triangle, 175, 178
  described, 168
  dissection of, 179
Anastomosis
  cruciate, in quadriceps femoris dissection, 306
  in digestive tract, 156–157
  in intercostal spaces, 92
Anatomical crown, versus clinical crown in teeth examination, 28–29
Anatomical position, 3, 81
  planes and directional terms used, *88*
Anatomy
  defined, 3
  developmental, 3
  functional, 3
  morphology relationship to, 81
  regional, 3–5
    laboratory review of, 8
  standardized descriptions of, normal body position for. *See* Anatomical position
  systematic, 5–7
    laboratory review of, 8
Anconeous, palpation of, 270
Angle of the mandible, 26, 37
Angle of the mouth, palpation and, 11
Ankle
  bones of, 310
  dissection of
    inferomedial view, *320*
    lateral aspect, *326*
    medial view, *320*
    posteromedial view, *320*
    dorsolateral view, *326*
    lateral view, *326*
Ankle joint, dissection of, 332–333
Anococcygeal ligament, 290
Anosmia, 186
ANS. *See* Autonomic nervous system (ANS)

Numbers in italic indicate figures.

INDEX

Ansa hypoglossi, in carotid triangle dissection, 57
Antagonist
  deep flexor leg muscles as, 312
  muscle action and, 225
Antecubital fossa, 264
  brachial artery division in, 269, 271, 274
  dissection of, 275
Anterior, defined, 3
Anterior interventricular artery, in heart dissection, 114
Anterior interventricular sulcus, in heart dissection, 114
Anterior nasal spine, 28
Anterior superior iliac spine (ASIS)
  in abdominal examination, 126
  origin of sartorius, 284–285
Anterior talofibular ligaments, in ankle joint dissection, 332–333
Anterior tibial artery, in calf dissection, 324
Antibodies, 6
Anular ligament, in elbow dissection, 276
Aorta. *See also* Ascending aorta; Descending aorta
  branches of, 166
  as relation of left lung, 97
  in mediastinum dissection, 107
  in mediastinum examination, 99
Aortic valve, blood flow control by, 109
Apex beat, 108
Aponeurosis
  bicipital, dissection of, 275
  extensor carpi ulnaris muscle and, 269–270
  of external oblique muscle of abdomen, 126
  palatine, pharynx dissection and, 66–67
  palmar, 268
  dissection of, 276
Appendices epiploicae, colon and, 157
Appendix, vermiform, 149, 151
Aqueous humor, in eyeball dissection, 210
Arachnoid granulations, in cranial cavity dissection, 195
Arachnoid mater membrane, in cranial cavity dissection, 195, 198
  in lumbar puncture, 125
Arch
  of foot, 313, *320*
  palatoglossal. *See* Palatoglossal fold
  palatopharyngeal, 61
  palmar, deep arterial, 276
  superciliary, 183
  zygomatic, 20, 26
Arcuate eminence, in inner ear dissection, 222
Arcuate line, in abdominal wall dissection, 132
Areola, in nipple examination, 84

Areolar glands, in breast examination, 85
Arm. *See also* Forearm
  anatomist usage of term, 4
  anterior aspect, *252, 258, 264*
  axilla and upper portions of, *82*
  blood vessels of, 260
  bony landmarks of, 255–256
  dissection of, 262–266
  functional anatomy and movements, 261
  lateral aspect, *264*
  muscles of, 256–257, 260
  nerves of, 260–261
  posterior aspect, *244, 252*
  skeletal structures of, *258*
  structures of, *252*
Armpit. *See* Axilla
Arterial arcades
  of ileum, 154
  of jejunum, 154
Artery(ies). *See also* individually named arteries
  in cardiovascular system, 7
  central, of retina, in eyeball dissection, 207
  of forearm, hand and wrist, 271, 274
  of body wall, in neurovascular plane, 86
Articular disk, ulnocarpal, 277
  of temporomandibular joint, 40
Articularis genu muscle, 322
Articular tubercle, in temporomandibular joint examination, 40
Aryepiglottic fold, 71
  in laryngopharynx dissection, 73
Arytenoid cartilages, 71
  vocal process of, in larynx dissection, 73
Ascending aorta
  blood flow through, 114
  in mediastinum examination, 99
Ascending colon, 134, 149
Ascending lumbar veins, 166
Ascending pharyngeal artery, dissection of, 66
ASIS. *See* Anterior superior iliac spine (ASIS)
Atlas
  head and neck movements involving, 246
  and nasopharynx relationship, 61
Atrioventricular bundle (of His), 118
Atrioventricular node, 118
Atrioventricular valves. *See* Left atrioventricular valve; Right atrioventricular valve
Atrium, of heart. *See* Left atrium; Right atrium
Auditory ossicles, dissection of, 219
Auditory tube
  dissection of, 66
  in nasal cavity dissection, 63

  and nasal cavity relationship, 60
  pharyngeal opening of, 60–61
Auricles
  in heart dissection, 114
  left, lateral aspect of, *216*
Auricular vein, posterior, dissection of, 55
Auriculotemporal nerve, 193
  in temporomandibular joint dissection, 48
Auscultation
  of heart sounds, 109
  of lower abdomen, 150
  in lung examination, 95
  of heart valves, 109–110
Autonomic nervous system (ANS), 6, 101, *104*
  cranial nerves and, 106
  sympathetic and parasympathetic parts, 106
Axilla
  examination of, 85
  superior inlet to, 227
  right, anterior aspect of, *258*
  suspensory ligament of, 87
Axillary artery
  in anterior triangle of neck dissection, 232
  in arm, 260
Axillary nerve, in arm dissection, 263, 266
Axis, head rotation and, 246
Azygos vein, as relation of right lung, 97

Back, *164*
  movements of, 246–247
  muscles of, *161*, 243, *244*, 246. *See also* individually named muscles
  upper limb, dissection of, 251, 254
  posterior aspect of, dissection, 248–249
  upper, dorsal view of, *90*
Basilic vein, 260
Basi-occipital bone, 61
Basisphenoid bone, 61
Biceps brachii muscle, 257
Biceps femoris, 305
  sciatic nerve relationship to, 297
Bicipital aponeurosis, dissection of, 275
Bicipital groove, 257
  in arm dissection, 262
  arm muscles and, 257
Bile canaliculi, in gallbladder dissection, 146
Bile duct
  in gallbladder dissection, 146
  in liver dissection, 142
Bile duct mesentery. *See* Lesser omentum

Numbers in italic indicate figures.

Bladder
    identification by percussion, 169
    in pelvis dissection, 171
Blood flow, to and from heart chambers, 108–109
Blood supply, to breast, 87
Blood vessels. *See* individually named arteries and veins
Bone marrow, 6
Bony labyrinth. *See* Inner ear
Bony pelvis, 169
    identification of, 4
Bony prominences
    of arm, elbow and shoulder, 255–256
    of back and neck, 241–242
    of chest wall, 85
    of forearm, wrist and hand, 267–268
    of leg and foot, 309–311
    of lower limbs, 281–282
    of oral vestibule, 28
Bowel sounds, 150
Brachial artery, 260
    division in antecubital fossa, 269, 271, 274
Brachialis muscle, 257, 260
Brachial plexus
    in anterior triangle of neck dissection, 233
    arm and, 260
    in neck posterior triangle dissection, 237
    nerves derived from, 274
Brachiocephalic trunk, 98–99
    in mediastinum dissection, 107
Brachiocephalic veins, 100
    in superior mediastinum examination, 100
Brachioradialis muscle, 269
Brain, inferior aspect, *196*
Breast. *See* Mammary gland
Breast bone. *See* Sternum
Bregma, 183
Bronchopulmonary segments, in lung dissection, 97
Bronchus, in lung dissection, 97
Buccinator muscle, 19
    examination of, 27–28
Buccoalveolar sulci, 10
Buccolabial sulcus, 10
Bulbar fascia, in eyeball dissection, 207
Bulbospongiosus muscle, 180
Bulla ethmoidalis, in nasal cavity dissection, 63
Bursa
    deltoid (subacromial), 263
    infrapatellar, 322
    omental, in liver dissection, 142
Buttocks. *See* Gluteal region

Cadaver, attitude towards, 8
Calcaneocuboid joint, movement of, 333
Calcaneofibular ligaments, in ankle joint dissection, 332–333
Calcaneum, 309
    sustentaculum tali, process of, 310

Calf
    dissection of, 322–323
        anterior aspect, 324–325
        left, posterior view, *316*
    muscles of, clinical examination of
        deep, 312–313
        superficial, 312
Calotte, 218
Calyces, in urinary pelvis dissection, 163, 166
Canaliculus, lacrimal, 201
Canal of Schlemm, in eyeball dissection, 210
Canthus, in eyelid examination, 201
Capillaries, in cardiovascular system, 7
Cardiac chambers, 108, 114. *See also* individually named chambers
Cardiac function, vagal control of, 100
Cardiac loop, embryological, 111
Cardiac plexus, identification of, 118
Cardiac veins, in heart dissection, 114
Cardiovascular system, in systematic anatomy, 6–7
Carotid artery
    in carotid triangle
        dissection, 55
        examination, 53
    common. *See* Common carotid artery
    external. *See* External carotid artery
    internal. *See* Internal carotid artery
    superior thyroid branch of external, in carotid triangle dissection, 57
    temporomandibular joint dissection and external, 48
Carotid nerve, internal, 233
Carotid sheath
    in carotid triangle dissection, 56
    in muscular triangle dissection, 76
Carotid triangle, 53–54
    dissection of, 55–57
Carpal bones, palpation of, 267
Carpal tunnel syndrome, flexor retinaculum and, 267
Carpometacarpal joint, dissection of, 277
Cartilaginous joints
    in long bones, 255
    inter vertebral, 247
Cartilaginous tissue, of nose, 59
Caruncula
    lacrimalis, 201
    sublingual, 29
Caudal, defined, 69
Caval venous system, 123
Cavernous sinus, in cranial cavity dissection, 199–200
Cecum, identification of, 149, 134
Celiac artery, 146
    branches of, 156
    dissection of, 151, 156
Celiac plexus, 166
Cellular immunity, described, 6
Central nervous system (CNS), 5, *104*
    gray matter in (versus ganglion in PNS), 106
    and peripheral nervous system, 101, *104*

peripheral nervous system and, 26
reflexes and, 130
spinal segment of, *104*
Cephalic vein, 260
Cerebral hemispheres, 186
Cerebrospinal fluid (CSF), sample acquisition, 125
Cerumen, in external ear dissection, 218
Cervical fascia, deep, dissection of, 55, 76
Cervical ganglia, in anterior triangle of neck dissection, 233
Cervical nerve, descending, in carotid triangle dissection, 56–57
Cervical plexus
    dissection of, 55
    in neck posterior triangle, 230
Cervical vertebrae
    first. *See* Atlas
    head and neck movements involving, 246
    identification of, 4
    lateral view of, *244*
Cervix
    in pelvis dissection, 171
    vaginal examination of, 169
Chambers
    cardiac, 108, 114. *See also* individually named chambers
    vitreous, in eyeball dissection, 207
Cheekbone. *See* Zygomatic bone
Chest wall
    anterior, inner surface of, *90*
    examination of, 81
Cholecystitis, and referred pain, 146
Chordae tendinae, in heart dissection, 115, 118
Chorda tympani, 192
    in relation to temporomandibular joint, 48
Ciliary body, components in eyeball dissection, 211
Ciliary ganglion
    in orbit dissection, 213
    pupils and, 190
    as example of a parasympathetic ganglion, 106
Ciliary muscle, in eyeball dissection, 213
Ciliary nerves, 190–191
    in orbit dissection, 210
Ciliary ring, in eyeball dissection, 210, 211
Circle of Willis, 233
Circumduction
    of lower limb, 284
    of upper limb, 261
Circumflex humeral artery, in arm dissection, 266
Clavicle
    dissection of, 122–124
    palpation of, 4, 84
    thorax and, 81
Clavipectoral fascia, in chest wall dissection, 87
Clinical crown, anatomical crown versus, in teeth examination, 28–29

Clitoris, 7
  dissection of, 180
  in female perineum examination, 178
CNS. *See* Central nervous system (CNS)
Coccygeal plexus, 290
Coccygeal vertebrae, identification of, 4
Coccygeus muscle, 290
  separation from iliococcygeus muscle, 291
Cochlea, duct of, in inner ear, 215
Colic arteries, course of, 156
Collar bone. *See* Clavicle
Collateral ligaments, of knee, 310
Colon
  dissection of, 157
  examination of, 149
  internal and external appearance of, *152*
  loop of, *160*
Color vision, testing of, 187
Common carotid artery
  in carotid triangle examination, 53
  as relation of lung, 97
  in mediastinum dissection, 107
  in mediastinum examination, 98, 99
  in trachea examination, 70
Common extensor tendon, 270
Common facial vein, in carotid triangle dissection, 56
Common peroneal nerve
  in calf dissection, 325
  in hamstring dissection, 297, 300
  in leg examination, 311
Conducting tissue, of the heart, 118
Condyle, in temperomandibular joint examination, 39–40
Condyloid process, of mandible, 37
Cones, of retina, 207
Conjoined tendon of inguinal canal, 132
Conjunctiva, 201
Connector neuron, 86
Contractions
  cardiac, 118
  peristaltic, 150
Conus elasticus, in larynx dissection, 73
Coracoacromial ligament, 263
Coracobrachialis muscle, 257
  in arm dissection, 262
Cornea, 201
  in eyeball dissection, 210
  in eyeball examination, 204
Corneal spur, in eyeball dissection, 210
Corocoid process, 85
Corona, in penis examination, 178
Coronal plane, described, 4
Coronal suture, 4
Coronary arteries, in heart dissection, 114
Coronary ligament, in liver dissection, 139, 142
Coronary sinus, in heart dissection, 114, 115
Coronary sulcus, in heart dissection, 114

Coronoid process, 28
  in mandible examination, 37
  in temporal muscle examination, 38
Corpus cavernosum
  dissection of, 179
  in female, 178
  in male, 175
Corpus spongiosum
  dissection of, 179
  in penis examination, 175
Costal cartilage, 84
Costocervicalis muscle, 251
Costocervical trunk, in neck posterior triangle dissection, 239
Costomediastinal recess, 97
Cough impulse, internal inguinal ring and, 130
Cranial cavity, median view, *197*
Cranial fossae, 186
Cranial nerves. *See also* individually named nerves
  in anterior triangle of neck dissection, 234–235
  autonomic nervous system and, 106
  clinical examination of, 186–187, 190–194
  facial muscles and, 19. *See also* Facial nerve
  and PNS relationship, 5
Craniocaudal specialization, 183
Cranium, base of, internal aspect, *188*
Cremasteric reflex, test for, 175
Cremaster muscle
  in abdominal wall dissection, 132
  in scrotum examination, 175
Cricoid cartilage
  in muscular triangle examination, 54
  palpation of, 69–70
Cricoid lamina, in larynx dissection, 73
Cricothyroid ligament, defined, 76
Cricothyroid muscle
  as relation of inferior pharyngeal constrictor, 77
  in larynx dissection, 73
Cricovocal membrane, in larynx dissection, 76
Crista terminalis, in heart dissection, 115
Crossed reflex, pupils and, 190
Cruciate anastomosis, 306
Cruciate ligaments, in knee joint dissection, 322
CSF. *See* Cerebrospinal fluid (CSF)
Cuboid, 310
Cuneiforms, 310
Cupolae, of diaphragm, 120
Cusps
  of pulmonary valve, 115
  sinuses of, 118
  of tricuspid valve, 115
Cystic duct, in gallbladder dissection, 146

Deep fascia
  cervical, dissection of, 55, 76
  of face, 43
  transverse, of leg, 322–323
Deep palmar arch, 276
Deep peroneal nerve, in calf dissection, 324, 325
Deglutition
  mechanisms of, 70
  muscles involved in, 61–62, 70–71
Deltoid bursa, in arm dissection, 263
Deltoid muscle, 85
Deltopectoral triangle, 85
Dens, head rotation and, 246
Dental arcade, in oral cavity examination, 9
Descending aorta
  abdominal portion
    examination of, 149
    surface projection to anterior abdominal wall, 159
  in mediastinum examination, 99
  thoracic portion
    in inferior mediastinum examination, 100
    as relation of left lung, 97
    in mediastinum dissection, 107
    surface projection of, 120–121
Descending colon, 149
Developmental anatomy, 3
Diaphragm
  examination by dissection, 124
  surface drawing of, 120
Diaphragm, pelvic, 168–169
Diaphysis, 255
Diastole, ventricular, 109
Digastric muscles
  dissection of, 57–58
  palpation of, 49, 52
Digastric triangle, 49, 52
  dissection of, 57–58
  submandibular salivary gland in, 52–53
Digestive system, in systematic anatomy, 5
Digestive tract
  blood supply of, 157
  divisions of, 151
  palpable portion of, 149
Digestive tube, 5
  pancreas and liver development from, 143, 146
Dilator muscles, of pupil, in eyeball dissection, 210
Disk(s)
  intraarticular, in temperomandibular joint, 40
  invertebral, 247
  optic, in eyeball dissection, 207
  "slipped," 292
Dissection
  description of technique, xv–xvi
  skinning techniques, *16*
Distal, defined, 4

Numbers in italic indicate figures.

INDEX **339**

Distal phalanges, 268
Domes. See Cupolae
Dorsal, defined, 14, *88*
Dorsal interossei muscles, dissection of, 332
Dorsalis pedis artery, 313
　dissection of, 325
Dorsal mesogastrium, in spleen dissection, 142
Dorsal rami, 86
Dorsal root ganglion, 86
"Dorsiflexion," 309
Douglas, pouch of. See Recto-uterine pouch
Ductus arteriosus, 107
Duodenal papilla, 146
Duodenohepatic ligaments, in liver dissection, 142
Duodenojejunal flexure, 136, 150
Duodenum, 135
　dissection of, 147
　laboratory examination of, 151
　and stomach mobility compared, 139
　surface drawing of, 136
　　on posterior abdominal wall, 159
Dura mater, in cranial cavity dissection, 195

Ear. See also External ear; Inner ear; Middle ear
　clinical examination of, 214–215
　dissection of, 218–222
Eardrum. See Tympanum
Efferent impulses, 6
Ejaculatory duct, in pelvic dissection, 172
Elbow
　blood vessels of, 260
　bony landmarks of, 255–256
　dissection of, 262–266, 276
　functional anatomy and movements, 261
　nerves of, 260–261
　posterior aspect, 252
Elevation, of scapula, 261
Embryological development
　abdominal, 138–139, 150
　heart, 111
　respiratory system, 69
Endocrine glands, 7
Endocrine-metabolic system, in systematic anatomy, 7
Endoderm, digestive system and, 5
Epididymis, 175
Epigastric arteries
　in abdominal examination, 126
　in chest wall dissection, 92
Epigastric region, of abdomen, 126
Epiglottis, 71
　in laryngopharynx examination, 68–69
Epiphyseal disk, 255
Epiphysis, 255
Epiploic foramen, 142

Epithelial cells, 9
Erector spinae. See Sacrospinalis muscles
Esophageal aperture, 135
Esophagus, 121
　abdominal portion, 135
　dissection of, 148
　examination by dissection, 123–124
　as relation of trachea, 70
Ethmoidal sinus, in nasal cavity
　dissection, 63
　examination, 59
Evagination, defined, 127
Eversion, 311
Excretion, defined, 7
Exocrine, defined, 9
Extension
　defined, 225
　of thumb, 268
Extensor muscles
　in calf, 324, 325
　of forearm, 269–270
　of leg
　　digitorum brevis, 313
　　digitorum longus, 311
Extensor reflex, test for, 274
Extensor retinaculum, 267
　dissection of, 276
　hallucis longus, 311
External auditory meatus
　in Frankfurt plane position, 3
External carotid artery
　in carotid triangle examination, 53
　in neck posterior triangle dissection, 237
　temporomandibular joint dissection and, 48
External ear
　clinical examination of, 214
　dissection of, 218
External jugular vein, dissection of, 55
Extorsion movement, of eye, 206
Extrapyramidal nervous pathways, 314
Extremity, defined
　lower, 4–5
　upper, 4
Extrinsic muscles, of tongue, 18
　dissection of, 24–25
Eye
　anterior chamber of, 210
　anterior quadrant, cut surface view, *208*
　clinical examination of, 201
　medial angle, structures of, *202*
　posterior chamber of, 210
　tarsal glands and, 201, *202*
Eyeball
　anterior half, interior aspect of, *208*
　clinical examination of, 204–205
　dissection of, 207–211
　muscle control of, 187
Eyebrows, clinical examination of, 201
Eyelashes, clinical examination of, 201
Eyelids, clinical examination of, 201

Face
　muscles of, 18–19
　　actions and palpation to identify, 19
　superficial dissection of, *16*
　superficial structures of, 20
Facial artery
　in carotid triangle examination, 53–54
　dissection of, 25
　in maxilla examination, 37
Facial nerve, 19
　branches of, 20
　dissection of, 25
　in inner ear dissection, 219
　in clinical examination, 191–192
Facial vein, common, in carotid triangle dissection, 56
Falx cerebri, in cranial cavity dissection, 195
Fascia
　bulbar, in eyeball dissection, 207
　clavipectoral, in chest wall dissection, 87
　deep. See Deep fascia
　inguinal. See Inguinal fascia
　parotideomasseteric, dissection of, 43
　popliteus, in leg dissection, 318
　pretrachial, in thyroid gland dissection, 76
　prevertebral, in anterior triangle of neck dissection, 233
　spermatic. See Spermatic fascia
　temporal, in temporalis muscle dissection, 42
　thoracolumbar, 167, 243
　transversalis
　　in abdominal wall dissection, 132
　　in peritoneal cavity dissection, 163
Fat pad, infrapatellar, in knee joint dissection, 322
Femoral artery
　in abdominal examination, 126
　in quadriceps femoris dissection, 305
Femoral hernia, 301
Femoral nerve, 167, 288
Femoral point, in abdominal examination, 126
Femoral triangle, 301
　dissection of, *302*
Femoral vein, 288
　in thigh dissection, 301
Femur
　lateral supracondylar line of, 292
　palpation of, 282
　superior end dissection of, *302*
Fenestra cochleae, of middle ear, 215
Fenestra vestibuli
　of middle ear, 215
　dissection, 219
Fibrous joints, 39
Fibrous pericardium, 101
Filiform papillae, 14, *16*
Fingernails, in integumentary system, 7
Fingers
　creases at base of, 270

Fingers (cont.)
　extensor muscles of, 269–270
　flexor muscles to, 268
Flat tendon. See Aponeurosis
Flexion
　defined, 225
　of thumb, 268, 271
Flexor muscles
　in foot dissection
　　digiti minimi brevis, 329
　　digitorum accessorius, 328
　　digitorum brevis, 328
　　digitorum longus, 328–329
　　hallucis brevis, 329
　of forearm, 268–269
　of leg
　　deep, 312–313
　　dissection, 323
　　superficial, 312
Flexor reflex, test for, 274
Flexor retinaculum
　of wrist, 267
　　dissection of, 276
　of ankle, dissection of, 323
Flexor sheath, 276
Flexor tendons
　carpi radialis, 268
　carpi ulnaris, 268
　hallucis longus, in foot dissection, 329
　of hand and forearm, 268–269
　　dissection of, 276
　pollicus, thumb and, 268
Floating ribs. See Free ribs
Foliate papillae, 15, *16*
Foot
　arch of, inferomedial view, *320*
　bony landmarks of, 309–311
　dissection of, 325–332
　　lateral aspect, *326*
　dorsolateral view, *326*
　functions of, 313
　inversion of, left and right, *326*
　lateral view, *326*
　muscles of, 311–313
　　layers, 328–329, *330*, 332
　plantar surface, *330*
　　dissection showing superficial nerves, *330*
　inferomedial dissection, *320*
　vessels and nerves of, 313–314
Foramen(ina)
　cecum, 15
　epiploic, 142
　of the facial skeleton, 20
　infraorbital, maxilla examination and, 37
　intervertebral, 86
　magnum
　　as relation of nasopharynx, 61
　　as relation of temperomandibular joint, 40
　mandibular, 28
　　as relation of temperomandibular joint, 40–41

obturator, in pelvis dissection, 173
ovale
　of heart, 115
　of base of skull, 46–47
sciatic, in lower limb dissection, 290
sphenopalatine, in nasal cavity dissection, 63
spinosum, as relation of temporomandibular joint dissection, 48
Forearm
　anterior aspect, *264*
　arteries and nerves of, 271, 274
　bony prominences of, 267–268
　defined, 4
　dissection of, 275–277
　dorsal aspect in pronation, *264*
　extensor muscles of, 269–270
　flexor muscles of, 268–269
　lateral aspect, *264, 272*
　skeleton of, *258*
　structures of, *252*
　ventral aspect, *272*
Fossa ovalis, in heart dissection, 115
Fovea centralis, in retina examination, 205
Foveolae, palatine, 30
Frankfurt plane, described, 3
Free gingival groove, 29
Free marginal gingiva, examination of, 29
Free ribs, 84
　palpation of, 85
Frenulum linguae, 15, 29
Frontal bone, 37
Frontal plane, described, 4
Frontal sinus, in nasal interior examination, 59
Functional anatomy, 3
Fundus, of uterus manual vaginal examination and, 169
Fungiform papillae, 14–15, *16*

Gallbladder
　dissection of, 146
　surface marking of, 136
Ganglion(ia)
　cervical, in anterior triangle of neck dissection, 233
　described, 6, 33
　dorsal root, 86
　geniculate, in inner ear dissection, 222
　otic, in medial pterygoid dissection, 48
　pterygopalatine, in nasal cavity dissection, 63
　submandibular, 33
　sympathetic, 106, *104*
　vagus nerve and, 101, 106
Gastric arteries, 156
Gastrocnemius muscle
　dissection of, 318
　popliteal fossa and, 309

Gastroduodenal artery, 156
Gastroepiploic artery, 156
Gastrohepatic ligaments, in liver dissection, 142
Gastrosplenic ligament, in spleen dissection, 142
Gemellus muscles, in gluteal region dissection, and piriformis relationship, 296
Genial tubercles, 18
Geniculate ganglion, in inner ear dissection, 222
Genioglossus, 18
　dissection of, 24
Geniohyoid muscle, 18
Genitalia
　in females, 130
　in males, 127
Genital system, in systematic anatomy, 6–7
Gingiva, 9
　examination of, 28–29
Glabella, 183
Gland(s). See also individually named glands
　defined, 9
　of the tongue, 15
Glans, in penis examination, 178
Glossoepiglottic folds, 71
Glossopharyngeal nerve
　in clinical examination, 193
　in oropharynx dissection, 72
　and taste bud innervation, 192
Gluteal arteries, in pelvis dissection, 173
Gluteal fold, identification of, 168
Gluteal line
　anterior, 293
　middle, 293
　posterior, 292
Gluteal region, dissection of, 291–293, 296
　posterior view, *294*
Gluteal tuberosity, gluteus maximus muscle and, 286–287
Gluteus maximus muscle, 281
　in gluteal region dissection, 292–293, 296, 297
　and piriformis relationship, 297
　palpation of, 286–287
　and related structures, posterior view, *294*
Gluteus medius muscle, 287
　in gluteal region dissection, 292–293
Gluteus minimus muscle, 287
　in gluteal region dissection, 292–293
Gonadal vein, in peritoneal cavity dissection, 163
Gracilis muscle
　clinical examination, 285
　tendon in leg dissection, 318
Gray matter
　defined, 106
　described, 6
　versus ganglion, 86

Numbers in italic indicate figures.

Greater horns of hyoid bone, palpation of, 4
Greater occipital nerve, dissection of, 248
Greater omentum
 and transverse mesocolon, 138
Greater sciatic foramen, in lower limb dissection, 290
Greater trochanter, of femur, 282
Great veins, examination by dissection, 111, 114
Great vessels, of mediastinum, 100
Gubernaculum
 in male, 127
 in pelvic dissection, 172
 homologue in female, 130
Gum. *See* Gingiva
Gyri, of cerebral hemispheres, 186

Hair, in integumentary system, 7
Hamate bone, hook of, 267
Hamstring muscles
 adductor magnus origin in, 286
 clinical examination of, 287
 dissection of, 297, 300
Hamulus
 laboratory examination of, 32
 palpation of, 27
Hand
 arteries and nerves of, 271, 274
 bony prominences of, 267–268
 clinical examination of, 270–271
 dissection of, 275–277
 dorsal aspect in pronation, *264*
 right, lateral aspect, *272*
 skeleton of, *258*
 ventral aspect, *272*
"Handedness," scoliosis and, 241
Hard palate, 10, 30
Haustrations
 of colon, 149, 157
Head
 movements of, 246–247
 palpation of, 4
Hearing, testing of, 193
Heart. *See also* Cardiac
 apex of
  apex beat and, 108
  auscultation of, 110
  in mediastinum palpation, 99
 in cardiovascular system, 6
 chambers of, 108, 114
 diaphragmatic surface and base view, *112*
 in embryological development, 111
 named surfaces of, 114
 percussion examination of, 108
 removal technique, 114
 and sinoatrial node role, 118
 sternocostal aspect of, *112*
 and sternum position, 99
Heart rhythm, sinoatrial node role in, 118
Heart sounds, 99
 auscultation of, 109

Heel bone. *See* Calcaneum
Hematic system, in systematic anatomy, 6
Hemopoietic organ, spleen as, 136
Hepatic artery, 156
Hepatic ducts, in gallbladder dissection, 146
 defined, 130
Hernia, femoral, 301
 inguinal, 130
Hilar structures, of left kidney, *164*
Hilus, of lung, 96
Hip joint, 284
 movements, and developmental rotation of lower limb, 286
 muscles acting on, 284–287
His, atrioventricular bundle of, 118
Homology
 defined, 130, 281
 of labium majus and scrotum, 130
 of round ligaments of ovary and uterus and gubernaculum, 172
 of male and female external genitalia, 178
 of upper and lower limbs, 281
 of upper limbs of human and wings of bird, 281
 of bones of hand and foot, 310
Horizontal plane, described, 4
Humerus
 examination of, 255
 lesser tubercle of, 85
 neck posterior triangle and, 227
Humoral immunity, described, 6
Hyaline cartilage, 247
Hyoid bone, palpation of, 4, 49, 52
Hypochondriac regions, of abdomen, 126
Hypoglossal nerve
 in carotid triangle dissection, 56
 in clinical examination, 193
 tongue muscles and, 15
Hyoglossus, 18
 dissection of, 24
Hypophysis, in cranial cavity dissection, 198
Hypothenar eminence
 dissection of, 276
 palpation of, 267, 271

Ileocecal junction, 150
Ileocolic artery, 156
Ileum
 arterial arcades of, *154*
 dissection of, 157
 identification of, 150
 mucosal surface of, *154*
Iliac crest, in abdominal examination, 125
Iliac spine. *See* Anterior superior iliac spine (ASIS); Posterior superior iliac spine (PSIS)
Iliac vein, in pelvis dissection, 173
Iliococcygeus muscle, 173
 separation from coccygeus muscle, 291

Iliofemoral ligament, 307
Iliohypogastric nerve
 in abdominal wall dissection, 131
 in gluteal region dissection, 292
Ilioinguinal nerves, in abdominal wall dissection, 131
Iliolumbar ligament, 283
Iliopsoas muscle, 306
Iliopubic eminence, 169
Iliotibial tract, in gluteal region dissection, 292–293
Immunocompetent lymphocytes, palatine tonsil and, 68
Immunolymphatic system
 spleen in, 136
 in systematic anatomy, 6
Impulses, carried by nerve cells, 6
Incus
 dissection of, 219
 in ear examination, 214
Index finger, extensor muscle of, 270
Inferior, defined, 3
Inferior alveolar nerve, 28
 in medial pterygoid dissection, 47
Inferior epigastric artery
 in abdominal examination, 126
 in abdominal wall dissection, 132
 in chest wall dissection, 92
Inferior laryngeal artery, in inferior pharyngeal constrictor dissection, 77
Inferior lobe, of lung, 95
Inferior mediastinum
 anterior, contents of, 121
 posterior, components of, 121
Inferior mesenteric artery, dissection of, *152*, 156
Inferior oblique muscle of eye
 eyeball control by, 187
Inferior oblique muscle of head, 249
Inferior pharyngeal constrictor, dissection of, 77
Inferior pubic ramus, palpation of, 168
Inferior rectus muscle of eye
 eyeball control by, 187
 in orbit dissection, 213
Inferior thyroid artery
 in anterior triangle of neck dissection, 232, 233
 in thyroid gland dissection, 76
Inferior tibiofibular joint, in foot dissection, 332
Inferior vena cava
 in cardiac loop, 111
 diaphragm examination and, 121
 in heart dissection, 115
 in liver dissection, 142
 and right atrium, 108
Infrahyoid muscles, 54
Infraorbital foramen, 20
 maxilla examination and, 37
Infraorbital margin, maxilla examination and, 37
Infraorbital nerve, dissection of, 42
Infrapatellar fat pad, 322
Infrapatellar synovial fold, 322
Infraspinatus muscle

Infundibulum, in nasal cavity dissection, 63
Inguinal canal, 127
Inguinal fascia, in abdominal wall dissection
  external, 131
  internal, 132
Inguinal (iliac) regions
  of abdomen, 126
  sexual differences observable in, 127
Inguinal ring
  external, 127
  internal, 127
    palpation of, 130
Inion, 183
Inner ear
  clinical examination of, 215
  dissection of, 219, 222
    anterolateral aspect, *220*
    posterolateral aspect, *220*
    posteromedial aspect, *220*
    superior aspect, *220*
Innervation, visceral versus somatic, 101
Integument, 7
Integumentary system, in systematic anatomy, 7
Intercondylar notch, of femur, 284
Intercostal muscles
  in chest wall dissection, 87
  dissection of, 87
  layers of, 81, *82*
  role in respiration, 81, 84
Intercostal nerves, 86
  in abdominal wall dissection, 131
Intercostal space
  dissection in, 87
  right anterior, medial dissection of, *82*
Intercristal plane, 125
Interdental papillae, examination of, 29
Intermediate tendon, of digastric muscle, 49
Internal auditory meatus, in cranial cavity dissection, 198
Internal carotid artery
  in carotid triangle examination, 53
  in cranial cavity dissection, 198
Internal carotid nerve, in anterior triangle of neck dissection, 233
Internal iliac artery, 290–291
Internal jugular vein
  carotid triangle dissection and, 55
  in superior mediastinum examination, 100
Internal thoracic artery, in chest wall dissection, 92
Interosseous membrane, of leg, 332
Interosseous muscles
  of foot, dissection of, 332
  of palm, 271
    dissection of, 276
Interspinal muscles, dissection of, 251

Intertransverse muscles, dissection of, 251
Intertubercular plane, 125
Intervertebral foramen, 86
Intorsion movement, of eye, 206
Intraarticular disk, in temperomandibular joint examination, 40
Intrinsic muscles
  of larynx, dissection of, 73
  of tongue, 18
    dissection of, 24–25
    swallowing and, 70
Invagination
  defined, 127
  of digestive tract into peritoneal sac, 138–139
  of heart into pericardial sac, 111
  of lungs into pleural sacs, 93
Iridocorneal angle, in eyeball dissection, 210
Iris, 201
  in eyeball dissection, 210
  in eyeball examination, 204
Ischial spine
  in lower limb dissection
    palpation of, 290
    and piriformis relationship, 296
  rectal examination of, 170
Ischial tuberosity, palpation of, 168, 171
Ischiocavernosus muscle, 180
Ischiofemoral ligament, 307
Ischiopubic ramus, palpation of, 171

Jejunum
  arterial arcades of, *154*
  dissection of, 157
  identification of, 150
  mucosal surface of, *154*
Joint capsule
  defined, 39
  in elbow dissection, 276
Joint(s)
  acromioclavicular, 256
  ankle, dissection of, 332–333
  calcaneocuboid, movement of, 333
  carpometacarpal, 277
  cartilaginous, 39
    back movements and, 247, 255
  classification of, 39
  hip, 284
  inferior tibiofibular, in foot dissection, 332
  knee. *See* Knee joint
  metatarsophalangeal, dissection of, 333
  midcarpal, 277
  midtarsal, 313
  radiocarpal, 277
  sacroiliac, 283–284
  sternoclavicular, lung extent and, 95
  subtaler, movement of, 333
  synovial, 246–247

  talocalcaneonavicular, movement of, 333
  temporomandibular, 39–41, 48
Jugular vein
  external, dissection of, 55
  internal
    carotid triangle dissection and, 55
    in superior mediastinum examination, 100

Kidney(s), 7
  dissection of, 163, 166
  identification of, 158–159
  left
    hilar structures relationship, *164*
    posterior relations, *164*
  palpation of, 159
  quadratus lumborum muscle relationship to, 163
  surface markings for, 159
Killer cells, 6
Knee
  and femur palpation, 282
  lateral and posterolateral aspects, *298*
  medial and posteromedial aspects, *298*
Knee cap. *See also* Patella, dissection of, 322
Knee joint
  anterior view, *298*
  clinical examination of, 308–309
  dissection of, 319, 322
  meniscus and, 319
Knuckles, of hand, 268
Kyphosis, 241

Labial arteries
  facial artery and, 20
  in maxilla examination, 37
  palpation of, 11
Labial commissure, 11, 28
  dissection of, 32
Labial glands, 11, *12*
Labia majora, 130, 178
Labia minora, 178
Labioalveolar sulci, 10
Labioscrotal fold, 172
Labium majus. *See* Labia majora
Lacrimal canaliculus, in eyelid examination, 201
Lacrimal gland, 201, 204
  cranial nerves and, 191
Lacrimal groove, 201
Lacrimal papilla, in eyelid examination, 201
Lactiferous ducts
  in breast examination, 85
  dissection of, 87
Large bowel. *See* Large intestine

Numbers in italic indicate figures.

Large intestine
    surface markings of, 149
Laryngeal nerves
    in inferior pharyngeal constrictor dissection, 77
    recurrent. See Recurrent laryngeal nerve
    in thyroid gland dissection, 76
Laryngeal prominence, in larynx examination, 69
Laryngopharynx
    dissection of, 72–73
    examination of, 68–69
Larynx, 14, 61
    dissection of, 73, 74, 76
    examination of, 69–70
    muscles involved in, 70–71
    transverse section, 64
Lateral, defined, 3
Lateral collateral ligament, of knee, 310
Lateral cutaneous nerve of the thigh in gluteal region dissection, 292
Lateral (lumbar) regions, of abdomen, 126
Lateral plantar nerve, in foot dissection, 329
Lateral pterygoid muscle, 38–39
    dissection of, 46–47
Lateral rectus muscle of eye
    eyeball control by, 187
    in orbit dissection, 213
Lateral temperomandibular ligament, 40
    in temporomandibular joint dissection, 48
Latissimus dorsi muscle, 243
    axilla and, 85
    dissection of, 249–250
    in shoulder and arm, 256–257
Left atrioventricular valve, 109
    in left ventricle dissection, 118
Left atrium
    blood flow and, 109
    dissection of, 118
    opened view, 112
Left ventricle
    blood flow and, 109
    dissection of, 114–115, 118
    interior view, 112
Leg
    anatomist usage of term, 4–5
    bony landmarks of, 309–311
    dissection of, 315, 318–319
        posterior aspect, 316
    muscles of, 311–313
        flexor (calf) muscles, 312–313
    right
        inferomedial view, 320
        posteromedial aspect, 316
    vessels and nerves of, 313–314
Lens, in eyeball dissection, 210
Lesser omentum
    liver and, 141
    in liver dissection, 142, 146
Lesser sciatic foramen, in lower limb dissection, 290
Lesser trochanter, of femur, 282

Levator muscles
    ani. See Pelvic diaphragm
    costarum, dissection of, 251
    palpebrae superioris, in orbit dissection, 212
    prostate, 173
    scapulae, in neck posterior triangle, 231
    veli palatini
        dissection of, 66
        in nasal cavity examination, 60–61
Lienorenal ligament, in spleen dissection, 143
Ligament(s). See also individually named ligaments
    in ankle joint dissection, 332–333
    cruciate, in knee joint dissection, 322
    described, 39
    elbow dissection and, 276
    long plantar, 329
    palmar, 277
Ligamentum arteriosum
    in mediastinum dissection, 107
Ligamentum nuchae, 241
Ligamentum teres, in abdominal wall dissection, 133
Ligamentum venosum, fissure for, 143
Linea alba, in abdominal examination, 126
Linea semilunaris
    in abdominal examination, 126
    in abdominal wall dissection, 131
Lingua. See Tongue
Lingual artery
    in carotid triangle examination, 53
    pulse of, 15
Lingual frenulum, 29
Lingual nerve, 28
    in medial pterygoid dissection, 47
Lingual tonsil, 15
    and palatine tonsil relationship, 68
Lingual vein, location of, 15
Lingula, 41
Lips
    laboratory review of, 21
    mucous membrane lined portion, 11
    as oral cavity boundary, 9
    red (prolabial) portion, 11
    regions of, 10
    skin covered portion, 10–11
    upper, tubercle of, 10, 12
Liver
    dissection of, 139, 143, 146
    lesser omentum, 141
    palpation of, 139, 142
    right anterolateral view, 141
    superior aspects, 144
    surface marking of, 136
    visceral surface of, 141
Lobes
    of cerebral hemispheres, 186
    of lung, 95
        dissection of, 97
    pyramidal, in thyroid gland dissection, 76

Longissimus capitis muscle
    in back dissection, 251
    dissection of, 248
Longissimus cervicis muscle
    in back dissection, 251
    dissection of, 248
Longitudinal muscles, of pharyngeal muscular coat, 61
Long plantar ligament, in foot dissection, 329
Long saphenous vein, 288
    in leg and foot, 313
    in thigh dissection, 301
Lordosis, 241
Lower abdominal organs
    examination procedures for, 149, 150
Lower extremity (lower limb), defined, 4–5
    anterior aspect, 302
    anteromedial aspect, 302
    bony prominences of, 281–282
    development of, 166–167
    dissection of, 289–300
    lateral aspect, 294, 316
    and trunk, articulation between, 283
    and upper limbs, homology of, 281
Lower limb girdle, 281–282
Lumbar arteries, in abdomen, 151
Lumbar plexus, nerve identification, 167
Lumbar puncture, position for, 125
Lumbar triangle, in back dissection, 250
Lumbar veins, ascending, 166
Lumbar vertebrae, identification of, 4
Lumbrical muscles
    of foot, 329
    of hand, 271
Lumbrosacral trunk, in lower limb dissection, 289–290
Lung contour, determination of, 95
Lung(s)
    apex of, 94
    auscultation in examination of, 95
    borders of, 95
    divisions of, 95
    function of, 93
    lateral aspects of, 90
    mobilization by hand dissection, 96
    percussion examination of, 94
    respiratory mechanism examination and, 94
"Lung sounds," 95
Lymph glands (Lymph nodes), 6
    upper limb dissection and, 266
Lymph nodes, submental, 54
Lymphocytes, 6
    immunocompetent, palatine tonsil and, 68
    producing organs for, 6
Lymph vessels, 6

Macula lutea, in retina examination, 205
Macular vision, 187
Malleolus, lateral, 310
Malleus, 214
    dissection of, 219

Mammary gland
  blood supply to, 87
  described, 85, *88*
  dissection of, 86–87
  in integumentary system, 7
  mobility of, *88*
Mandible
  angle of, 26
  infratemporal region dissection and, *44*
  mylohyoid muscle relationship to, 4
  in oral cavity examination, 9–10
  palpation of, 37
  ramus of, 26, 28
Mandibular division, of trigeminal nerve, 191–192
Mandibular fossa, 40
Mandibular nerve, 20
  buccal branch, in medial pterygoid dissection, 48
Manubrium of sternum, in chest wall examination, 84
Masseteric nerve, in masseter muscle dissection, 43
Masseteric vessels, in masseter muscle dissection, 43
Masseter muscle
  elevation of, *44*
  in maxilla examination, 37
  palpation of, 26, 38
  superficial dissection of, 43, *46*
Masticatory apparatus. *See also* individually named muscles; Temporomandibular joint
  clinical examination of, 37–41
  defined, 37
  dissection of, 42–48
  primary muscles of, 38–39
Mastoid air cells, and middle ear, 215
Mastoid process, identification of, 49
Maxilla
  in oral cavity examination, 9–10
  palatine process of, 10
  palpation of, 37, 37–38
Maxillary artery
  in lateral pterygoid dissection, *46*
  middle meningeal branch, in temporomandibular joint dissection, 48
Maxillary division, of trigeminal nerve, 191
Maxillary nerve, 20
Maxillary sinus
  in nasal cavity dissection, 63
  in nasal clinical examination, 59
Maxillary vein, in lateral pterygoid dissection, *46*
Meatus
  external auditory
    in Frankfurt plane position, 3
    lateral aspect, *216*
  nasal, examination of, 59
Medial, defined, 3
Medial circumflex artery, of thigh, 306
Medial collateral ligament, of knee, 310

Medial malleolus, in ankle dissection, 325, 332
Medial pterygoid muscle, 38
  dissection of, 47–48
Medial rectus muscle of eye, eyeball control by, 187
Median lingual sulcus, 14
Median nerve
  derivation of, 274
  dissection of, 277
  recurrent branch dissection, 276
  wrist and, 269
Median plane. *See* Sagittal plane
Median spinal furrow, 125, 158
Mediastinum. *See also* Inferior mediastinum; Superior mediastinum
  dissection of, 101
  great vessels of, 100
  left aspect of, *102*
  location of, 98
  in lung dissection, 96–97
  clinical examination of
    inferior portion, 99
    superior portion, 98–99
  right aspect of, *102*
Meniscus, in knee joint dissection, 319
Mental foramen, 20
Mental nerve, dissection of, 42
Mentolabial sulcus, 10
Mesenteric arteries, dissection of, 151, 156. *See also* Inferior mesenteric artery; Superior mesenteric artery
Mesenteric veins, 156
Mesentery
  in abdominal wall dissection, 134
  bile duct. *See* Lesser omentum
  blood vessels courses and, 156
  defined, 150
  sigmoid colon and, 150
Mesocardium, 111
Mesosalpinx, 172
Mesovarium, 172
Metabolism, defined, 7
Metacarpal bones, 267–268
Metaphysis, 255
Metatarsals, 310
Metatarsophalangeal joint, dissection of, 333
Midaxillary line, 85
Midcarpal joint, dissection of, 277
Midclavicular line, lung extent and, 94
Middle ear
  clinical examination of, 214–215
  dissection of, 218–219
  lateral aspect of medial wall of, *216*
  medial aspect of lateral wall of, *216*
  and nasal cavity relationship, 60
Middle lobe, of lung, 95
Middle phalanges, 268
Midtarsal joint, 313
Mitral valve. *See* Left atrioventricular valve
Mixed spinal nerve, defined, 86
Mons pubis, 178

Morphology
  and anatomy relationship, 81
  defined, 3
  surfaces and directions, terms for, *88*
Mouth
  angle of the, labial arterial pulse palpation and, 11
  floor of. *See also* Sublingual region
    dissection of, 22
    sagittal section, 22
  roof of. *See also* Hard palate; Soft palate
    dissection of, 33, 36
    examination of, 30–31
Mucocutaneous junction, on lips, 11, *12*
Mucoperichondrium, elevation of, 63
Mucoperiosteum
  defined, 33, 36
  elevation of, 63
Mucous membrane
  on lips, 11
  of oral cavity, 9
Muscles. *See also* individually named muscles; Tendons
  of the back, *161*
  facial, 18–19
    dissection of, 25
  hip joint and, 284–287
  of mastication, 38–39
  of oral cavity, 10
  of limbs and body wall as somatic structures, 6
  of tongue, 18
Muscular system, in systematic anatomy, 6
Muscular triangle
  larynx and, dissection of, 76
  of neck, 54
  swallowing and, 70–71
Musculocutaneous nerve, 260–261
  in arm dissection, 262
Musculophrenic artery, in chest wall dissection, 92
Musculoskeletal system, 6
Mydriatics, side effects of, 205
Mylohyoid line, 28, 52
Mylohyoid muscle, palpation of, 4, 52

Nasal alae, palpation of, 59
Nasal cavity
  dissection of, 63
  examination of, 60–62
  sagittal section, 22
Nasal choanae, in nasal cavity examination, 60
Nasal conchae, examination of, 59
Nasal meatuses, examination of, 59
Nasal septum
  dissection of, 63
  speculum examination of, 59–60
Nasal spine, anterior, 28
Nasolabial sulcus, 10

Numbers in italic indicate figures.

Nasolacrimal duct, 201
  in nasal cavity dissection, 63
  in nasal examination, 59
Nasopalatine nerves, in nasal cavity dissection, 63
Nasopharynx
  described, 61
  dissection of, 66
  sagittal section, 22
Neck
  anterior aspect, 82
  anterior triangles of, 50, 225–226
    dissection, 232–235
    palpation of, 49
      digastric triangle, 49, 52
  movements of, 246–247
  palpation of, 4
  posterior aspect, 241
    palpation of muscles, 242–243
  posterior triangle of, 226–227, 230–231
    anterior view, 228
    boundaries and relations, 228
    dissection, 236–237, 240
    posterolateral views, 238
  superficial dissection of, 16
Nelaton's line, 176, 282
Neurovascular plane, of vessels and arteries, 86
Nerve cell body, 5
  neurons and, 86
Nerve cell processes, 5
Nerve cells
  fibers. See Nerve cell processes
  impulses carried by, 6
Nerves. See individually named nerves
Nervous system, in systematic anatomy, 5–6
Neurocranium, 14. See also Skull
  clinical examination of, 183, 186
  laboratory examination of, 195, 198–200
Neurons, and nerve cell bodies, 86
Nipple
  dissection of, 87
  location of
    in female, 84
    in male, 84
Normal anatomical position, 3
  directional parameters employed in, 3
  Franfurt plane in, 3
Nose
  external palpation, 59
  internal examination, 59
  nasal septum examination, 59–60
Nostrils, described, 59
Notch(es), sciatic, 290
Nuchal line, superior, 183
"Numb bum syndrome," 292

Oblique muscles
  in abdominal wall dissection, 131
    capitis superior, in posterior neck dissection, 249

of eye
  eyeball control by, 187
  inferior, in orbit dissection, 213
  superior, in orbit dissection, 212, 213
Oblique pericardial sinus, in heart dissection, 111, 114
Obturator artery, in pelvis dissection, 173
Obturator foramen, in pelvis dissection, 173
Obturator muscles
  adductor muscles and, 286
  external, 287
  internal, 287
    in gluteal region dissection, 296
  in thigh, 306–307
Obturator nerve, 167
  in thigh, 306
Occipital artery
  in digastric triangle dissection, 57–58
  in posterior neck dissection, 248
Occipital crest, external, 183
Occipital eminence, of neurocranium, 183
Occipital triangle, of neck, 226
Oculomotor nerve
  in cranial cavity dissection, 198
  in clinical examination, 187, 190
  in orbit dissection, 212
Olecranon process, of ulna, 256
Olfactory epithelium, in nasal roof examination, 60
Olfactory nerve, in clinical examination, 186
Olfactory tract, in cranial cavity dissection, 198
Omental bursa, in liver dissection, 142
Omentum
  greater, 134, 138
  lesser, 141, 142, 146
Omohyoid muscle
  in carotid triangle dissection, 55–56
  in carotid triangle examination, 53
  in neck posterior triangle, 227
Ontogeny, defined, 14, 69
Ophthalmic artery, in orbit dissection, 212
Ophthalmic division, of trigeminal nerve, 190–191
Ophthalmic nerve, 20
Opposition, of thumb, 271
Optic chiasma, 186
Optic disk
  examination of, 204, 205
  in eyeball dissection, 207
Optic groove, 186
Optic nerve, 186
  anterior extent of, 208
  in cranial cavity dissection, 198
  in clinical examination, 186
Oral cavity
  examination of, 9–10
  laboratory review of, 21
  in maxilla examination, 37–38
  sagittal section, 34
  structures of, 12

Oral cavity proper
  defined, 10
  laboratory session, 36
  lateral boundary of, 28
Oral fissure, 11
Oral region, structures of, 12
Oral vestibule. See Vestibule
Ora serrata, in eyeball dissection, 210
Orbicularis oris, 11, 19
Orbiculus ciliaris, in eyeball dissection, 210
Orbit
  anterolateral view, 202
  clinical examination of, 205–206
  dissection of, 212–213
  in Frankfurt plane position, 3
  left, 184
  maxilla examination and, 37
  right
    deep view, 208
    superior aspect, 208
Orbital septum, in orbit dissection, 213
Oropharynx
  dissection of, 72
  examination of, 68
Ossification, centers of, 255
Otic ganglion, 193
  in medial pterygoid dissection, 48
Ovary, 7
  round ligament of, 172
Oxygenation, of blood, heart chambers and, 108–109

Palate, defined, 10. See also Hard palate; Soft palate
Palatine aponeurosis, in pharynx dissection, 66–67
Palatine bone, 10
  horizontal plate of, 30
Palatine foveolae, 30
Palatine glands, 30
Palatine nerves, in nasal cavity dissection, 63
Palatine raphe, 30
Palatine tonsil
  extent of, 68
  and lingual tonsil relationship, 68
Palatoglossal fold
  in oral cavity, 10
  in pharynx examination, 61
  and pterygomandibular ligament proximity, 27
  tongue examination and, 15
Palatoglossus muscle
  in oral cavity, 10, 30
  in pharynx dissection, 67
  tongue and, 17, 18
Palatopharyngeal arch, in pharynx examination, 61
Palatopharyngeal sphincter. See Passavant, ridge of
Palatopharyngeus muscle, 61
  in pharynx dissection, 66, 67
Palm
  creases of, 270

Palm (*cont.*)
  dissection of, 276
  interosseous muscles of, 271
  ventral aspect of, *272*
Palmar aponeurosis, 268
  dissection of, 276
Palmar arch, deep, 276
Palmar ligaments, 277
Palmaris longus tendon, 268
  dissection of, 276
Palpebrae, clinical examination of, 201
Pancreas
  dissection of, 138, 143, 146–147
  surface marking of, 136, 159, 162
Pancreaticoduodenal arteries, 156
Papillae
  duodenal, 146
  incisive, 30
  interdental, 29
  lacrimal, 201
  in oral vestibule, 26, 27. *See also* Parotid papilla
  sublingual, 29
  of tongue, 14–15, *16*
Paranasal sinuses, 59
Pararectal fossa, in pelvic dissection, 172
Parasaggital plane, described, 3
Parosympathetic part of autonic nervous system, 106, 104
Parasympathetic fibers, of facial nerve, 192
Parasympathetic nerves, in pelvis dissection, 173
Parathyroid glands, 76
Paravesical fossa, in pelvic dissection, 172
Parietal peritoneum, in abdominal wall dissection, 133
Parietal pleura, 93
  in chest wall dissection, 87
Parosmia, 186
Parotid duct, *16*, 20
  dissection of, 25, 32
  in oral vestibule, 26
  palpation of, 26–27
Parotideomasseteric fascia, dissection of, 43
Parotid gland, 20
  dissection of, 32
  in oral vestibule, 26
  palpation of, 26–27
Parotid papilla, 26, 27
  dissection of, 32
Passavant, ridge of, 62, 66
Patella, in knee joint dissection, 322
Patellar ligament, 308
Pectinate muscles, in heart dissection, 115
Pectineal line
  in abdominal wall dissection, 132
  pelvis and, 169
Pectineus muscle
  and hip joint flexion, 285
  in thigh dissection, 301, 305

Pectoralis major muscle
  axilla and, 85
  dissection of, 87
  and female breast, 85
  palpation of, 84–85
Pectoralis minor muscle
  in chest wall dissection, 87
Pectoral nerves, dissection of, 87
Pelvic brim, and division of pelvis, 169
Pelvic diaphragm
  described, 168–169
  pelvic organs and, 169
  in pelvis dissection, 173–174
Pelvic inlet, 169
Pelvic organs, clinical examination, 169–170
Pelvis. *See also* Bony pelvis
  in abdominal examination, 125
  arteries and veins of, 173
  bony landmarks, 168
  dissection of, 171–174
  organs of, 169–170
  regional definition, 168–169
  right side of, *164*
Penis, 7, 175, 178
  anterior aspect, *176*
  erectile components, 179–180
  posterior aspect, *176*
Percussion
  cardiac dullness to, 99
  in heart examination, 108
  of lower abdomen, 150
  in lung examination, 94
  lung resonance to, 99
  of upper abdominal organs, 135
Perforating cutaneous nerve, in gluteal region dissection, 292
Pericardial sinuses, 111
Pericardium
  dissection of, 111
  fibrous, heart and, 101
  serous. *See* Serous pericardium
Peridontal sulcus, 29
Perineal membrane, in penis dissection, 179–180
Perineum
  anatomical structures of
    female, *176*, 178
    male, 175, *176*, 178
  dissection of, 179–180
  identification of, 4
Periosteum, mucosa and, 33
Peripheral nerves, 5
  and CNS relationship, 5
Peripheral nervous system (PNS), 5
  central nervous system and, 26, 101, *104*
  and extrapyramidal pathway connections, 314
  segmental nerves of, *104*
Peripheral vision, 187
Peristaltic contractions, 150
Peritoneal sac, in abdominal wall dissection, 134

Peritoneum, upper abdominal organs relationship with, 138
Peroneal compartment of leg, clinical examination of, 311–312
Peroneal nerve
  common. *See* Common peroneal nerve
  deep and superficial, in calf dissection, 324–325
Peroneus muscles, 311–312
  in calf dissection
    brevis, 324–325
    longus, 324–325
    tertius, 324
Per rectum (PR) examination, 170
Per vaginum (PV) examination, 169
Petrosal nerves
  in cranial cavity dissection, 200
  greater, 192, 193
  lesser, 193
Petrotympanic fissure, 192
Phalanges
  of fingers, 268
  of toes, 310
Pharyngeal artery, ascending, dissection of, 66
Pharyngeal constrictors, 61
  inferior, and related structures, 77
  middle, dissection of, 72
  superior, in pharynx dissection, 66
  swallowing and, 62
Pharyngeal raphe, 27, 61
Pharyngeal recess, in nasal cavity examination, 60
Pharyngeal tonsil, 61
Pharyngeal tube, lumen of, 69
Pharyngeal tubercle, 61
Pharyngeal venous plexus, 234
Pharynx
  in anterior triangle of neck dissection, 233–234
  oral part of, 11. *See also* Oropharynx
  parts of, 14, *22*
  sagittal section, *22*
Philtrum, 10, *12*
Phrenic nerve
  in mediastinum dissection, 106, 107
  in neck posterior triangle dissection, 123, 237, 239
Phylogeny, defined, 14, 69
Pia mater, in cranial cavity dissection, 195
Piriform fossa, in laryngopharynx dissection, 73
Piriformis muscle, 287
  in gluteal region dissection, relationships at inferior border, 296–297
  in lower limb dissection, 290
Pisiform, palpation of, 267
Pituitary gland. *See* Hypophysis
Planes. *See* individually named planes
Plantar aponeurosis, in foot dissection, 328
Plantar arch, in foot dissection, 325

Numbers in italic indicate figures.

Plantar ligaments
  calcaneocuboid, 333
  calcaneonavicular, 332
Plantaris muscle
  Achilles tendon and, 318
  in calf, 312
  Plantar interosseous muscle, 332
Plantar reflex, 314
Platysma muscle, 94
  fiber identification in, 55
Pleura
  cervical dome of, neck anterior triangle and, 226
  visceral and parietal, 93
Pleural sac
  lung invagination and, 93
  recesses of, in lung dissection, 97
Plica(e)
  circulares, of duodenal mucosa, 146
  fimbriata, 15
  semilunaris, 201
  sublingual, 29, 30
PNS. See Peripheral nervous system (PNS)
Popliteal artery
  in knee examination, 309
  in quadriceps femoris dissection, 305
Popliteal fossa, 287, 298
  clinical examination of, 309
  in hamstring dissection, 297
Popliteal vein, in knee examination, 309
Popliteus fascia, in leg dissection, 318
Popliteus muscle, in leg dissection, 318
Portal vein, 156
Portal venous system
  esophageal, 123–124
  in liver dissection, 142
Position sense, trigeminal nerve maxillary division and, 191
Posterior, defined, 3
Posterior atlantooccipital membrane, in posterior neck dissection, 249
Posterior auricular vein, dissection of, 55
Posterior cutaneous nerve of thigh, in gluteal region dissection, 292
Posterior gluteal line, identification of, 292
Posterior lingual glands, 15
Posterior superior iliac spine (PSIS), in abdominal examination, 126
Posterior talofibular ligaments, in ankle joint dissection, 332–333
Pouch
  of Douglas, 172
  rectovesical, in pelvic dissection, 172
  vesicouterine, in pelvic dissection, 172
PR. See Per rectum (PR) examination
Precordium, palpation of, 108
Pretrachial fascia, in thyroid gland dissection, 76
Prevertebral fascia, in anterior triangle of neck dissection, 233
Prevertebral muscles, in anterior triangle of neck dissection, 234
Primary cartilaginous joint, defined, 247

Profunda femoris artery, in quadriceps femoris dissection, 305
Pronation
  defined, 256
  forearm in, dissection of, 275–276
  hand, wrist and forearm in, dorsal aspect, 264
  and supination relationships, 267
Pronator quadratus muscle, 269
Pronator teres muscle, 269
Proprioception, sensory nerves and, 274
Prostate gland
  pelvic examination and, 170
  in pelvis dissection, 171
Prostatic urethra, in pelvic dissection, 172
Protraction, of scapula, 261
Proximal, defined, 4
Proximal phalanges, 268
PSIS. See Posterior superior iliac spine (PSIS)
Psoas major muscle, 163
Pterion, of skull, 183
Pterygoid muscle, medial and lateral. See Lateral pterygoid muscle; Medial pterygoid muscle
Pterygoid process, medial and lateral lamina of, 27
Pterygomandibular fold, 27
  and medial pterygoid muscle palpation, 38
Pterygomandibular ligament, 27
  laboratory examination of, 32
  in lateral pterygoid dissection, 47
Pterygopalatine fossa, in nasal cavity dissection, 63
Pterygopalatine ganglion, in nasal cavity dissection, 63, 66
Pubic crest, in abdominal examination, 126
Pubic hair, sexual differences in, 127
Pubic symphysis. See Symphysis pubis
Pubic tubercle, in abdominal examination, 126
Pubofemoral ligament, 307
Puborectalis muscle, 173
Pudendal artery, in pelvis dissection, 173
Pudendal canal, 291
Pudendal cleft, 178
Pudendal nerve
  in pelvis dissection, 173
  in perineum dissection, 180
Pulmonary arteries, in lung dissection, 96
Pulmonary trunk, blood flow through, 109, 114
Pulmonary valve
  blood flow control by, 109, 114
  cusp identification, 115
Pulmonary veins, in lung dissection, 96
Pulse
  angle of the mouth, palpation of, 11
  lingual artery, 15
Punctum lacrimale, in eyelid examination, 201

Pupil
  and cranial nerve function, 190
  dilator muscles of, in eyeball dissection, 210
  in eyeball examination, 204
PV. See Per vaginum (PV) examination
Pyloris, 135
Pyramidal lobe, in thyroid gland dissection, 76
Pyramids, of the kidney, 166
Pyriform. See Piriform fossa; Piriformis muscle

Quadratus femoris muscle, 287
  in gluteal region dissection, 296
  and piriformis relationship, 296
Quadratus lumborum muscle
  back muscles and, 246
  identification of, 158
  kidney relationship to, 163
  surface markings for, 159
Quadratus plantae muscle, in foot dissection, 328
Quadriceps femoris muscles, 286
  dissection of, 304–307
Quadriceps tendon, and knee joint extension, 285

Radial artery, 269, 271, 274
  dissection of, 276
Radial collateral ligament, in elbow dissection, 276
Radial nerve, 261
  in arm dissection, 263
  derivation of, 274
  dissection of, 277
Radiocarpal joint, dissection of, 277
Radius
  examination of, 256
  palpation of, 267
Ramus
  dorsal, 86
  of the mandible, 26, 28
Raphe
  palatine, 30
  pharyngeal, 27, 61
  pterygomandibular. See Pterygomandibular ligament
Rectal artery, 156
  in pelvis dissection, 173
Rectal veins, in pelvis dissection, 173
Recto-uterine pouch, in pelvic dissection, 172
Rectovesical pouch, in pelvic dissection, 172
Rectum, 134
  pelvic examination and, 170
Rectus muscles
  abdominis
    in abdominal examination, 126
    in chest wall dissection, 87
  capitis posterior, in posterior neck dissection, 249

Rectus muscles (cont.)
  eyeball control by, 187
  in orbit dissection, 213
  femoris
    in quadriceps femoris dissection, 304
    in thigh dissection, 301
Rectus sheath, in abdominal wall dissection, 131
Recurrent laryngeal nerve
  in inferior pharyngeal constrictor dissection, 77
  in mediastinum dissection, 107
  in thyroid gland dissection, 76
  in neck, 122–123
Referred pain, gallbladder and, 146
Reflex arc, 26, 86, 274, 314
Reflexes
  abdominal, 130
  and nervous system integrity, 130
  plantar, 314
  of upper limb, testing, 274
Regional anatomy, 3–5
  body regions defined, 81
  laboratory review of, 8
Renal arteries, laboratory identification of, 163
Renal system, in systematic anatomy, 7
Renal veins, laboratory identification of, 163
Respiration
  abdominal, male versus female, 127
  accessory muscles of, 94
  intercostal muscles role in, 81, 84
  mechanisms of, lungs and, 94
  quiet versus deep inspiration, 95
Respiratory distress, 94
Respiratory excursion, adequacy of, 94
Respiratory movement, palpation of, 94
Respiratory system
  laryngopharynx and, 69
  in systematic anatomy, 6
Retina
  in eyeball dissection, 207, 210
    central artery, 207
  in eyeball examination, 204–205
Retromandibular vein
  dissection of, 55
  in lateral pterygoid dissection, 46
Retroperitoneal organs, development in embryo, 138–139
Rhomboid muscles, 243
  dissection of, 250
Rib cage, 81
Ribs, palpation of, 84. *See also* Free ribs
Right atrioventricular valve, 109, *116*
  auscultation of, 109–110
  cusp identification, 115
Right atrium
  blood flow and, 108–109
  dissection of, 115
  opened view, *116*

and ventricular wall thickness compared, 118
Right ventricle
  and atrial wall thickness compared, 118
  blood flow and, 109
  dissection of, 114–115
  opened view, *116*
  outflow tract, *116*
Rima glottidis, 71
  in larynx dissection, 73
Rods, in eyeball, 207
Rotation
  at hip joint, 284
  of scapula, 261
  of upper limb, 85
Rotator muscles, dissection of, 251

Sacculus, in inner ear dissection, 222
Sacral vertebrae, identification of, 4
Sacroiliac joints, 283–284
  lower limb articulation with trunk, 283
Sacrospinalis muscles, 246
  dissection of, 251
  identification of, 158
Sacrospinus ligament, 283–284
Sacrotuberous ligament, 283
Sacrum, in abdominal examination, 126
Sagittal plane, described, 3
Sagittal suture, on skull, 3
Saliva, 11
Salivary glands, 26, 27. *See also* individually named glands
  classification of, 53
Salpingopharyngeus muscle, 61
  in pharynx dissection, 66
  tubal elevation and, 60
Saphenous vein
  long. *See* Long saphenous vein
  short, of leg and foot, 313
Sartorius muscle
  and hip joint flexion, 284–285
  as relation of quadriceps femoris dissection, 304–305
  tendon in leg dissection, 318
  in thigh dissection, 301, 304
Scalene triangle, 230
Scalenus anterior muscle
  homology with abdominal wall muscle, 126
  neck posterior triangle and, 227
  in upper limb girdle dissection, 122
Scalenus medius muscle, neck posterior triangle and, 227
Scalenus posterior muscle, neck posterior triangle and, 227
Scaphoid bone, tubercle of, 267
Scapula. *See also* Shoulder
  clinical examination of, 241–242
  dissection of, 122–124
  glenoid surface of, 255
  movements of, 261

Scapular line, lung extent and, 94
Schlemm, canal of, in eyeball dissection, 210
Sciatic nerve
  in gluteal region dissection, 296
  in hamstring dissection, 297, 300
    biceps femoris relationship to, 297
Sciatic notch, 290
Sclera, in eyeball examination, 204
Scoliosis, 241
Scrotum, 7, 127, 175
Seam. *See* Raphe; Suture
Sebaceous glands, on lips, 11
Secondary cartilaginous joint, defined, 247
Secretion, defined, 7. *See also* Exocrine, Endocrine
Secretomotor nerve impulses, auriculotemporal nerve and, 48
Segmentation
  morphological characteristics of, 81
  pattern of, 127
Semicircular canals, of inner ear, 215
Semilunar hiatus, in nasal cavity dissection, 63
Semimembranosus muscles, 287
  dissection, 297
Seminal vesicles
  in pelvic dissection, 172
  pelvic examination and, 170
Semispinalis capitis muscle
  dissection of, 251
  in neck posterior triangle dissection, 237
Semispinalis cervicis muscle
  dissection of, 251
  in posterior neck dissection, 249
Semitendinosus muscles, 287
  dissection, 297
Sensory cells, in reflex arc, 26
Sensory nerves
  integrity of, testing for, 191, 274
  tongue and, 26
Sensory retina, in eyeball dissection, 207, 210
Serous pericardium, 111
  parietal layer of, 118–119
Serratus anterior muscle
  in breast examination, 85
  in chest wall dissection, 87
Serratus posterior muscle, 243
  dissection of, 250–251
Sesamoid bone, patella as, 285
Sex organs, external and internal, 7
Sexual differences
  in abdominal wall, 127, 130
  in inguinal region, 127
Short plantar ligament, in foot dissection, 333
Short saphenous vein
  of leg and foot, 313
Shoulder
  anterior aspect, *82, 252*
  blood vessels of, 260

Numbers in italic indicate figures.

bony landmarks of, 255–256
dissection of, 262–266
functional anatomy and movements, 261
muscles of, 256
nerves of, 260–261
posterior aspect, *244*, *252*
scapula examination, 241–242
skeletal structures of, *258*
Shoulder girdle, movements of, 261
Sigmoid colon, 149–150
laboratory identification of, 134, 151
Sinoatrial node, site identification for, 118
Sinus arrhythmia, 100
Sinus(es)
cranial venous
cavernous, 199–200
superior sagittal, 195
of the heart-valve cusps, 118
of nasal cavity, 59
Skeletal system, in systematic anatomy, 6
Skin. *See* Integument
Skull. *See also* Neurocranium
anterior aspect, *202*
from below, view of, *188*
immature, *184*
from below, view of, *188*
left half, posterior aspect, *184*
lines on, 3–4
major portions of, 14
right lateral aspect, *184*
"Slipped disks," 292
Small bowel. *See* Small intestine
Small intestine
components of, 134, 150
SNS. *See* Somatic nervous system (SNS)
Soft palate, 10, 30
muscle dissection, 66–67
Somatic nervous system (SNS), 6, *104*
and muscular system, 6
Somatic structures, in body wall and limbs, 5, 6
Spermatic cord, 127
Spermatic fascia, in abdominal wall dissection
external, 131
internal, 132
Sperm cells, 7
Sphenoidal sinus
in nasal cavity dissection, 63
in nasal clinical examination, 59
Sphenoid bone, sphenomandibular ligament and, 40
Sphenoid spine, and sphenomandibular ligament, 40
Sphenomandibular ligament, 40
in medial pterygoid dissection, 47
Spheno-occipital synchondrosis, 39, 247
nasopharynx roof and, 61
Sphenopalatine foramen, in nasal cavity dissection, 63
Sphincter
anal, rectal examination and, 170
mouth opening and, 11

urethral, 180
vaginae, 173
Spinal accessory nerve, in neck posterior triangle, 230
dissection, 236
Spinal cord, as part of CNS, 5
Spinal curvatures, 241
Spinal nerves, 5
Splanchnic nerve, 166
in pelvis dissection, 173
Spleen, 6
dissection of, 147–148
laboratory examination of, 142–143
surface marking of, 136
Splenic artery, branches and arteries of, *160*
Splenic flexure, ascending colon and, 149
Splenic vein
inferior mesenteric vein and, 156
tributaries of, *160*
Splenius capitis muscle
dissection of, 248, 251
in neck posterior triangle dissection, 237
Splenius cervicis muscle, 246
dissection of, 251
Spring ligament, in foot dissection, 332
Stapedius muscle, in middle ear dissection, 219
Stapes
dissection of, 219
in ear examination, 214
Statoacoustic nerve
in inner ear dissection, 219
in clinical examination, 192–193
Sternoclavicular joint, lung extent and, 95
Sternocleidomastoid muscles, 49
as accessory muscles of respiration, 94
activation of, 242
dissection of, 57–58
Sternocostalis muscles, in chest wall dissection, 87
Sternohyoid muscles, swallowing and, 70–71
Sternomanubrial plane, in chest wall examination, 84
Sternothyroid muscles, 54
swallowing and, 70
Sternum, 4
heart position and, 99
palpation of, 84
surface marking of, 120
Stomach
dissection of, 148
and duodenum mobility compared, 139
greater curvature of, 135
lesser curvature of, 135–136
mobilization of, 138
percussion of, 135
surface marking of, 135
Strap muscles, 54, 70
dissection of, 76

Styloglossus muscle, of tongue, 18
Stylohyoid bone, in digastric triangle examination, 52
Stylohyoid ligament, in oropharynx dissection, 72
Styloid processes, identification of, 18
Stylomandibular ligament, 40
in temporomandibular joint dissection, 48
Stylopharyngeus muscle, 61
in oropharynx dissection, 72
Subarachnoid space, lumbar puncture and, 125
Subscapularis muscle, 257
Subclavian arteries
as relation of lung, 97
in mediastinum dissection, 107
in mediastinum examination, 99
Subclavian triangle, of neck, 227
Subclavian vein
in superior mediastinum examination, 100
in neck posterior triangle dissection, 237
Subcostal plane, 125
Sublingual plica, 29, 30
Sublingual region
dissection of, 32–33, 34
examination of, 29–30
Submandibular duct, 29–30
Submandibular ganglion, 33
Submandibular gland
in digastric triangle, 52–53
dissection of, 57
palpation of, 30
secretion from, 29
in sublingual region dissection, 33
Submandibular triangle. *See* Digastric triangle
Submental lymph nodes, 54
Submental triangle, 54
Suboccipital nerve, in posterior neck dissection, 249
Suboccipital triangle, in posterior neck dissection, 249
Subscapular artery, in arm dissection, 266
Subscapularis muscle, in back dissection, 250
Subtaler joint, movement of, 333
Sulcus terminalis, of tongue, 15
in heart dissection, 115
Superciliary arches, 183
Superficial peroneal nerve, in calf dissection, 325
Superior, defined, 3
Superior alveolobuccal sulcus, 38
Superior epigastric artery
in abdominal examination, 126
in chest wall dissection, 92
Superior laryngeal nerve, in thyroid gland dissection, 76
Superior lobe, of lung, 95
Superior mediastinum
dissection of, 122–124
structures of, 121

350 INDEX

Superior mesenteric artery
  dissection of, *152*, 156
  in pancreas dissection, 146
Superior nuchal line, 183
Superior oblique muscle of eye
  eyeball control by, 187
  in orbit dissection, 212, 213
Superior rectus muscle of eye, eyeball control by, 187
Superior sagittal sinus, in cranial cavity dissection, 195
Superior thyroid branch, of external carotid artery, in carotid triangle dissection, 57
Superior vena cava, 100
  as relation of lung, 97
  and right atrium, 108
Supination
  of palm, 271
  and pronation relationships, 267
Supinator muscle, 270
  lateral aspect, *258*
Supraclavicular fossa
  identification of, 93
  lung extent and, 93
  in mediastinum examination, 99
Supraglenoid tubercle, in arm dissection, 263
Supraorbital foramen, 20
Suprapleural membrane, neck anterior triangle and, 226
Suprarenal gland, 159
  dissection of, 163, 166
Suprascapular artery, in anterior triangle of neck dissection, 232
Suprascapular notch, in neck posterior triangle, 227
Supraspinatus muscle, 254, 256
Supraspinous ligaments, 242
Suprasternal notch
  in mediastinum examination, 98
  trachea and, 70
Sustentaculum tali, 310
Sutures
  of skull
    coronal, 4
    sagittal, 3
    zygomaticomaxillary, 37
Swallowing. *See* Deglutition
Sympathetic part of Autonomic Nervous System, 106, *104*
Sympathetic trunk, 106
  in anterior triangle of neck dissection, 232–233
  in mediastinum, 123
Symphysis, as secondary cartilaginous joint, 247
Symphysis menti, mandible examination and, 37
Symphysis pubis, 284
  in abdominal examination, 125
  lower limb articulation with trunk and, 283
  palpation of, 171

Synapse, defined, 33
Synchondrosis, spheno-occipital. *See* Spheno-occipital synchondrosis
Synovial fluid, 39
Synovial fold, infrapatellar, 322
Synovial joints, 39
Synovial joints, back movements and intervertebral, 246–247
Synovial membrane
  of deltoid bursa, 263
  of elbow joint, 276
Systematic anatomy, 5–7
  laboratory review of, 8
Systole, ventricular, 109

Taenia coli, colon and, 157
Talocalcaneal joint. *See* Subtaler joint
Talocalcaneonavicular joint, movement of, 333
Talofibular ligaments, in ankle joint dissection, 332
Talus bone, 310
Tarsal bones, of ankle, 310
Tarsal glands, 201, *202*
  in eyelid dissection, 213
Tarsal plate, in eyelid dissection, 213
Taste, facial nerve as mediator of, 192
Taste buds, 26, 192
Taurus tubarius, 60
  in nasal cavity dissection, 63
Teeth
  anatomical parts, 28–29
  in oral cavity examination, 9
Temporal artery, dissection of, 25
Temporal bone, 26
  petrous part, in nasal cavity examination, 60
Temporal fascia, in temporalis muscle dissection, 42
Temporal fossa, in temporalis muscle dissection, 42
Temporalis muscle, 38
  dissection of, 42–43
  deep, 43, 46
  elevation of, *44*
Temporal line, in temporalis muscle dissection, 42
Temporal nerves, in temporalis muscle dissection, 42, 46
Temporal vessels, in temporalis muscle dissection, 42, 46
Temporomandibular joint, 39–41
  dissection of, 48
Tendon(s), 267. *See also* individually named tendons
  Achilles, 318, 322
  central, of diaphragm, 120
  conjoined, in inguinal canal dissection, 132

  flat. *See* Aponeurosis
  flexor. *See* Flexor tendons
  of forearm, 268–269
    dissection of, 276
  in hamstring dissection, 297
  intermediate of digastric muscle, palpation of, 49
  in leg dissection, 318–319
  tibialis posterior, 312, 323
"Tennis elbow," and common extensor tendon, 270
Tensor fasciae latae muscle, 287
  in gluteal region dissection, 292
Tensor veli palatini muscle, 61, *64*, 66
Tentorium cerebelli, in cranial cavity dissection, 195
Teres major muscle, 243, 254
  neck posterior triangle and, 227
  in shoulder and arm, 256
Terminology, in regional anatomy descriptions, 3–4
Testis, 7
  palpation of, 175
Thenar eminence
  dissection of, 276
  palpation of, 267
Thigh
  cutaneous nerves of, in gluteal region dissection, 292
  dissection of, anterior aspect, 301, 304
  vessels and nerves of, 287–288
Thoracic artery
  internal, in chest wall dissection, 92
  lateral, in arm dissection, 266
Thoracic cavity
  identification of, 93
  volume change in respiration, 94
Thoracic duct, dissection of, 106–107
Thoracic inlet, 122
  and neck anterior triangle, 226
Thoracic respiration, 94
Thoracic vertebrae, identification of, 4
Thoracodorsal nerve, in back dissection, 250
Thoracolumbar fascia, 167, 243
Thorax
  dissection of, 86
  identification on skeleton, 4
  palpation of, 81
  right side, anterior view, *102*
  ventral view of contents, *90*
Thumb
  dissection of, 277
  movements of, 268, 271
Thymus gland, in upper limb girdle dissection, 122
Thyrocervical trunk
  in anterior triangle of neck dissection, 232
  in neck posterior triangle dissection, 239
Thyroglossal duct, 15
  in thyroid gland dissection, 76

Numbers in italic indicate figures.

Thyrohyoid membrane
  in laryngopharynx dissection, 73
  in thyroid gland dissection, 76
Thyrohyoid muscles, 54
  swallowing and, 70
Thyroid cartilage
  in carotid triangle examination, 53
  and laryngopharynx, 69
  in larynx examination, 69
  palpation of, 4
Thyroid gland
  dissection of, 76
  palpation of, 71
Tibial condyles, and related structures, superior surface of, *316*
Tibialis anterior muscle, 311
  tendon of, 324
Tibialis posterior tendon, 312, 323
Tibial nerve
  in hamstring dissection, 297
  in knee examination, 309
Tongue. *See also* Lingual
  dissection of, *22*
  dorsal surface examination, 14
  dorsal surface of, *16*
  extrinsic muscles of, 17, 18
  inferior surface examination of, 15
  intrinsic muscles of, 17, 18
    swallowing and, 70
  laboratory review of, 21, 24–26
  in oral cavity, 10
  portions of, 11
Tonsil
  lingual, 15
  palatine, 68
  pharyngeal, 61
Tonsillar cleft, in oropharynx examination, 68
Tonsillar crypts, in oropharynx examination, 68
Torsion movements, of eye, 206
Trabeculae carneae, in heart dissection, 115, 118
Trachea, 70
  identification of, 93
  in mediastinum examination, 98
Tracheal rings, 70, 93
Tractus spiralis foraminosus, in inner ear, 215
  dissection, 219
Transpyloric plane, surface projection of, 120
Transverse acetabular ligament, in quadriceps femoris dissection, 307
Transverse cervical artery, in anterior triangle of neck dissection, 232
Transverse colon, 149
  as relation of stomach, 138
Transverse mesocolon, as relation of stomach, 138
Transverse palatine rugae, 30
Transverse pericardial sinus, in heart dissection, 111
Transverse plane, described, 4

Transversospinalis muscles, 246
  dissection of, 251
Transversus thoracis. *See* Sternocostalis muscle
Trapezius muscle, 243
  dissection of, 249–250
  palpation of, 242
Trapezium bone
  in wrist, 267
Triangular ligament, left, in liver dissection, 142
Triceps, 260
  in arm dissection, 262
Tricuspid valve. *See* Right atrioventricular valve
Trigeminal nerve, 20
  in cranial cavity dissection, 198
  in clinical examination, 190–191
Tripartite erectile body, of penis, 178
Trochlear nerve
  in cranial cavity dissection, 198
  in clinical examination, 187
  in orbit dissection, 212
Trunk
  axilla and upper portions of, *82*
  in "bent over rowing" motion, *128*
  limbs and, 4
  and lower limb, articulation between, 283
  lumbosacral, in lower limb dissection, 289–290
  neck, shoulder and upper portion of anterior aspect, *82*
  right lateral aspect, *82*
  posterior aspect, *128*
  pulmonary. *See* Pulmonary trunk
  right posterolateral aspect, *128*
Tubal elevation, auditory tube and, 60
Tubercle
  articular, in temperomandibular joint examination, 40
  epiglottal, 71
  of fifth metatarsal, 310
  genial, 18
  of humerus
    greater, 255
    lesser, 85, 255
  of iliac crest, 125
  pharyngeal, 61
  pubic, in abdominal examination, 126
  supraglenoid, in arm dissection, 263
  of upper lips, 10, *12*
Tunica albuginea, in penis dissection, 179
Tympanic cavity, 193
Tympanic membrane
  dissection of, 219
  laboratory examination of, 218
  in nasal cavity examination, 60
Tympanic plate, in temporomandibular joint dissection, 48
Tympanic plexus, 193
Tympanum, 214–215. *See also* Middle ear

Ulna
  examination of, 256
  palpation of, 267
Ulnar artery, 269, 271, 274
  dissection of, 276
Ulnar collateral ligament, in elbow dissection, 276
Ulnar nerve, 261
  in arm dissection, 262, 266
  derivation of, 274
  dissection of, 277
Umbilical artery
  obliterated
    in abdominal wall dissection, 133
  in pelvic dissection, 173
Umbilical region, of abdomen, 126
Umbilical vein, obliterated, in abdominal wall dissection, 133
Umbilicus, in abdominal examination, 125
Uncinate process, of pancreas and, 136
Upper abdominal organs
  percussion of, 135
  and peritoneum relationship, 138
Upper extremity, defined, 4
Upper limb (upper extremity)
  anterior aspect, *82*
  development of, 166–167
  functional anatomy and movements of, 261
  lateral and medial aspects, *264*
  and lower limbs, homology of, 281
  muscles, dissection of, 251, 254
  reflex integrity in, 274
Upper limb girdle, dissection of, 122.
Urachus, 133
Ureters
  laboratory identification of, 163
  in pelvic examination, 169
Urethra
  in pelvis dissection, 171
  prostatic, in pelvic dissection, 172
  sphincter of, 180
Urogenital diaphragm, 180
Urogenital system, 7
Urogenital triangle, 175
  described, 168
  dissection of, 179–180
  in female, 178
  in male, 175
Uterine tubes
  in female genitalia, 7
  in pelvis dissection, 171–172
Uterus, 7
  bimanual vaginal examination of, 169
  in pelvis dissection, 171
  round ligament of, 130, 172
Utriculus, in inner ear dissection, 222
Uvula, 30

"Vagal tone," 110
Vagina, 7
  and pelvic examination, 169
  in pelvis dissection, 171

Vagus nerve, 100
  in mediastinum dissection, 101
  in clinical examination, 193
  in neck and mediastinum, 122
Vallate papillae, 15, *16*
Vallecula, 71
Valves
  of heart, 108–110. *See also* individually named valves
    dissection of, 115
  of thigh veins, competence of, 288
Varicose veins, and valve competence, 288
Vas deferens
  in pelvic dissection, 172
  and testis examination, 175
Vastus muscles, in quadriceps femoris dissection, 304
Veins. *See* individually named veins
Venous lacunae, in cranial cavity dissection, 195
Venous systems, caval and portal, 123–124
Ventral, defined, 14, *88*
Ventral rami of mixed spinal nerves, 86, *104*
Ventral rami, in limb plexuses, 230
Ventricles, cerebrospinal fluid and, 125
Ventricular diastole, 109
Ventricular systole, 109
Vermiform appendix, colon and, 157
Vertebrae
  articulating facets of, back movements and, 246
  cervical. *See* Cervical vertebrae
  identification of, 4. *See also* individually named vertebrae
  sacral, 126
  segmental sequence of, 94
Vertebral artery
  in cranial cavity dissection, 199
  in neck posterior triangle dissection, 239
Vertebral column
  anterior abdominal wall markings on, 159
  CNS relationship with, 5
Vertebra prominens, 4
  in thoracic wall examination, 94
Vesicouterine pouch, in pelvic dissection, 172
Vestibular nerve function, tests for, 193
Vestibule of oral cavity
  bony structures of, 28
    laboratory session, 36
  described, 9–10
  dissection of, 32
  examination of, 26–27
  structures of, *12*
Vestibulocochlear nerve. *See* Statoacoustic nerve
Villi, in jejunum, 157
Viscera
  of digestive system, 5
  of respiratory system, 6
Visceral nervous system, 6
Visceral pleura, 93
Viscerocranium, 14
Viscus, of lung, 93
Vision, field of, test for, 186–187
Visual acuity, testing of, 186
Vitreous chamber, in eyeball dissection, 207
Vitreous humor, in eyeball dissection, 207
Vocal cords, in larynx dissection, 73, 76
Vocal folds, 71
Vocal process, of arytenoid cartilage, in larynx dissection, 73, 76
Voice box. *See* Larynx

Voluntary nervous system. *See* Somatic nervous system (SNS)
Vomer, in nasal cavity examination, 60

Waldeyer's lymphatic ring, 68
White matter, described, 5
Willis, circle of, 233
Wrist
  arteries and nerves of, 271, 274
  bony prominences of, 267–268
  clinical examination of, 270–271
  dissection of, 275–277
  dorsal aspect in pronation, *264*
  skeleton of, *258*
  structural features of, *272*
  tendon identification in, 276
  ventral aspect, *272*

Xiphisternal plane, in chest wall examination, 84
Xiphisternum, in chest wall examination, 84
Xiphoid process. *See* Xiphisternum

Zonular fibers, in eyeball dissection, 210
Zygoma. *See* Zygomatic bone
Zygomatic arch, 20
  palpation of, 26
Zygomatic bone
  maxilla examination and, 37
  palpation of, 26
Zygomatic nerve, 191
Zygomaticomaxillary suture, 37

Numbers in italic indicate figures.